9.230

50%
DISC

PARTIAL DIFFERENTIAL EQUATIONS

PARTIAL
DIFFERENTIAL EQUATIONS

P. R. GARABEDIAN, *Professor of Mathematics, New York University*

John Wiley & Sons, Inc., New York · London · Sydney

Preface

The present book has been prepared primarily as a text for a graduate course in partial differential equations. However, it also includes in the later chapters material that might have been developed in a research monograph. In both respects it represents a statement of my ideas about the subject thought out over a period of years of teaching at Stanford University and New York University.

Since one of my principal objectives has been to achieve an adequate treatment of the more subtle aspects of the theory of partial differential equations within the framework of a single volume, I have found it necessary to assume a knowledge of perhaps the most important elementary technique, namely, the method of separation of variables and Fourier analysis. Such a step is made feasible by the extensive literature already available in that field, which has become part of the standard curriculum for courses in mathematical physics at both the undergraduate and graduate levels. Besides the technique of separation of variables, the reader should be acquainted with the elements of ordinary differential equations and with the theory of functions of a complex variable. Given these prerequisites, I have attempted insofar as possible to present the subject in terms of simple illustrative examples, and I have leaned toward constructive approaches in preference to purely existential arguments.

The book has been written for engineers and physicists as well as for mathematicians. Thus I have hoped to bring to the attention of a wider audience modern mathematical methods that have proved to be useful in the applications. An entire chapter is devoted to the nonlinear partial differential equations of fluid dynamics, which are a source of some of the most interesting initial and boundary value problems of mathematical physics. This chapter ends with a section on magnetohydrodynamics. Finally, a special chapter on difference equations not only serves as a guide to the numerical solution of partial differential equations but also furnishes a convenient review of the general theory.

The central theme of the presentation has of necessity been existence and uniqueness theorems. However, I have placed the emphasis on constructive procedures and have imposed unnecessarily strong hypotheses in order to simplify the analysis and bring out those features that are of significance for the applications. One limitation of this treatment is the small place given to methods based primarily on *a priori* estimates. Moreover, I have failed to discuss nonlinear problems in the calculus of variations and related linear elliptic equations with merely measurable coefficients. On the other hand, I have devoted special attention to the theory of analytic partial differential equations in the complex domain.

It has been my intention to avoid unsupported references to results whose proof is beyond the scope of the book. I preferred to include such material in the form of exercises following the sections to which it is relevant. Thus the exercises vary in difficulty from quite elementary examples of the theory to more substantial problems on the level of thesis topics. This broad range of problems seems justified as an indication of the variety of questions that a mathematician encounters in practice. Of course, the harder exercises have in general been listed after the easy ones, although no precise rule has been strictly observed.

Enough alternate ways of treating the principal theorems about partial differential equations are suggested to provide a textbook that might serve in courses of widely differing description. One selection of material that I tried out in a two-semester course at New York University is the following: Sections 1.1, 1.2, 3.1, 3.2, 3.5, 4.1, 4.2, 4.4, 5.1, 6.1, 6.2, 7.1, 7.2, 7.3, 8.1, 8.2, 9.2, 11.1, 12.1 and 12.3. The course was paced for second year graduate students.

I am indebted to the Office of Naval Research, the Atomic Energy Commission, and the Sloan Foundation for support while writing this book. It is a pleasure to acknowledge, too, the contributions of colleagues and students. In particular, E. Rodemich compiled a set of notes from lectures delivered by me at Stanford University, and J. Kazdan made detailed criticisms of the first draft of the manuscript. C. D. Hill revised Chapters 10 and 15 extensively, and R. Sacker composed the first draft of Section 15.4. Many helpful comments were offered by C. Morawetz, B. Friedman, P. Lax, J. Douglas, Jr., J. Berkowitz, G. Deem and V. L. Chuckrow, who read various portions of the manuscript. Invaluable assistance with typing and editing was given by C. Engle and

P. Hunt; and C. Bass and B. Prine prepared an excellent set of pen and ink drawings for the figures.

PAUL R. GARABEDIAN

New York City
December 1963

Contents

1

The Method of Power Series

1. INTRODUCTION

The subject matter of the present book is largely motivated by a desire to solve the partial differential equations that arise in mathematical physics. Many problems of continuum mechanics can be formulated as relationships among various unknown functions and their partial derivatives, and such relationships constitute the partial differential equations in which we shall be primarily interested. For these equations we shall have to investigate the existence and uniqueness of solutions, their construction, and the description of their properties.

To start with, let us review a few elementary concepts. We define the *order* of a partial differential equation to be the order of the highest derivatives to appear in the equation, and in practice it will virtually always be a finite number. Since we refer to partial derivatives, it is of course understood that there are at least two independent variables involved, which may or may not occur explicitly in the equation. On the other hand, ordinary differential equations for functions of just one independent variable actually form a special case of our topic, and they serve on occasion as a simple model from which to deduce certain aspects of the broader theory by generalization. As is customary in most treatments of partial differential equations, we shall assume that the reader is already acquainted with the fundamentals about ordinary differential equations.

By way of illustration, consider the most general partial differential equation of, say, the second order, for one unknown function u of two independent variables x and y, which can be expressed in the form

$$F(x, y, u, u_x, u_y, u_{xx}, u_{xy}, u_{yy}) = 0.$$

1

Often the notation

$$u_x = p, \quad u_y = q, \quad u_{xx} = r, \quad u_{xy} = s, \quad u_{yy} = t$$

is introduced and the equation is written as follows:

(1.1) $$F(x, y, u, p, q, r, s, t) = 0.$$

If the given function F is linear in the quantities u, p, q, r, s and t, we call the partial differential equation (1.1) itself *linear*. If, more generally, F is a polynomial of degree m in the highest order derivatives r, s and t, we say that the equation is of *degree m*. Such properties of a partial differential equation as its order, its degree, or the number of independent variables involved have an important influence on the family of its solutions. With regard to such qualitative features, the theory we have to describe will be considerably more complicated than that for ordinary differential equations.

Of course, a single partial differential equation for one unknown function, such as (1.1), might possess many solutions, and one of the first questions to arise concerns their multiplicity and the auxiliary data that might serve to distinguish them from one another in a unique way. In this connection we can recall the familiar result that the general solution of an ordinary differential equation of order n depends on n arbitrary constants of integration. We shall attempt in the next section to extend this statement to a partial differential equation of order n for a function of k independent variables by indicating that the general solution ought to depend on n arbitrary functions of $k - 1$ independent variables. However, the situation is by no means as simple as it might appear at first sight, and a variety of restrictions must be imposed before such a straight-forward theorem can be asserted (cf. Section 2).

Some insight into what will be needed to assure the existence and uniqueness of solutions can be gained by referring to some of the classical examples of partial differential equations of mathematical physics and by examining the initial conditions or boundary conditions naturally associ-ated with them through the context in which they arise. Consider, to begin with, the equation of the vibrating string,

(1.2) $$u_{xx} - u_{tt} = 0,$$

whose solution $u = u(x, t)$ represents the infinitesimal displacement at time t of a point of the string distant x units from a fixed origin. Typical auxiliary data which might be given in a physical problem involving the wave equation (1.2) are the initial values

(1.3) $$u(x, 0) = f(x), \qquad u_t(x, 0) = g(x)$$

of the displacement u and velocity u_t of the string at the time $t = 0$. Formulas (1.2) and (1.3) describe an *initial value problem* quite analogous to initial value problems of the theory of ordinary differential equations, and we might expect on physical grounds that these requirements suffice to determine the motion $u(x, t)$ of the string uniquely.

Another reasonable question for the wave equation (1.2) is encountered when the initial data (1.3) are only available along a finite length

$$0 \leq x \leq l$$

of string which is, however, pinned down at its ends $x = 0$ and $x = l$. The latter physical requirement is formulated mathematically by imposing the additional boundary conditions

(1.4) $$u(0, t) = 0, \qquad u(l, t) = 0$$

on the displacement u for every choice $t \geq 0$ of the time t.

It is helpful to recall briefly (cf. Churchill 1) how the mixed initial and boundary value problem (1.2), (1.3), (1.4) can be solved by means of Fourier series. The expressions

(1.5) $$u = \sin \frac{n\pi x}{l} \cos \frac{n\pi t}{l}, \quad u = \sin \frac{n\pi x}{l} \sin \frac{n\pi t}{l}, \qquad n = 1, 2, \ldots,$$

are found (cf. Exercise 3 below) to comprise those solutions of (1.2) and (1.4) that have the form of a function of x alone multiplied by a function of t alone. Through linear combination of the particular solutions (1.5) we construct a Fourier expansion

(1.6) $$u(x, t) = \sum_{n=1}^{\infty} \left(a_n \sin \frac{n\pi x}{l} \cos \frac{n\pi t}{l} + b_n \sin \frac{n\pi x}{l} \sin \frac{n\pi t}{l} \right)$$

for the answer u to our problem. The Fourier coefficients a_n and b_n are given in terms of the initial data (1.3) by the integral formulas

(1.7) $$a_n = \frac{2}{l} \int_0^l f(x) \sin \frac{n\pi x}{l} \, dx, \quad b_n = \frac{2}{n\pi} \int_0^l g(x) \sin \frac{n\pi x}{l} \, dx,$$

and when f and g vanish at $x = 0$ and at $x = l$ and are sufficiently differentiable, the infinite series (1.6) converges (cf. Exercise 4 below) and fulfills (1.2), (1.3) and (1.4).

A quite different example is presented by the heat equation

(1.8) $$u_{xx} - u_t = 0$$

for the temperature $u = u(x, t)$ in a rod, considered as a function of the distance x measured along the rod and of the time t. Thermodynamics would suggest that the initial values

(1.9) $$u(x, 0) = f(x)$$

of the temperature should be sufficient to specify the distribution $u(x, t)$ at all later times $t > 0$. Here we are faced with a new situation, however, because the problem we have been led to involves only the one initial condition (1.9).

As before we may consider a finite rod of length l, too, with the temperatures

(1.10) $$u(0, t) = 0, \quad u(l, t) = 0$$

assigned at its ends $x = 0$ and $x = l$. The method of separation of variables is, of course, applicable to the mixed initial and boundary value problem (1.8), (1.9), (1.10). The separated solutions of (1.8) and (1.10) are now given by the expression

$$u = e^{-n^2\pi^2 t/l^2} \sin \frac{n\pi x}{l}, \qquad n = 1, 2, \dots .$$

Hence the answer to the problem can be represented by a Fourier series

(1.11) $$u(x, t) = \sum_{n=1}^{\infty} a_n e^{-n^2\pi^2 t/l^2} \sin \frac{n\pi x}{l}$$

convergent for $t > 0$. It is easy to verify that the coefficients a_n are once again defined by the rule (1.7). Observe that the explicit form of the answer (1.11) provides confirmation of the correctness of our mathematical formulation of the original heat conduction problem that it solves.

More obscure from the point of view of what we know about initial value problems for ordinary differential equations are the boundary conditions naturally associated with Laplace's equation

(1.12) $$u_{xx} + u_{yy} = 0.$$

A useful physical interpretation of u here is that of the potential of a two-dimensional electrostatic field. One of the standard problems of electrostatics is to determine u inside a specific region of the (x,y)-plane, for example, inside the unit circle

$$x^2 + y^2 < 1,$$

when its value on the boundary, in this case the circumference

$$x^2 + y^2 = 1,$$

are prescribed. Note how both a different geometry and a lesser number of arbitrary functions in the auxiliary data distinguish the latter boundary value problem for Laplace's equation, which is called the *Dirichlet problem*, from the initial value problem (1.2), (1.3) for the wave equation.

In order to find solutions of Laplace's equation inside the unit circle by the procedure of separating variables, we introduce the preliminary transformation

$$x = r \cos \theta, \qquad y = r \sin \theta$$

to polar coordinates. In the new coordinate system equation (1.12) becomes

(1.13) $$u_{rr} + \frac{1}{r} u_r + \frac{1}{r^2} u_{\theta\theta} = 0,$$

where

$$u = u(r, \theta)$$

is viewed as a function of r and θ. The advantage of such a transformation is that it leads to the convenient formulation

(1.14) $$u(1, \theta) = f(\theta)$$

of our boundary condition at the periphery of the unit circle.

We recall (cf. Churchill 1) that separated solutions of (1.13) which are appropriately bounded in their dependence on r and periodic in their dependence on θ are defined by the products

$$u = r^n \cos n\theta, \quad u = r^n \sin n\theta, \qquad n = 0, 1, \ldots,$$

of powers of r times trigonometric functions of θ. Thus we seek to represent the answer to the boundary value problem (1.13), (1.14) as a power series of the form

(1.15) $$u(r, \theta) = \frac{a_0}{2} + \sum_{n=1}^{\infty} (a_n r^n \cos n\theta + b_n r^n \sin n\theta),$$

where the expressions

$$a_n = \frac{1}{\pi} \int_0^{2\pi} f(\theta) \cos n\theta \, d\theta, \qquad b_n = \frac{1}{\pi} \int_0^{2\pi} f(\theta) \sin n\theta \, d\theta$$

are to be substituted for the Fourier coefficients a_n and b_n. Although (1.15) does furnish a solution of our problem in the special case of the unit circle, it becomes apparent that we shall have to develop new techniques of broader scope in order to handle even quite analogous questions for more complicated regions whose geometry is not associated with any workable coordinate system.

We shall suppose that the reader is already well acquainted with the above method of separation of variables and Fourier analysis by means of which the physical problems we have just outlined have been solved explicitly in terms of definite integrals and infinite series. Therefore this familiar and fundamental technique will not be a subject for further discussion in the present book. Here we shall be interested in more general methods and procedures which can be applied to a wider class of partial differential equations that arise in practice. Thus we shall investigate what can be done with the multitude of harder problems for which no answer in closed form is to be expected or for which it is a quite subtle matter to find an explicit solution.

EXERCISES

1. The general solution of a partial differential equation of the first order in two independent variables involves one arbitrary function of one variable. Establish this result for the special equation

$$u_x + u_y = 0.$$

2. Rearrange the expression (1.6) for the solution of the mixed initial and boundary value problem (1.2), (1.3), (1.4) so that it has the form

$$u(x, t) = \phi(x + t) + \psi(x - t).$$

Interpret the result as a general solution of the wave equation and use it to solve the initial value problem (1.2), (1.3) in the absence of any boundary conditions.

3. Derive formula (1.5) by looking for solutions of the wave equation (1.2) that have the form
$$u(x, t) = X(x)T(t)$$

and by finding ordinary differential equations for the factors X and T as functions of their respective arguments x and t. Give a physical interpretation of the results.

4. Discuss the convergence of the Fourier series (1.6), (1.11) and (1.15). Make a comparison of their respective properties as functions of the variables t and r.

2. THE CAUCHY-KOWALEWSKI THEOREM

Let us return to the question we have already brought up concerning the number and type of arbitrary functions that should occur in the general solution of a partial differential equation of given order for a function of a specific number of independent variables. For this study a possible tool which comes to mind is expansion of the solution in power series and identification of the coefficients of the series through substitution into the equation and substitution into auxiliary initial conditions. Of course such an approach, which is familiar for ordinary differential equations, will confine us to equations and solutions that have convergent power series representations. This assumption of analyticity excludes many of the most significant situations in the theory of partial differential equations, but as a first step toward the general theory the method of power series expansion will turn out to have its merits.

In order to describe the power series method for the solution of partial differential equations, we shall apply it to the particular equation (1.1). To be more specific about the requirement that the equation be of the second order, we assume that it can be solved for the second derivative r, which means that it can be written in the form

$$(1.16) \qquad r = G(x, y, u, p, q, s, t),$$

where G is an analytic function of its arguments x, y, u, p, q, s and t in, say, a neighborhood of the origin and has therefore a convergent Taylor series expansion in such a neighborhood. We impose on the solution u of (1.16) initial conditions of the type

$$(1.17) \qquad u(0, y) = f(y), \qquad p(0, y) = g(y),$$

where f and g are arbitrary analytic functions of the real variable y in some neighborhood of the origin and are supposed to vanish together with their derivatives f', f'' and g' at $y = 0$, because this fits in with our assumption about the region of regularity of the right-hand side of (1.16). Of course, a more general situation in which f and g have convergent Taylor series representations about any specific point $y = y_0$ and in which G is regular near a correspondingly more general set of values x_0, y_0, u_0, p_0, q_0, s_0 and t_0 of its arguments can be reduced to the present simpler case by means of a translation to the new coordinates $X = x - x_0$ and $Y = y - y_0$. Moreover, introduction of a new unknown of the form

$$U = u - u_0 - (x - x_0)p_0 - (y - y_0)q_0 - (x - x_0)^2 r_0/2$$
$$- (x - x_0)(y - y_0)s_0 - (y - y_0)^2 t_0/2$$

can always be exploited to make G zero at the origin, an additional simplification that we shall therefore adopt here, too. By further transformation of variables we could even achieve initial conditions along an arbitrary analytic curve rather than just along the y-axis, as in (1.17), but this would lead to a more careful analysis of the significance of our assumption that the partial differential equation (1.16) has been solved for r, and we therefore postpone discussion of such matters until Chapter 3.

Our attack on the analytic initial value problem embodied in (1.16) and (1.17) consists in calculating all the partial derivatives of the solution u at the origin by successive differentiation of the initial conditions and of the partial differential equation and then in using the results to write down a Taylor series expansion for u. The success of this approach hinges on the convergence of the series, which we shall attempt to establish by constructing a *majorant*, or series of larger, positive terms, based on the observation that no minus signs are introduced by the formal computation of the derivatives. Although we could proceed directly with this program, a better insight into the determination of the majorant is gained if we digress for a moment to reduce the problem to a more convenient one for a suitable canonical system of partial differential equations of the first order.[1] In such a simplified formulation it will be easier to see that the method of power series applies to any system of analytic partial differential equations in the same number of functions of two or more independent variables, a case that is, of course, as important to know about as the above example.

Many situations arise in which it is helpful to convert a partial differential equation of higher order into a canonical system of the first order. Such a reduction therefore has considerable significance in its own right. Thus the reader should view the conversion process presented below as a separate lemma for future reference, noting that it is not actually essential for the construction of power series solutions. Although we confine our attention to the special case of equation (1.1), the technique of reduction to be developed here is of broader scope and applies to equations of any order in any number of independent variables.

Our procedure is to interpret all the quantities x, y, u, p, q, r, s and t as new unknown functions

$$(1.18)\quad u_1 = x,\ u_2 = y,\ u_3 = u,\ u_4 = p,\ u_5 = q,\ u_6 = r,\ u_7 = s,\ u_8 = t$$

of two new independent variables ξ and η. The introduction of u_1 and u_2 is a device to avoid explicit appearance of the independent variables later on. This succeeds because we contrive to make ξ and η coincide with x and y by imposing on $u_1(\xi, \eta)$ and $u_2(\xi, \eta)$ the pair of first order partial differential equations

$$(1.19)\qquad\qquad \frac{\partial u_1}{\partial \xi} = \frac{\partial u_2}{\partial \eta}\,, \qquad \frac{\partial u_2}{\partial \xi} = 0,$$

[1] Cf. Courant-Hilbert 1.

subject to the initial conditions

(1.20) $\qquad\qquad u_1(0, \eta) = 0, \qquad u_2(0, \eta) = \eta.$

Direct integration with respect to ξ of the differential equation for u_2, which simply states that u_2 is independent of ξ, followed by a comparison with the corresponding initial condition at $\xi = 0$, gives, indeed,

$$y = u_2 \equiv \eta.$$

Thereupon we see that

(1.21) $\qquad\qquad \dfrac{\partial u_1}{\partial \xi} = \dfrac{\partial u_2}{\partial \eta} = 1,$

so that a similar integration for u_1 yields the result

$$x = u_1 \equiv \xi.$$

The various requirements

$$u_x = p, \quad p_x = r, \quad q_x = p_y, \quad s_x = r_y, \quad t_x = s_y,$$

which merely assert that the mixed partial derivatives of u are independent of the order in which the differentiations are performed, provide us with a system of five additional partial differential equations

(1.22) $\qquad \dfrac{\partial u_3}{\partial \xi} = u_4 \dfrac{\partial u_2}{\partial \eta}, \quad \dfrac{\partial u_4}{\partial \xi} = u_6 \dfrac{\partial u_2}{\partial \eta}, \quad \dfrac{\partial u_5}{\partial \xi} = \dfrac{\partial u_4}{\partial \eta},$

$$\dfrac{\partial u_7}{\partial \xi} = \dfrac{\partial u_6}{\partial \eta}, \quad \dfrac{\partial u_8}{\partial \xi} = \dfrac{\partial u_7}{\partial \eta}$$

among the eight unknowns $u_1, \ldots, u_8$, since differentiation with respect to ξ and η is equivalent to differentiation with respect to x and y. Here we have exploited the relation (1.21) to obtain expressions on the right that are in every case linear and homogeneous in the first partial derivatives of the functions u_k with respect to η. Finally, differentiation of (1.16) with respect to x yields

$$r_x = G_x + G_u u_x + G_p p_x + G_q q_x + G_s s_x + G_t t_x,$$

whence, using (1.21) and (1.22), we derive the eighth equation

(1.23) $\qquad \dfrac{\partial u_6}{\partial \xi} = G_x \dfrac{\partial u_2}{\partial \eta} + u_4 G_u \dfrac{\partial u_2}{\partial \eta} + u_6 G_p \dfrac{\partial u_2}{\partial \eta} + G_q \dfrac{\partial u_4}{\partial \eta}$

$$+ G_s \dfrac{\partial u_6}{\partial \eta} + G_t \dfrac{\partial u_7}{\partial \eta},$$

where G_x, G_u, G_p, G_q, G_s and G_t are to be understood as functions of the new arguments $u_1, \ldots, u_5$, u_7 and u_8 in place of x, y, u, p, q, s and t.

In the above manner we obtain a canonical system (1.19), (1.22), (1.23) of quasi-linear partial differential equations of the first order for the unknowns u_k of the general form

$$(1.24) \qquad \frac{\partial u_j}{\partial \xi} = \sum_{k=1}^{m} a_{jk}(u_1, \ldots, u_m) \frac{\partial u_k}{\partial \eta}, \qquad j = 1, \ldots, m,$$

with $m = 8$. We say that the system is *quasi-linear* because all the partial derivatives appear linearly, and we describe it as *canonical* because it has been solved for the partial derivatives $\partial u_k / \partial \xi$ in terms of the quantities u_k and $\partial u_k / \partial \eta$. Notice that the coefficients a_{jk} that occur here do not involve ξ or η explicitly and are analytic functions of their arguments u_k in a neighborhood of the origin.

Initial conditions for the u_k, in addition to (1.20), are derived by differentiating (1.17) a suitable number of times with respect to y and by evaluating the partial differential equation (1.16) itself at $x = 0$. We obtain in this way

$$(1.25) \qquad u_3(0, \eta) = f(\eta), \quad u_4(0, \eta) = g(\eta), \quad u_5(0, \eta) = f'(\eta),$$

$$u_7(0, \eta) = g'(\eta), \quad u_8(0, \eta) = f''(\eta)$$

and

$$(1.26) \qquad u_6(0, \eta) = G(0, \eta, f(\eta), g(\eta), f'(\eta), g'(\eta), f''(\eta)).$$

Thus the initial conditions are all of the form

$$(1.27) \qquad u_j(0, \eta) = h_j(\eta), \qquad j = 1, \ldots, m,$$

where the prescribed functions h_j are analytic in η and vanish at the origin,

$$(1.28) \qquad h_j(0) = 0, \qquad j = 1, \ldots, m.$$

We have now established that the original initial value problem (1.16), (1.17) for a single partial differential equation of the second order can be reduced to one for a canonical system (1.24), subject to initial conditions of the type (1.27), (1.28). It is, of course, necessary to verify that our steps can be retraced and that a solution of the new problem also solves the old one. We have already indicated why (1.19) and (1.20) imply that $x = \xi$ and $y = \eta$, but we must also prove from the differential equations (1.22), (1.23) and the initial conditions (1.25), (1.26) that p, q, r, s and t, as defined by (1.18), are actually appropriate partial derivatives of a function u fulfilling the partial differential equation (1.16). A sample case will serve to show why this is true. Thus we shall be content below merely to check that

(1.29) $q = u_y$.

From (1.18), (1.21) and (1.22) we obtain, indeed,

$$\frac{\partial u_5}{\partial \xi} = \frac{\partial u_4}{\partial \eta}, \quad u_4 = \frac{\partial u_3}{\partial \xi}, \quad u = u_3, \quad q = u_5,$$

and therefore, since ξ is identical with x and η with y,

$$\frac{\partial}{\partial x}\left(q - \frac{\partial u}{\partial y} \right) = 0.$$

Integrating with respect to x we find that

$$q - u_y = C(y),$$

where $C(y)$ is a constant of integration depending only on y and not on x. But according to the initial conditions (1.25) we must have

$$C(y) = u_5(0, y) - \frac{\partial}{\partial y} u_3(0, y) = f'(y) - f'(y) = 0$$

and this suffices to establish the desired result (1.29).

Notice here that while we can always reduce a partial differential equation of high order for a single unknown function to a system of several first order equations by introducing an appropriate set of derivatives as new unknowns, the return from this system to the original equation may depend on the application of initial conditions. A larger number of independent variables distinguishes the present situation from that encountered in the theory of ordinary differential equations and leads to elimination processes that cannot be reversed without the use of auxiliary information. As a further example of such phenomena consider a pair of first order partial differential equations

$$F_j(x, y, u, u_x, u_y, v, v_x, v_y) = 0, \quad j = 1, 2,$$

for the determination of two unknown functions u and v. These can in general be differentiated to provide enough relations so that one of the unknowns and its derivatives can be eliminated. But differentiation of each equation just once with respect to each of the independent variables yields only four new equations in addition to the original two and that does not usually suffice to furnish an equation for one unknown by elimination of the other together with its five partial derivatives of the first and second orders. Therefore we have to differentiate again, getting altogether twelve relations from which to eliminate one function plus its nine derivatives of order three or less. The latter procedure leaves us, however, with two relations for the remaining function and its first, second and third partial derivatives, and we cannot expect either of these two third order partial differential equations alone to be equivalent to the original pair of the first order. It is well to keep such complications in mind when transforming the equation (1.16) into a canonical system like (1.24).

Having convinced ourselves that the range of applicability of the canonical system (1.24) is adequate, we now return to the main problem of solving it by the method of power series, which was, after all, our original purpose. More specifically, we seek Taylor series expansions about the origin for the unknowns u_k. To find the coefficients in the expansions we merely calculate the partial derivatives of the functions u_k at the origin. For them we develop recursion formulas that express partial derivatives involving any number μ of differentiations with respect to the variable ξ in terms of partial derivatives involving at most $\mu - 1$ such differentiations.

The partial differential equations (1.24) are themselves recursion formulas of the type we have described. The rest of the recursion formulas that are needed can be derived by applying the differential operator $\partial^{\mu+\nu-1}/\partial\xi^{\mu-1}\partial\eta^\nu$ to both sides of (1.24), a procedure that gives the result

$$(1.30) \qquad \frac{\partial^{\mu+\nu}u_j}{\partial\xi^\mu\partial\eta^\nu} = \sum_{k=1}^{m} \frac{\partial^{\mu+\nu-1}}{\partial\xi^{\mu-1}\partial\eta^\nu}\left[a_{jk}(u_1,\ldots,u_m)\frac{\partial u_k}{\partial\eta}\right].$$

By application of the rules for differentiating a function of a function, a sum, and a product, none of which involves minus signs, the derivatives on the right can be worked out in detail. Proceeding thus, we obtain a polynomial in the partial derivatives of the a_{jk} and the u_k whose coefficients are exclusively positive integers. The recursion formulas for the successive calculation of the partial derivatives of the u_k at the origin therefore do not contain any minus signs, a fact that will enable us to introduce majorants which establish the convergence of the Taylor series for the unknowns u_k.

A preliminary step in our construction is to compute all the derivatives of the u_k with respect to η alone by differentiating the initial conditions (1.27). Thus we obtain

$$(1.31) \qquad \frac{\partial^\nu u_j(0,0)}{\partial\eta^\nu} = h_j^{(\nu)}(0), \qquad j = 1, \ldots, m.$$

Next we insert (1.31) into (1.30), with $\mu = 1$, to derive expressions for all partial derivatives of the form $\partial^{\nu+1}u_j/\partial\xi\partial\eta^\nu$. Then we can substitute the latter results into (1.30) once more, but with $\mu = 2$, to find the derivatives $\partial^{\nu+2}u_j/\partial\xi^2\partial\eta^\nu$, and so on. In such a manner we determine in succession all the partial derivatives of the functions u_k without ever encountering a subtraction. This remarkable fact will now be exploited to establish the convergence of the formal Taylor series expansions

$$(1.32) \qquad u_k(\xi, \eta) = \sum_{\mu, \nu=0}^{\infty} \frac{1}{\mu!\,\nu!} \frac{\partial^{\mu+\nu}u_k(0, 0)}{\partial\xi^\mu\partial\eta^\nu} \xi^\mu\eta^\nu$$

for the unknowns u_k.

Because the given functions $a_{jk}(u_1, \ldots, u_m)$ and $h_j(\eta)$ are analytic at the origin, their Taylor series representations there must converge for special arguments of the form

$$u_1 = u_2 = \cdots = u_m = \eta = \rho$$

when ρ is a sufficiently small positive number. For such a point of convergence there must exist an upper bound M on the absolute values of all the terms of the Taylor series. Hence at the origin we have

$$(1.33) \qquad \left| \frac{\partial^{n_1+n_2+\cdots+n_m}a_{jk}}{\partial u_1^{n_1}\cdots\partial u_m^{n_m}} \right| \leq \frac{n_1!\,n_2!\cdots n_m!\,M}{\rho^{n_1+n_2+\cdots+n_m}},$$

$$(1.34) \qquad |h_j^{(n)}(0)| \leq \frac{n!\,M}{\rho^n}.$$

The question that we must ask is how large the coefficients on the right in the expansions (1.32) of the unknowns u_k can become in view of the restrictions (1.33) and (1.34). Due to the absence of minus signs in the recursion formulas (1.30) and (1.31) for the coefficients, they are evidently largest when all the derivatives of the given functions a_{jk} and h_j are positive and coincide with the upper bounds defined by (1.33) and (1.34). Therefore we can estimate them by investigating only this least favorable situation, which is advantageous not just because it avoids cancellations, but rather because it turns out to be vastly simpler than the general case.

Actually, we prefer to treat an even cruder example with

$$(1.35) \qquad \frac{\partial^{n_1+n_2+\cdots+n_m}a_{jk}}{\partial u_1^{n_1}\cdots\partial u_m^{n_m}} = \alpha_{n_1\cdots n_m} \frac{n_1!\,n_2!\cdots n_m!\,M}{\rho^{n_1+n_2+\cdots+n_m}},$$

$$(1.36) \qquad h_j(0) = 0,$$

$$(1.37) \qquad h_j^{(n)}(0) = \frac{n!\,M}{\rho^n}, \qquad n \geq 1,$$

where $\alpha_{n_1 \cdots n_m}$ is the coefficient of $u_1^{n_1} \cdots u_m^{n_m}$ in the formal expansion of the polynomial $(u_1 + \cdots + u_m)^{n_1 + \cdots + n_m}$. Since most of the factors $\alpha_{n_1 \cdots n_m}$ exceed unity, while none is smaller than that, this results in unnecessarily large estimates of the Taylor series coefficients in the solution (1.32). However, it leads to geometric series and provides us with a comparison problem that can be handled explicitly.

The essential idea now is to construct a majorant for the series (1.32) by solving in closed form the special case of the canonical initial value problem (1.24), (1.27), (1.28) defined by (1.35), (1.36) and (1.37). We can sum the particular Taylor series for a_{jk} and h_j that occur here to obtain the results

$$a_{jk} = M \sum_{n=0}^{\infty} \left[\frac{u_1 + \cdots + u_m}{\rho} \right]^n = \frac{M\rho}{\rho - u_1 - \cdots - u_m}$$

and

$$h_j = M \sum_{n=1}^{\infty} \frac{\eta^n}{\rho^n} = \frac{M\eta}{\rho - \eta}.$$

Note that the expressions on the right are independent of the indices j and k. Thus to prove convergence of the general series (1.32) we have only to study the simple canonical system

$$(1.38) \qquad \frac{\partial u_j}{\partial \xi} = \frac{M\rho}{\rho - u_1 - \cdots - u_m} \sum_{k=1}^{m} \frac{\partial u_k}{\partial \eta}, \qquad j = 1, \ldots, m,$$

subject to the initial conditions

$$(1.39) \qquad u_j(0, \eta) = \frac{M\eta}{\rho - \eta}, \qquad j = 1, \ldots, m,$$

for its solution defines the desired majorant.

Next it is important to observe that there can be at most one analytic solution of any analytic initial value problem of the type we have discussed here, since the recursion formulas for the Taylor series coefficients involved can yield only one answer. We shall make use of this uniqueness theorem to identify the formal series solution of (1.38) and (1.39) with an analytic solution, represented in terms of elementary functions, which we shall find by a more direct integration procedure.

A preliminary remark is that, according to (1.38) and (1.39), the u_j all have identical first derivatives with respect to ξ and all satisfy the same initial condition at $\xi = 0$. Therefore they must all be equal, and they reduce to a single function u solving the simplified initial value problem

$$(1.40) \qquad \frac{\partial u}{\partial \xi} = \frac{mM\rho}{\rho - mu} \frac{\partial u}{\partial \eta},$$

$$(1.41) \qquad u(0, \eta) = \frac{M\eta}{\rho - \eta}.$$

A special device can be exploited to solve equation (1.40) in closed form (cf. Exercise 2.1.1[2]). Observe that if we introduce the expression

$$v = \frac{mM\rho\xi}{\rho - mu} + \eta,$$

we can write (1.40) in the more readily understandable form

$$(1.42) \qquad \frac{\partial u}{\partial \xi} \frac{\partial v}{\partial \eta} - \frac{\partial u}{\partial \eta} \frac{\partial v}{\partial \xi} = 0,$$

since the contributions to the derivatives of v involving differentiation of u cancel out in this particular combination of terms. The partial differential equation (1.42) states that the Jacobian of the pair of functions u and v with respect to the variables ξ and η is identically zero. This means that u and v are functionally dependent, and hence the general solution of (1.42) may be presented in the form

$$(1.43) \qquad u = \phi(v) = \phi\left(\frac{mM\rho\xi}{\rho - mu} + \eta\right),$$

where ϕ is an arbitrary function of the one variable v. The initial condition (1.41) serves to establish that

$$\phi(\eta) = \frac{M\eta}{\rho - \eta},$$

and therefore we derive from (1.43) the final implicit relation

(1.44)

$$(m\rho - m\eta)u^2 - (\rho^2 + mM\eta - \rho\eta - mM\rho\xi)u + M\rho\eta + mM^2\rho\xi = 0$$

defining u as a function of ξ and η.

The desired solution of (1.40) and (1.41) is given by the smaller root

$$(1.45) \quad u = [2m(\rho - \eta)]^{-1} [\rho^2 + mM\eta - \rho\eta - mM\rho\xi$$
$$- \sqrt{(\rho^2 + mM\eta - \rho\eta - mM\rho\xi)^2 - 4mM\rho(\rho - \eta)(\eta + mM\xi)}]$$

of the quadratic (1.44), which is chosen because it vanishes at the origin. Since the expressions in the denominator and under the radical sign in (1.45) do not vanish when $\xi = \eta = 0$, this formula yields a convergent Taylor series expansion for u about that point. By its very construction

[2] This notation is used to refer to Exercise 1 at the end of Section 1 in Chapter 2.

the latter series must be the same as the formal solution of (1.40) and (1.41), or, rather, of (1.38) and (1.39), which was defined before by successive application of recursion formulas. Hence for small positive choices of ξ and η it is a convergent series of positive terms which are at least as large in absolute value or, in other words, which *majorize* the corresponding terms of the formal Taylor series solution (1.32) of the general initial value problem (1.24), (1.27), (1.28). Indeed, the special problem (1.38), (1.39) was designed to provide just such a majorant. It follows that the expansions (1.32) converge in all cases for small enough values of ξ and η. Thus we have completed our proof that for analytic data a_{jk} and h_j the canonical initial value problem (1.24), (1.27), (1.28) has a unique analytic solution in some sufficiently small neighborhood of the origin. The existence and uniqueness of an analytic solution of the original analytic initial value problem given by (1.16) and (1.17) are a direct consequence of this statement. We summarize the principal result in the form of a theorem which we shall refer to as the *Cauchy-Kowalewski theorem*.

THEOREM. *About any point at which the given matrix A of coefficients a_{jk} and the prescribed column vector h of functions h_j are analytic a neighborhood can be found where there exists a unique vector u with analytic components u_k solving the initial value problem*

$$u_\xi = A(u)u_\eta, \quad u(0, \eta) = h(\eta).$$

Despite its apparent generality, the Cauchy-Kowalewski theorem has a number of significant limitations. Not only is it restricted to problems involving exclusively analytic functions, but even in such cases the actual computation of series coefficients by means of recursion formulas can turn out to be too tedious for practical application. Furthermore, the series may not converge in the full region where the solution is needed in a specific example. For these and other more theoretical reasons we shall find presently that we want to investigate more direct and effective methods for the integration of partial differential equations. However, the Cauchy-Kowalewski theorem does show that within the class of analytic solutions of analytic equations the number of arbitrary functions required for a general solution is equal to the order of the equation and that the arbitrary functions involve one less independent variable than the number occurring in the equation.

EXERCISES

1. Use reduction to a canonical system analogous to (1.24) to prove the Cauchy-Kowalewski theorem for partial differential equations of arbitrary order in more than two independent variables.

2. Formulate a majorizing problem that permits one to establish the Cauchy-Kowalewski theorem for (1.16) directly, without resort to the reduction to a canonical system.

3. Modify the initial value problem (1.16), (1.17) so that the initial conditions are given on an arbitrary line $x = x_0$, and assume that the data are analytic at points x_0, y_0 which are allowed to range over a prescribed rectangle. Verify that the Taylor series solution about x_0, y_0 converges in a circle of fixed size around that point whose radius can be chosen independently of the parameters x_0, y_0.

4. Show that if (1.16) is Laplace's equation $r = -t$, the Cauchy-Kowalewski solution of the initial value problem defined by (1.16) and (1.17) is equivalent to performing an analytic continuation of the function $f'(y) + ig(y)$ to complex values $y - ix$ of its argument by means of a power series representation.

5. Given a solution of the system (1.19), (1.22), (1.23) satisfying the initial conditions (1.20), (1.25), (1.26), show in complete detail that the formulas (1.18) define a corresponding solution of the initial value problem (1.16), (1.17).

2

Equations of the First Order

1. CHARACTERISTICS

The simplest partial differential equations to treat are those of the first order for the determination of just one unknown function. It turns out, in fact, that the integration of such a partial differential equation can be reduced to a family of initial value problems for a system of ordinary differential equations. This is a quite satisfactory result because of the substantial knowledge of ordinary differential equations that is available to us.[1] For examples and applications of the theory of first order equations that we shall now develop, the reader is referred ahead to Sections 4, 5 and 6. Moreover, anyone who is not interested in the present topic may proceed directly to the later chapters, which do not depend on it in an essential way.

We begin our discussion by studying partial differential equations of the form

$$(2.1) \qquad a(x, y, u)u_x + b(x, y, u)u_y = c(x, y, u)$$

in two independent variables x and y, with coefficients a, b and c that are not necessarily analytic functions of their arguments. The results we shall obtain are easily generalized afterward to the case of $n > 2$ independent variables (cf. Exercise 2 below). Because it is given by an expression that is linear in the partial derivatives u_x and u_y, we call the equation (2.1) *quasi-linear*. When c is linear in u and when a and b do not depend on u at all, we may go further and say that the equation (2.1) is *linear*.

We shall investigate the quasi-linear equation (2.1) by examining in detail what it says about the geometry of the surface $u = u(x, y)$ defined

[1] Cf. Ince 1.

by the solution. It is a familiar fact that at a point x, y, u of such a surface the vector $(u_x, u_y, -1)$ has the direction of the normal. This is essentially the significance of the relation

$$(2.2) \qquad du = u_x \, dx + u_y \, dy.$$

Equation (2.1) simply states that the scalar product of the vector $(u_x, u_y, -1)$ with the vector (a, b, c) vanishes. Thus (a, b, c) is perpendicular to the normal and must lie in the tangent plane of the surface $u = u(x, y)$. We can therefore interpret the first order partial differential equation (2.1) geometrically as a requirement that any solution surface $u = u(x, y)$ through the point with coordinates x, y, u must be tangent there to a prescribed vector, namely, to the one with the components $a(x, y, u)$, $b(x, y, u)$, $c(x, y, u)$.

Within a specific surface $u = u(x, y)$ solving (2.1) we can consider the field of directions defined by the tangential vectors (a, b, c). This field of directions is composed of the tangents of a one-parameter family of curves in that surface, called *characteristics*, which are determined by the system of ordinary differential equations

$$(2.3) \qquad \frac{dx}{a} = \frac{dy}{b} = \frac{du}{c}.$$

Indeed, the differential equations (2.3) state that the increment (dx, dy, du) is parallel to the vector (a, b, c) and lies in the tangent plane of the surface $u = u(x, y)$, which must therefore be generated by a one-parameter family of solutions of (2.3). Moreover, any one-parameter family of solutions of the pair of ordinary differential equations (2.3) sweeps out a surface whose equation $u = u(x, y)$ must solve (2.1), since a surface so generated necessarily has at each of its points x, y, u a tangent plane containing the corresponding vector (a, b, c).

A more analytical way to see that the integral curves of (2.3) sweep out solutions of (2.1) is suggested by the relation (2.2). Let the surface $u = u(x, y)$ consist of curves satisfying (2.3). Then we can select the quantities dx, dy, du occurring in (2.2) so that they are proportional to a, b, c. Consequently it is feasible to multiply (2.2) by a factor which converts it into the partial differential equation (2.1). Thus u has the desired property of solving (2.1).

There are actually only two independent ordinary differential equations in the system (2.3); therefore its solutions comprise in all a two-parameter family of curves in space. What we have established is that any one-parameter subset of these curves generates a solution of the first order quasi-linear partial differential equation (2.1). Such a one-parameter subset is defined by a single relation between the two arbitrary constants occurring in the general solution of (2.3) or, in other words, by an arbitrary function of one independent variable.

The manner in which the arbitrary function appears can be formulated more specifically in terms of an *initial value problem*, or *Cauchy problem*, that asks for a solution of (2.1) passing through a prescribed curve

$$(2.4) \qquad x = x(t), \quad y = y(t), \quad u = u(t)$$

in space. The latter problem can be solved in the small by considering for each value of t the integral curve of the system of ordinary differential equations (2.3) whose initial values are defined by (2.4). The local existence and uniqueness of the required solution of (2.3) follow from the theory of ordinary differential equations.[2] If we introduce a parameter s along these integral curves, which might, for example, be the arc length measured from (2.4), we obtain from our construction a surface

$$(2.5) \qquad x = x(s, t), \quad y = y(s, t), \quad u = u(s, t)$$

in parametric form. According to what has been said earlier, the non-parametric representation $u = u(x, y)$ of (2.5) should yield the desired solution of (2.1) passing through the initial curve (2.4). In this context note that we can replace (2.3) by the system

$$(2.6) \qquad \frac{dx}{ds} = a(x, y, u), \quad \frac{dy}{ds} = b(x, y, u), \quad \frac{du}{ds} = c(x, y, u),$$

in which the parameter s plays the role of an independent variable but does not in general stand for arc length

The only difficulty that might occur in our solution of the initial value problem defined by (2.1) and (2.4) arises from inverting (2.5) to express s, t and u as functions of the independent variables x and y. Such a step goes through for sufficiently small values of s whenever the Jacobian

$$(2.7) \qquad J = x_s y_t - x_t y_s = \begin{vmatrix} x_s & y_s \\ x_t & y_t \end{vmatrix}$$

differs from zero all along the initial curve (2.4). However, when J vanishes, we find that

$$\frac{x_t}{y_t} = \frac{x_s}{y_s} = \frac{a}{b},$$

in view of (2.6). Hence if there exists a solution $u = u(x, y)$ of (2.1)

[2] Cf. Ince 1.

through a curve (2.4) on which $J = 0$, we can write

$$\frac{u_t}{c} = \frac{u_x x_t + u_y y_t}{a u_x + b u_y} = \frac{y_t}{b} = \frac{x_t}{a}.$$

This means that in the case of a vanishing Jacobian our initial value problem (2.1), (2.4) can be solved only when the initial curve (2.4) is itself a solution of the system of ordinary differential equations (2.3). Clearly the solution is not unique here, since a curve of the latter type can be imbedded in a variety of one-parameter families of solutions of (2.3), each of which generates a solution of (2.1).

In connection with the matter of uniqueness, we mention again that any curve solving the system of ordinary differential equations (2.3) is called a *characteristic* of the first order quasi-linear partial differential equation (2.1). We have seen that each solution of (2.1) consists of a one-parameter family of characteristics. These can be arranged in a unique fashion so that the solution passes through a prescribed curve (2.4) if the condition

(2.8) $ay_t - bx_t \neq 0$

assuring us that the Jacobian (2.7) will differ from zero is fulfilled. When the requirement (2.8) fails, the initial curve (2.4) must itself be restricted to coincide with a characteristic, and then there are infinitely many solutions of the initial value problem. It follows in particular that if two solutions of (2.1) intersect at an angle, the curve along which they are joined must be a characteristic, for through this curve the solution of the initial value problem is not unique. To visualize the situation here, simply observe that the characteristics of (2.1) comprise a two-parameter family of curves which exhaust uniquely every point in space where the coefficients a, b and c are defined.

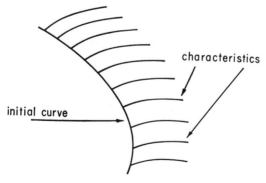

FIG. 1. Solution of the initial value problem by a one-parameter family of characteristics.

The way in which the characteristics feature in our construction of a solution of the first order partial differential equation (2.1) exhibits them as those curves along which jumps in the derivatives of the solution can occur. Thus any singularity of the initial data (2.4) propagates in the (x,y)-plane along the projection there of a relevant characteristic curve (2.3), which is itself sometimes called a characteristic. In connection with the propagation of singularities, notice that a given arc of the data (2.4) only influences the values of the solution u at points on the characteristics through that arc. The projection of these characteristics onto the (x,y)-plane is referred to as the *range of influence* of the arc of initial data in question. Similarly, we can describe the portion of the initial curve cut out by characteristics traced back from a specific section of the solution as the *domain of dependence* of that section. Due to the nonlinearity of (2.1), the notions of a domain of dependence and of a range of influence have only local significance. Also, the geometry of the characteristics can become quite complicated in the large.

It is noteworthy that our solution of (2.1) by reduction to the system of ordinary differential equations (2.3) in no way requires analyticity of the partial differential equation or the initial data involved. In that respect it differs radically from the power series procedure given in Section 1.2.[3] The method based on characteristics provides a much greater insight into the nature of the solution and indicates clearly how it depends on the initial data. Such an approach has the advantage for application to specific examples that it leaves unfinished only the integration of the ordinary differential equations (2.3). Thus it furnishes a quite satisfactory analysis of the problem at hand. Furthermore, we now see without any hypothesis of analyticity that the general solution of a quasi-linear partial differential equation of the first order for a function of two independent variables depends on a single arbitrary function of one variable.

EXERCISES

1. Derive the general solution (1.43) of the first order equation (1.40) by integrating the corresponding system of ordinary differential equations (2.3) for the characteristics.

2. Show that the solution of the nonlinear equation

$$u_x + u_y = u^2$$

passing through the initial curve

$$x = t, \quad y = -t, \quad u = t$$

---

[3] This notation is used to refer to Section 2 of Chapter 1.

becomes infinite along the hyperbola

$$x^2 - y^2 = 4.$$

3. Explain why there are no solutions of the linear equation

$$u_x + u_y = u$$

which pass through the straight line

$$x = t, \quad y = t, \quad u = 1.$$

4. Solve the quasi-linear first order partial differential equation

(2.9) $$\sum_{j=1}^{n} a_j(x_1, \ldots, x_n, u) \frac{\partial u}{\partial x_j} = c(x_1, \ldots, x_n, u)$$

in n independent variables by a reduction to the system of ordinary differential equations

(2.10) $$\frac{dx_1}{a_1} = \frac{dx_2}{a_2} = \cdots = \frac{dx_n}{a_n} = \frac{du}{c}$$

for an n-parameter family of characteristic curves. Show that any $(n-1)$-parameter family of solutions of (2.10) generates an n-dimensional surface in the $(n + 1)$-dimensional space with Cartesian coordinates $x_1, \ldots, x_n, u$ which, when represented in the non-parametric form $u = u(x_1, \ldots, x_n)$, furnishes a solution of (2.9). Prove that the general solution of (2.9) depends on one arbitrary function of $n - 1$ variables.

5. Using the developments of Exercise 4, construct a solution of (2.9) through the $(n-1)$-dimensional initial manifold

(2.11) $$u = u(t_1, \ldots, t_{n-1}), \quad x_j = x_j(t_1, \ldots, t_{n-1}), \quad j = 1, \ldots, n,$$

when the requirement

(2.12) $$\begin{vmatrix} a_1 & \cdots & a_n \\ \dfrac{\partial x_1}{\partial t_1} & \cdots & \dfrac{\partial x_n}{\partial t_1} \\ \cdot & \cdots & \cdot \\ \dfrac{\partial x_1}{\partial t_{n-1}} & \cdots & \dfrac{\partial x_n}{\partial t_{n-1}} \end{vmatrix} \neq 0$$

analogous to (2.8) is fulfilled. Show that both existence and uniqueness of the solution may fail if this determinant vanishes.

6. If two different solutions of the first order partial differential equation (2.9) intersect in a manifold (2.11) of dimension $n - 1$, prove that this manifold

must be characteristic in the sense that the matrix

$$\begin{pmatrix} a_1 & \cdots & a_n & c \\ \dfrac{\partial x_1}{\partial t_1} & \cdots & \dfrac{\partial x_n}{\partial t_1} & \dfrac{\partial u}{\partial t_1} \\ \cdot & \cdot & \cdot & \cdot \\ \dfrac{\partial x_1}{\partial t_{n-1}} & \cdots & \dfrac{\partial x_n}{\partial t_{n-1}} & \dfrac{\partial u}{\partial t_{n-1}} \end{pmatrix}$$

has a rank less than n.

7. Under the hypothesis (2.12), establish that the solution of the initial value problem defined by (2.9) and (2.11) depends continuously on the expressions (2.11) for the data.

2. THE MONGE CONE

We turn our attention to the general nonlinear partial differential equation of the first order

$$(2.13) \qquad F(x, y, u, p, q) = 0$$

for one unknown function u of two independent variables x and y. In order to be sure that the partial derivatives $p = u_x$ and $q = u_y$ actually appear in the equation, we make the formal assumption that

$$F_p^2 + F_q^2 \neq 0.$$

Our objective will be to reduce (2.13) to a system of ordinary differential equations analogous to (2.3) for a family of characteristic curves. However, the situation is now considerably more complicated than before and calls for a larger number of ordinary differential equations involving p and q as unknowns in addition to x, y and u.

From the geometrical point of view adopted in Section 1, the partial differential equation (2.13) represents a functional relationship between p and q that confines the possible normals of solutions through any point x_0, y_0, u_0 to a one-parameter family of directions. These may be described by a vector whose components $p_0(\alpha)$, $q_0(\alpha)$, -1 are functions of a single variable α and are defined so that x_0, y_0, u_0, $p_0(\alpha)$ and $q_0(\alpha)$ fulfill the relation (2.13). The associated family of feasible tangent planes

$$(2.14) \qquad u - u_0 = (x - x_0)p_0(\alpha) + (y - y_0)q_0(\alpha)$$

to solutions through x_0, y_0, u_0 have as their envelope[4] a cone which is

----

[4] Cf. Courant 1.

called the *Monge cone* for (2.13) at that point. The geometrical significance of the first order partial differential equation (2.13) is that any solution surface through a point in space must be tangent there to the corresponding Monge cone. Hence such a surface is covered with Monge cones whose vertices just touch it.

At each point of a solution u of (2.13) one generator of the Monge cone lies in the tangent plane and defines a unique direction within the surface $u = u(x, y)$. The field of directions thus specified comprises the tangents of a one-parameter family of curves sweeping out the solution. These curves are known once again as *characteristics* of the partial differential equation (2.13).

In the quasi-linear case (2.1) which we have already studied, the Monge cone degenerates to the single vector (a, b, c) and the characteristics form a two-parameter family of curves in space that are independent of the particular solutions of the partial differential equation involved. For the nonlinear equation (2.13) the situation tends to be more complicated because many characteristics pass through a given point in space, each arising from a different solution of (2.13). Thus only the Monge cone can be determined without knowledge of more than the coordinates of the point in question. We proceed, nevertheless, to derive a system of ordinary differential equations for the characteristics. However, we shall describe them now as *strips*, rather than curves, because we must include with x, y and u the quantities p and q as dependent variables along each characteristic. These quantities determine a tangent plane at each point of space and thus sweep out an infinitesimal strip.

To find ordinary differential equations for the characteristics, we must associate with a solution of (2.13) through any point x_0, y_0, u_0 the generator

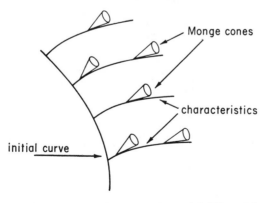

FIG. 2. Geometry of the solution of a nonlinear partial differential equation of the first order.

of the Monge cone there which lies in the tangent plane of the solution. The Monge cone in question is the envelope of the one-parameter family of planes (2.14). It can therefore be represented analytically by eliminating α from that relation and from the equation

$$(2.15) \qquad (x - x_0)p_0'(\alpha) + (y - y_0)q_0'(\alpha) = 0$$

obtained from it by performing a partial differentiation with respect to the parameter α. For each fixed choice of α, the two simultaneous linear equations (2.14) and (2.15) for x, y and u describe a line which is a generator of the Monge cone. We can avoid mentioning $p_0'(\alpha)$ and $q_0'(\alpha)$ here if we take into account the formula

$$(2.16) \qquad F_p p_0'(\alpha) + F_q q_0'(\alpha) = 0,$$

which follows from a differentiation of the identity

$$F(x_0, y_0, u_0, p_0(\alpha), q_0(\alpha)) = 0$$

with respect to α. Indeed, (2.16) shows that (2.15) can be replaced by

$$(2.17) \qquad \frac{x - x_0}{F_p} = \frac{y - y_0}{F_q}.$$

The condition (2.17) alone determines the projection onto the (x,y)-plane of the generator of the Monge cone that lies in the tangent plane of our solution $u = u(x, y)$ of (2.13) through the point x_0, y_0, u_0. Of course, (2.17) presupposes a knowledge of relevant values of $p = u_x$ and $q = u_y$, too. Since the above generator has the direction of a characteristic in the surface $u = u(x, y)$, we deduce from (2.17) that this characteristic must satisfy the ordinary differential equation

$$(2.18) \qquad \frac{dx}{F_p} = \frac{dy}{F_q}.$$

We need to derive similar differential equations for u, p and q as well as x and y along the characteristic, since all five quantities appear as arguments of F_p and F_q.

From the familiar expression

$$du = p\,dx + q\,dy$$

for the differential of the function u we obtain

$$(2.19) \qquad \frac{du}{pF_p + qF_q} = \frac{p\,dx + q\,dy}{pF_p + qF_q} = \frac{dx}{F_p} = \frac{dy}{F_q},$$

in view of (2.18). To find expressions for dp and dq, we differentiate (2.13) with respect to x and y, getting

$$F_x + F_u p + F_p p_x + F_q q_x = 0,$$
$$F_y + F_u q + F_p p_y + F_q q_y = 0.$$

Since $p_y = q_x$ and since F_p and F_q are connected with dx and dy by (2.18), we deduce that

$$(2.20) \qquad \frac{dp}{F_x + pF_u} = -\frac{p_x\,dx + p_y\,dy}{p_x F_p + p_y F_q} = -\frac{dx}{F_p}.$$

Similarly we have

$$(2.21) \qquad \frac{dq}{F_y + qF_u} = -\frac{q_x\,dx + q_y\,dy}{q_x F_p + q_y F_q} = -\frac{dy}{F_q}.$$

Formulas (2.18) to (2.21) may be summarized in the final system of four ordinary differential equations

$$(2.22) \qquad \frac{dx}{F_p} = \frac{dy}{F_q} = \frac{du}{pF_p + qF_q} = -\frac{dp}{F_x + pF_u} = -\frac{dq}{F_y + qF_u}$$

for the determination of x, y, u, p and q along a characteristic strip.

Introducing an appropriate parameter s, we can express the system of ordinary differential equations (2.22) for the characteristics of the first order partial differential equation (2.13) in the more tangible form

$$(2.23) \qquad \begin{aligned} &\frac{dx}{ds} = F_p, \quad \frac{dy}{ds} = F_q, \quad \frac{du}{ds} = pF_p + qF_q, \\ &\frac{dp}{ds} = -F_x - pF_u, \quad \frac{dq}{ds} = -F_y - qF_u \end{aligned}$$

analogous to (2.6). We obtain in this manner a system of five equations to replace the four defined by (2.22). The extra arbitrary constant thus incorporated in the solution merely represents a translation of the parameter s, since when five specific functions of s solve (2.23), so also do the same five functions of $s - s_0$, with s_0 fixed. We have to deal therefore with a family of solutions of (2.23) involving in an essential way only four parameters. It will even turn out that the relevant characteristics of (2.13) depend on just three parameters, since they correspond to solutions of (2.23) along which F must vanish identically.

In order to establish that the latter requirement can be met simply by choosing one parameter appropriately, we prove that

$$(2.24) \qquad F(x, y, u, p, q) \equiv \text{const.}$$

along any solution of (2.23). Indeed, according to (2.23) we have

$$\frac{dF}{ds} = F_x \frac{dx}{ds} + F_y \frac{dy}{ds} + F_u \frac{du}{ds} + F_p \frac{dp}{ds} + F_q \frac{dq}{ds}$$

$$= F_x F_p + F_y F_q + F_u(pF_p + qF_q)$$

$$- F_p(F_x + pF_u) - F_q(F_y + qF_u) = 0.$$

Because of this property of the function F we say that it is an *integral* of the system (2.23). Such an integral will vanish identically along a solution of (2.23) if we select the initial values of p and q so that it vanishes for a fixed choice of s. Evidently we thus dispose of precisely one arbitrary constant in the solution.

Our analysis shows that each solution of the partial differential equation (2.13) is swept out by a one-parameter subset of the three-parameter family of characteristic curves defined by (2.23) and the auxiliary condition $F = 0$. The next problem is to establish the converse result that a one-parameter family of characteristics of this type passing through a pre-scribed initial curve

$$x = x(t), \quad y = y(t), \quad u = u(t)$$

yields a solution of (2.13) at least in the small.

The initial condition, which is of the same form as (2.4), must be elaborated in the present situation to include the two requirements

(2.25) $u'(t) = px'(t) + qy'(t),$

(2.26) $F(x(t), y(t), u(t), p, q) = 0.$

Formulas (2.25) and (2.26) determine p and q along the curve (2.4) as functions

(2.27) $p = p(t), \quad q = q(t)$

of t such that the solutions

(2.28) $x = x(s, t), \quad y = y(s, t), \quad u = u(s, t), \quad p = p(s, t), \quad q = q(s, t)$

of (2.23) with initial values at $s = 0$ given by (2.4) and (2.27) sweep out a surface for which p and q are the first partial derivatives of u and for which (2.13) is fulfilled. That (2.13) will be satisfied follows immediately from (2.26) and the statement (2.24) that F is an integral of the system (2.23). Our principal difficulty is therefore with the verification that p and q, as defined by (2.28), are actually the partial derivatives of u with respect to x and y.

To prove this lemma we exploit a device which appears in a variety of useful forms in the theory of partial differential equations. Let

$$U = \frac{\partial u}{\partial s} - p\frac{\partial x}{\partial s} - q\frac{\partial y}{\partial s},$$

$$V = \frac{\partial u}{\partial t} - p\frac{\partial x}{\partial t} - q\frac{\partial y}{\partial t}.$$

What we really need to show is that

$$U \equiv V \equiv 0.$$

The first three equations of the system (2.23) imply directly that U vanishes for all s and t. We proceed to derive a linear homogenous ordinary differential equation for V which permits us to conclude from the uniqueness theorem for an associated initial value problem that it is identically zero, too.

Observe that

$$(2.29) \qquad \frac{\partial V}{\partial s} - \frac{\partial U}{\partial t} = \frac{\partial p}{\partial t}\frac{\partial x}{\partial s} - \frac{\partial p}{\partial s}\frac{\partial x}{\partial t} + \frac{\partial q}{\partial t}\frac{\partial y}{\partial s} - \frac{\partial q}{\partial s}\frac{\partial y}{\partial t},$$

since all the terms that might appear on the right involving mixed partial derivatives with respect to s and t cancel out. If we use (2.23) to eliminate from (2.29) the derivatives of x, y, p and q with respect to s, and if we take into account the relation $U \equiv 0$, we find that

$$\frac{\partial V}{\partial s} = F_p\frac{\partial p}{\partial t} + F_q\frac{\partial q}{\partial t} + F_x\frac{\partial x}{\partial t} + F_y\frac{\partial y}{\partial t} + F_u\left(p\frac{\partial x}{\partial t} + q\frac{\partial y}{\partial t}\right)$$

$$= \frac{\partial F}{\partial t} - F_u\left(\frac{\partial u}{\partial t} - p\frac{\partial x}{\partial t} - q\frac{\partial y}{\partial t}\right).$$

Since $F \equiv 0$, this leads to the linear homogeneous ordinary differential equation

$$(2.30) \qquad \frac{\partial V}{\partial s} + F_u V = 0$$

for V, in which we may view the coefficient F_u as a known function. From the initial condition (2.25) we see that V vanishes at $s = 0$. Hence we can deduce from the uniqueness of the solution of initial value problems for (2.30) that V must coincide everywhere with the trivial solution $V = 0$.

For the solution (2.28) of the Cauchy problem given by (2.23), (2.4) and (2.27) we have shown that F, U and V all vanish identically. However, this is not quite sufficient to establish that (2.28) defines a solution of

(2.13) in parametric form. It remains to be proved, in particular, that we can replace s and t as the independent variables by x and y. Thus to complete the final steps of the demonstration we have to require that the Jacobian

$$J = \begin{vmatrix} x_s & y_s \\ x_t & y_t \end{vmatrix}$$

differ from zero all along the prescribed initial curve (2.4), just as in the quasi-linear case. Then we can change variables and we can compare the standard relations

$$\frac{\partial u}{\partial s} - u_x \frac{\partial x}{\partial s} - u_y \frac{\partial y}{\partial s} = 0, \qquad \frac{\partial u}{\partial t} - u_x \frac{\partial x}{\partial t} - u_y \frac{\partial y}{\partial t} = 0$$

with the fact that U and V vanish in order to verify that p and q are identical with u_x and u_y. The latter result is seen to follow because both pairs of quantities satisfy the same two linear equations whose determinant J is not zero. Thus (2.28) does provide a solution u of (2.13) at least locally.

By virtue of the first two equations of the system (2.23), we can put the condition $J \neq 0$ into the convenient form

$$(2.31) \qquad \begin{vmatrix} F_p & F_q \\ \dfrac{\partial x}{\partial t} & \dfrac{\partial y}{\partial t} \end{vmatrix} \neq 0,$$

where the arguments of F_p and F_q are subject to (2.25) and (2.26). Our discussion shows that under the restriction (2.31) there exists locally a solution of the first order partial differential equation (2.13) passing through the prescribed curve (2.4). The solution is unique for a fixed determination of the roots p and q of the nonlinear relations (2.25) and (2.26). It can be constructed from a one-parameter family of characteristic strips (2.28) which are obtained by integrating the system of five ordinary differential equations (2.23), with initial data given by (2.4) and (2.27). In particular, the general solution of (2.13) involves in an essential way only one arbitrary function of a single variable, since that is enough to describe the initial curve (2.4) when it is confined to lie, for example, in one of the coordinate planes $x = $ const. or $y = $ const.

The exceptional case where the determinant in (2.31) vanishes identically along the initial curve (2.4) still merits our attention. The vanishing of this determinant means that (2.4) itself satisfies the ordinary differential equation (2.18). In order to lie on a solution u of (2.13) possessing derivatives of the second order, the curve (2.4) would therefore also have to fulfill the requirements (2.19), (2.20) and (2.21). Thus it would have to be a characteristic. We conclude that in the exceptional case at hand our initial value problem may have no smooth solutions, that is, solutions with continuous second derivatives, at all. However, if the initial curve has a vanishing determinant J and happens to be a characteristic as well, the problem will have infinitely many solutions consisting of one-parameter families of characteristics amongst which the given one is imbedded. Finally, we observe that when the determinant J is zero, but (2.4) is not a characteristic, our construction based on passing suitable characteristics through (2.4) might still furnish a solution of (2.13) exhibiting the initial curve as a singularity along its edge which the characteristics touch in such a way that certain second derivatives of u no longer exist (cf. Exercise 3 below). In this situation we call (2.4) a *focal curve*, or *caustic*, for the solution of (2.13).

Our results show that any curve of intersection of two solutions of (2.13) which bifurcate smoothly in the sense that both have the same tangent planes along their intersection must necessarily be a characteristic. For the proof, merely note that along such a curve the solution of the initial value problem corresponding to a fixed determination of the roots p and q of (2.25) and (2.26) is not unique. However, when the equation (2.13) is not quasi-linear, it may have two solutions intersecting at an angle along a curve that is not a characteristic, for the possibility of putting these two solutions through the same curve can arise from a transition from one branch of the roots of the nonlinear relation (2.26) for p and q to another.

We have explained the structure of the solution of a first order partial differential equation in two independent variables as a one-parameter family of characteristic curves. Usually the family is to be selected from a set of characteristics depending on three parameters, but in the quasi-linear case this number diminishes to two. Furthermore, the number of ordinary differential equations needed to find the characteristics reduces in the quasi-linear case from the four given by (2.22) to the two given by (2.3). We have seen that a more complicated situation regarding the determination of p and q and involving the Monge cone arises in the general case. Finally, the characteristics appear geometrically as bifurcations of solutions, and as such they may always be interpreted as the loci along which discontinuities in the higher derivatives of a solution can propagate.

EXERCISES

1. Prove that the conoidal surface generated by the one-parameter family of characteristic strips through any fixed point x_0, y_0, u_0 in space is a solution of (2.13). These strips are the solutions of the system (2.23) with initial values

x, y, u, p and q such that $x = x_0, y = y_0, u = u_0$ and

$$F(x_0, y_0, u_0, p, q) = 0.$$

2. Investigate the possibility of finding more than one (or no) solution of the initial value problem (2.13), (2.4) because the relations (2.25), (2.26) for p and q possess more than one (or no) solution.

3. Give a detailed discussion of focal curves. In particular, explain why second derivatives of u are needed to deduce (2.20) and (2.21) from (2.18). Show that any focal curve is the envelope of a one-parameter family of characteristics.

4. Solve the equation

$$u^2(1 + p^2 + q^2) = 1$$

by the method of characteristics. In particular, determine the conoidal solutions. Find a solution possessing a focal curve.

5. Given the values of a solution u of (2.13) along a prescribed curve

$$x = x(t), \qquad y = y(t)$$

in the (x,y)-plane, consider the problem of calculating in succession p and q and all the higher partial derivatives of u there from this data and from the partial differential equation itself (cf. Section 1.2). Show that the hypothesis (2.31) is needed in order to ensure that such a process can be performed.

6. Given a solution $u = u(x_1, \ldots, x_n)$ of the partial differential equation

(2.32) $$F(x_1, \ldots, x_n, u, p_1, \ldots, p_n) = 0$$

in n independent variables x_i, where

$$p_i = \frac{\partial u}{\partial x_i},$$

show how it is composed of integrals of the system of $2n + 1$ ordinary differential equations

(2.33) $$\frac{dx_i}{ds} = F_{p_i}, \quad \frac{du}{ds} = \sum_{j=1}^{n} p_j F_{p_j}, \quad \frac{dp_i}{ds} = -F_{x_i} - p_i F_u, \quad i = 1, \ldots, n,$$

for a $(2n - 1)$-parameter family of characteristic strips satisfying the auxiliary condition $F = 0$. Establish that an $(n - 1)$-parameter family of solutions of (2.33) defined by initial data at $s = 0$ of the form

$$u = u(t_1, \ldots, t_{n-1}), \quad x_i = x_i(t_1, \ldots, t_{n-1}), \quad p_i = p_i(t_1, \ldots, t_{n-1}),$$

where $p_1, \ldots, p_n$ are supposed to fulfill the requirements

$$\frac{\partial u}{\partial t_i} = \sum_{j=1}^{n} p_j \frac{\partial x_j}{\partial t_i}, \qquad i = 1, \ldots, n - 1,$$

$$F(x_1(t_1, \ldots, t_{n-1}), \ldots, x_n(t_1, \ldots, t_{n-1}), u(t_1, \ldots, t_{n-1}), p_1, \ldots, p_n) = 0,$$

generates a solution of (2.32) provided that

$$
\begin{vmatrix}
F_{p_1} & \cdots & F_{p_n} \\
\dfrac{\partial x_1}{\partial t_1} & \cdots & \dfrac{\partial x_n}{\partial t_1} \\
\cdot & \cdot \cdot \cdot & \cdot \\
\dfrac{\partial x_1}{\partial t_{n-1}} & \cdots & \dfrac{\partial x_n}{\partial t_{n-1}}
\end{vmatrix} \neq 0.
$$

Discuss the exceptional case in which this determinant vanishes identically on the initial manifold.

7. On how many arbitrary functions of how many parameters does the general solution of (2.32) really depend?

8. Give a generalization of Exercise 1 for the case of the equation (2.32) in $n > 2$ independent variables.

3. THE COMPLETE INTEGRAL

We have seen that the general solution of the first order partial differential equation

$$(2.13) \qquad\qquad F(x, y, u, p, q) = 0$$

is an expression involving an arbitrary function of one variable. This is a natural extension of the result that the general solution of an ordinary differential equation of the first order involves one arbitrary constant. However, we shall now be interested in solutions

$$(2.34) \qquad\qquad u = \phi(x, y; a, b)$$

of (2.13) that depend on two parameters a and b rather than on one arbitrary function. An expression of this kind is called a *complete integral* of the equation (2.13) when the parameters a and b occur in a truly independent fashion. We shall proceed to show how we can deduce from such a complete integral relations defining the general solution and relations describing the characteristics. On the other hand, no systematic rule determining the complete integral is available.

The basic theorem which makes the complete integral (2.34) significant is that the envelope of any family of solutions of the first order equation (2.13) depending on some parameter must again be a solution. Our assertion is easily understood geometrically because equation (2.13) is merely a statement about the tangent plane of a solution. Consequently if a surface has the same tangent plane as a solution at some point in space, then it also satisfies the equation there. The envelope of a family of solutions is touched at each of its points by one of these solutions, and thus it must itself be a solution.

To obtain the general solution of (2.13) from the complete integral, we have only to prescribe the second parameter b, say, as an arbitrary function

$$b = b(a)$$

of the first parameter a. We then consider the envelope of the one-parameter family of solutions so defined. This envelope is described by the two simultaneous equations

(2.35) $$u = \phi(x, y; a, b(a)),$$

(2.36) $$\phi_a(x, y; a, b(a)) + \phi_b(x, y; a, b(a))b'(a) = 0,$$

of which the second is obtained from the first by performing a partial differentiation with respect to a. Elimination of the parameter a between the equations (2.35) and (2.36) would yield a single expression for the envelope that would involve the arbitrary function $b(a)$ and would also represent a solution of (2.13). It would therefore be a general solution of (2.13). Hence we can say that the two relations (2.35) and (2.36) together define the general solution.

A more analytical demonstration that the simultaneous equations (2.35) and (2.36) describe a solution $u = u(x, y)$ of the partial differential equation (2.13) proceeds as follows. The second of these equations can be understood to define a as a function of x and y. Then we can write

$$u_x = \phi_x + [\phi_a + \phi_b b'(a)]\, a_x,$$
$$u_y = \phi_y + [\phi_a + \phi_b b'(a)]\, a_y.$$

Since the quantity in square brackets vanishes by virtue of (2.36), we obtain

(2.37) $$u_x = \phi_x, \qquad u_y = \phi_y.$$

The five numbers x, y, ϕ, ϕ_x and ϕ_y are known to fulfill (2.13) because ϕ, considered as a function of x and y, is one of the solutions of (2.13). However, since (2.35) and (2.37) establish that the above values are identical with the set x, y, u, u_x and u_y for each particular choice of a, we conclude that u is also a solution, as indicated.

In addition to forming the general solution (2.35), (2.36) of (2.13), we can construct the so-called *singular solution* by finding the envelope of the full two-parameter family of surfaces defined by the complete integral (2.34). This envelope is given by the three relations

(2.38) $\quad u = \phi(x, y; a, b), \quad \phi_a(x, y; a, b) = 0, \quad \phi_b(x, y; a, b) = 0,$

of which the last two can be solved to eliminate a and b provided that

$$\begin{vmatrix} \phi_{aa} & \phi_{ab} \\ \phi_{ba} & \phi_{bb} \end{vmatrix} \neq 0.$$

Whenever the above determinant differs from zero, so that the procedure we have just outlined yields a genuine result, we obtain a singular solution (2.38) of the partial differential equation (2.13) that cannot be found by specialization of the general solution. Its nature is similar to that of the singular solution of an ordinary differential equation of the first order.[5]

It is important to observe that the characteristics of (2.13) can be found from the complete integral (2.34). We have pointed out earlier that any curve where two solutions of (2.13) bifurcate smoothly has to be a characteristic. Now for a fixed value of the parameter a, the two relations (2.35) and (2.36) represent the intersection of a special solution defined by (2.35) alone and of the envelope solution obtained by eliminating a from (2.35) and (2.36). The two solutions in question are tangent along their intersection, which therefore cannot be the result of a transition from one branch of (2.26) to another and must, indeed, be a characteristic. With this in mind our next aim will be to determine all the characteristics of (2.13) by considering, for fixed choices of the parameters a and b, appropriate systems of simultaneous relations like (2.35) and (2.36).

We shall discuss first the case in which the partial differential equation (2.13) involves the argument u explicitly, so that

$$F_u \neq 0.$$

We note that for a specific choice of a the relation (2.36) implies that the partial derivatives ϕ_a and ϕ_b remain in a fixed ratio. Therefore we are led to introduce a parameter σ along the characteristic defined by (2.35) and (2.36) so that these equations take the form

(2.39) $\quad u = \phi(x, y; a, b), \quad \phi_a(x, y; a, b) = \lambda\sigma, \quad \phi_b(x, y; a, b) = \mu\sigma,$

where σ remains variable, but where a, b, λ and μ are constants satisfying the requirement

$$\lambda + \mu b'(a) = 0.$$

[5] Cf. Ince 1.

Using our previous reasoning only as a motivation, we shall proceed to establish directly that the strips

$$x = x(\sigma), \quad y = y(\sigma), \quad u = u(\sigma), \quad p = p(\sigma), \quad q = q(\sigma)$$

given by the relations (2.39) and

$$(2.40) \qquad p = \phi_x(x, y; a, b), \qquad q = \phi_y(x, y; a, b),$$

along which σ features as the basic independent variable, are precisely the characteristics of (2.13) and solve the system of ordinary differential equations (2.22), provided that

$$(2.41) \qquad \begin{vmatrix} \phi_{ax} & \phi_{ay} \\ \phi_{bx} & \phi_{by} \end{vmatrix} \neq 0.$$

In the case $F_u \neq 0$ the requirement (2.41) makes explicit the independent roles played by the parameters a and b which justify calling (2.34) a complete integral of (2.13).

Since multiplication of λ and μ by a common factor only serves to change the scale of the variable σ, we see that the family of curves (2.39), (2.40) depends on three distinct parameters, namely, a, b and λ/μ. This is precisely the number needed to describe the characteristics of (2.13). We can solve the last pair of equations (2.39) for x and y as functions

$$x = x(\sigma), \qquad y = y(\sigma)$$

of the variable σ because the relevant Jacobian (2.41) differs from zero. When we differentiate those equations with respect to σ, we find that

$$(2.42) \qquad \phi_{ax}\frac{dx}{d\sigma} + \phi_{ay}\frac{dy}{d\sigma} = \lambda, \qquad \phi_{bx}\frac{dx}{d\sigma} + \phi_{by}\frac{dy}{d\sigma} = \mu.$$

On the other hand, the partial differential equation (2.13) is satisfied identically in a and b by the complete integral (2.34). Therefore partial differentiation of (2.13) with respect to the parameters a and b yields

$$(2.43) \qquad F_p\phi_{xa} + F_q\phi_{ya} = -F_u\phi_a, \qquad F_p\phi_{xb} + F_q\phi_{yb} = -F_u\phi_b.$$

We may view (2.42) and (2.43) as pairs of simultaneous linear equations for the determination of $dx/d\sigma$, $dy/d\sigma$ and of F_p, F_q. We notice that they have the same non-vanishing determinant (2.41). Since (2.39) implies that the right-hand sides λ, μ and $-F_u\phi_a$, $-F_u\phi_b$ are proportional, we conclude from the uniqueness of the solution of such simultaneous linear equations that

(2.44)
$$\frac{dx}{d\sigma} = -\frac{F_p}{\sigma F_u}, \quad \frac{dy}{d\sigma} = -\frac{F_q}{\sigma F_u}.$$

Let us introduce a new independent variable s along the curve (2.39), (2.40) which is defined by the ordinary differential equation

$$\frac{ds}{d\sigma} = -\frac{1}{\sigma F_u}.$$

Then (2.44) assumes the form

$$\frac{dx}{ds} = F_p, \quad \frac{dy}{ds} = F_q$$

equivalent to (2.18) or, in other words, to the first two equations (2.23). The three remaining ordinary differential equations (2.23) for the characteristics now follow for the strip given by (2.39) and (2.40) in exactly the same way that (2.22) followed from (2.18), for this strip is imbedded in the special solution $u = \phi$ of (2.13) and that is precisely the information required for the final step in the proof that the characteristic differential equations are satisfied.

The foregoing explanation of how the characteristics are to be found from a complete integral is only valid under the hypothesis $F_u \neq 0$, for F_u appears in the denominator in (2.44). When the partial differential equation (2.13) has the form

(2.45)
$$F(x, y, p, q) = 0$$

not involving u explicitly, it becomes desirable to reduce the complete integral (2.34) to an expression of the type

(2.46)
$$u = \phi(x, y; a) + b.$$

Such a step is feasible because addition of any constant b to a solution ϕ of (2.45) furnishes another solution. The general solution of (2.44) can now be derived by eliminating a from the two equations

$$u = \phi(x, y; a) + b(a), \quad \phi_a(x, y; a) + b'(a) = 0.$$

If we assign arbitrary constant values to a, b and $\lambda = -b'(a)$ here, it can be seen that the relations

(2.47)
$$u = \phi(x, y; a) + b, \quad \phi_a(x, y; a) = \lambda,$$

(2.48)
$$p = \phi_x(x, y; a), \quad q = \phi_y(x, y; a)$$

define a characteristic strip satisfying the system of ordinary differential equations (2.22), provided only that ϕ_{ax} and ϕ_{ay} do not vanish simultaneously. Indeed, differentiation of the second equation (2.47) gives

$$(2.49) \qquad \phi_{ax}\,dx + \phi_{ay}\,dy = 0,$$

and we may also substitute (2.46) into (2.45) and take a partial derivative with respect to the parameter a to derive

$$(2.50) \qquad F_p\,\phi_{xa} + F_q\,\phi_{ya} = 0.$$

The two relations (2.49) and (2.50) can hold simultaneously only when (2.18) is fulfilled. Now (2.22) follows in the usual way from the fact that the strip (2.47), (2.48) lies on a solution (2.46) of (2.45).

For examples of partial differential equations of the first order where some special device permits us to determine a complete integral, the theory we have developed above can be quite useful in describing the general solution or the characteristics. Moreover, the problem of passing a solution of (2.13) through a prescribed curve in space can be handled by considering a one-parameter family of solutions from the complete integral which just touch the given curve, and then rolling them along the curve to generate an envelope that furnishes the answer. On the other hand, when the system of ordinary differential equations (2.23) for the characteristic strips can be solved in closed form, it is, of course, an easy matter to find a complete integral. For this purpose it would suffice, indeed, to introduce the conoidal solution composed of all characteristics through a given point in space and then to allow that point to vary over an appropriate parameter surface (cf. Exercise 2.1).

EXERCISES

1. Prove that
$$u = ax + by + f(a, b)$$
is a complete integral of Clairaut's equation
$$(2.51) \qquad u = px + qy + f(p, q).$$
Show that the characteristics of (2.51) are straight lines.

2. Show that
$$(2.52) \qquad u = \sqrt{1 - (x - a)^2 - (y - b)^2}$$
is a complete integral of
$$(2.53) \qquad u^2(1 + p^2 + q^2) = 1.$$

THE COMPLETE INTEGRAL 39

Use (2.52) to find the characteristics, the general solution, and the singular solution of this partial differential equation (cf. Exercise 2.4).

3. Given any two-parameter family of surfaces (2.34), prove that there is a corresponding partial differential equation of the first order for which it is the complete integral.

4. Show that the singular solution of (2.13) satisfies the three simultaneous equations

$$F(x, y, u, p, q) = 0, \quad F_p(x, y, u, p, q) = 0, \quad F_q(x, y, u, p, q) = 0,$$

from which it can be found by eliminating p and q. Apply this result to the equation (2.53).

5. The complete integral of the first order partial differential equation (2.32) for a function u of n independent variables x_i is a solution

$$(2.54) \qquad u = \phi(x_1, \ldots, x_n; a_1, \ldots, a_n)$$

depending on n different parameters a_i, with

$$\begin{vmatrix} \phi_{a_1 x_1} & \cdots & \phi_{a_1 x_n} \\ \cdot & \cdot & \cdot & \cdot & \cdot \\ \phi_{a_n x_1} & \cdots & \phi_{a_n x_n} \end{vmatrix} \neq 0$$

in the case $F_u \neq 0$. Show that the general solution of (2.32) can be found by setting

$$a_i = a_i(t_1, \ldots, t_{n-1}), \qquad i = 1, \ldots, n,$$

and introducing the envelope of (2.54) with respect to the $n - 1$ new parameters $t_1, \ldots, t_{n-1}$, which is defined by eliminating them from (2.54) and the relations

$$\sum_{j=1}^{n} \frac{\partial \phi}{\partial a_j} \frac{\partial a_j}{\partial t_i} = 0, \qquad i = 1, \ldots, n - 1.$$

Prove that the characteristic strips (2.33) for (2.32) are given by (2.54) and the equations

$$\phi_{a_i}(x_1, \ldots, x_n; a_1, \ldots, a_n) = \lambda_i \sigma, \qquad i = 1, \ldots, n,$$

$$p_i = \phi_{x_i}(x_1, \ldots, x_n; a_1, \ldots, a_n), \qquad i = 1, \ldots, n,$$

with σ variable, but with the parameters $a_1, \ldots, a_n$ and $\lambda_1, \ldots, \lambda_n$ fixed.

6. For a partial differential equation

$$(2.55) \qquad F(x_1, \ldots, x_n, p_1, \ldots, p_n) = 0$$

of the first order in n variables with u missing, a complete integral may be expressed in the form

$$(2.56) \qquad u = \phi(x_1, \ldots, x_n; a_1, \ldots, a_{n-1}) + b$$

analogous to (2.46). Show that the general solution can be obtained by setting

$$b = b(a_1, \ldots, a_{n-1})$$

and considering the envelope defined by (2.56) and the equations

$$\phi_{a_i}(x_1, \ldots, x_n; a_1, \ldots, a_{n-1}) + b_{a_i}(a_1, \ldots, a_{n-1}) = 0, \qquad i = 1, \ldots, n-1.$$

Assuming that the matrix

$$\begin{pmatrix} \phi_{a_1 x_1} & \cdots & \phi_{a_1 x_n} \\ \cdot & \cdot \cdot \cdot \cdot & \cdot \\ \phi_{a_{n-1} x_1} & \cdots & \phi_{a_{n-1} x_n} \end{pmatrix}$$

is of rank $n - 1$, prove that (2.56) and the relations

$$(2.57) \qquad \phi_{a_i}(x_1, \ldots, x_n; a_1, \ldots, a_{n-1}) = \lambda_i, \qquad i = 1, \ldots, n-1,$$

$$(2.58) \qquad p_i = \phi_{x_i}(x_1, \ldots, x_n; a_1, \ldots, a_{n-1}), \qquad i = 1, \ldots, n,$$

with $a_1, \ldots, a_{n-1}$ and $\lambda_1, \ldots, \lambda_{n-1}$ and b held fixed, define a solution of the characteristic system of ordinary differential equations (2.33) for (2.55).

7. Show that a partial differential equation (2.32) in n independent variables can be reformulated as one of the special form (2.55) in $n + 1$ independent variables by interpreting the coordinate of the unknown u as a new independent variable x_{n+1} and by replacing the solutions $u(x_1, \ldots, x_n)$ of (2.32) by solutions

$$\phi = x_{n+1} - u(x_1, \ldots, x_n)$$

of the new equation. Establish conversely that the locus of zeros of a solution ϕ of the latter type produces a function u satisfying the original equation (cf. Exercise 4.1 below).

4. THE EQUATION OF GEOMETRICAL OPTICS

As an application of the theory of partial differential equations of the first order that we have developed in Sections 2 and 3, we shall study in detail the special equation

$$(2.59) \qquad p^2 + q^2 = 1,$$

which arises in geometrical optics.[6] The solutions $u = u(x, y)$ of (2.59) serve to describe two-dimensional phenomena of geometrical optics. In this context the level curves

$$(2.60) \qquad u(x, y) = \text{const.}$$

[6] Cf. Keller-Lewis-Seckler 1, Kline 1.

represent wave fronts, while the characteristics represent light rays (cf. Section 6.1). Note that (2.59) is an example with the property $F_u \equiv 0$.

The system of five ordinary differential equations (2.23) for the characteristics of (2.59) assumes the form

$$(2.61) \quad \frac{dx}{ds} = 2p, \quad \frac{dy}{ds} = 2q, \quad \frac{du}{ds} = 2p^2 + 2q^2, \quad \frac{dp}{ds} = 0, \quad \frac{dq}{ds} = 0.$$

It follows that p and q are constants along any characteristic and that

$$x = 2ps + \text{const.}, \quad y = 2qs + \text{const.}, \quad u = 2s + \text{const.}$$

there. Hence the characteristics are straight lines. In view of the restriction (2.59) on p and q, they are precisely those lines which make an angle of 45° with the u-axis.

It is interesting to study how the characteristics sweep out a solution u of (2.59). Observe that according to the first two relations (2.61), the projection of any characteristic of (2.59) onto the (x,y)-plane is a line in the direction of the vector (p, q), which is, of course, normal to the level curves (2.60). In the language of geometrical optics, we can assert that each light ray is perpendicular to all the wave fronts that it cuts. In particular, any two level curves (2.60) of a solution of (2.59) have identical normals. Furthermore, the distance between these curves measured along the normals must be constant, because u differs from the arc length along the characteristics only by the factor $\sqrt{2}$. The physical significance of the latter remark is that the wave fronts move at a constant rate in the direction of their normals.

A geometrical construction of the solution of (2.59) passing through a prescribed curve in the (x,y)-plane is obtained by moving all points of the curve in the normal direction at a fixed rate and simultaneously raising, or lowering, them by an equal amount. The trajectories that are described in this way by individual points are simply the characteristics which sweep out the solution in question. Two answers are found, one from raising the points and the other from lowering them; they correspond physically to wave fronts which move forward or backward. Analytically, it is the quadratic character of the partial differential equation (2.59) that leads to two solutions of the initial value problem.

With the information now at hand, it would be an easy matter to write down by inspection formulas for the general solution of (2.59). We prefer, however, to resort to the method of the complete integral for that purpose, which leads in a natural way to a convenient formulation of the answer.

To find a complete integral of (2.59) we use the technique of separation of variables and seek solutions of the special form

$$(2.62) \qquad\qquad u = f(x) + g(y).$$

Substitution of (2.62) into (2.59) gives

$$f'(x)^2 + g'(y)^2 = 1.$$

Therefore $f'(x)$ and $g'(y)$ must be constants, and we can write

$$f'(x) = \cos a, \qquad g'(y) = \sin a.$$

We conclude that

$$(2.63) \qquad\qquad u = x \cos a + y \sin a + b$$

is a complete integral of the equation of geometrical optics. The particular solutions described by (2.63) merely comprise the family of planes that cut the (x,y)-plane at an angle of 45°.

To find the general solution of (2.59), we introduce an arbitrary function $b = b(a)$ into (2.63) and examine the envelope of the resulting one-parameter family of planes. This is obtained by elimination of the parameter a from the two equations

$$(2.64) \qquad\qquad u = x \cos a + y \sin a + b(a),$$

$$(2.65) \qquad\qquad 0 = -x \sin a + y \cos a + b'(a),$$

which therefore represent the desired general solution. The characteristics are found by fixing a in (2.64) and (2.65) and allowing $b(a)$ and $b'(a)$ to assume arbitrary constant values. They are once more seen to be lines inclined at an angle of 45° with the u-axis because they are now given as intersections of the vertical planes (2.65) with perpendicular planes (2.64) making an angle of 45° with the horizontal.

The formulas (2.64) and (2.65) for a general solution of (2.59) fit in elegantly with our earlier remarks about the solution through a prescribed curve in the (x,y)-plane. In fact, if we put $u = 0$ in (2.64), it becomes the equation in normal form of a line in the (x,y)-plane whose distance from the origin is $-b(a)$ and whose normal makes an angle a with the x-axis. As the parameter a varies, this line envelopes the initial curve $u = 0$ in the (x,y)-plane, whereas (2.65) describes the corresponding normals. Thus in order to determine a solution of (2.59) passing through a given curve in the (x,y)-plane, we have only to express the distance $-b$ from the origin to the tangent of that curve as a function of the angle a between the normal and the x-axis and then insert the particular function $b(a)$ so obtained into the representation (2.64), (2.65).

We have presented a detailed account of the structure of the solutions of (2.59). Each of them is generated by a one-parameter family of characteristic straight lines that we have interpreted as light rays. It is natural to inquire about the extent of such a solution in the large. This leads immediately to a discussion of the envelope of the one-parameter family of characteristics involved, which is called a *focal curve*, or *caustic*.

The characteristics generating a solution of (2.59) can be represented by the equations (2.64) and (2.65) for various values of the parameter a. The second equation (2.65) was derived from the first one (2.64) by performing a partial differentiation with respect to a aimed at determining the envelope of the one-parameter family of planes (2.64). To find the envelope of the one-parameter family of characteristic lines defined by both (2.64) and (2.65), the appropriate procedure is to make a further partial differentiation with respect to a, which yields

$$(2.66) \qquad 0 = -x \cos a - y \sin a + b''(a).$$

Elimination of the parameter a from the three simultaneous equations (2.64), (2.65) and (2.66) gives two equations that determine the desired focal curve. In optics this would appear, of course, as a bright streak due to the condensation of light rays. Note that the tangents of the focal curve are, conversely, the characteristics that sweep out the original solution of (2.59) from which we started.

There are solutions of (2.59) which exhibit no focal curve and hence can be extended into the large indefinitely. Indeed, the planes (2.63) have the desired property. At the other extreme, the solution composed of all characteristics through a fixed point x_0, y_0, u_0 in space has a focal curve which degenerates to that single point (cf. Exercise 2.1). This solution, which is of special interest, is the cone

$$(2.67) \qquad (u - u_0)^2 = (x - x_0)^2 + (y - y_0)^2.$$

It can be used to form a complete integral. From the point of view of geometrical optics it represents light rays emanating from a single source, with circular wave fronts.

EXERCISES

1. Show that the partial differential equation

$$\phi_x^2 + \phi_y^2 = \phi_u^2$$

in three independent variables, x, y and u, can be reduced to (2.59).

2. Determine the characteristics, the complete integral and the general solution of the equation

(2.68) $$u_x^2 + u_y^2 + u_z^2 = 1,$$

which has to do with geometrical optics in space.

3. In analogy with (2.64), (2.65) and (2.66), define the notion of a caustic surface for (2.68).

4. Represent the general solution of (2.59) as the surface swept out by the tangents of any focal curve in space satisfying the ordinary differential equation

$$du^2 = dx^2 + dy^2.$$

5. Construct a solution of (2.59) through the curve (2.4) by considering the envelope generated by the cone (2.67) as its vertex x_0, y_0, u_0 traverses that curve. Show that this initial value problem has two solutions when

$$u'(t)^2 < x'(t)^2 + y'(t)^2$$

and none when

$$u'(t)^2 > x'(t)^2 + y'(t)^2.$$

6. To describe diffraction of a plane electromagnetic wave by a cylinder, one may introduce solutions

$$U = e^{ikx} + Ae^{iku}$$

of the reduced wave equation

$$U_{xx} + U_{yy} + k^2 U = 0$$

that are appropriate for large frequencies k. An asymptotic solution U is found[7] by choosing for u a function that satisfies the first order equation (2.59) of geometrical optics (cf. Exercise 5.2.8). In order to achieve the boundary condition $U = 0$ on, say, the parabolic cylinder

$$y = x^2,$$

it is convenient to impose the requirements $u = x$ and $A = -1$ there. In these circumstances determine u so that it fulfills (2.59) everywhere but does not coincide with the incident plane wave solution x. What are the incident and reflected light rays to be associated here with x and u, respectively?

5. THE HAMILTON-JACOBI THEORY

We have developed at length a technique for reducing the solution of a first order partial differential equation to a set of initial value problems for an associated characteristic system of ordinary differential equations.

[7] Cf. Keller-Lewis-Seckler 1.

We shall now investigate the representation of systems of ordinary differential equations arising in mechanics in a canonical form that exhibits them as the characteristic system for a certain partial differential equation of the first order known as the *Hamilton-Jacobi equation*. This approach permits us to use a complete integral of the latter equation in order to discuss the solutions of the ordinary differential equations of mechanics. Since large systems of ordinary differential equations occur in problems of mechanics involving many degrees of freedom, we shall treat our subject in the context of first order partial differential equations with an arbitrary finite number of independent variables. Actually it will turn out to be an equation like (2.55), in which the unknown function u does not appear explicitly, that is relevant here.

We suppose that a mechanical system with n degrees of freedom is given and that its motion can be described from the knowledge of expressions for certain *generalized coordinates* $x_1, \ldots, x_n$ as functions of the time t. We shall assume that the system is *conservative*, which means that the external forces can be derived from a potential V depending only on $x_1, \ldots, x_n$. The kinetic energy, which is a quadratic form in the velocities

$$\dot{x}_i = \frac{dx_i}{dt}, \qquad i = 1, \ldots, n,$$

with possibly variable coefficients, will be denoted by T. The difference between the kinetic energy T and the potential energy V is the so-called *Lagrangian*

$$L = L(t, x_1, \ldots, x_n; \dot{x}_1, \ldots, \dot{x}_n) = T - V$$

of the system. It may be a function of the time t, the coordinates x_i and the velocities $\dot{x}_i$.

We refer to the literature on mechanics[8] for a derivation of *Hamilton's principle*. It states that equilibrium is achieved for a motion $x_i = x_i(t)$ that makes the integral

$$(2.69) \qquad I = \int_{t_0}^{t_1} L(t, x_1, \ldots, x_n; \dot{x}_1, \ldots, \dot{x}_n) \, dt$$

stationary amongst all trajectories joining a pair of fixed terminal points

$$(2.70) \qquad x_i(t_0) = y_i, \quad x_i(t_1) = z_i, \qquad i = 1, \ldots, n.$$

Euler's equations for this problem in the calculus of variations[9] yield the

---

[8] Cf. H. Goldstein 1.
[9] Cf. Courant 1.

Lagrange differential equations of motion

$$(2.71) \qquad \frac{d}{dt} L_{\dot{x}_i} - L_{x_i} = 0, \qquad i = 1, \ldots, n,$$

of the mechanical system. We intend to transform the latter equations into a canonical form similar to the characteristic system (2.33).

It is convenient to introduce the *Legendre transformation*

$$(2.72) \qquad p_i = L_{\dot{x}_i}(t, x_1, \ldots, x_n; \dot{x}_1, \ldots, \dot{x}_n),$$

which takes $\dot{x}_1, \ldots, \dot{x}_n$ into new variables $p_1, \ldots, p_n$, called *generalized momenta*, for each fixed choice of $t, x_1, \ldots, x_n$. We need to assume that the Jacobian of this transformation differs from zero, in other words that

$$\begin{vmatrix} L_{\dot{x}_1\dot{x}_1} & \cdots & L_{\dot{x}_1\dot{x}_n} \\ \cdot & \cdot \quad \cdot \quad \cdot & \cdot \\ L_{\dot{x}_n\dot{x}_1} & \cdots & L_{\dot{x}_n\dot{x}_n} \end{vmatrix} \neq 0,$$

in order to be sure that we can solve (2.72) for the velocities $\dot{x}_i, \ldots, \dot{x}_n$ as functions of the momenta $p_1, \ldots, p_n$, and, of course, $t, x_1, \ldots, x_n$. We are then in a position to define the *Hamiltonian H* by means of the formula

$$(2.73) \qquad H + L = \sum_{j=1}^{n} p_j \dot{x}_j.$$

We concieve of it as a function

$$H = H(t, x_1, \ldots, x_n; p_1, \ldots, p_n)$$

of the new variables t, x_i and p_i. Because the kinetic energy T is a quadratic form in the velocities $\dot{x}_i$, direct computation shows that the Hamiltonian is simply the total energy

$$H = T + V$$

of the mechanical system.

As an elementary example of the general theory we recall the equation of motion

$$m\ddot{x} = -gm$$

of a particle of mass m falling freely under the influence of gravity g. In this case the Lagrangian is

$$L = \tfrac{1}{2}m\dot{x}^2 - gmx,$$

where the term $m\dot{x}^2/2$ represents kinetic energy and the term gmx represents potential energy. The Hamiltonian

$$H = \tfrac{1}{2}m\dot{x}^2 + gmx,$$

which gives the total energy of the particle, is seen to be a constant of the motion.

It requires only a straightforward calculation to reformulate the differential equations (2.71) in terms of the Hamiltonian H instead of the Lagrangian L. Taking partial derivatives of (2.73) with respect to x_i and p_i and remembering that the velocities $\dot{x}_i, \ldots, \dot{x}_n$ are now functions of these variables, we obtain

$$(2.74) \qquad H_{x_i} + L_{x_i} + \sum_{j=1}^{n} L_{\dot{x}_j} \frac{\partial \dot{x}_j}{\partial x_i} = \sum_{j=1}^{n} p_j \frac{\partial \dot{x}_j}{\partial x_i},$$

$$(2.75) \qquad H_{p_i} + \sum_{j=1}^{n} L_{\dot{x}_j} \frac{\partial \dot{x}_j}{\partial p_i} = \sum_{j=1}^{n} p_j \frac{\partial \dot{x}_j}{\partial p_i} + \dot{x}_i.$$

The expressions under the summation signs on both sides of (2.74) and (2.75) cancel by virtue of (2.72), and we can eliminate L_{x_i} on the left in (2.74) by means of the relation

$$L_{x_i} = \frac{d}{dt} L_{\dot{x}_i} = \frac{dp_i}{dt}.$$

After simplification, (2.75) and (2.74) reduce to

$$(2.76) \qquad \frac{dx_i}{dt} = H_{p_i}, \quad \frac{dp_i}{dt} = -H_{x_i}, \qquad i = 1, \ldots, n.$$

These $2n$ ordinary differential equations of the first order for $x_1, \ldots, x_n$, $p_1, \ldots, p_n$ as functions of t, which we shall use to replace the n second order Lagrange equations (2.71), are referred to as *Hamilton's canonical equations*.

The reader will note an analogy between (2.76) and the system of ordinary differential equations (2.33) for the characteristics of the partial differential equation (2.32). In fact, consider an auxiliary unknown function u of the $n + 1$ independent variables $t, x_1, \ldots, x_n$ and put

$$p = \frac{\partial u}{\partial t}, \quad p_i = \frac{\partial u}{\partial x_i}, \qquad i = 1, \ldots, n.$$

The analogy suggests that we introduce the first order partial differential equation

$$(2.77) \qquad p + H(t, x_1, \ldots, x_n; p_1, \ldots, p_n) = 0$$

for u, which is called the *Hamilton-Jacobi equation*. The system (2.33) defining the characteristics of the Hamilton-Jacobi equation turns out to be

$$(2.78) \qquad \frac{dt}{ds} = 1, \quad \frac{dx_i}{ds} = H_{p_i}, \quad \frac{du}{ds} = p + \sum_{j=1}^{n} p_j H_{p_j},$$

$$\frac{dp}{ds} = -H_t, \quad \frac{dp_i}{ds} = -H_{x_i}, \qquad i = 1, \ldots, n.$$

Since the first equation (2.78) implies that

$$s = t + \text{const.},$$

the only essential equations here are those already occurring in the system (2.76). Indeed, it is not necessary to retain the differential equations for u or for p, since they do not appear explicitly as arguments of the Hamiltonian function H and could always be found independently after $x_1, \ldots, x_n$ and $p_1, \ldots, p_n$ have been determined.

From Section 2 we recall that $p + H$ has to be an integral of the characteristic system (2.78). It follows that every solution of (2.76) is a genuine characteristic of the Hamilton-Jacobi equation, since we can always pick the initial value of p to satisfy (2.77). Thus the problem of solving Lagrange's equations of motion is altogether equivalent to the question of determining the characteristics of the first order partial differential equation (2.77).

Of special interest is the case where the Hamiltonian H is independent of the time t. Then

$$\frac{dp}{ds} = -H_t = 0,$$

so that p is a constant of the motion $x_i = x_i(t)$. Because $p + H$ is an integral of the system (2.78), we conclude that

$$H = T + V = \text{const.},$$

which is seen to be a statement of the law of conservation of energy.

The characteristics of the Hamilton-Jacobi equation can be found from a complete integral

$$u = \phi(t, x_i, \ldots, x_n; a_1, \ldots, a_n) + b.$$

Indeed, according to (2.57) and (2.58) they are defined by the relations

$$(2.79) \qquad \phi_{a_i}(t, x_1, \ldots, x_n; a_1, \ldots, a_n) = \lambda_i, \qquad i = 1, \ldots, n,$$

$$(2.80) \qquad p_i = \phi_{x_i}(t, x_1, \ldots, x_n; a_1, \ldots, a_n), \qquad i = 1, \ldots, n,$$

for fixed choices of the parameters $a_1, \ldots, a_n$ and $\lambda_1, \ldots, \lambda_n$. The n equations (2.79) alone are sufficient to determine the coordinates x_i that describe the motion of our mechanical system as functions of the time t. In this context the a_i and λ_i play the role of the constants of integration of the Lagrangian system of ordinary differential equations (2.71). The solution (2.79) of (2.71) presents itself in an exceptionally elegant form whenever the complete integral ϕ can be found explicitly. However, the determination of the complete integral is in general a prohibitively difficult problem; thus it can only be treated adequately for especially simple examples.

EXERCISES

1. Consider the stationary value of the integral I defined by (2.69) and (2.70) as a function of the upper limit of integration t_1 and of the associated coordinates $z_1, \ldots, z_n$. With t and $x_1, \ldots, x_n$ interpreted now as these variables and with $p_1, \ldots, p_n$ defined as usual in terms of the velocities $\dot{x}_1, \ldots, \dot{x}_n$, show that $\partial I/\partial x_i = p_i$ and show that I satisfies the Hamilton-Jacobi partial differential equation (2.77).[10]

2. Apply the Hamilton-Jacobi theory to study the motion of a particle in space under the influence of gravity.

3. Generalize the Hamilton-Jacobi theory to show that it applies to the motion of charged particles in an external electromagnetic field. Observe that the potential V depends on the velocities $\dot{x}_i$ in this example.[11]

4. Consider the possibility that in an application of Hamilton's principle the integral

$$I = \int_{t_0}^{t_1} (T - V)\, dt$$

might be minimized by a trajectory achieving equilibrium. Show that this is the case for the motion of a bead constrained to rotate on a frictionless circular wire located in a horizontal plane.

6. APPLICATIONS

As an application of the Hamilton-Jacobi theory we shall study the motion of n particles, with masses m_i and coordinates x_i, y_i, z_i, that are under the influence of the forces of mutual attraction described by Newton's law of gravitation. This is the classical n-body problem.

[10] Cf. Courant-Hilbert 1.
[11] Cf. H. Goldstein 1.

The kinetic energy of the system of n particles is defined to be

$$T = \frac{1}{2} \sum_{j=1}^{n} m_j (\dot{x}_j^2 + \dot{y}_j^2 + \dot{z}_j^2),$$

and its potential energy, according to Newton's law of gravitation, is

$$V = -\sum_{i<j} \frac{Km_i m_j}{r_{ij}},$$

where

$$r_{ij} = \sqrt{(x_i - x_j)^2 + (y_i - y_j)^2 + (z_i - z_j)^2}$$

indicates the distance between the ith and the jth particles, and where K is a gravitational constant. Since

$$L = T - V,$$

the Lagrange differential equations of motion (2.71) of the system of particles are given by

$$(2.81) \qquad m_i \ddot{x}_i + \sum_{j \neq i} Km_i m_j \frac{x_i - x_j}{r_{ij}^3} = 0, \qquad i = 1, \ldots, n,$$

together with analogous equations for $\ddot{y}_i$ and $\ddot{z}_i$.

Introducing the Hamiltonian

$$H = T + V,$$

we see that the Hamilton-Jacobi partial differential equation (2.77) associated with (2.81) takes the form

(2.82)

$$u_t + \sum_{i=1}^{n} \frac{u_{x_i}^2 + u_{y_i}^2 + u_{z_i}^2}{2m_i} = \sum_{i<j} \frac{Km_i m_j}{\sqrt{(x_i - x_j)^2 + (y_i - y_j)^2 + (z_i - z_j)^2}}.$$

A complete integral of (2.82) is known only in the case $n = 2$ of the motion of two bodies. To find it we make the preliminary change of variables

$$\bar{x} = \frac{m_1 x_1 + m_2 x_2}{m_1 + m_2}, \quad \bar{y} = \frac{m_1 y_1 + m_2 y_2}{m_1 + m_2}, \quad \bar{z} = \frac{m_1 z_1 + m_2 z_2}{m_1 + m_2},$$

$$\xi = r \sin \theta \cos \phi = x_2 - x_1,$$

$$\eta = r \sin \theta \sin \phi = y_2 - y_1,$$

$$\zeta = r \cos \theta \qquad = z_2 - z_1.$$

Thus the point $\bar{x}, \bar{y}, \bar{z}$ lies at the center of gravity of the two particles involved and r, θ and ϕ are the spherical coordinates of the displacement vector (ξ, η, ζ) joining them. An elementary calculation shows that in terms of the new variables $\bar{x}, \bar{y}, \bar{z}, r, \theta$ and ϕ, the Hamilton-Jacobi equation (2.82) becomes

$$(2.83) \quad u_t + \frac{1}{2}\frac{u_{\bar{x}}^2 + u_{\bar{y}}^2 + u_{\bar{z}}^2}{m_1 + m_2}$$

$$+ \frac{1}{2}\left(\frac{1}{m_1} + \frac{1}{m_2}\right)\left(u_r^2 + \frac{1}{r^2}u_\theta^2 + \frac{1}{r^2 \sin^2 \theta}u_\phi^2\right) = \frac{Km_1m_2}{r}.$$

A complete integral of (2.83) can be obtained by the method of separation of variables. If we seek solutions of the form

$$u = f_1(t) + f_2(\bar{x}) + f_3(\bar{y}) + f_4(\bar{z}) + f_5(r) + f_6(\theta) + f_7(\phi),$$

we find upon substituting into (2.83) that

$$f_1'(t) + \frac{1}{2}\frac{f_2'(\bar{x})^2 + f_3'(\bar{y})^2 + f_4'(\bar{z})^2}{m_1 + m_2}$$

$$+ \frac{1}{2}\left(\frac{1}{m_1} + \frac{1}{m_2}\right)\left[f_5'(r)^2 + \frac{1}{r^2}\left\{f_6'(\theta)^2 + \frac{1}{\sin^2 \theta}f_7'(\phi)^2\right\}\right] = \frac{Km_1m_2}{r}.$$

It follows by the usual argument that the desired complete integral is

$$(2.84)$$

$$u = -\frac{1}{2}\left(\frac{a^2 + b^2 + c^2}{m_1 + m_2} + \frac{m_1 + m_2}{m_1m_2}\alpha\right)t + a\bar{x} + b\bar{y} + c\bar{z} + \beta\phi$$

$$+ \int\sqrt{\gamma^2 - \frac{\beta^2}{\sin^2 \theta}}\,d\theta + \int\sqrt{\alpha + \frac{2Km_1^2m_2^2}{m_1 + m_2}\frac{1}{r} - \frac{\gamma^2}{r^2}}\,dr + \text{const.},$$

where a, b, c, α, β and γ are arbitrary constants.

According to (2.79) we can obtain the various possible trajectories of the two bodies under consideration by putting the partial derivatives of the complete integral (2.84) with respect to the parameters a, b, c, α, β and γ equal to fixed values. Applying this rule to the partial derivatives with respect to a, b and c and solving the resulting equations, we find that

$$\bar{x} = \frac{at}{m_1 + m_2} + \text{const.}, \quad \bar{y} = \frac{bt}{m_1 + m_2} + \text{const.},$$

$$\bar{z} = \frac{ct}{m_1 + m_2} + \text{const.}$$

Consequently the center of gravity $\bar{x}, \bar{y}, \bar{z}$ of the two particles moves along a straight line at a uniform rate, as was to be expected, of course. On the other hand, partial differentiation of (2.84) with respect to β yields the relation

$$\sqrt{\gamma^2 - \beta^2} \sin \theta \cos (\phi - \text{const.}) = \beta \cos \theta$$

between θ and ϕ, which implies that the orbit of one particle lies in a plane of fixed inclination through the other particle and through the center of gravity $\bar{x}$, $\bar{y}$, $\bar{z}$. There is no loss of generality if we assume henceforth that the plane in question is vertical, so that $\beta = 0$.

To find the equation of the orbit of one particle about the other in the plane of the polar coordinates r and θ, we set the partial derivative with respect to γ of the complete integral (2.84) equal to a constant λ. The formula

$$\theta - \lambda + \sin^{-1} \frac{\dfrac{\gamma^2}{r} - \dfrac{Km_1^2 m_2^2}{m_1 + m_2}}{\sqrt{\dfrac{K^2 m_1^4 m_2^4}{(m_1 + m_2)^2} + \alpha \gamma^2}} = 0$$

results. It represents a conic section with one focus at the origin and with the eccentricity

$$\epsilon = \sqrt{1 + \alpha \gamma^2 \frac{(m_1 + m_2)^2}{K^2 m_1^4 m_2^4}} \ .$$

Thus the trajectory of one of our two particles moving around the other as a frame of reference describes an ellipse if $\alpha < 0$, a parabola if $\alpha = 0$, or a hyperbola if $\alpha > 0$, and the second particle is always located at a focus. An analogous statement is valid for the orbit of either of the particles about their center of gravity, since this is described by a vector that differs only by a specified scale factor from the vector (ξ, η, ζ) joining the two particles. Finally, the expression

$$t = \frac{m_1 m_2}{m_1 + m_2} \int \frac{dr}{\sqrt{\alpha + \dfrac{2Km_1^2 m_2^2}{m_1 + m_2}\dfrac{1}{r} - \dfrac{\gamma^2}{r^2}}} + \text{const.}$$

for the time as a function of position can be deduced from the partial derivative of the complete integral with respect to α.

A quite different application of the Hamilton-Jacobi theory arises from the problem of finding the geodesics in n-dimensional Euclidean space for a metric whose arc length element ds is given by a quadratic form

(2.85)
$$ds^2 = \sum_{i,j=1}^{n} A_{ij}\, dx_i\, dx_j.$$

A knowledge of these geodesics will be of service to us in Section 5.2, for they are related to the so-called *bicharacteristics* of a linear partial differential equation of the second order.

We assume that the coefficients A_{ij} are functions of the Cartesian coordinates $x_1, \ldots, x_n$ and that the corresponding matrix is symmetric,

$$A_{ij} = A_{ji}, \qquad i, j = 1, \ldots, n.$$

The *geodesics* are curves $x_i = x_i(t)$ which minimize, or at least make stationary, the arc length integral

(2.86)
$$I = \int_{t_0}^{t_1} \sqrt{\sum_{i,j=1}^{n} A_{ij} \dot{x}_i \dot{x}_j}\, dt$$

between fixed limits

$$y_i = x_i(t_0), \quad z_i = x_i(t_1), \qquad i = 1, \ldots, n.$$

Our aim will be to show that as a function of the coordinates $z_1, \ldots, z_n$ of its upper limit, the *geodesic distance* I satisfies the first order partial differential equation

(2.87)
$$\sum_{i,j=1}^{n} a_{ij} I_{z_i} I_{z_j} = 1,$$

where the coefficients a_{ij} are the elements of the inverse matrix specified by the relations

$$\sum_{k=1}^{n} a_{ik} A_{kj} = \delta_{ij} = \begin{cases} 0, & i \neq j, \\ 1, & i = j. \end{cases}$$

The Lagrangian L for the extremal problem defining the geodesic distance is obviously

$$L = \sqrt{Q},$$

where Q is the quadratic form

$$Q = \sum_{i,j=1}^{n} A_{ij} \dot{x}_i \dot{x}_j.$$

It does not fulfill the requirement

$$\begin{vmatrix} L_{\dot{x}_1 \dot{x}_1} & \cdots & L_{\dot{x}_1 \dot{x}_n} \\ \cdot & \cdot & \cdot & \cdot & \cdot \\ L_{\dot{x}_n \dot{x}_1} & \cdots & L_{\dot{x}_n \dot{x}_n} \end{vmatrix} \neq 0$$

needed in connection with the Legendre transformation (2.72) because it is homogeneous of degree one in the quantities $\dot{x}_i$. However, this difficulty

can be overcome by writing the Euler equations (2.71) associated with the
integral (2.86) in terms of Q to obtain

(2.88)
$$\frac{d}{dt}\frac{Q_{\dot{x}_i}}{\sqrt{Q}} - \frac{Q_{x_i}}{\sqrt{Q}} = 0, \qquad i = 1, \ldots, n.$$

We are then led to take the parameter t along the geodesics to be the arc
length s defined by (2.85), measured from the point $y_1, \ldots, y_n$, so that
$Q \equiv 1$.

With the special choice of t just made we observe that (2.88) reduces to

(2.89)
$$\frac{d}{dt} Q_{\dot{x}_i} - Q_{x_i} = 0, \qquad i = 1, \ldots, n.$$

These are the Euler equations which state that the integral

(2.90)
$$J = \int_0^s Q \, dt$$

is stationary. Now fix the terminal values $y_1, \ldots, y_n$ and $z_1, \ldots, z_n$ of
$x_1, \ldots, x_n$, but allow the upper limit of integration s to vary. The
extremals of (2.90) come to depend on s and on the variable of integration
t according to the simple rule

$$x_i = x_i\left(\frac{I}{s} t\right)$$

for a change of scale, since Q is homogeneous of degree two in the
quantities $\dot{x}_1, \ldots, \dot{x}_n$. Therefore the constant value of Q becomes I^2/s^2
along each such extremal and

(2.91)
$$J = \frac{I(z_1, \ldots, z_n)^2}{s}.$$

The Hamilton-Jacobi theory applies to the system of ordinary differential
equations (2.89). The Hamiltonian is given by

(2.92)
$$H = Q = \frac{1}{4} \sum_{i,j=1}^n a_{ij} p_i p_j,$$

since (2.72) takes the form

$$p_i = Q_{\dot{x}_i} = 2 \sum_{j=1}^n A_{ij} \dot{x}_j, \qquad i = 1, \ldots, n,$$

so that

$$\dot{x}_j = \frac{1}{2} \sum_{i=1}^n a_{ji} p_i, \qquad j = 1, \ldots, n,$$

$$\sum_{i=1}^n p_i \dot{x}_i = 2 \sum_{i,j=1}^n A_{ij} \dot{x}_i \dot{x}_j = \frac{1}{2} \sum_{i,j=1}^n a_{ij} p_i p_j.$$

It follows from Exercise 5.1 that J satisfies the Hamilton-Jacobi equation

$$(2.93) \qquad J_s + \frac{1}{4} \sum_{i,j=1}^{n} a_{ij}(z_1, \ldots, z_n) J_{z_i} J_{z_j} = 0.$$

If we substitute (2.91) into (2.93) and then put $s = 1$, we obtain the desired partial differential equation (2.87) for the geodesic distance I. In terms of the quantity

$$\Gamma = I^2,$$

which will play an important role in Section 5.2, the result is

$$(2.94) \qquad \sum_{i,j=1}^{n} a_{ij} \Gamma_{z_i} \Gamma_{z_j} = 4\Gamma.$$

Observe that Γ is the conoidal solution (cf. Exercise 2.1) of (2.94) generated by all the characteristics of that equation passing through the point $y_1, \ldots, y_n$, which are, of course, geodesics of the metric (2.85).

In terms of the Hamiltonian (2.92) the geodesics satisfy the canonical system of $2n$ ordinary differential equations

$$(2.95) \qquad \frac{dx_i}{ds} = \frac{1}{2} \sum_{j=1}^{n} a_{ij} p_j, \quad \frac{dp_i}{ds} = -\frac{1}{4} \sum_{j,k=1}^{n} \frac{\partial a_{jk}}{\partial x_i} p_j p_k,$$

together with appropriate initial conditions of the form

$$(2.96) \qquad x_i(0) = y_i, \quad p_i(0) = q_i.$$

Note that a change of scale caused by substituting ρs for s does not alter the geometry of the geodesics, provided that we replace $p_1, \ldots, p_n$ at the same time by $p_1/\rho, \ldots, p_n/\rho$ in the system of differential equations (2.95). Therefore the solutions $x_1, \ldots, x_n$ of (2.95) can be expressed as functions

$$x_i = x_i(sq_1, \ldots, sq_n; y_1, \ldots, y_n)$$

exclusively of the initial values $y_1, \ldots, y_n$ and of the products

$$(2.97) \qquad \theta_i = sq_i, \quad i = 1, \ldots, n.$$

It is sometimes convenient to introduce $\theta_1, \ldots, \theta_n$ as new coordinates to take the place of $x_1, \ldots, x_n$, after making a fixed choice of the parameters $y_1, \ldots, y_n$. At the point $y_1, \ldots, y_n$ we have

$$\frac{\partial x_i}{\partial \theta_j} = \frac{1}{2} a_{ij}$$

in view of (2.95), whence the Jacobian

$$\frac{\partial(x_1, \ldots, x_n)}{\partial(\theta_1, \ldots, \theta_n)} = \frac{1}{2^n} \begin{vmatrix} a_{11} & \cdots & a_{1n} \\ \cdot & \cdot & \cdot & \cdot & \cdot \\ a_{n1} & \cdots & a_{nn} \end{vmatrix}$$

differs from zero there. Consequently such a transformation of variables is feasible in a neighborhood of the vertex $y_1, \ldots, y_n$.

EXERCISES

1. Use the Hamilton-Jacobi equation to determine the motion of the center of gravity of $n > 2$ particles acted on exclusively by the forces of mutual attraction.

2. Discuss the partial differential equation (2.87) in the case of the Euclidean metric

$$ds^2 = dx^2 + dy^2 + dz^2.$$

3. Show that if we insert $x_1, \ldots, x_n$ to replace $z_1, \ldots, z_n$ as the arguments of Γ, then

$$\frac{\partial \Gamma}{\partial x_i} = sp_i.$$

4. Using the notation of Exercise 3, prove that the square Γ of the geodesic distance I is given in terms of the coordinates $\theta_1, \ldots, \theta_n$ by the quadratic form

$$\Gamma = \frac{1}{4} \sum_{i,j=1}^{n} a_{ij}\theta_i\theta_j.$$

Show that if the coefficients A_{ij} and a_{ij} are analytic functions of the arguments $x_1, \ldots, x_n$, then Γ is regular in a neighborhood of its vertex $y_1, \ldots, y_n$. Verify that the lowest order terms in the Taylor series expansion for Γ there are

$$(2.98) \qquad \Gamma = \sum_{i,j=1}^{n} A_{ij}(y_1, \ldots, y_n)(x_i - y_i)(x_j - y_j) + \cdots.$$

5. Use the geodesic distance I between two points $y_1, \ldots, y_n$ and $z_1, \ldots, z_n$ to set up a complete integral of the partial differential equation (2.87).

6. Find the geodesics on a sphere.

3

Classification of Partial
Differential Equations

1. REDUCTION OF LINEAR EQUATIONS IN TWO INDEPENDENT VARIABLES TO CANONICAL FORM

The theory of partial differential equations of the second order is a great deal more complicated than that for equations of the first order, and it is actually much more typical of the subject as a whole. Within this framework we shall see that considerably better results can be achieved for equations of the second order in two independent variables than for equations in space of higher dimensions. Moreover, equations that are linear in the unknown function and its partial derivatives are by all means the easiest to handle. Thus we prefer to begin our discussion of partial differential equations of higher order by studying the equation

$$(3.1) \qquad au_{xx} + 2bu_{xy} + cu_{yy} + du_x + eu_y + fu = g$$

in two independent variables. The coefficients a, b, c, d, e, f and g are assumed here to be functions of x and y only, whence the equation is *linear*.

The difficulties that can be expected to arise in treating even the linear equation (3.1) are foreshadowed by the variety of initial value and boundary value problems that we mentioned in connection with the wave equation (1.2), the heat equation (1.8), and Laplace's equation (1.12) in Chapter 1. However, we shall establish that these three cases are in a certain sense typical of what occurs in the general theory, inasmuch as we shall be able, whenever there is no degeneracy, to reduce (3.1) by means of a change of variables

$$(3.2) \qquad \xi = \xi(x, y), \quad \eta = \eta(x, y)$$

to one of the three canonical forms

$$(3.3) \qquad u_{\xi\xi} - u_{\eta\eta} + \cdots = 0,$$

$$(3.4) \qquad u_{\xi\xi} + \cdots = 0,$$

$$(3.5) \qquad u_{\xi\xi} + u_{\eta\eta} + \cdots = 0.$$

The dots indicate in the present context terms involving u and its first derivatives u_x and u_y but not its second derivatives. It is significant that the transformation (3.2) enables us to simplify equation (3.1) to the extent indicated.

The reason we refer to (3.3), (3.4) and (3.5) as *canonical forms* is that they correspond to particularly simple choices of the coefficients of the second partial derivatives of u to which we shall be able to ascribe an invariant meaning. The structure of these coefficients, which is suggested by the analytic geometry of quadratic equations, turns out to have a connection with the question already raised at the beginning of Section 1.2 concerning solvability of the equation (1.1) for $r = u_{xx}$ as a function of u and its other derivatives. Indeed, the transformation (3.2) will lead us to the discovery of special loci known as *characteristics* along which the equation (3.1) provides only an incomplete expression for the second derivatives of u.

Let us formulate our main result more precisely. We intend to prove that we can achieve the canonical form (3.3) if and only if

$$(3.6) \qquad b^2 - ac > 0,$$

that we can achieve the canonical form (3.4) if and only if

$$(3.7) \qquad b^2 - ac = 0,$$

while a, b and c do not all vanish simultaneously, and that we can achieve the canonical form (3.5) if and only if

$$(3.8) \qquad b^2 - ac < 0.$$

We say that the partial differential equation (3.1) is of the *hyperbolic type* in the case (3.6), of the *parabolic type* in the case (3.7), and of the *elliptic type* in the case (3.8). The type of the equation (3.1) will turn out to be decisive in establishing the kind of initial conditions or boundary conditions that serve in a natural way to determine a solution uniquely. Thus we can use as a guide in such matters the examples of the wave equation (hyperbolic), the heat equation (parabolic), and Laplace's equation (elliptic), which present themselves in canonical form to start with.

In order to work out the reduction of (3.1) to canonical form by means of the transformation of coordinates (3.2), we need to express the derivatives of u with respect to x and y in terms of derivatives with respect to ξ and η. We can facilitate these calculations by adopting once and for all the convention, exploited already in formulas (3.3), (3.4) and (3.5), that terms involving only u and its first derivatives are to be abbreviated merely by inserting dots to replace them. The advantage of such a notation stems from the fact that in much of our theory it will be exclusively the terms involving second derivatives that matter. Thus we write

$$u_x = u_\xi \xi_x + u_\eta \eta_x, \qquad u_y = u_\xi \xi_y + u_\eta \eta_y$$

and

$$(3.9) \qquad u_{xx} = u_{\xi\xi}\xi_x^2 + 2u_{\xi\eta}\xi_x\eta_x + u_{\eta\eta}\eta_x^2 + \cdots,$$

$$(3.10) \qquad u_{xy} = u_{\xi\xi}\xi_x\xi_y + u_{\xi\eta}\xi_x\eta_y + u_{\xi\eta}\xi_y\eta_x + u_{\eta\eta}\eta_x\eta_y + \cdots,$$

$$(3.11) \qquad u_{yy} = u_{\xi\xi}\xi_y^2 + 2u_{\xi\eta}\xi_y\eta_y + u_{\eta\eta}\eta_y^2 + \cdots,$$

where in (3.9), for example, the dots stand for $u_\xi \xi_{xx} + u_\eta \eta_{xx}$.

From (3.9), (3.10) and (3.11) we derive the identity

$$au_{xx} + 2bu_{xy} + cu_{yy} = Au_{\xi\xi} + 2Bu_{\xi\eta} + Cu_{\eta\eta} + \cdots,$$

where

$$(3.12) \qquad A = a\xi_x^2 + 2b\xi_x\xi_y + c\xi_y^2,$$

$$(3.13) \qquad B = a\xi_x\eta_x + b\xi_x\eta_y + b\xi_y\eta_x + c\xi_y\eta_y,$$

$$(3.14) \qquad C = a\eta_x^2 + 2b\eta_x\eta_y + c\eta_y^2.$$

It follows that in terms of the new variables ξ and η the partial differential equation (3.1) assumes the form

$$(3.15) \qquad Au_{\xi\xi} + 2Bu_{\xi\eta} + Cu_{\eta\eta} + \cdots = 0$$

with coefficients A, B and C defined by (3.12), (3.13) and (3.14).

It is easy to verify that

$$(3.16) \qquad B^2 - AC = (b^2 - ac)(\xi_x\eta_y - \xi_y\eta_x)^2$$

by substituting the expressions (3.12), (3.13) and (3.14) directly for A, B and C. The factor that is squared on the right is recognized to be the Jacobian

$$(3.17) \qquad \frac{\partial(\xi, \eta)}{\partial(x, y)} = \begin{vmatrix} \xi_x & \xi_y \\ \eta_x & \eta_y \end{vmatrix}$$

of the transformation (3.2). Clearly we should confine our attention to locally one-to-one transformations whose Jacobians are different from zero. Under such changes of coordinates formula (3.16) shows that the sign of the discriminant $b^2 - ac$ remains invariant, since the squared factor on the right must be positive. According to our definitions (3.6), (3.7) and (3.8) of hyperbolic, parabolic and elliptic partial differential equations (3.1), we conclude that *the type of such an equation cannot be altered by a real change of variables* (3.2). Of course, to be significant any classification of second order partial differential equations as to type would have to exhibit an invariance of this kind.

We turn our attention to the problem of selecting the transformation (3.2) so that (3.15) will reduce to one of the canonical forms (3.3), (3.4) or (3.5). Consider first the hyperbolic case (3.6). If we attempt to make B vanish, we are led according to (3.13) to a partial differential equation

$$a\xi_x\eta_x + b\xi_x\eta_y + b\xi_y\eta_x + c\xi_y\eta_y = 0$$

of the first order for both of the functions $\xi = \xi(x, y)$ and $\eta = \eta(x, y)$ simultaneously. It would be far simpler to try to make A or C vanish, since that would yield a first order equation for either ξ or η alone stemming from (3.12) or (3.14). This suggests that we introduce the alternate canonical form

(3.18) $$u_{\xi\eta} + \cdots = 0$$

for equations (3.1) of the hyperbolic type (3.6), which turns out anyway to be more convenient than (3.3) for a variety of applications.

To achieve (3.18) it is necessary to select ξ and η so that both A and C vanish. In view of (3.12) and (3.14) we must therefore choose them to be independent solutions of the first order partial differential equation

(3.19) $$a\xi_x^2 + 2b\xi_x\xi_y + c\xi_y^2 = 0.$$

We shall analyze the solutions of (3.19) by investigating their level curves

(3.20) $$\xi(x, y) = \text{const.}$$

Such a procedure is suggested by the ordinary differential equation (2.23) for the characteristics of (3.19), which reduces here to

$$\frac{d\xi}{ds} = pF_p + qF_q = 2(a\xi_x^2 + 2b\xi_x\xi_y + c\xi_y^2) = 0$$

and thus shows that those characteristics coincide with the loci (3.20).

Along each level curve (3.20) we have

$$\xi_x \, dx + \xi_y \, dy = 0,$$

or

(3.21) $$\frac{dy}{dx} = -\frac{\xi_x}{\xi_y}.$$

We can divide (3.19) by ξ_y^2 and then eliminate ξ_x and ξ_y by means of (3.21) to derive

(3.22) $$a \, dy^2 - 2b \, dx \, dy + c \, dx^2 = 0.$$

The quadratic (3.22) represents an ordinary differential equation for the level curves (3.20) in the (x,y)-plane, since the coefficients a, b and c are functions of x and y only. Solving it for dy/dx, we find that

(3.23) $$\frac{dy}{dx} = \frac{b(x, y) \pm \sqrt{b(x, y)^2 - a(x, y)c(x, y)}}{a(x, y)}.$$

Because of the hyperbolicity requirement (3.6), formula (3.23) actually furnishes two quite distinct ordinary differential equations that correspond to choosing the plus sign or the minus sign in front of the square root.

We can express the integrals of the two differential equations (3.23) in the form

(3.24) $$\phi(x, y) = \text{const.}, \qquad \psi(x, y) = \text{const.},$$

with the constants of integration appearing on the right. The functions ϕ and ψ that occur on the left yield the desired solutions

(3.25) $$\xi = \phi(x, y), \qquad \eta = \psi(x, y)$$

of the partial differential equation (3.19) because their level curves define two different families of solutions of (3.22), which is equivalent to (3.19) by virtue of (3.20) and (3.21). Note that any smooth function of $\phi(x, y)$ could be used to replace $\phi(x, y)$ and that any smooth function of $\psi(x, y)$ could be used to replace $\psi(x, y)$ in the formulas (3.24) for the integrals of (3.23). Thus an arbitrary function of one variable could be introduced into each of the solutions (3.25) of (3.19) that we have found.

We can verify that the two functions ξ and η defined by (3.25) are independent by appealing to the hypothesis (3.6) that (3.1) is of the hyperbolic type. Indeed, if ξ results from taking the plus sign in (3.23) and η results from taking the minus sign there, then

(3.26) $$\frac{\xi_x}{\xi_y} - \frac{\eta_x}{\eta_y} = -2\frac{\sqrt{b^2 - ac}}{a} \neq 0.$$

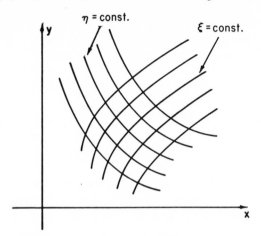

FIG. 3. Characteristics of a hyperbolic equation.

Hence the Jacobian (3.17) differs from zero. Degenerate cases in which $\xi_y = 0$ or $\eta_y = 0$ here are easy to avoid. If $\xi_y = 0$, for example, then $\xi_x = 0$ also, because of (3.19), and it is possible to replace ξ by an appropriate function of ξ to eliminate such a situation. Incidentally, when the division by a required for (3.23) becomes meaningless because a vanishes, we can interchange the roles of x and y so that c appears in the denominator there instead. Moreover, if c vanishes, too, then we can omit the change of variables (3.2) altogether, since we introduced it in the first place only to make the coefficients A and C in (3.15) disappear. Thus in the hyperbolic case (3.6) we are always assured of the existence, at least in the small, of coordinates ξ and η in terms of which (3.1) reduces to

$$Bu_{\xi\eta} + \cdots = 0.$$

We have only to divide out the coefficient B, which differs from zero by virtue of (3.16), in order to achieve the desired canonical form (3.18).

The condition (3.26) merely states that the level curves

(3.27) $\xi(x, y) = \text{const.}, \qquad \eta(x, y) = \text{const.}$

which furnish our new coordinate network have distinct slopes at each point of the (x,y)-plane. These two one-parameter families of curves, which are defined in the hyperbolic case by the quadratic ordinary differential equation (3.22), are called the *characteristics* of the second order partial differential equation (3.1). In the (ξ,η)-plane, where equation (3.1) appears in the form (3.18) with the second derivatives $u_{\xi\xi}$ and $u_{\eta\eta}$ missing, the characteristics become lines parallel to the coordinate axes. We shall attempt next to identify them as loci that are exceptional for initial value problems in the same sense as are the characteristics of a partial differential equation of the first order.

In the case of analytic coefficients a, b, c, d, e, f and g, we may refer back to the power series solution of the second order equation (1.16) presented in Chapter 1. In order to solve the initial value problem (1.17) there with u and its first derivatives prescribed along the y-axis, we had to assume that the equation (1.16) had been solved for the second derivative $r = u_{xx}$. This was necessary because the partial differential equation in question was to be used to furnish recursion formulas expressing partial derivatives of u of a high order with respect to x in terms of derivatives of lower orders in that variable. Such a procedure would fail for the equation (3.18), since r is missing from it, and since it consequently imposes an ordinary differential equation on the functions

$$f(\eta) = u(0, \eta), \qquad g(\eta) = u_\xi(0, \eta)$$

of η that are to be given along the initial line $\xi = 0$, instead of serving as a recursion formula for $u_{\xi\xi}$ or for the remaining higher derivatives. Therefore we maintain that it would not be reasonable, at least within the framework of analytic equations, to pose an initial value problem for (3.18) with u and its normal derivative u_ξ assigned along the η-axis or, for that matter, along any of the lines $\xi = $ const. Neither would it be correct to pose a similar problem with data prescribed on one of the lines $\eta = $ const., since the second derivative $u_{\eta\eta}$ is also absent from (3.18) and thus the roles played by ξ and η are interchangeable.

It is in the sense just described that we say that both of the families of characteristics (3.27) are exceptional for the initial value problem for a second order equation (3.1) of the hyperbolic type. Our statements in this respect are as valid in the (x,y)-plane as they are in the (ξ,η)-plane, since assigning u and its normal derivative along any curve in one of those planes is equivalent to assigning the same quantities along the corresponding curve in the other plane. The significance of the characteristics is thus similar for first order equations and for second order hyperbolic equations with the obvious difference that there are two families of characteristics in the latter case and only one in the former.

It is quite natural to call the new variables ξ and η defined by (3.25) *characteristic coordinates* for (3.1). The canonical form (3.18) is the one that occurs when we introduce these characteristic coordinates as new independent variables. We have still to show that by a further transformation we can bring (3.1) into the canonical form (3.3) in the hyperbolic case. To find out how that should be done, we use the theory of characteristic coordinates to reduce (3.3) to the form (3.18) and then examine the inverse transformation.

The analysis just described amounts to putting the wave equation

(3.28) $$u_{xx} - u_{yy} = 0$$

into the canonical form (3.18). The differential equation (3.23) becomes

$$\frac{dy}{dx} = \pm 1$$

here, and its solutions (3.24) are the lines

(3.29) $$x - y = \text{const.,} \quad x + y = \text{const.}$$

It is convenient, and certainly permissible, to divide the functions on the left in (3.29) by $\sqrt{2}$ and to express the resulting transformation (3.25) to characteristic coordinates ξ and η as a rotation

(3.30) $$\xi = \frac{x - y}{\sqrt{2}}, \qquad \eta = \frac{x + y}{\sqrt{2}}$$

through $-45°$. The inverse of (3.30) is also a rotation, this time through $+45°$. We conclude that in order to bring (3.1) into the canonical form (3.3) in the general hyperbolic case, we have only to superimpose on the change of variables (3.25) an additional rotation through $-45°$ leading to the new coordinates

$$\xi_1 = \frac{\phi(x, y) - \psi(x, y)}{\sqrt{2}}, \qquad \eta_1 = \frac{\phi(x, y) + \psi(x, y)}{\sqrt{2}},$$

which yield the desired reduction.

We observe from (3.29) that the characteristics of the wave equation (3.28) are the two families of straight lines that make angles of $45°$ with the x-axis and the y-axis. Furthermore, since the special transformation (3.30) of (3.28) to the canonical form (3.18) is linear, the terms in u or its first derivatives that we have abbreviated with dots actually drop out altogether here. Hence we find that the precise expression for the wave equation in the plane of the characteristic coordinates is

(3.31) $$u_{\xi\eta} = 0.$$

The form of the general solution of the wave equation (3.28) can easily be deduced from (3.31). In fact, (3.31) asserts that u_ξ does not depend on η, so that we can write

(3.32) $$u_\xi = \phi'(\xi).$$

But (3.32) shows that $u - \phi$ does not depend on ξ, and therefore

$$u = \phi(\xi) + \psi(\eta).$$

If we suppress the inessential factors of $\sqrt{2}$ in the transformation (3.30), which are awkward in the present context, we thus find for the general solution of the wave equation (3.28) the final expression

$$(3.33) \qquad u = \phi(x - y) + \psi(x + y)$$

involving two arbitrary functions ϕ and ψ of one variable (cf. Exercise 1.1.2).

Formula (3.33) represents a strict solution u of the wave equation provided merely that ϕ and ψ have continuous second derivatives. We observe that jumps in the higher derivatives of ϕ and ψ result in corresponding discontinuities of the partial derivatives of the wave function u. The particular form of (3.33) shows that discontinuities of the latter kind occur only along the lines (3.29), which comprise the characteristics of (3.28). Since these discontinuities might be viewed as *wave fronts*, we conclude that the characteristics are to be interpreted physically as the possible loci along which a wave front can advance.

We change the topic and discuss next equations (3.1) of the parabolic type (3.7). In order to achieve the canonical form (3.4) here, we want to select the function η in the transformation (3.2) so that the coefficient C occurring in (3.15) will vanish. According to (3.14), we should construct η as a solution of the first order partial differential equation

$$(3.34) \qquad a\eta_x^2 + 2b\eta_x\eta_y + c\eta_y^2 = 0.$$

Equation (3.34) is, of course, the same as (3.19), and it can therefore be reduced to (3.22) and then to (3.23). Now, however, (3.23) defines only one ordinary differential equation because of the hypothesis

$$b^2 - ac = 0$$

that (3.1) is of the parabolic type. Consequently we obtain just a single one-parameter family of solutions

$$\psi(x, y) = \text{const.}$$

analogous to (3.24). We set

$$(3.35) \qquad \eta = \psi(x, y)$$

and proceed to show that it suffices to choose as the remaining coordinate ξ any function independent of η in order to arrive at the canonical form (3.4).

To prove this, we observe once again that substitution of (3.7) into (3.16) gives

$$(3.36) \qquad\qquad B^2 - AC = 0$$

no matter how the transformation (3.2) is chosen. Since the rule (3.35) specifies η in such a way that $C = 0$, we deduce from (3.36) that $B = 0$, too, despite the arbitrariness in our choice of ξ. Furthermore, we suppose that the coefficients a, b and c do not all vanish simultaneously, whence A must differ from zero and can be divided out of (3.15) to bring equation (3.1) into the required canonical form (3.4).

In the parabolic case, the ordinary differential equation (3.22) defines only one family of level curves

$$(3.37) \qquad\qquad \eta(x, y) = \text{const.}$$

that we can describe as characteristics of the second order partial differential equation (3.1). However, these characteristics are once more to be identified as exceptional curves for the initial value problem for (3.1), since they are straight lines parallel to the ξ-axis in the (ξ,η)-plane, where equation (3.1) appears in the canonical form (3.4) with the second derivative $u_{\eta\eta}$ absent. When the coefficients a, b, c, d, e, f and g are analytic, a detailed verification of the latter irregularity property of the characteristics (3.37) can be based on the power series method of Chapter 1, but since it is patterned on our explanation of the same phenomenon for hyperbolic equations we do not repeat it here.

If the coefficient E of the partial derivative u_η in the canonical representation (3.4) of a partial differential equation (3.1) of the parabolic type (3.7) does not vanish identically, the equation has some of the features of the heat equation. On the other hand, when u_η does not appear at all in (3.4), we are left with an ordinary differential equation of the second order for u as a function of ξ alone, and η plays more the role of a parameter than that of an independent variable. Although such a situation should be viewed as degenerate, it can be helpful for the study of pecularities in the behavior of u along a characteristic, since it permits a dependence of u on η that can be quite arbitrary.

We have postponed until now reduction of the partial differential equation (3.1) to the canonical form (3.5) in the elliptic case (3.8) because it is in general much more difficult to handle. For this purpose it is necessary, in view of (3.12), (3.13) and (3.14), to solve simultaneously the coupled pair of partial differential equations

$$(3.38) \qquad a(\xi_x^2 - \eta_x^2) + 2b(\xi_x\xi_y - \eta_x\eta_y) + c(\xi_y^2 - \eta_y^2) = 0,$$

$$(3.39) \qquad a\xi_x\eta_x + b\xi_x\eta_y + b\xi_y\eta_x + c\xi_y\eta_y = 0$$

of the first order, which state that $A = C$ and $B = 0$. We obtain instead a single complex equation more analogous to (3.19) if we multiply (3.39) by $2i$ and then add the answer to (3.38). We are thus led to the complex identity

$$(3.40) \quad a(\xi_x + i\eta_x)^2 + 2b(\xi_x + i\eta_x)(\xi_y + i\eta_y) + c(\xi_y + i\eta_y)^2 = 0.$$

From (3.40) it follows that

$$(3.41) \qquad \frac{\xi_x + i\eta_x}{\xi_y + i\eta_y} = - \frac{b + i\sqrt{ac - b^2}}{a}.$$

If we multiply (3.41) on both sides by $\xi_y + i\eta_y$ and consider the real and imaginary parts of the result separately, we obtain two linear equations for ξ_x and ξ_y in terms of η_x and η_y that can be solved for those derivatives to yield

$$(3.42) \qquad \xi_x = \frac{c\eta_y + b\eta_x}{\sqrt{ac - b^2}}, \qquad \xi_y = - \frac{a\eta_x + b\eta_y}{\sqrt{ac - b^2}}.$$

We call this system of linear partial differential equations of the first order for ξ and η the *Beltrami equations*. We can differentiate the first of them with respect to y and the second with respect to x and then subtract to eliminate ξ. Thus the single equation

$$(3.43) \qquad \frac{\partial}{\partial x} \frac{a\eta_x + b\eta_y}{\sqrt{ac - b^2}} + \frac{\partial}{\partial y} \frac{c\eta_y + b\eta_x}{\sqrt{ac - b^2}} = 0$$

of the second order for η alone is obtained.

The difficulty here is that in general equation (3.43) is not significantly easier to solve than the original equation (3.1), since for both of them the coefficients of the second derivatives of the unknown function are essentially the same. It can be shown by deeper methods from the theory of conformal mapping that when mild hypotheses about the coefficients a, b and c are fulfilled, the desired solutions ξ and η of the Beltrami system (3.42) do exist even in the large (cf. Ahlfors 2). Alternately, solutions of (3.43) can be found in the small by means of an appropriate scheme of successive approximations (cf. Section 5.3). However, we prefer at present to avoid lengthy discussion of these more subtle aspects of the question by confining ourselves to the case where a, b and c are analytic functions of their arguments x and y. The more general problem will be touched on again in Sections 5.3 and 8.3.

Formula (3.41) suggests analytic continuation into the domain of complex values of the independent variables x and y as a device for the determination of ξ and η when the coefficients a, b and c are analytic. Although the ordinary differential equation (3.22) for the characteristics has no real solutions in the elliptic case,

$$b^2 - ac < 0,$$

we can nevertheless consider the equation

(3.44)
$$\frac{dy}{dx} = \frac{b + i\sqrt{ac - b^2}}{a}$$

in the complex domain. It will be our aim here to explain how complex analytic solutions of (3.44) provide a pair of real functions ξ and η satisfying the Beltrami system (3.42).

By the method of Section 1.2 we can construct a convergent power series solution

(3.45)
$$y = \Phi(x, K)$$

of (3.44) depending on a constant of integration K that can be taken, for example, to be the value y_0 of y for some fixed real choice x_0 of x. The analytic function Φ that appears in (3.45) will in general be complex-valued, and, of course, it is necessary to view its argument x as a complex variable in the present context. By solving equation (3.45) for the constant of integration K we can also express our solution of (3.44) in the implicit form

(3.46)
$$\phi(x, y) = K.$$

Differentiation of (3.46) gives in the usual way

$$\phi_x \, dx + \phi_y \, dy = 0$$

for each fixed choice of K. Therefore even when its arguments x and y are real, ϕ has to satisfy the identity

(3.47)
$$\frac{\phi_x}{\phi_y} = -\frac{b + i\sqrt{ac - b^2}}{a},$$

which is equivalent to (3.44). Our final step is to break ϕ down into its real and imaginary parts for real values of x and y and to identify them with ξ and η. Thus we set

(3.48)
$$\phi(x, y) = \xi(x, y) + i\eta(x, y)$$

in the real (x,y)-plane. We verify right away from (3.47) that the functions

ξ and η defined in this manner satisfy (3.41). Hence they yield at least locally a change of variables (3.2) fulfilling the requirements (3.38) and (3.39) that $A = C$ and $B = 0$ in (3.15). We have now only to divide (3.15) through by A to achieve the desired canonical form (3.5) of (3.1) in the elliptic case.

It should be emphasized that an equation (3.1) of the elliptic type has no real characteristics, since the quadratic forms (3.19) and (3.22) are positive-definite when the discriminant $b^2 - ac$ is negative. This property of an elliptic equation has the peculiar consequence in the case of analytic coefficients that when initial data are prescribed analytically along any analytic curve, a regular solution of the corresponding initial value problem for (3.1) exists in the small (cf. Section 1.2).

Our remarks thus far indicate to some extent the wide differences among hyperbolic, parabolic and elliptic linear partial differential equations in two independent variables. In the theory that follows we shall find that such distinctions concerning the type of an equation play an essential role, especially in matters related to the dependence of a solution on auxiliary initial or boundary data. Of interest in addition to the cases that we have already mentioned here are problems where the equation to be treated is of mixed type in the sense that it is elliptic in one region but hyperbolic in an adjacent region. Situations.of this kind crop up in the applications, too, adding to the difficulties of classifying and reducing to a convenient canonical form the variety of partial differential equations that arise in mathematical physics (cf. Chapter 12).

EXERCISES

1. Determine the type of each of the following partial differential equations and reduce them to canonical form:

$$u_{xx} + 2u_{xy} + u_{yy} + u_x - u_y = 0,$$

$$u_{xx} + 2u_{xy} + 5u_{yy} + 3u_x + u = 0,$$

$$3u_{xx} + 10u_{xy} + 3u_{yy} = 0.$$

2. Find the general solution of the last equation given in Exercise 1.
3. Give a detailed derivation of the identity (3.16).
4. Use the method of Section 1.2 to establish the existence of a convergent power series solution (3.45) of the analytic ordinary differential equation (3.44).
5. Find the characteristics of the *Tricomi equation*

$$(3.49) \qquad\qquad yu_{xx} + u_{yy} = 0$$

in the lower half-plane $y < 0$. Put this equation into canonical form in the upper half-plane $y > 0$.

6. Show that the transformations which leave the canonical form (3.18) for hyperbolic equations invariant are given by

$$\xi_1 = \phi(\xi), \qquad \eta_1 = \psi(\eta),$$

where ϕ and ψ are arbitrary functions. Show that conformal transformations leave the equation (3.5) in canonical form. Establish a relationship between these two results based on the complex substitution

(3.50) $\xi = x + iy, \qquad \eta = x - iy.$

7. By using the complex substitution (3.50) to reduce Laplace's equation

$$u_{xx} + u_{yy} = 0$$

formally to the hyperbolic canonical form (3.18), deduce the formula

$$u(x, y) = \tfrac{1}{2}\phi(x + iy) + \tfrac{1}{2}\overline{\phi(x + iy)}$$

expressing any harmonic function u as the real part of some analytic function ϕ of the complex variable $x + iy$.

8. A fundamental theorem about conformal mapping states that any sufficiently smooth, simply connected, two-dimensional curved surface can be mapped conformally onto a plane region. More generally, any two-dimensional Riemannian manifold defined by a positive-definite metric

(3.51) $ds^2 = a\,dy^2 - 2b\,dx\,dy + c\,dx^2$

over the (x,y)-plane can be transformed by a substitution (3.2) onto the (ξ,η)-plane conformally in the sense that we have

$$ds^2 = \lambda(\xi, \eta)(d\xi^2 + d\eta^2)$$

in terms of the new variables ξ and η, where λ is a positive function (cf. Ahlfors 2). Use this result to establish the existence of a solution of the Beltrami equations (3.42). In the above circumstances we call ξ and η *conformal coordinates* on the Riemannian manifold given by (3.51).

2. EQUATIONS IN MORE INDEPENDENT VARIABLES

In the case of more than two independent variables it is usually not possible to reduce a linear partial differential equation of the second order to a simple canonical form. However, we shall discuss here some aspects of the problem of classifying such equations, and we shall take up in detail equations with constant coefficients, for which a satisfactory canonical form can be achieved. The reader who would prefer to postpone studying this more involved topic can proceed directly to Chapter 4, where characteristic coordinates are applied to solve initial value problems for equations of the hyperbolic type.

Consider the linear partial differential equation

$$(3.52) \qquad \sum_{i,j=1}^{n} a_{ij} \frac{\partial^2 u}{\partial x_i \partial x_j} + \sum_{i=1}^{n} b_i \frac{\partial u}{\partial x_i} + cu = d,$$

whose coefficients a_{ij}, b_i and c and whose right-hand side d are supposed to be functions of the independent variables $x_1, \ldots, x_n$ only. We assume without loss of generality that the matrix of coefficients a_{ij} is symmetric, and we investigate what happens to such an equation under a transformation of coordinates

$$(3.53) \qquad \xi_i = \xi_i(x_1, \ldots, x_n), \qquad i = 1, \ldots, n,$$

with Jacobian

$$\frac{\partial(\xi_1, \ldots, \xi_n)}{\partial(x_1, \ldots, x_n)} \neq 0.$$

To facilitate our calculations, we continue to make use of the convention that dots stand for terms involving only u and its first derivatives. We have

$$(3.54) \qquad \frac{\partial u}{\partial x_i} = \sum_{j=1}^{n} \frac{\partial u}{\partial \xi_j} \frac{\partial \xi_j}{\partial x_i}, \qquad i = 1, \ldots, n,$$

$$(3.55) \qquad \frac{\partial^2 u}{\partial x_i \partial x_j} = \sum_{k,l=1}^{n} \frac{\partial^2 u}{\partial \xi_k \partial \xi_l} \frac{\partial \xi_k}{\partial x_i} \frac{\partial \xi_l}{\partial x_j} + \cdots, \qquad i, j = 1, \ldots, n.$$

Therefore, in terms of the new variables $\xi_1, \ldots, \xi_n$, (3.52) becomes

$$(3.56) \qquad \sum_{k,l=1}^{n} A_{kl} \frac{\partial^2 u}{\partial \xi_k \partial \xi_l} + \cdots = 0,$$

where

$$(3.57) \qquad A_{kl} = \sum_{i,j=1}^{n} a_{ij} \frac{\partial \xi_k}{\partial x_i} \frac{\partial \xi_l}{\partial x_j}, \qquad k, l = 1, \ldots, n.$$

Since we have only the n functions $\xi_1, \ldots, \xi_n$ at our disposal, plus the possibility of dividing out some factor in (3.56), we cannot hope to impose more than $n + 1$ conditions on the $n(n + 1)/2$ independent coefficients A_{kl}. This is not enough freedom to achieve a reasonable canonical form in most situations where $n \geq 3$. If $n = 4$, for example, we cannot expect in general to eliminate all the mixed second derivatives in (3.56), since that leads to the overdetermined system of six homogeneous partial differential equations

$$A_{12} = A_{13} = A_{14} = A_{23} = A_{24} = A_{34} = 0$$

of degree two for the four unknowns ξ_1, ξ_2, ξ_3 and ξ_4.

Although a canonical form for (3.52) with mixed derivatives omitted fails in space of higher dimension when we demand that it be valid in some full region, we are able to obtain such a form for the equation at each isolated point. For the sake of simplicity we take the point in question to be the origin, and we restrict our attention to linear transformations

$$(3.58) \qquad \xi_i = \sum_{j=1}^{n} f_{ij} x_j, \qquad i = 1, \ldots, n,$$

of coordinates, since nothing of significance would be achieved if we were to include higher order terms in a Taylor series development on the right here. Under the change of variables (3.58), formula (3.57) for the transformation of the coefficients of the second derivatives in (3.52) becomes

$$(3.59) \qquad A_{kl} = \sum_{i,j=1}^{n} a_{ij} f_{ki} f_{lj}, \qquad k, l = 1, \ldots, n.$$

We have chosen to consider (3.59) only at the origin so that not only the elements f_{ij} but also the a_{ij} and A_{ij} may be thought of as constants rather than as functions of $x_1, \ldots, x_n$.

After a suitable interchange of the order in which the indices appear, (3.59) is recognized as the rule for transforming the quadratic form

$$(3.60) \qquad Q = \sum_{i,j=1}^{n} a_{ij} \lambda_i \lambda_j$$

naturally associated with the symmetric matrix of elements a_{ij}. It therefore follows from standard theorems in linear algebra concerning the diagonalization of quadratic forms[1] that the coefficients f_{ij} can always be selected in a non-trivial way so that

$$(3.61) \qquad A_{kl} = 0, \qquad k \neq l.$$

Even more significant for our analysis is the observation that the invariants with respect to linear transformation[2] of the two quadratic forms associated with the matrix a_{ij} and the matrix A_{ij} must be identical. This

[1] Cf. Birkhoff-MacLane 1.
[2] Cf. Birkhoff-MacLane 1.

suggests how to classify the partial differential equation (3.52) in a fashion that is invariant under changes of coordinates such as (3.58) or even (3.53). Thus we now find ourselves in a position to define what is meant by the *type* of an equation (3.52) of the second order in more than two independent variables, extending the results of Section 1.

When the matrix of elements A_{kl} is diagonal, which means that (3.61) holds, the numbers $A_{11}, \ldots, A_{nn}$ differ from the eigenvalues of the matrix of coefficients a_{ij} in (3.52) only by positive factors. We say that (3.52) is of the *elliptic type* if and only if all these eigenvalues are different from zero and have the same sign. If the sign in front of (3.52) is chosen so that $a_{11} > 0$, the requirement that the equation is elliptic signifies that the quadratic form (3.60) is positive-definite. The situation usually encountered is that the quadratic form (3.60) remains positive-definite throughout some region of values of $x_1, \ldots, x_n$, and in such circumstances we call (3.52) elliptic in that region. Laplace's equation

$$(3.62) \qquad u_{x_1 x_1} + \cdots + u_{x_n x_n} = 0$$

is the classical example of an equation of the elliptic type in n independent variables.

When the eigenvalues of the matrix of coefficients a_{ij} all differ from zero and all have the same sign except for precisely one of them, we say that (3.52) is of the *normal hyperbolic type* or simply that it is of the *hyperbolic type*. The most familiar example of a hyperbolic equation in n variables is the wave equation

$$(3.63) \qquad u_{x_1 x_1} - u_{x_2 x_2} - \cdots - u_{x_n x_n} = 0,$$

where x_1 stands for the time t. When the eigenvalues are all different from zero, but there are at least two of them of each sign, we call (3.52) *ultra-hyperbolic*. This situation can only occur when $n \geq 4$, the simplest special case being the equation

$$u_{x_1 x_1} + u_{x_2 x_2} = u_{x_3 x_3} + u_{x_4 x_4}$$

in four independent variables. Ultrahyperbolic equations do not arise in a natural way in mathematical physics, and satisfactory initial or boundary value problems for them are not known.

If any of the eigenvalues of the matrix of coefficients a_{ij} vanish, we define (3.52) to be of the *parabolic type*. The heat equation

$$(3.64) \qquad u_{x_1 x_1} + \cdots + u_{x_{n-1} x_{n-1}} - u_{x_n} = 0$$

in $n - 1$ space variables $x_1, \ldots, x_{n-1}$ and one time variable $x_n = t$ is the most important parabolic equation. The theory of parabolic equations is concerned principally with examples that have in common with (3.64) the property that they appear in the canonical form

$$(3.65) \qquad \frac{\partial u}{\partial x_n} = \sum_{i,j=1}^{n-1} a_{ij} \frac{\partial^2 u}{\partial x_i \partial x_j} + \sum_{i=1}^{n-1} b_i \frac{\partial u}{\partial x_i} + cu,$$

where the differential operator on the right is supposed to be elliptic in the sense that the quadratic form in $n - 1$ variables defined by the matrix of coefficients a_{ij} is positive definite. Degenerate parabolic equations in which partial differentiations with respect to some of the independent variables do not occur at all are also conceivable. Finally, the example

$$u_t = u_{xy}$$

involving three independent variables x, y and t is amusing because it does not fit into any of these more easily understood categories.

Although a diagonalized canonical form for (3.52) is in general out of the question in the case of variable coefficients, when the a_{ij} are constants a satisfactory result can be achieved by introducing an appropriate linear transformation of variables (3.58). Indeed, we have observed from (3.59) that it is possible to diagonalize the coefficients A_{kl} at an isolated point by means of (3.58), and, of course, when the coefficients a_{ij} are constant the reduction thus obtained is valid everywhere. By also choosing the scale of the new coordinates $\xi_1, \ldots, \xi_n$ in the right way we can therefore always bring an equation (3.52) with constant coefficients a_{ij} into the canonical form

$$(3.66) \qquad \sum_{i=1}^{n} A_i \frac{\partial^2 u}{\partial \xi_i^2} + \sum_{i=1}^{n} B_i \frac{\partial u}{\partial \xi_i} + Cu = D,$$

where the numbers A_i are all either 1, -1 or 0.

If $b_1, \ldots, b_n$ and c are constants and $d = 0$, then $B_1, \ldots, B_n$ and C are constants, too, and $D = 0$, as can be seen from (3.54), (3.55) and (3.58). This is the case of the most general homogeneous linear partial differential equation of the second order with constant coefficients. A further simplification can be introduced here by making the substitution

$$(3.67) \qquad v = u \exp\left(\frac{B_1 \xi_1}{2A_1} + \cdots + \frac{B_n \xi_n}{2A_n}\right),$$

provided that none of the A_i vanish. For v we deduce from (3.66) the equation

$$\sum_{i=1}^{n} A_i \frac{\partial^2 v}{\partial \xi_i^2} + \left(C - \frac{1}{4} \sum_{i=1}^{n} \frac{B_i^2}{A_i}\right) v = 0,$$

in which terms involving the first derivatives are absent. In particular, the most general homogeneous elliptic equation (3.52) with constant coefficients can be reduced by the above devices to the canonical form

$$(3.68) \qquad u_{\xi_1\xi_1} + \cdots + u_{\xi_n\xi_n} + \lambda u = 0,$$

where λ is a suitable constant. Similarly, the most general homogeneous hyperbolic equation with constant coefficients can be reduced to the analogous form

$$(3.69) \qquad u_{\xi_1\xi_1} - u_{\xi_2\xi_2} - \cdots - u_{\xi_n\xi_n} + \lambda u = 0,$$

where, again, λ denotes a constant. The canonical forms (3.68) and (3.69) for equations of the elliptic type and of the hyperbolic type with constant coefficients are useful when explicit solutions are being sought by the method of separation of variables.

As was the case for two independent variables, the question of classifying partial differential equations of the second order in space of higher dimension is related to the notion of a *characteristic surface*. A characteristic of (3.52) is defined to be any level surface

$$(3.70) \qquad \xi(x_1, \ldots, x_n) = \text{const.}$$

on which the differential relation

$$(3.71) \qquad \sum_{i,j=1}^{n} a_{ij} \frac{\partial \xi}{\partial x_i} \frac{\partial \xi}{\partial x_j} = 0$$

holds. If ξ is taken to be one of the new coordinates ξ_i given by (3.53), then according to (3.57) the corresponding diagonal coefficient A_{ii} drops out of (3.56) along the characteristic (3.70). Hence the partial differential equation (3.52) does not provide a formula for the second derivative $u_{\xi\xi}$ along (3.70) but represents instead a relation within that characteristic among tangential derivatives of u and of its normal derivative u_ξ. A characteristic surface (3.70) of equation (3.52) thus exhibits the property with which the name has been associated before of being exceptional for an initial value problem in which we seek a solution of (3.52) with the data u and u_ξ prescribed on it.

Observe that all the level surfaces of a solution ξ of the equation (3.71) are characteristics. On the other hand, we can convert (3.71) into a partial differential equation of the first order analogous to (3.22) for a function of only $n - 1$ variables that corresponds to a single characteristic by solving the implicit equation (3.70) for one of the coordinates x_i in terms of the others. If (3.52) is of the hyperbolic type, the conoidal solutions of this first order equation (cf. Exercise 2.2.1) comprise a family of characteristics that are of special interest. Consider, for example, any function

$$x_1 = \phi(x_2, \ldots, x_n)$$

representing a characteristic of the wave equation (3.63). It follows from (3.71) that ϕ must satisfy the equation

$$(3.72) \qquad \phi_{x_2}^2 + \cdots + \phi_{x_n}^2 = 1,$$

which provided the topic of Section 2.4. The conoidal solutions of (3.72) are evidently generated by translation of the ordinary cone

$$x_1 = \pm\sqrt{x_2^2 + \cdots + x_n^2},$$

whose level surfaces describe spherical wave fronts.

For a more general hyperbolic equation (3.52) the particular nature of the quadratic form (3.60) ensures that the characteristic conoids have near their vertices the structure of standard cones of two sheets dividing the surrounding space into three parts (except for the case of two independent variables). On the other hand, an equation of the elliptic type is associated with a quadratic form (3.60) that is positive-definite, so that the corresponding first order partial differential equation (3.71) can have no non-trivial real solutions. Consequently elliptic equations possess no real characteristics whatever. Finally, for a parabolic equation in the canonical form (3.65) the characteristics are simply the one-parameter family of planes $x_n = $ const., since for them the relation (3.70) implies that the first partial derivatives of ξ with respect to $x_1, \ldots, x_{n-1}$ all vanish.

We have now seen some of the complications that can arise in classifying and putting into canonical form linear partial differential equations of the second order in any number of independent variables. If we pass to equations of still higher order, these questions become quite troublesome even for the case of constant coefficients. Except for the remarks about systems in Section 5 below we shall not investigate such matters here, but we do wish to mention the *biharmonic equation*

$$(3.73) \qquad u_{xxxx} + 2u_{xxyy} + u_{yyyy} = 0$$

in two independent variables. Equation (3.73) is to be described as elliptic because it has no real characteristics. It will serve occasionally in what follows as an example of the phenomena that are encountered in the theory of partial differential equations of higher order.

EXERCISES

1. Perform a reduction of any hyperbolic equation (3.1) with constant coefficients a, b, c, d, e and f to the canonical form

$$v_{\xi\eta} + \lambda v = 0$$

in the homogeneous case $g = 0$. Find out under what circumstances the constant λ vanishes.

2. Show that the most general homogeneous partial differential equation (3.52) with constant coefficients can be reduced by a linear transformation like (3.58) and an exponential substitution like (3.67) to the canonical form

$$v_{\xi_1\xi_1} + \cdots + v_{\xi_k\xi_k} - v_{\eta_1\eta_1} - \cdots - v_{\eta_l\eta_l} + \lambda v_t + \mu v = 0,$$

where λ and μ are constants and where it is conceivable that $k + l < n - 1$.

3. Show that any partial differential equation (3.52) of the normal hyperbolic type can be brought locally into a canonical form

(3.74) $$\frac{\partial^2 u}{\partial t^2} = \sum_{i,j=1}^{n-1} A_{ij} \frac{\partial^2 u}{\partial \xi_i \partial \xi_j} + \cdots$$

where the differential operator on the right is elliptic.

4. Show that all the characteristic surfaces for the wave equation

$$u_{tt} = u_{xx} + u_{yy}$$

can be found by solving the equation (2.59) of geometrical optics.

5. Explain why we say that the biharmonic equation (3.73) is of the elliptic type by showing that it has no real characteristics.

6. Discuss the conoidal solutions of the first order equation for the characteristics of a second order partial differential equation with constant coefficients. In particular, show that each characteristic conoid for an ultrahyperbolic equation consists of only one sheet dividing space into just two parts.

7. Show that when $\lambda > 0$ the linear combinations

$$u = \alpha r^{-(n-2)/2} J_{(n-2)/2}(\sqrt{\lambda} r) + \beta r^{-(n-2)/2} Y_{(n-2)/2}(\sqrt{\lambda} r)$$

of Bessel functions[3] are the solutions of the partial differential equation (3.68) which depend on the distance

$$r = \sqrt{(\xi_1 - \eta_1)^2 + \cdots + (\xi_n - \eta_n)^2}$$

alone. Find the corresponding solutions of the general homogeneous elliptic

[3] Cf. Churchill 1.

equation (3.52) with constant coefficients and show that they are of the form

$$(3.75) \qquad u = \frac{U}{\Gamma^{(n-2)/2}} + V \log \Gamma + W,$$

with U, V and W regular at the point $x_i = y_i$ and with

$$\Gamma = \sum_{i,j=1}^{n} A_{ij}(x_i - y_i)(x_j - y_j),$$

where the A_{ij} are defined (cf. Exercise 2.6.4) to be the elements of the matrix inverse to that of the coefficients a_{ij}. Obtain analogous solutions of (3.52) in the hyperbolic case.

3. LORENTZ TRANSFORMATIONS AND SPECIAL RELATIVITY

Thus far the emphasis has been on altering and simplifying the coefficients a_{ij} of the linear second order partial differential equation (3.52) by means of a transformation of the independent variables. A quite different question would be to determine all the transformations of coordinates which leave invariant the form of a specific partial differential equation, such as Laplace's equation or the wave equation. Problems of this kind have genuine significance for equations with constant coefficients, since they lead to groups of linear transformations that are important in geometry and in mathematical physics.[4]

As our first example we ask for the transformations that leave the Laplace equation

$$u_{x_1 x_1} + \cdots + u_{x_n x_n} = 0$$

invariant. In the plane case $n = 2$ it is only combinations of conformal mappings and reflections that need be considered, but we now require that they do not introduce any factor in front of the Laplace operator. Hence we are restricted to translations, rotations and reflections, since the modulus of the derivative of any analytic function providing a conformal mapping of the desired kind must be constant, and the analytic function itself must therefore be linear. In space of dimension $n \geq 3$ the absence of conformal mappings other than translations, orthogonal transformations, magnifications and inversions makes it even easier to conclude that our attention can be limited to the translations and orthogonal transformations. We omit translations from further consideration because their properties are quite elementary. With this understood, it is the group of orthogonal

[4] Cf. Petrovsky 1.

transformations that should provide the answer to our problem. We recall that orthogonal transformations are of the form

(3.76) $$\xi_i = \sum_{j=1}^{n} f_{ij}x_j, \qquad i = 1, \ldots, n,$$

with

(3.77) $$\sum_{i=1}^{n} f_{ki}f_{li} = \delta_{kl} = \begin{cases} 0, & k \neq l, \\ 1, & k = l, \end{cases}$$

where δ_{kl} is the Kronecker delta. It is evident from formula (3.59) that (3.77) is, indeed, the condition for the substitution (3.76) to leave the form of Laplace's equation unchanged.

A similar analysis, which we shall not go into in detail, suggests that it is only translations and certain linear transformations (3.76) that leave the wave equation

$$u_{x_1 x_1} - u_{x_2 x_2} - \cdots - u_{x_n x_n} = 0$$

invariant. Once again we discard the translations as trivial, so that our interest will be in determining the group of linear transformations that do not alter the form of the wave equation. These are known as the *Lorentz transformations*. They are characterized within the group (3.76) by the requirements

(3.78) $$f_{11}^2 - \sum_{i=2}^{n} f_{1i}^2 = 1,$$

(3.79) $$f_{k1}f_{l1} - \sum_{i=2}^{n} f_{ki}f_{li} = -\delta_{kl}, \qquad k, l = 1, \ldots, n; \ k + l > 2.$$

According to (3.59), formulas (3.78) and (3.79) give precisely the conditions for the invariance of the wave equation under the change of coordinates (3.76). It is to be observed that they differ from (3.77) principally in reversals of sign before terms not involving the subscript 1. In particular, any transformation that leaves x_1 unaltered and is orthogonal with respect to the variables $x_2, \ldots, x_n$ all by themselves must satisfy (3.78) and (3.79). Our immediate objective will be to exploit this remark in order to establish a representation for the Lorentz transformations on n variables in terms of orthogonal transformations on $n - 1$ variables.

We start by finding the Lorentz transformations when $n = 2$. Adopting the notation $x_1 = x$, $x_2 = y$, we accomplish this by performing a preliminary rotation (3.30) through $-45°$ to reduce the one-dimensional wave equation (3.28) to the canonical form

(3.80) $$2u_{\xi\eta} = 0,$$

which differs from (3.31) only by a factor of 2 that we are not allowed to divide out in the present context. Our problem is, then, to determine the substitutions that leave (3.80) invariant in the strict sense that not even the factor of 2 appearing there should be altered. The answer is furnished by the special case

$$\xi' = \alpha\xi, \qquad \eta' = \frac{1}{\alpha}\eta$$

of the change of characteristic coordinates

$$\xi' = \phi(\xi), \qquad \eta' = \psi(\eta),$$

where α is a parameter different from zero. When we reverse the rotation through $-45°$ that led to (3.80) and return to our original notation, we obtain the general representation

(3.81) $$\xi_1 = x_1 \operatorname{ch} \theta + x_2 \operatorname{sh} \theta, \quad \xi_2 = x_1 \operatorname{sh} \theta + x_2 \operatorname{ch} \theta$$

for the Lorentz transformations on two variables, which involves only a single parameter

$$\theta = -\log|\alpha|.$$

We shall prove that the Lorentz transformations in n-dimensional space can be constructed by composition of Lorentz transformations of the special type (3.81) on the first two variables x_1 and x_2 with orthogonal transformations on the last $n - 1$ variables $x_2, \ldots, x_n$. Any such combination of transformations is, of course, a Lorentz transformation itself, since the composition of any two transformations leaving the wave equation invariant has that same property.

Suppose now that we are given an arbitrary Lorentz transformation (3.76). If not all the coefficients $f_{12}, \ldots, f_{1n}$ vanish, we can make a preliminary orthogonal transformation on $x_2, \ldots, x_n$ alone such that

$$\xi_2 = \frac{f_{12}x_2 + \cdots + f_{1n}x_n}{\sqrt{f_{12}^2 + \cdots + f_{1n}^2}}$$

replaces x_2. Therefore it suffices to consider the case where f_{1i} vanishes for $i = 3, \ldots, n$ and (3.78) reduces to

$$f_{11}^2 - f_{12}^2 = 1.$$

Here we can write

$$f_{11} = \operatorname{ch} \theta, \qquad f_{12} = \operatorname{sh} \theta,$$

which means that the formulas

$$\xi_1 = f_{11}x_1 + f_{12}x_2, \quad \xi_2 = f_{12}x_1 + f_{11}x_2, \quad \xi_i = x_i, \qquad i = 3, \ldots, n,$$

define a special Lorentz transformation like (3.81) on x_1 and x_2 alone.

If we suppose that the above reductions have already been performed, we have left merely to treat a situation where

$$f_{11} = \pm 1, \qquad f_{12} = \cdots = f_{1n} = 0.$$

Under the latter hypothesis we deduce from (3.79) with $k = 1$ and $l = 2, \ldots, n$ that

$$f_{21} = \cdots = f_{n1} = 0,$$

too. The remainder of the conditions (3.79) then show that (3.76) provides an orthogonal transformation on the variables $x_2, \ldots, x_n$ alone. Therefore any Lorentz transformation consists of a special one of the form (3.81) involving only the first two coordinates, preceded and followed by orthogonal transformations of the last $n - 1$ coordinates. This is the representation theorem for Lorentz transformations that we set out to establish.

A direct corollary of our theorem is that the *adjoint*

$$\eta_i = \sum_{j=1}^{n} f_{ji} x_j, \qquad i = 1, \ldots, n,$$

of any Lorentz transformation (3.76) is again a Lorentz transformation, since the orthogonal transformations and the plane transformations (3.81) enjoy this property. Thus (3.78) and (3.79) must be equivalent to the relations

$$f_{11}^2 - \sum_{i=2}^{n} f_{i1}^2 = 1,$$

$$f_{1k} f_{1l} - \sum_{i=2}^{n} f_{ik} f_{il} = -\delta_{kl}, \qquad k, l = 1, \ldots, n; \ k + l > 2.$$

It follows that under a Lorentz transformation (3.76) we have

(3.82) $$\xi_1^2 - \xi_2^2 - \cdots - \xi_n^2 = x_1^2 - x_2^2 - \cdots - x_n^2.$$

The invariance of the quadratic form (3.82) is related to the invariance of the characteristic conoids of the wave equation under Lorentz transformations, since the cone

$$x_1^2 - x_2^2 - \cdots - x_n^2 = 0$$

is such a characteristic fulfilling (3.72). Observe that any pair of points $x_1, \ldots, x_n$ and $\xi_1, \ldots, \xi_n$ satisfying (3.82) correspond under some Lorentz transformation, for they can be brought by orthogonal transformations of their last $n - 1$ coordinates into the plane of the first two coordinates, where a substitution (3.81) can be arranged to provide the desired correspondence.

The notion of a Lorentz transformation is fundamental in the *special theory of relativity*. Let us work under the simplified assumption that we observe in nature only solutions of the wave equation

$$(3.83) \qquad \frac{1}{c^2} u_{tt} - u_{xx} - u_{yy} - u_{zz} = 0,$$

where c denotes the *speed of light*. The basic hypothesis of special relativity[5] is that c has the same value when measured in any pair of coordinate systems that move uniformly with respect to each other at a speed slower than c. This justifies using units for which $c = 1$ in (3.83), so that we can set

$$ct = x_1, \quad x = x_2, \quad y = x_3, \quad z = x_4$$

there. When a new coordinate frame $\xi_1, \ldots, \xi_4$ is introduced whose origin $\xi_2 = \xi_3 = \xi_4 = 0$ moves with uniform velocity in the frame given by $x_1, \ldots, x_4$, it is physically just as acceptable as that frame if the two are connected by a linear transformation leaving the wave equation (3.83) unchanged in form. We have seen that such transformations comprise the Lorentz group; thus we are not able to prefer one coordinate system over another in the special theory of relativity when they are related by a Lorentz transformation.

The consequences of the above freedom in our choice of coordinates can be studied in terms of the special substitutions (3.81) involving only two variables, since our representation theorem shows that the general case can be obtained from this one merely by a change of orientation. One of the interesting observations to be made is that under a Lorentz transformation the distance among the space variables is not preserved, a fact which is apparent directly from (3.81). Linked with the latter phenomenon is our inability to distinguish x_1 from ξ_1 as a more natural time variable. Indeed, the quadratic form (3.82) is the only geometrical invariant of the group of Lorentz transformations.

The concepts that a direction $(x_1, \ldots, x_4)$ with

$$x_1^2 > x_2^2 + x_3^2 + x_4^2$$

is *time-like* and that a direction $(x_1, \ldots, x_4)$ with

$$x_1^2 < x_2^2 + x_3^2 + x_4^2$$

[5] Cf. Sommerfeld 2.

is *space-like* are invariant, and we can describe the intermediate characteristic

$$x_1^2 = x_2^2 + x_3^2 + x_4^2$$

meaningfully as a *light cone*. This enables us to say that two *events* described by the coordinates x_1', x_2', x_3', x_4' and x_1'', x_2'', x_3'', x_4'' occur at different times whenever the vector $(x_1'' - x_1', \ x_2'' - x_2', \ x_3'' - x_3', \ x_4'' - x_4')$ joining them has a time-like direction, or to say that they occur at different positions in space whenever that vector has a space-like direction. In the former case there is always a Lorentz transformation to a new coordinate system $\xi_1, \ldots, \xi_4$ in which both events correspond to the same values of ξ_2, ξ_3 and ξ_4, so that we cannot distinguish their positions in space. In the latter case, on the other hand, a Lorentz transformation to new coordinates $\xi_1, \ldots, \xi_4$ exists such that both events correspond to a single value of ξ_1 and may be conceived of as occurring simultaneously. Our conclusion is that the metric

$$ds^2 = dx_1^2 - dx_2^2 - dx_3^2 - dx_4^2 = c^2 \, dt^2 - dx^2 - dy^2 - dz^2$$

based on the invariant quadratic form (3.82) furnishes the only tool available in special relativity to measure time intervals. It is remarkable that the principle of relativity thus appears as a natural consequence of our search for transformations leaving the partial differential equation (3.83) unchanged.

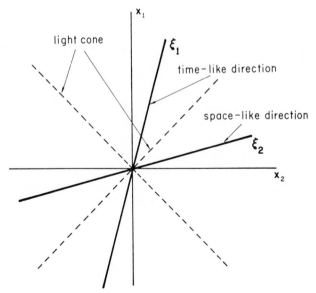

FIG. 4. A Lorentz transformation in the plane.

EXERCISES

1. Show that if $u = u(x_1, \ldots, x_n)$ is a solution of Laplace's equation, then the *Kelvin transformation*, or *inversion*,

$$(3.84) \qquad v = \frac{1}{R^{n-2}} u\left(\frac{x_1}{R^2}, \ldots, \frac{x_n}{R^2}\right), \qquad R^2 = x_1^2 + \cdots + x_n^2,$$

furnishes another solution $v = v(x_1, \ldots, x_n)$.

2. Extend the construction (3.84) of harmonic functions to the case of the wave equation.

3. Explain how the imaginary substitution $x_1' = ix_1$ on the first coordinate x_1 can be used to relate the group of orthogonal transformations (3.76) on n variables to the Lorentz group. In particular, compare (3.81) with the group of rotations in the plane.

4. Show that *Maxwell's equations*

$$(3.85) \qquad \frac{\partial B}{\partial t} + \nabla \times E = 0, \qquad \frac{\partial E}{\partial t} - \nabla \times B = 0,$$

$$(3.86) \qquad \nabla \cdot B = 0, \qquad \nabla \cdot E = 0$$

for the electric field vector $E = (E_1, E_2, E_3)$ and the magnetic field vector $B = (B_1, B_2, B_3)$ in a vacuum are invariant under Lorentz transformations (3.76) with $n = 4$ and

$$x_1 = t, \quad x_2 = x, \quad x_3 = y, \quad x_4 = z.$$

It is to be understood[6] that the components of the electric and magnetic field vectors E' and B' in the new coordinate system $\xi_1, \ldots, \xi_4$ are defined by the tensor rule

$$F_{ij}' = \sum_{k,l=1}^{4} f_{ik} f_{jl} F_{kl}, \qquad i,j = 1, \ldots, 4,$$

where

$$F_{12} = E_1, \qquad F_{13} = E_2, \quad F_{14} = E_3, \quad F_{23} = B_3,$$

$$F_{24} = -B_2, \quad F_{34} = B_1, \quad F_{ij} = -F_{ji}, \qquad i,j = 1, \ldots, 4,$$

and where identical relations for the corresponding primed quantities hold.

5. Prove that each component E_i or B_i of a solution of the Maxwell equations (3.85), (3.86) satisfies the wave equation (3.83) with $c = 1$.

6. Deduce from (3.85) that

$$\frac{\partial}{\partial t}(\nabla \cdot B) = \frac{\partial}{\partial t}(\nabla \cdot E) = 0.$$

[6] Cf. Sommerfeld 2.

4. QUASI-LINEAR EQUATIONS IN TWO INDEPENDENT VARIABLES

A nonlinear partial differential equation of the second order

(3.87) $$F(x_i, u, u_{x_i}, u_{x_i x_j}) = 0$$

for a function u of n independent variables $x_1, \ldots, x_n$ can be classified in an invariant way according to rules analogous to those we have developed for linear equations. However, its type turns out to depend in general on the particular solution involved and not just on the values of the independent variables at hand, for we define the type at each point in space in terms of the signs of the eigenvalues of the matrix of elements

$$a_{ij} = \frac{\partial F}{\partial u_{x_i x_j}},$$

which are now functions of u and its derivatives. Since the precise specifications are identical with those given in Section 2 for the linear case, where the coefficients a_{ij} are functions of $x_1, \ldots, x_n$ only, we need not repeat them here. As an example we merely observe that (3.87) is said to be elliptic in a region, and for a solution, such that the quadratic form

$$Q = \sum_{i,j=1}^{n} \frac{\partial F}{\partial u_{x_i x_j}} \lambda_i \lambda_j$$

is either positive-definite or negative-definite.

It is futile to attempt to reduce a nonlinear equation in more than two independent variables to a useful canonical form, but much progress can be made for partial differential equations in the plane. Of special importance are equations

(3.88) $$au_{xx} + 2bu_{xy} + cu_{yy} + d = 0$$

with coefficients a, b, c and d depending exclusively on x, y, u, u_x and u_y, that is, with

$$a = a(x, y, u, u_x, u_y), \qquad b = b(x, y, u, u_x, u_y),$$

and so on. We call such an equation *quasi-linear* because it is linear in the derivatives of the highest order. We shall describe here how characteristic coordinates can be introduced for (3.88) in the hyperbolic case

$$b^2 - ac > 0$$

by a procedure quite closely related to the one used to reduce the linear hyperbolic equation (3.1) to the canonical form (3.18). The principal new feature that makes the quasi-linear equation more complicated than the linear one in this connection is that the differential equations for the characteristics occur in such a way that they are necessarily coupled with the partial differential equation (3.88) itself in the truly quasi-linear case, and therefore we cannot solve for the characteristics separately.

For the sake of variety, we introduce the characteristics of (3.88) in a somewhat different fashion from that used for (3.1), although the two methods work equally well in both cases. We set out to find curves

$$(3.89) \qquad x = x(\sigma), \qquad y = y(\sigma)$$

in the (x,y)-plane along which a knowledge of u, $p = u_x$ and $q = u_y$ is *not* sufficient to determine, together with the partial differential equation (3.88), all the second derivatives u_{xx}, u_{xy} and u_{yy}.

Along a curve (3.89) where p and q are prescribed, we know the values of the differentials.

$$(3.90) \qquad dp = p_x \, dx + p_y \, dy = u_{xx} \, dx + u_{xy} \, dy,$$

$$(3.91) \qquad dq = q_x \, dx + q_y \, dy = u_{xy} \, dx + u_{yy} \, dy.$$

Since the arguments x, y, u, p and q of the coefficients a, b, c and d are given along (3.89), we can view (3.88), (3.90) and (3.91) as a system of three simultaneous linear equations for the determination of the three unknowns u_{xx}, u_{xy} and u_{yy}. The situation in which these linear equations cannot be solved is described by the relation

$$\begin{vmatrix} a & 2b & c \\ dx & dy & 0 \\ 0 & dx & dy \end{vmatrix} = 0$$

stating that the determinant of the system vanishes. It furnishes the ordinary differential equation

$$(3.92) \qquad a \, dy^2 - 2b \, dx \, dy + c \, dx^2 = 0$$

for the characteristics of (3.88), which is equivalent to (3.22) in the case when a, b and c do not depend on u, p and q.

For a quasi-linear equation (3.88) in which a, b and c involve p and q in an essential way, (3.92) has no meaning unless a solution u of (3.88) is specified. However, when u is given in a non-trivial way, (3.92) defines two families of characteristic curves for that solution in the hyperbolic case $b^2 - ac > 0$. We denote by ξ and η, respectively, parameters along these two families of curves that are chosen so that ξ is constant along each characteristic on which η varies and η is constant along each characteristic on which ξ varies. The variables ξ and η are once more called *characteristic coordinates*.

The quadratic (3.92) breaks down into two linear factors, each of which furnishes an ordinary differential equation for one of the families of characteristics. The possibility of making a rotation in the (x,y)-plane allows us to assume without loss of generality that $ac \neq 0$, whence we can write these differential equations in the form

$$(3.93) \qquad y_\xi - \lambda_+ x_\xi = 0, \qquad y_\eta - \lambda_- x_\eta = 0,$$

where

$$\lambda_+ = \frac{b + \sqrt{b^2 - ac}}{a}, \qquad \lambda_- = \frac{b - \sqrt{b^2 - ac}}{a}.$$

The key to our treatment of the quasi-linear case is that we think of (3.93) as a pair of simultaneous partial differential equations for x and y as functions of the new coordinates ξ and η rather than as ordinary differential equations for the characteristics of (3.88). Since u, p and q appear as arguments of the coefficients λ_+ and λ_- in (3.93), we should consider them together with x and y as unknown functions of ξ and η. Our problem is to derive three additional partial differential equations of the first order for these functions to complete the system (3.93).

If we re-examine the three linear equations (3.88), (3.90) and (3.91) and take into account the vanishing of their determinant along a characteristic, we notice that the rank of the matrix

$$M = \begin{pmatrix} a & 2b & c & d \\ dx & dy & 0 & -dp \\ 0 & dx & dy & -dq \end{pmatrix}$$

must be two when dx, dy, dp and dq are measured in a characteristic. When we put equal to zero the determinant of the matrix obtained by striking out the second column of M, we find

$$(3.94) \qquad a\, dy\, dp + c\, dx\, dq + d\, dx\, dy = 0.$$

This can be evaluated along the characteristics on which ξ varies, and after division by ay_ξ it yields

$$(3.95) \qquad p_\xi + \lambda_- q_\xi + \frac{d}{a} x_\xi = 0$$

because of the first equation (3.93) and the obvious relation $c = a\lambda_+\lambda_-$. Similarly, along the characteristics on which η varies (3.94) reduces to

$$(3.96) \qquad\qquad p_\eta + \lambda_+ q_\eta + \frac{d}{a} x_\eta = 0.$$

We have been able to divide out the factors ay_ξ and ay_η in passing from (3.94) to (3.95) and (3.96) because the assumption $ac \neq 0$ ensures that they do not vanish.

When u does not appear explicitly as an argument of the coefficients a, b, c and d, the four partial differential equations (3.93), (3.95), (3.96) form by themselves a system adequate for the determination of x, y, p and q as functions of the characteristic coordinates ξ and η. If a, b, c or d does depend on u, we can use the familiar differential relation

$$du = p \, dx + q \, dy$$

to derive the two further partial differential equations

$$(3.97) \qquad\qquad u_\xi - px_\xi - qy_\xi = 0,$$

$$(3.98) \qquad\qquad u_\eta - px_\eta - qy_\eta = 0,$$

either of which can be added to the system to provide a fifth equation for the five unknowns x, y, p, q and u. Notice that (3.97) and (3.98) are not truly independent, since

$$(3.99) \quad \frac{\partial}{\partial \xi}(u_\eta - px_\eta - qy_\eta) - \frac{\partial}{\partial \eta}(u_\xi - px_\xi - qy_\xi)$$

$$= p_\eta x_\xi + q_\eta y_\xi - p_\xi x_\eta - q_\xi y_\eta$$

$$= (p_\eta + \lambda_+ q_\eta)x_\xi - (p_\xi + \lambda_- q_\xi)x_\eta = 0.$$

We have seen how to reduce the quasi-linear equation (3.88) to the characteristic system

$$y_\xi - \lambda_+ x_\xi = 0, \qquad y_\eta - \lambda_- x_\eta = 0,$$

$$p_\xi + \lambda_- q_\xi + \frac{d}{a} x_\xi = 0, \qquad p_\eta + \lambda_+ q_\eta + \frac{d}{a} x_\eta = 0,$$

$$u_\xi - px_\xi - qy_\xi = 0$$

or

$$u_\eta - px_\eta - qy_\eta = 0$$

of five equations of the first order. It is still necessary to investigate under what circumstances a solution x, y, u, p, q of this canonical system generates a solution u of the original equation (cf. Section 2.2).

Suppose that we are given a solution of the canonical system with the values of x, y, u, p and q prescribed on an initial line

(3.100) $$\xi + \eta = \text{const.}$$

Since new characteristic coordinates can be introduced by replacing ξ with any function of ξ and by replacing η with any function of η, and since the initial curve should not be itself a characteristic $\xi = \text{const.}$ or $\eta = \text{const.}$, there is no actual loss of generality in taking it to be of the special form (3.100). We may confine our attention to initial data such that

$$du = p\,dx + q\,dy, \qquad x_\xi x_\eta \neq 0$$

on (3.100). From an integration of the identity (3.99) it follows that if we take (3.97) to be the last equation of our system, then (3.98) has to hold everywhere, too. Moreover, the Jacobian

$$x_\xi y_\eta - x_\eta y_\xi = (\lambda_- - \lambda_+)x_\xi x_\eta = -2\frac{\sqrt{b^2 - ac}}{a}\,x_\xi x_\eta$$

differs from zero, whence we can replace ξ and η as the independent variables by x and y. When this is done, (3.97) and (3.98) show that $p = u_x$ and $q = u_y$.

It remains to establish that the function $u = u(x, y)$ satisfies the partial differential equation (3.88). From (3.93) and rules of the elementary calculus we obtain

$$\eta_x + \lambda_+\eta_y = 0, \qquad \xi_x + \lambda_-\xi_y = 0.$$

Consequently (3.95) and (3.96) yield

$$u_{xx} = p_\xi\xi_x + p_\eta\eta_x$$

$$= -\left(\lambda_- q_\xi + \frac{d}{a}x_\xi\right)\xi_x - \left(\lambda_+ q_\eta + \frac{d}{a}x_\eta\right)\eta_x$$

$$= -(\lambda_+ + \lambda_-)(q_\xi\xi_x + q_\eta\eta_x) - \lambda_+\lambda_-(q_\xi\xi_y + q_\eta\eta_y) - \frac{d}{a},$$

whence (3.88) follows because

$$\lambda_+ + \lambda_- = \frac{2b}{a}, \qquad \lambda_+\lambda_- = \frac{c}{a}.$$

Our conclusion is that any quasi-linear hyperbolic partial differential equation (3.88) of the second order in two independent variables x and y is equivalent to a canonical system of five partial differential equations of the first order for five unknown functions x, y, u, p and q of the characteristic coordinates ξ and η. The characteristic system has the special property that in each equation partial derivatives with respect to only one of the two independent variables occur. Let us differentiate with respect to η each of the equations involving derivatives with respect to ξ only, while differentiating with respect to ξ each equation involving derivatives with respect to η only. We thus obtain five linear equations for the mixed second derivatives of x, y, u, p and q whose right-hand sides are quadratic expressions in the first derivatives. The determinant

$$\begin{vmatrix} -\lambda_+ & 1 & 0 & 0 & 0 \\ -\lambda_- & 1 & 0 & 0 & 0 \\ \dfrac{d}{a} & 0 & 0 & 1 & \lambda_- \\ \dfrac{d}{a} & 0 & 0 & 1 & \lambda_+ \\ -p & -q & 1 & 0 & 0 \end{vmatrix} = -4\frac{b^2 - ac}{a^2}$$

of these five linear equations differs from zero. Therefore we can solve them for the mixed derivatives to obtain a canonical system of five second order equations

(3.101) $x_{\xi\eta} + \cdots = 0, \quad y_{\xi\eta} + \cdots = 0, \quad u_{\xi\eta} + \cdots = 0,$

$$p_{\xi\eta} + \cdots = 0, \quad q_{\xi\eta} + \cdots = 0$$

analogous to (3.18), where the dots now stand for nonlinear combinations of all of the unknowns and of their first derivatives. In a certain sense we can view (3.101) as a generalization of the canonical form (3.18) for linear hyperbolic equations. Of course, if a, b and c do not depend on u, p and q, but d does, the original canonical form (3.18) is still adequate.

The canonical system just described is far simpler to handle than the quasi-linear equation (3.88) of the second order from which it arose. In this connection it turns out that characteristic coordinates provide a powerful tool for the integration of both linear and nonlinear partial differential equations in two independent variables (cf. Chapter 4). We shall refer to the important collection of techniques associated with them as the *method of characteristics*.

A better appreciation of the significance of the characteristic system (3.93) to (3.97), in which each equation involves partial differentiations in one direction only, can be gained if we discuss for a moment the question of classification of more general quasi-linear systems of partial differential equations of the first order in two independent variables. Consider the system

(3.102) $$\frac{\partial u_i}{\partial x} + \sum_{j=1}^{n} a_{ij}\frac{\partial u_j}{\partial y} + b_i = 0, \qquad i = 1, \ldots, n,$$

where the coefficients a_{ij} and b_i are functions of x and y and of the n unknowns $u_1, \ldots, u_n$. We attempt to transform it into a new system such that each equation involves differentiations in one direction only.

To achieve our objective we need to form n independent linear combinations

$$\sum_{j=1}^{n} v_{kj}\frac{\partial u_j}{\partial x} + \sum_{i,j=1}^{n} v_{ki}a_{ij}\frac{\partial u_j}{\partial y} + \sum_{i=1}^{n} v_{ki}b_i = 0$$

of the n equations (3.102) such that

$$\sum_{i=1}^{n} v_{ki}a_{ij} = \lambda_k v_{kj}, \qquad k, j = 1, \ldots, n.$$

Whenever this can be accomplished, (3.102) assumes the characteristic form

(3.103) $$\sum_{j=1}^{n} v_{kj}\left(\frac{\partial u_j}{\partial x} + \lambda_k \frac{\partial u_j}{\partial y}\right) + \sum_{i=1}^{n} v_{ki}b_i = 0, \qquad k = 1, \ldots, n.$$

Note that each equation (3.103) involves differentiation in a single direction defined by one of the n ordinary differential equations

(3.104) $$\frac{dy}{dx} = \lambda_k, \qquad k = 1, \ldots, n.$$

The reduction of (3.102) to the canonical form (3.103) succeeds if and only if the matrix of coefficients a_{ij} has n real characteristic roots λ_k satisfying the polynomial equation

$$\begin{vmatrix} a_{11} - \lambda & a_{12} & \cdots & a_{1n} \\ a_{21} & a_{22} - \lambda & \cdots & a_{2n} \\ \cdot & \cdot \cdot \cdot \cdot \cdot \cdot \cdot \cdot & & \cdot \\ a_{n1} & a_{n2} & \cdots & a_{nn} - \lambda \end{vmatrix} = 0$$

and has a full set of n linearly independent characteristic vectors $(v_{k1}, \ldots, v_{kn})$. When this is the case we say that the quasi-linear system of partial differential equations (3.102) is of the *hyperbolic type*. We call the n families of curves defined by the ordinary differential equations (3.104) its *characteristics.* On the other hand, if all the roots λ_k are complex, so that the characteristics (3.104) are imaginary, we say that the system (3.102) is of the *elliptic type*.

The system (3.93) to (3.97) derived from the second order equation (3.88) is seen to be in the characteristic form (3.103). The ordinary differential equations (3.104) for the characteristics played a less important role there because we introduced the pair of characteristic coordinates ξ and η and thus succeeded in incorporating them in the system itself as equations (3.93). The device of resorting to characteristic coordinates is only available when there are no more than two distinct families of characteristics at hand. Thus it cannot in general be used for systems of n equations when $n > 2$. However, the relationship between the theory of the quasi-linear equation (3.88) of the second order and the theory of quasi-linear systems (3.102) of the first order is apparent and should be helpful for an understanding of the situation as a whole.

We shall not pursue the study of hyperbolic systems further here because that constitutes the main topic of Section 5. It will be convenient there to review the whole subject in matrix notation, for we shall see that matrices and vectors provide a natural language in which to describe the principal ideas involved without excessive computation.

EXERCISES

1. Show that the most general nonlinear hyperbolic equation

$$F(x, y, u, p, q, r, s, t) = 0$$

of the second order in two independent variables can be reduced to a characteristic system analogous to (3.93) to (3.97) consisting of eight partial differential equations of the first order for the eight unknowns x, y, u, p, q, r, s and t as functions of a pair of characteristic coordinates ξ and η.

2. Evaluate the expressions that are abbreviated by dots in (3.101).

3. Show that every analytic solution of an analytic quasi-linear equation (3.88) can be transformed in the elliptic case

$$b^2 - ac < 0$$

into a solution of a canonical system of the form

$$(3.105) \qquad x_{\alpha\alpha} + x_{\beta\beta} + \cdots = 0, \qquad y_{\alpha\alpha} + y_{\beta\beta} + \cdots = 0,$$
$$u_{\alpha\alpha} + u_{\beta\beta} + \cdots = 0, \qquad p_{\alpha\alpha} + p_{\beta\beta} + \cdots = 0,$$
$$q_{\alpha\alpha} + q_{\beta\beta} + \cdots = 0$$

analogous to (3.101), where the new coordinates α and β are related to characteristic coordinates ξ and η by the complex substitution (cf. Section 16.2)

$$\xi = \alpha + i\beta, \qquad \eta = \alpha - i\beta.$$

How can it be seen that the hypothesis of analyticity is inessential?

4. Use the transformation of an elliptic equation (3.88) to the canonical form (3.105) to reduce the equation

$$(1 + q^2)r - 2pqs + (1 + p^2)t = 0$$

of minimal surfaces (cf. Section 15.4) to the system of three Laplace equations

$$x_{\alpha\alpha} + x_{\beta\beta} = 0, \quad y_{\alpha\alpha} + y_{\beta\beta} = 0, \quad u_{\alpha\alpha} + u_{\beta\beta} = 0,$$

subject to the side conditions

$$x_\alpha^2 + y_\alpha^2 + u_\alpha^2 = x_\beta^2 + y_\beta^2 + u_\beta^2,$$

$$x_\alpha x_\beta + y_\alpha y_\beta + u_\alpha u_\beta = 0$$

comparable to (3.38) and (3.39). Show that α and β can be interpreted as *conformal coordinates* (cf. Exercise 1.8) on the minimal surface $u = u(x, y)$.

5. Discuss the type of the *Monge-Ampère equation*

$$rt - s^2 = 1$$

and reduce it to a canonical form.

6. For a hyperbolic equation (3.88) with characteristic coordinates ξ and η defined by (3.93), establish the identity

$$\begin{vmatrix} x_{\xi\eta} & y_{\xi\eta} & u_{\xi\eta} \\ x_\xi & y_\xi & u_\xi \\ x_\eta & y_\eta & u_\eta \end{vmatrix} = \frac{(x_\xi y_\eta - x_\eta y_\xi)^2}{2\sqrt{b^2 - ac}} d.$$

7. Show that an arbitrary linear combination of (3.97) and (3.98) can be used to complete the characteristic system (3.93) to (3.96) for the five unknowns x, y, u, p and q.

8. Show that the characteristics of a hyperbolic system of two equations

$$a_{i1}\frac{\partial u_1}{\partial x} + a_{i2}\frac{\partial u_2}{\partial x} + b_{i1}\frac{\partial u_1}{\partial y} + b_{i2}\frac{\partial u_2}{\partial y} + c_i = 0, \qquad i = 1, 2,$$

for two unknowns are defined by the ordinary differential equation

$$\begin{vmatrix} a_{11} & a_{12} & b_{11} & b_{12} \\ a_{21} & a_{22} & b_{21} & b_{22} \\ dx & 0 & dy & 0 \\ 0 & dx & 0 & dy \end{vmatrix} = 0,$$

and prove that corresponding characteristic coordinates ξ and η can be introduced (cf. Section 14.2).

9. Reduce any elliptic system (3.102) of $n = 2m$ linear partial differential equations such that the associated characteristic vectors $(v_{k1}, \ldots, v_{kn})$ are linearly independent to the canonical form

$$\frac{\partial U_k}{\partial x} - A_k \frac{\partial V_k}{\partial y} + B_k \frac{\partial U_k}{dy} + C_k = 0, \qquad k = 1, \ldots, m,$$

$$\frac{\partial V_k}{\partial x} + A_k \frac{\partial U_k}{\partial y} + B_k \frac{\partial V_k}{\partial y} + D_k = 0, \qquad k = 1, \ldots, m,$$

where C_k and D_k depend on the new unknowns $U_1, \ldots, U_m, V_1, \ldots, V_m$ linearly.

10. Reduce the general system of n nonlinear partial differential equations

$$(3.106) \quad F_i\left(x, y, u_1, \ldots, u_n, \frac{\partial u_1}{\partial x}, \ldots, \frac{\partial u_n}{\partial x}, \frac{\partial u_1}{\partial y}, \ldots, \frac{\partial u_n}{\partial y}\right) = 0,$$

$$i = 1, \ldots, n,$$

of the first order to a quasi-linear system of $3n$ equations by introducing the partial derivatives

$$p_i = \frac{\partial u_i}{\partial x}, \qquad q_i = \frac{\partial u_i}{\partial y}$$

as additional unknowns. What does it mean to say that (3.106) is of the elliptic type or to say that it is of the hyperbolic type?

5. HYPERBOLIC SYSTEMS

The classification of systems of partial differential equations is most easily described when the equations are formulated in matrix notation. Here we shall be concerned primarily with first order linear or quasi-linear systems of the hyperbolic type. To begin with, let u stand for an m-dimensional column vector whose components are unknown functions u_j of the two independent variables x and t. We denote by A a square matrix with m rows and m columns of elements a_{ij} depending in a prescribed fashion on x and t, and possibly on u, too. Similarly, the letter B will be used to indicate a column vector whose m components are given functions of x, t and u. Our intention is to discuss under what circumstances it is appropriate to say that the system of partial differential equations

$$(3.107) \qquad u_t + Au_x + B = 0$$

is of the hyperbolic type (cf. Section 4).

A *characteristic* of the system (3.107) is a curve along which the values of
u, combined with the equations (3.107) themselves, do not suffice to
determine the normal derivative of u. Let s be the arc length along such a
curve, and let ν be the unit normal. The chain rule for differentiation
yields the relations

$$u_x = u_s s_x + u_\nu \nu_x = u_s x_s + u_\nu x_\nu,$$

and

$$u_t = u_s s_t + u_\nu \nu_t = u_s t_s + u_\nu t_\nu.$$

Hence (3.107) is equivalent to

$$(3.108) \qquad (t_\nu I + x_\nu A)\frac{\partial u}{\partial \nu} + (t_s I + x_s A)\frac{\partial u}{\partial s} + B = 0,$$

where I stands for the identity matrix.

On a characteristic the determinant of the matrix of coefficients
multiplying the normal derivative $\partial u/\partial \nu$ in (3.108) must vanish, which
means that we must have

$$(3.109) \qquad |A - \lambda I| = 0,$$

with

$$\lambda = -\frac{t_\nu}{x_\nu} = \frac{x_s}{t_s}.$$

It follows that every characteristic curve for the system (3.107) is a solution
of the ordinary differential equation

$$(3.110) \qquad \frac{dx}{dt} = \lambda,$$

where λ is one of the characteristic roots of the polynomial equation
(3.109).

With the characteristic root λ we can associate at least one characteristic
row vector[7] v of the matrix A such that

$$v(A - \lambda I) = 0.$$

Multiplying (3.108) through on the left by v, we derive for u the scalar
equation

$$v(t_s I + x_s A)\frac{\partial u}{\partial s} + vB = 0,$$

which has the property that every derivative appearing in it is directed

[7] Cf. Birkhoff-MacLane 1.

along the tangent to the characteristic curve (3.110). If all the characteristic roots $\lambda_1, \ldots, \lambda_m$ of the matrix A are real, and if, in addition, A has a full set of m linearly independent (real) characteristic vectors $v_1, \ldots, v_m$, then the partial differential equations (3.107) can be transformed into a canonical system

$$(3.111) \qquad v_k \left(\frac{\partial t_k}{\partial s} I + \frac{\partial x_k}{\partial s} A \right) \frac{\partial u}{\partial s} + v_k B = 0, \qquad k = 1, \ldots, m,$$

of m distinct equations each involving differentiation in just one characteristic direction $(\partial t_k / \partial s, \partial x_k / \partial s)$ defined by

$$(3.112) \qquad \frac{\partial x_k}{\partial s} \bigg/ \frac{\partial t_k}{\partial s} = \lambda_k.$$

It is in precisely this situation that we call the system (3.107) *hyperbolic* (cf. Section 4).

For a *linear* system (3.107) of the hyperbolic type, the characteristic curves (3.112) and the square matrix

$$P = \begin{pmatrix} v_1 \\ \cdot \\ \cdot \\ \cdot \\ v_m \end{pmatrix}$$

composed of the characteristic row vectors $v_1, \ldots, v_m$ are all independent of the particular solution u under consideration, since the matrix A is a function of x and t alone. Consequently it turns out that we are able to reduce (3.107) to an especially simple canonical form involving the diagonal matrix

$$\Lambda = PAP^{-1} = \begin{pmatrix} \lambda_1 & 0 & \ldots & 0 \\ 0 & \lambda_2 & \ldots & 0 \\ \cdot & \cdot & \cdot & \cdot \\ 0 & 0 & \ldots & \lambda_m \end{pmatrix}.$$

However, we have to confine our attention to the special case where the roots $\lambda_1, \ldots, \lambda_m$ are all distinct, so that no multiple characteristics appear and P can be differentiated with respect to x and t. See Exercise 7 below for the analogous reduction of a quasi-linear system.

Let us introduce the new unknown

$$U = Pu.$$

Multiplying (3.107) on the left by P, we obtain[8]

$$(3.113) \qquad U_t + \Lambda U_x + \Gamma = 0,$$

where

$$\Gamma = P\frac{\partial P^{-1}}{\partial t}U + PA\frac{\partial P^{-1}}{\partial x}U + PB = \Gamma_1 U + \Gamma_2,$$

and where P^{-1} stands for the inverse of the (non-singular) matrix P. Observe that each of the m scalar equations

$$(3.114) \qquad \frac{\partial U_k}{\partial t} + \lambda_k\frac{\partial U_k}{\partial x} + c_k = 0$$

comprising the canonical system (3.113) is solved for the derivative of a single component U_k of U in the direction of a corresponding characteristic curve

$$(3.115) \qquad \frac{dx}{dt} = \lambda_k.$$

The components c_k of Γ are, of course, known functions of x, t and U.

The simplest sufficient condition for the matrix A to have real characteristic roots and a full set of m linearly independent characteristic vectors is that it be symmetric. When that is so, we call the system (3.107) *symmetric hyperbolic*. Cruder techniques than the above method of characteristics are available for the treatment of symmetric hyperbolic systems. They will not only enable us to overcome the difficulties connected with multiple characteristics in the symmetric case but will also furnish a means of handling problems in more than two independent variables. Thus we are now ready to study what becomes of our theory in space of higher dimension.

Consider the quasi-linear system of m partial differential equations

$$(3.116) \qquad Au_t = \sum_{j=1}^{n} A_j u_{x_j} + B$$

for an m-dimensional column vector u of unknown functions of the n space variables

$$x = (x_1, \ldots, x_n)$$

and of the time variable t. The square matrices A and A_j and the column vector B are supposed to consist of coefficients depending in a given way on x, t and u.

[8] Cf. Courant-Lax 1.

In order to motivate our classification of the system (3.116), we shall discuss a preliminary example in which B is set equal to zero, while the coefficients $A_1, \ldots, A_n$ and A are frozen to coincide everywhere with their values at some specific point. The method of separation of variables shows that in a situation of this kind exponential solutions of the form

$$(3.117) \qquad u = U e^{i(\lambda_1 x_1 + \cdots + \lambda_n x_n)} e^{i\lambda t}.$$

can be found, where $\lambda_1, \ldots, \lambda_n$ and λ are fixed parameters such that

$$(3.118) \qquad \left| \lambda A - \sum_{j=1}^{n} \lambda_j A_j \right| = 0,$$

and where U is a non-trivial m-dimensional column vector of constants with the property

$$(3.119) \qquad \lambda A U = \sum_{j=1}^{n} \lambda_j A_j U.$$

When $\lambda_1, \ldots, \lambda_n$ and λ are real, the expression (3.117) may be interpreted as the general term in a Fourier expansion. Thus to distinguish t as a time variable it is natural to ask that for all real choices of the parameters $\lambda_1, \ldots, \lambda_n$, every root λ of the characteristic equation (3.118) be real, too. In that case an oscillatory dependence of (3.117) on the space variables $x_1, \ldots, x_n$ results in a function u that neither grows nor decays exponentially with the time t. These remarks, together with our analysis of (3.107), lead us to say even in the case of variable coefficients that (3.116) is of the *hyperbolic type* when A is non-singular and when, for arbitrary assignment of the real parameters $\lambda_1, \ldots, \lambda_n$, the roots λ of the polynomial equation (3.118) are all real and have associated with them a full set of m linearly independent characteristic vectors U satisfying (3.119).

In the applications we shall be concerned with symmetric hyperbolic systems, which are those fulfilling the simplest condition sufficient to guarantee that they be of the hyperbolic type. To be precise, we call the system (3.116) *symmetric hyperbolic*, with the variable t *time-like*, if and only if all the coefficient matrices $A_1, \ldots, A_n$ and A are symmetric and, in addition, A is positive-definite (or negative-definite). In particular, A can be diagonalized by an auxiliary matrix P such that

$$PAP' = I,$$

where P' stands for the transpose of P. Hence the roots λ of (3.118) also solve the polynomial equation

$$(3.120) \qquad \left| \sum_{j=1}^{n} \lambda_j P A_j P' - \lambda I \right| = 0.$$

Thus whenever the parameters $\lambda_1, \ldots, \lambda_n$ are real, λ must be real, too, since it appears in (3.120) as the characteristic root of a real symmetric matrix. The necessary characteristic vectors also exist, which makes it clear that our symmetry hypotheses about the coefficients A and A_j do ensure that (3.116) is of the hyperbolic type.

A *characteristic* of the system (3.116) is any level surface

$$(3.121) \qquad \qquad \phi(x, t) = \text{const.}$$

on which the partial differential equation

$$(3.122) \qquad \qquad \left| \phi_t A - \sum_{j=1}^{n} \phi_{x_j} A_j \right| = 0$$

is satisfied (cf. Section 2). According to our analysis of the example (3.107), the requirement (3.122) asserts that the equations (3.116) do not suffice to determine the normal derivative of u along (3.121). For a symmetric hyperbolic system (3.116), surfaces on which the matrix

$$M = \phi_t A - \sum_{j=1}^{n} \phi_{x_j} A_j$$

is positive-definite (or negative-definite) also play a significant role (cf. Exercise 13 below). We say that they are *space-like* (cf. Sections 3.3 and 6.1). Notice that there are characteristics which appear as limits of space-like surfaces.

EXERCISES

1. What type would you ascribe to a single partial differential equation of the first order?

2. Show that (3.103) and (3.111) are equivalent.

3. Specify when the system

$$Au_t + Bu_x + C = 0$$

should be called hyperbolic, and put it into canonical form.

4. Show that the special form of the coefficient of u_t in (3.107) simply means that none of the lines $t = \text{const.}$ are characteristics of that equation.

5. Discuss the characteristics and the type of the system

$$u_x = v, \qquad u_t = v_x,$$

which is connected with the heat equation.

6. Verify that the characteristic system (3.93) to (3.97) associated with a second order quasi-linear equation (3.88) is of the hyperbolic type.

7. Through the introduction of m additional unknowns U_k given by the formula

$$U = Pu_t,$$

reduce (3.107) in the quasi-linear case where A depends on u to a diagonalized canonical system of $2m$ partial differential equations

$$\frac{\partial u_k}{\partial t} + \alpha_k = 0, \qquad \frac{\partial U_k}{\partial t} + \lambda_k \frac{\partial U_k}{\partial x} + \beta_k = 0$$

like (3.114), in which α_k and β_k are known functions of x, t, u and U. For the proof assume that the characteristic roots λ_k of A are all distinct and differ from zero.

8. Show that a reduction of (3.107) to the canonical form (3.114) is feasible even when multiple characteristics appear, provided that the matrix P of characteristic row vectors remains differentiable.

9. Supposing that c stands for a given function of ρ, put the quasi-linear system

$$u_t + uu_x + \frac{c^2}{\rho} \rho_x = 0, \qquad \rho_t + \rho u_x + u \rho_x = 0$$

into canonical form (cf. Section 14.2).

10. Explain in detail why the characteristic surfaces (3.121) of a symmetric hyperbolic system (3.116) ought to satisfy the scalar partial differential equation (3.122).

11. Verify that the Maxwell equations

$$\dot{B} + \nabla \times E = 0, \qquad \dot{E} - \nabla \times B = 0$$

form a symmetric hyperbolic system, and find their characteristics.

12. Why is the system (3.116) said to be of the *elliptic type* when the polynomial equation (3.118) for $\lambda_1, \ldots, \lambda_n$ and λ has no exclusively real non-trivial solutions?

13. For a symmetric hyperbolic system (3.116) with constant coefficients, show that the exponential solutions (3.117) corresponding to real choices of $\lambda_1, \ldots, \lambda_n$ and λ are complete within every space-like hyperplane.

14. Develop complex substitutions that convert an analytic system

$$u_t = \sum_{j=1}^{n} A_j u_{x_j} + B$$

of arbitrary type into the symmetric hyperbolic system

$$u_t = \sum_{j=1}^{n} \frac{A_j + \bar{A}_j'}{2} u_{x_j} + \sum_{j=1}^{n} \frac{A_j - \bar{A}_j'}{2i} u_{y_j} + B$$

for m complex-valued unknown functions of the $2n + 1$ independent variables $x_1, \ldots, x_n, y_1, \ldots, y_n$ and t (cf. Section 16.1).

4

Cauchy's Problem for Equations with Two Independent Variables

1. CHARACTERISTICS

In our discussions thus far we have frequently encountered the concept of an *initial value problem*. By this we meant the problem of determining the solution of an ordinary differential equation or of a partial differential equation with a number of data related to the order of the equation prescribed at an initial point or along an initial curve. Quite often in physical applications where the time t occurs in a natural way as one of the independent variables, the term *initial values* refers to the fact that the data are assigned at $t = 0$. However, this is by no means the only situation leading to initial value problems, because the data can be given just as reasonably along some curve as along the line $t = 0$. In such a context our problem is also called a *Cauchy problem* instead of an initial value problem, although the two names are actually synonymous.

We intend to study in detail the Cauchy problem for a partial differential equation of the second order in two independent variables. In this connection it will turn out that the appropriate equations to consider are those of the hyperbolic type. To start with we shall assume that our equation has already been put in the canonical form (3.18), although we shall allow the terms involving the unknown function and its first derivatives to be nonlinear. Thus we shall be interested in the equation

$$(4.1) \qquad u_{xy} = f(x, y, u, u_x, u_y),$$

where the function f on the right need not be analytic but must satisfy smoothness requirements in its dependence on the arguments $x, y, u, p = u_x$ and $q = u_y$ which will be specified later.

The Cauchy problem for (4.1) asks for a solution u with the property that prescribed values

$$u = u(\sigma), \quad p = p(\sigma), \quad q = q(\sigma)$$

of u, p and q are assumed along a given curve

(4.2) $$x = x(\sigma), \quad y = y(\sigma).$$

The data u, p and q must fulfill the compatibility condition

(4.3) $$\frac{du}{d\sigma} = p \frac{dx}{d\sigma} + q \frac{dy}{d\sigma}$$

along the initial curve (4.2) if the function u is to have p and q as its first partial derivatives. Therefore p and q cannot be assigned independently. It is actually the values of u and of its normal derivative $\partial u / \partial \nu$ that can be prescribed as arbitrary functions along (4.2). We shall find it convenient to refer to these quantities as the *Cauchy data*.

In Section 3.1 we have had occasion to observe that the Cauchy data for a solution of (4.1) cannot be given freely along one of the characteristics $x = $ const. or $y = $ const. of that equation. Along any line $y = $ const., for example, (4.1) represents an ordinary differential equation

(4.4) $$\frac{dq}{dx} = f\left(x, y, u, \frac{du}{dx}, q\right)$$

for u and q in their dependence on the variable x. It should be pointed out in this connection that the relation (4.4) is comparable to (3.95) and (3.96). Also of interest is the fact that even when the Cauchy data along a characteristic $y = $ const. happen to satisfy (4.4), we cannot expect the solution of the associated Cauchy problem to be unique.

For a verification of the last result consider the special equation

(4.5) $$u_{xy} = 0,$$

which was found in Section 3.1 to have the general solution

(4.6) $$u = \phi(x) + \psi(y),$$

where ϕ and ψ are arbitrary functions of one variable. Here (4.4) merely asserts that the values of $q = \psi'(y)$ prescribed along the initial characteristic $y = $ const. must be independent of x. However, the choice of ψ remains otherwise quite unspecified, since the rest of the requirements of the Cauchy problem at hand can be met by selecting the function ϕ appropriately.

It becomes quite clear from (4.4) that the characteristics are exceptional curves for the assignment of Cauchy data. Therefore we have to impose the hypothesis on the initial curve (4.2) that it should nowhere be tangent to a characteristic. Since the equation (4.1) appears in canonical form, its characteristics are simply the vertical lines $x = $ const. and the horizontal lines $y = $ const. Hence our hypothesis states that (4.2) has non-parametric representations

$$(4.7) \qquad\qquad y = y(x), \qquad x = x(y)$$

in terms of functions $y(x)$ and $x(y)$ that are strictly monotonic. Thus we can use either x or y as a parameter σ with which to represent the Cauchy data.

We shall find that it is convenient to represent p along (4.7) as a function $p = p(x)$ and to represent q there as a function $q = q(y)$, while using representations $u = u(x)$ and $u = u(y)$ of both kinds for u. In the case of the special partial differential equation (4.5) it is then possible to conclude that the arbitrary functions ϕ and ψ appearing in the general solution (4.6) must satisfy

$$\phi'(x) = p(x) = \frac{p(x) + u'(x) - q(y(x))y'(x)}{2},$$

$$\psi'(y) = q(y) = \frac{q(y) + u'(y) - p(x(y))x'(y)}{2},$$

where we have made a pertinent application of (4.3) to obtain the expressions on the right. It follows that the solution of the Cauchy problem for (4.5) is given explicitly by the formula

$$(4.8) \qquad u(x, y) = \frac{u(x) + u(y)}{2} + \frac{1}{2}\int_{x(y)}^{x} p(\xi)\,d\xi + \frac{1}{2}\int_{y(x)}^{y} q(\eta)\,d\eta,$$

which involves integrals evaluated along the initial curve (4.7).

It is easy to generalize the expression (4.8) for the solution of Cauchy's problem for (4.5) so that it includes the case of the inhomogeneous equation

$$(4.9) \qquad\qquad u_{xy} = g(x, y).$$

Because of the linearity of the problem we have only to insert an additional term that vanishes together with its first partial derivatives on the initial curve (4.7) and satisfies (4.9) identically. Such a term can be found by integrating the right-hand side g of (4.9) twice, first with respect to x and then with respect to y, starting both times at points on the curve (4.7).

The result is

$$u(x, y) = \int_{y(x)}^{y} \int_{x(\eta)}^{x} g(\xi, \eta) \, d\xi \, d\eta.$$

Combining this with the formula (4.8) associated with inhomogeneous initial values, we obtain the final result

$$(4.10) \qquad u(x, y) = \frac{u(x) + u(y)}{2} + \frac{1}{2} \int_{x(y)}^{x} p(\xi) \, d\xi$$

$$+ \frac{1}{2} \int_{y(x)}^{y} q(\eta) \, d\eta + \int_{y(x)}^{y} \int_{x(\eta)}^{x} g(\xi, \eta) \, d\xi \, d\eta.$$

It is possible, of course, to verify by direct substitution into the relevant equations that (4.10) does indeed yield the desired solution of the Cauchy problem for (4.9), provided merely that the data are continuously differentiable.

We can exploit formula (4.10) to derive an integrodifferential equation for the solution of Cauchy's problem in the general nonlinear case (4.1) which combines both the partial differential equation and the initial conditions in one single statement. To see how this is to be done we suppose for the moment that the solution u of the Cauchy problem is known to us. We thus conceive of the right-hand side of (4.1) as given. With that in mind we can interpret (4.1) as an inhomogeneous linear equation of the type (4.9) in which the term g on the right is a prescribed function of x and y. It follows that the representation (4.10) for u must be an identity applicable to the solution of our nonlinear Cauchy problem when we replace the integrand g by the function f appearing on the right in (4.1).

We conclude that the Cauchy problem for (4.1) is equivalent to the integrodifferential equation

$$(4.11) \qquad u(x, y) = \frac{u(P) + u(Q)}{2} - \frac{1}{2} \int_{P}^{Q} (q \, d\eta - p \, d\xi)$$

$$+ \iint_{D} f(\xi, \eta, u, u_{\xi}, u_{\eta}) \, d\xi \, d\eta.$$

Here we have introduced some changes in notation according to which D stands for the curvilinear triangle of integration in the (ξ, η)-plane bounded by the two characteristic lines $\xi = x$ and $\eta = y$ and by the initial curve (4.7). The letter P stands for the point $(x(y), y)$ and the letter Q for the point $(x, y(x))$, and the line integral on the right in (4.11) is to be evaluated between P and Q along the initial curve (4.7). Observe that it is an elementary matter to verify by direct substitution that a solution of

(4.11) solves the Cauchy problem for (4.1). However, note also that a convention must be adopted so that the double integral over D is taken with a plus sign only when (4.7) is monotonically decreasing and is preceded by a minus sign instead when (4.7) is monotonically increasing.

We postpone until Section 2 our construction of the solution of the integrodifferential equation (4.11) in order to discuss first various qualitative properties of the solution of Cauchy's problem that can be deduced from it. Let R denote the point (x, y). The characteristics through R play an important role in formula (4.11), since they serve to define the limits of integration there. It follows from (4.11), and also from (4.10) for the special equation (4.9), that the values of the given function f over the triangular region D bounded by the initial curve and the pair of characteristics through R, together with the Cauchy data along the arc from P to Q of the initial curve which is cut out by these characteristics, are sufficient to determine the solution u of the Cauchy problem at the point R. For this reason we call D, together with the arc of the initial curve bounding it, the *domain of dependence* of the solution u with respect to the point R.

The conclusion that only the portion of the Cauchy data between P and Q is needed to fix u at the point R gives considerably more insight into the nature of the solution of the Cauchy problem than could be provided by the power series method of Section 1.2. Our present construction based on characteristics is thus reminiscent of the treatment of initial value problems for an equation of the first order that was developed in Chapter 2. There the notion of a characteristic was also essential.

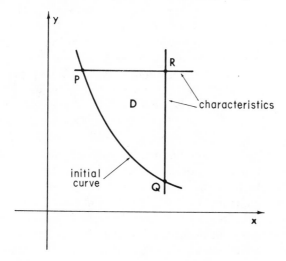

FIG. 5. Geometry of the solution of Cauchy's problem.

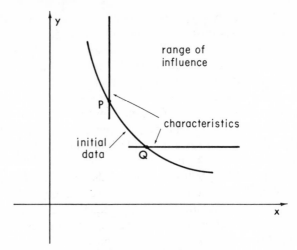

FIG. 6. Range of influence.

If we reverse our point of view, we can state that Cauchy data beyond the arc from P to Q of the initial curve (4.7) have no influence on the value u of the solution of the Cauchy problem at the point R. It is therefore natural to ask over what region of the plane the Cauchy data along a specific arc of the initial curve, such as that between P and Q, can affect the solution of the Cauchy problem. We call this region the *range of influence* of the arc in question. If we proceed to the opposite side of the initial curve from R, we find that the range of influence of the arc from P to Q is the infinite wedge bounded by that arc and by the extensions beyond P and Q of the characteristics through R. On the other hand, if we seek the range of influence on the same side of the initial curve as R, we see that it is bounded by the opposite pair of characteristics through P and Q, together, of course, with the initial arc between P and Q. Indeed, the domain of dependence associated with any point beyond the latter two characteristics does not extend far enough along the initial curve to reach P or Q.

Closely related to the notion of a range of influence is the question of the propagation of singularities of the Cauchy data. For the example (4.5) it can be verified by inspection that discontinuities in the data u, p and q or their derivatives generate corresponding discontinuities of the solution (4.8) along the characteristics $x = \text{const.}$ and $y = \text{const.}$ that pass through the points of the initial curve where the discontinuities in the data occur. We have seen before that the characteristics are those curves along which the Cauchy data do not determine the second derivatives of a solution uniquely. Hence they are precisely the curves where discontinuities in the second derivatives could appear.

More generally, if we try to calculate partial derivatives of the solution of a given order in terms of lower derivatives along some curve, we are led to a system of linear equations with the same determinant that furnished the ordinary differential equation (3.92) of the characteristics when it was the second derivatives that were sought. Thus discontinuities in any higher derivative of a solution can occur only along the characteristics. We conclude that for any hyperbolic equation, singularities in the Cauchy data correspond to singularities in the solution of Cauchy's problem along the characteristics through the points on the initial curve where the original discontinuities in the data appeared. In particular, examination of (4.11) shows that a discontinuity in some derivative of the Cauchy data at a point P on the initial curve propagates along the two characteristics through P. These characteristics are, of course, the ones that bound the range of influence corresponding to the degenerate arc consisting of P alone.

We give a physical interpretation of the above results for the special example (4.5), which is recognized to represent the equation

$$(4.12) \qquad u_{ss} = \frac{1}{c^2} u_{tt}$$

of the vibrating string in terms of characteristic coordinates

$$x = s + ct, \qquad y = s - ct.$$

It is natural to pick the line

$$x - y = 2ct = 0$$

as our initial curve. In the context of the vibrating string, the Cauchy data represent initial values of the displacement and of the velocity of the string (cf. Section 1.1). An initial disturbance at some point P spreads out at a uniform rate, covering a range of influence for that point in the (x,y)-plane. The disturbance travels along the string at the speed of sound c, and it can be viewed as a sound wave.

Discontinuities in any of the higher derivatives of u should thus be interpreted as wave fronts that travel up and down the string. Our analysis establishes that their trajectories in a space-time diagram are the characteristics of the wave equation (4.12). Moreover, the initial data along a fixed length of the string extending from a point P to a point Q determine the motion in a triangular domain of dependence bounded by the initial length of string and two characteristics through its endpoints. The physical significance of the latter statement is that subsequent displacements of the ends P and Q of the string cannot influence its motion in the triangle just described because their effect cannot travel back and forth faster than the speed of sound c. Thus it is because we have a finite signal speed c that each point in the space-time diagram cannot influence every other point. This causality law is connected, of course, with the hyperbolicity of the equation of motion.

The qualitative properties of the solution of the Cauchy problem for (4.1) that we have established in connection with the integrodifferential equation (4.11) are valid in the small for any hyperbolic partial differential equation which can be brought into such a canonical form. In the more general case the characteristics play the same important role that they have assumed in the present discussion, even though they are not usually straight lines. The triangular domain of dependence and the range of influence corresponding to a prescribed arc of Cauchy data may have curvilinear boundaries, but otherwise the geometry of the Cauchy problem is always similar locally to what we have described here.

For an elliptic equation the situation with regard to the Cauchy problem is altogether different, essentially because there are then no real characteristics. Mathematical physics does not provide in a natural or direct fashion Cauchy problems for equations of the elliptic type. In this connection it is instructive to examine an example devised by Hadamard[1] which shows that the Cauchy problem is actually not the right question to pose for elliptic equations, at least not in the real domain.

Consider the Laplace equation

$$(4.13) \qquad u_{xx} + u_{yy} = 0,$$

together with the specific Cauchy data

$$(4.14) \qquad u(0, y) = 0, \qquad u_x(0, y) = \frac{1}{n} \sin ny,$$

where $n > 0$. The method of separation of variables suggests the explicit form

$$(4.15) \qquad u(x, y) = \frac{1}{n^2} \operatorname{sh} nx \sin ny$$

for the solution of the Cauchy problem defined by (4.13) and (4.14). Let us study what happens as the parameter n becomes infinite. It is observed that the Cauchy data (4.14) approach zero uniformly, whereas for $x \neq 0$ the solution (4.15) oscillates between limits that increase indefinitely. Since zero is an obvious solution of (4.13) with vanishing Cauchy data, we conclude that for Laplace's equation the dependence of the solution of Cauchy's problem on the data is not in general continuous.

For any Cauchy problem describing a stable physical situation one would expect that small errors in the data should result in correspondingly small errors in the answer. If this did not turn out to be true, we would be

[1] Cf. Hadamard 3.

inclined to believe that the mathematical model of the physical problem had been badly formulated. Hadamard's example therefore leads us to set up rigid criteria for the Cauchy problem, or for any other boundary or initial value problem in the theory of partial differential equations, that justify calling it *correctly set*, or *well posed*. These criteria are intended to distinguish the problems that are reasonable and correspond to well formulated questions in mathematical physics from those that are peculiar and yield either a badly behaved solution or none at all.

To be precise, we define a boundary or initial value problem for a partial differential equation, or for a system of partial differential equations, to be *correctly set in the sense of Hadamard* if and only if its solution exists, is unique, and depends continuously on the data assigned. Much of the theory to be developed in what follows will be devoted to the investigation of specific problems to decide whether they are correctly set and, if so, to determine the structure of their solutions.

We have shown that the Cauchy problem (4.13), (4.14) is not well posed because its solution does not depend continuously on the Cauchy data in the limit as $n \to \infty$. This conclusion is drawn despite the fact that the Cauchy-Kowalewski theorem of Chapter 1 ensures the existence and uniqueness of a solution within the class of real analytic functions. Such a situation provides a clear indication of the inadequacy of the power series method. Therefore we must seek a more satisfactory way in which to establish that Cauchy's problem is at least correctly set for hyperbolic equations in the canonical form (4.1), a result that does not now appear to be entirely trivial.

Existence and uniqueness theorems for the Cauchy problem in the hyperbolic case are of basic importance because they are needed to confirm the correctness of our mathematical formulation of physical questions which arise in the applications, such as that of the vibrating string. The proofs of such theorems that we shall present in the next section are built around a procedure of successive approximations for the integrodifferential equation (4.11) due to Picard. It will be seen that our approach there also furnishes a rapidly convergent construction of solutions.

EXERCISES

1. Work out the notions of a domain of dependence and of a range of influence in the case of the Cauchy problem for a single linear partial differential equation

$$au_x + bu_y + cu + d = 0$$

of the first order (cf. Section 2.1).

2. Show that any solution u of equation (4.5) satisfies the difference equation

(4.16) $$u_1 - u_2 + u_3 - u_4 = 0,$$

where u_1, u_2, u_3 and u_4 are the values of u at the four successive corners of any rectangle whose edges are characteristics.

3. From (4.8) derive d'Alembert's formula

(4.17) $$u(s, t) = \frac{1}{2} u(s + ct, 0) + \frac{1}{2} u(s - ct, 0) + \frac{1}{2c} \int_{s-ct}^{s+ct} u_t(\sigma, 0)\, d\sigma$$

for the solution of the initial value problem for the wave equation (4.12) with data assigned at $t = 0$. What term should be added to take care of the case where (4.12) is inhomogeneous?

4. Show that the Cauchy problem for the heat equation

$$u_{xx} = u_t$$

with initial data u and u_t assigned at $t = 0$ is not correctly set but that the problem makes sense when u alone is prescribed at $t = 0$ for all values of x.

5. Describe curvilinear domains of dependence and ranges of influence for the partial differential equation

$$u_{xx} + y^2 u_{xy} - (1 + y^2)u_{yy} = 0.$$

6. Explain why we might describe (4.10) as a generalized, or weak, solution of Cauchy's problem for the inhomogeneous wave equation (4.9) when the prescribed functions $x(y)$, $y(x)$, $u(x)$, $u(y)$, $p(x)$, $q(y)$ and $g(x, y)$ are continuous but not differentiable.

7. Use (4.4) to show that except for one arbitrary constant the prescribed values of u along a characteristic are enough to fix the remaining Cauchy data $\partial u/\partial v$ there.

2. THE METHOD OF SUCCESSIVE APPROXIMATIONS

Our next aim is to establish that Cauchy's problem is well posed for the hyperbolic partial differential equation

$$u_{xy} = f(x, y, u, u_x, u_y)$$

when the Cauchy data are assigned on a monotonic curve (4.7). For that purpose we shall prove that a solution of the equivalent integrodifferential equation (4.11) exists, is unique, and depends continuously on the data.

There is no loss of generality if we assume that the Cauchy data u, p and

THE METHOD OF SUCCESSIVE APPROXIMATIONS   111

q vanish along the initial curve, since the general case associated with continuously differentiable data can be reduced to such a simplified problem by subtracting from the unknown function the solution (4.8) of the Cauchy problem for (4.5) with the same data. This substitution merely has the effect of altering the functional form of the right-hand side f of (4.1) in an elementary fashion. Thus it suffices to deal exclusively with the integrodifferential equation

$$(4.18) \qquad u(x, y) = \iint_D f(\xi, \eta, u, u_\xi, u_\eta) \, d\xi \, d\eta,$$

where, as indicated earlier, D is the triangular region bounded by the curve (4.7) and the two straight lines $\xi = x$ and $\eta = y$. Finally, there is no loss of generality if we suppose that the initial curve (4.7) is the straight line

$$(4.19) \qquad x + y = 0,$$

since when that is not so we can introduce new characteristic coordinates $x' = y(x)$, $y' = -y$ or $x' = -x$, $y' = x(y)$ to bring (4.7) into such a form without changing any of the essential features of (4.18).

Picard's procedure for solving (4.18) is to set up a sequence of successive approximations u_n defined by the formula

$$(4.20) \qquad u_{n+1}(x, y) = \iint_D f(\xi, \eta, u_n, p_n, q_n) \, d\xi \, d\eta,$$

where p_n and q_n denote the first partial derivatives of u_n. Although the first approximation u_0 could be selected more or less at random, it will be taken here to be

$$(4.21) \qquad u_0(x, y) = 0.$$

Our purpose is to establish that under suitable hypotheses about the integrand f, the limit

$$(4.22) \qquad u = \lim_{n \to \infty} u_n = \sum_{n=0}^{\infty} [u_{n+1} - u_n]$$

of the successive approximations u_n exists and satisfies the integrodifferential equation (4.18).

For the proof of convergence of the infinite series (4.22) we shall need to estimate differences of f corresponding to increments in the arguments u, p

and q. For these differences the mean value theorem yields the expression

$$(4.23) \quad f(\xi, \eta, u'', p'', q'') - f(\xi, \eta, u', p', q')$$

$$= (u'' - u')f_u + (p'' - p')f_p + (q'' - q')f_q,$$

where the partial derivatives f_u, f_p and f_q on the right are evaluated at a set of arguments intermediate between ξ, η, u', p', q' and ξ, η, u'', p'', q''. If $|f_u|$, $|f_p|$ and $|f_q|$ are bounded by some positive number M, we can conclude from (4.23) that

$$(4.24) \quad |f(\xi, \eta, u'', p'', q'') - f(\xi, \eta, u', p', q')|$$

$$\leq M[\, |u'' - u'| + |p'' - p'| + |q'' - q'|\,].$$

We prefer, however, to make the Lipschitz condition (4.24) itself our basic hypothesis about the integrand f. We shall suppose in addition that f is a continuous function of all its arguments, and we shall let K stand for an upper bound on $|f|$,

$$|f| \leq K.$$

We shall require that these assumptions be valid when the point (ξ, η) ranges over a region G containing our initial curve and when the remaining three arguments u, p and q of f are sufficiently small, say when

$$(4.25) \quad\quad\quad |u| < \epsilon, \quad |p| < \epsilon, \quad |q| < \epsilon.$$

If u_n, p_n, q_n and $u_{n-1}, p_{n-1}, q_{n-1}$ satisfy (4.25), the definition (4.20) of our successive approximations and the Lipschitz condition (4.24) lead to the inequality

$$(4.26) \quad |u_{n+1}(x, y) - u_n(x, y)|$$

$$\leq \iint\limits_{D} |f(\xi, \eta, u_n, p_n, q_n) - f(\xi, \eta, u_{n-1}, p_{n-1}, q_{n-1})|\, d\xi\, d\eta$$

$$\leq M \iint\limits_{D} [\,|u_n - u_{n-1}| + |p_n - p_{n-1}| + |q_n - q_{n-1}|\,]\, d\xi\, d\eta.$$

Furthermore, the formulas

$$(4.27) \quad\quad\quad p_{n+1}(x, y) = \int_{-x}^{y} f(x, \eta, u_n, p_n, q_n)\, d\eta,$$

$$(4.28) \quad\quad\quad q_{n+1}(x, y) = \int_{-y}^{x} f(\xi, y, u_n, p_n, q_n)\, d\xi$$

obtained by differentiating (4.20) with respect to x and with respect to y furnish in a similar way the inequalities

(4.29) $\quad |p_{n+1}(x, y) - p_n(x, y)|$

$$\leq M \int_{-x}^{y} [|u_n - u_{n-1}| + |p_n - p_{n-1}| + |q_n - q_{n-1}|] \, |d\eta|,$$

(4.30) $\quad |q_{n+1}(x, y) - q_n(x, y)|$

$$\leq M \int_{-y}^{x} [|u_n - u_{n-1}| + |p_n - p_{n-1}| + |q_n - q_{n-1}|] \, |d\xi|$$

for the differences of derivatives $p_{n+1} - p_n$ and $q_{n+1} - q_n$.

To exploit the similarity of the integrands in (4.26), (4.29) and (4.30), it is convenient to set

(4.31) $\quad U_n(t) = \max_{x+y=t} [|u_n(x, y) - u_{n-1}(x, y)| + |p_n(x, y) - p_{n-1}(x, y)|$

$$+ |q_n(x, y) - q_{n-1}(x, y)|],$$

with the points (x, y) restricted so that each of them generates a domain of dependence lying inside the region G specified above. To achieve an adequate global analysis of (4.1) in the linear case, we need to make exceptionally careful estimates of U_n here. Introducing new coordinates

$$\sigma = x - y, \qquad \tau = x + y$$

parallel and perpendicular to the initial line (4.19), we derive the inequality

$$\iint_D [|u_n - u_{n-1}| + |p_n - p_{n-1}| + |q_n - q_{n-1}|] \, d\xi \, d\eta$$

$$\leq \frac{1}{2} \int_0^{x+y} \int_{2x-\tau}^{\tau-2y} U_n(\tau) \, |d\sigma| \, |d\tau|$$

$$= \int_0^{x+y} |x + y - \tau| \, U_n(\tau) \, |d\tau|.$$

Hence (4.26), (4.29) and (4.30) can be combined to yield

(4.32) $\quad |u_{n+1}(x, y) - u_n(x, y)| + |p_{n+1}(x, y) - p_n(x, y)|$

$$+ |q_{n+1}(x, y) - q_n(x, y)| \leq M \int_0^{x+y} [2 + |x + y - \tau|] U_n(\tau) \, |d\tau|.$$

If we replace the left-hand side of (4.32) by its maximum value, we obtain the result

$$(4.33) \qquad U_{n+1}(t) \leq M(2 + T) \int_0^t U_n(\tau) \, |d\tau|$$

for $|t| \leq T$. This will provide the key to our convergence proof.

We restrict the point (x, y) to a neighborhood

$$(4.34) \qquad |x + y| < T$$

of the initial line (4.19) that is contained in G. From (4.20), (4.27) and (4.28) we have

$$|u_n(x, y)| \leq \frac{KT^2}{2}, \quad |p_n(x, y)| \leq KT, \quad |q_n(x, y)| \leq KT,$$

where K is our bound on $|f|$. Hence if T is taken to be sufficiently small, the quantities u_n, p_n, q_n fulfill (4.25) for all values of n. Because of (4.21) it follows in particular that

$$(4.35) \qquad U_1(t) \leq KT\left(2 + \frac{T}{2}\right) \leq KT(2 + T).$$

The hypotheses needed to assure the validity of the basic integral inequality (4.33) are now satisfied. Inserting (4.35) into (4.33), we find that

$$U_2(t) \leq KTM(2 + T)^2 t.$$

Another application of (4.33) then gives

$$U_3(t) \leq \frac{KTM^2(2 + T)^3 t^2}{2!}.$$

More generally, mathematical induction serves to establish that

$$(4.36) \qquad U_n(t) \leq \frac{KTM^{n-1}(2 + T)^n t^{n-1}}{(n - 1)!}.$$

Indeed, (4.36) reduces to (4.35) for $n = 1$, and for larger values of n it becomes a consequence of (4.33).

Comparison of (4.31) and (4.36) shows that the exponential series

$$(4.37) \qquad KT(2 + T) \, e^{MT(2+T)} = \sum_{n=0}^{\infty} \frac{KTM^n T^n (2 + T)^{n+1}}{n!}$$

is a majorant for the infinite series (4.22), as well as for its formal partial derivatives with respect to x and y, in the neighborhood (4.34) of the initial curve (4.19). It follows that these series converge uniformly. Hence we can let $n \to \infty$ and pass to the limit under the integral sign in (4.20) to establish that formula (4.22) defines a solution of the integro-differential equation (4.18). Any function u that can be represented as a double integral of the kind appearing in (4.18) must be twice continuously differentiable, and therefore we have found in (4.22) a solution u of Cauchy's problem for the hyperbolic partial differential equation (4.1).

In the nonlinear case the successive approximations u_n have only been shown to converge in a sufficiently small neighborhood (4.34) of the initial curve, since only in such a neighborhood could we expect the Lipschitz condition (4.24) to remain valid. However, for a linear equation

$$u_{xy} + a(x, y)u_x + b(x, y)u_y + c(x, y)u = 0$$

no restriction of this kind is needed, since the Lipschitz constant M on which our estimates are based depends in that case only on the size of the coefficients a, b and c and not on the range of values of the unknown and its derivatives. Because of the care we took to achieve a factorial in the denominator on the right in (4.36), the majorizing series (4.37) converges for all M and T. Hence we are able to conclude that for a linear equation (4.1) the successive approximations u_n actually converge in the large to a solution u of the Cauchy problem.

The uniqueness of the solution u of (4.18) is easy to deduce from the Lipschitz condition (4.24) and an estimate similar to (4.32). If u' is a second solution we have

$$|u'(x, y) - u(x, y)| \leq \iint_D |f(\xi, \eta, u', p', q') - f(\xi, \eta, u, p, q)|\, d\xi\, d\eta$$

$$\leq M \iint_D [\,|u' - u| + |p' - p| + |q' - q|\,]\, d\xi\, d\eta$$

in some neighborhood of the initial curve (4.19) so small that both solutions satisfy the requirement (4.25). Inequalities analogous to (4.29) and (4.30) are valid for the differences $p' - p$ and $q' - q$, too. It follows that

$$U(t) \leq M(2 + |t|) \int_0^t U(\tau)\, |d\tau|\,,$$

where

$$U(t) = \max_{x+y=t} [\,|u'(x, y) - u(x, y)|$$

$$+ |p'(x, y) - p(x, y)| + |q'(x, y) - q(x, y)|\,].$$

Therefore

(4.38) $\max_{|t| \le T} U(t) \le MT(2 + T) \max_{|t| \le T} U(t),$

which can be true for small T only if

$$U(t) \equiv 0.$$

The conclusion to be drawn is that u and u' are identical; this completes our proof that the solution of the Cauchy problem is unique.

In order to discuss the dependence of u on the Cauchy data, it suffices to study a situation where the integrand f in (4.18) depends on a parameter λ. Indeed, our reduction of the general integrodifferential equation (4.11) to the simpler form (4.18) was based on a substitution such that the Cauchy data became involved in the functional representation of f. Suppose that f is a continuous function of λ but that the constant M in the Lipschitz condition (4.24) and the bound K on $|f|$ can be chosen quite independently of λ. Our aim is to show that under these circumstances

$$u = u(x, y; \lambda)$$

must vary continuously with λ.

It follows from the uniqueness of the solution of the integrodifferential equation (4.18) that u can be expressed as an infinite series (4.22) of successive approximations. Formula (4.20) exhibits each individual approximation in terms of the preceding one as a continuous function of λ, and the form of the majorant (4.37) shows that the series (4.22) converges uniformly. Since a uniformly convergent series of continuous functions is itself continuous, we conclude that u does depend continuously on λ. Continuity of the derivatives p and q can be deduced by the same argument.

We have now compiled the information about existence, uniqueness and continuous dependence on parameters which is necessary to establish that the Cauchy problem for (4.1) is well posed. In addition, we shall prove that if the given function f has a continuous partial derivative f_λ with respect to the parameter λ, and if f_u, f_p and f_q vary continuously with λ, then the partial derivative u_λ of u with respect to λ exists, is continuous, and satisfies a linear hyperbolic partial differential equation obtained by differentiating (4.1) formally with respect to λ.

Consider the difference quotient

$$v(x, y; \lambda_1, \lambda_2) = \frac{u(x, y; \lambda_2) - u(x, y; \lambda_1)}{\lambda_2 - \lambda_1},$$

and observe that it satisfies the integrodifferential equation

$$(4.39) \qquad v(x, y; \lambda_1, \lambda_2) = \iint_D [f_\lambda + f_u v + f_p v_\xi + f_q v_\eta] \, d\xi \, d\eta$$

derived from (4.18). The partial derivatives f_λ, f_u, f_p and f_q occurring in the integrand are evaluated for a choice of the arguments λ, u, p and q intermediate between those corresponding to $\lambda = \lambda_1$ and those corresponding to $\lambda = \lambda_2$, in line with the mean value theorem. We view f_λ, f_u, f_p and f_q as known functions of ξ, η, λ_1 and λ_2, and we note that their limits as $\lambda_1 \to \lambda_2$ are well defined because u, p and q are continuous functions of λ. Thus the integrand in (4.39), conceived of for a fixed choice of λ_2 as a function of the six variables λ_1, ξ, η, v, v_ξ and v_η, depends continuously on the parameter λ_1 at the point $\lambda_1 = \lambda_2$.

From our previous theorem about the continuous variation with parameters of a solution u of the general integrodifferential equation (4.18), which also applied to the first derivatives p and q, we conclude that the difference quotients v, v_x and v_y have limits u_λ, p_λ and q_λ as $\lambda_1 \to \lambda_2$. Moreover, a passage to the limit in (4.39) shows that these partial derivatives with respect to λ fulfill

$$(4.40) \qquad u_\lambda(x, y; \lambda) = \iint_D [f_\lambda + f_u u_\lambda + f_p p_\lambda + f_q q_\lambda] \, d\xi \, d\eta.$$

Since f_λ, f_u, f_p and f_q are continuous functions of λ, it follows that u_λ is, too. The linear partial differential equation

$$u_{\lambda xy} = f_\lambda + f_u u_\lambda + f_p u_{\lambda x} + f_q u_{\lambda y}$$

for u_λ, which is what we originally set out to establish, now becomes a consequence of (4.40).

We have discussed the Cauchy problem for (4.1) at length. However, there are more complicated mixed boundary and initial value problems that can also be solved by Picard's method of successive approximations. The most interesting examples will be described briefly below.

Suppose that we give the values of u along a characteristic, say along the x-axis, and that we also prescribe u alone along a monotonically increasing curve (4.7). Assume for the sake of simplicity that this curve intersects the x-axis at the origin, and assume that the assigned values of u match continuously there. In the region of the first quadrant lying between the x-axis and the monotonic curve (4.7) we wish to determine a solution u of (4.1) subject to these mixed boundary conditions. We call the question so defined the *Goursat problem*. Actually, the solution of Goursat's problem ought to exist throughout the entire first quadrant, since the values of u

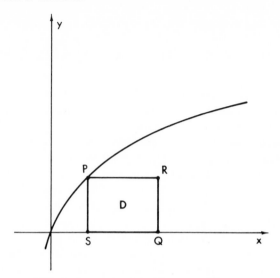

FIG. 7. Geometry of the Goursat problem.

below (4.7) furnish Cauchy data along that curve which lead to a Cauchy problem fixing the solution above it.

We shall describe as the *characteristic initial value problem* for (4.1) the degenerate case of the Goursat problem where the monotonic curve (4.7) reduces to the y-axis, which has the significance in this context of being the remaining characteristic through the origin. The characteristic initial value problem also has an interpretation as a limiting case of the Cauchy problem in which the initial curve, assumed here to be monotonically decreasing, collapses onto a pair of intersecting characteristics that we have taken in our formulation to be the coordinate axes. In the limit we omit the normal derivative $\partial u/\partial \nu$ from the Cauchy data because it is already determined by the values of u along the characteristics (cf. Exercise 1.7).

We can prove that the Goursat problem is well posed by transforming it into an integrodifferential equation resembling (4.11) and then solving by the technique of successive approximations. Let the point R with co-ordinates x and y lie in the region above the x-axis but below the curve (4.7). Let P denote the point $(x(y), y)$ on (4.7), let S denote the point $(x(y), 0)$, and let Q denote the point $(x, 0)$. We allow D now to stand for the rectangle with corners at R, P, S and Q, and we integrate (4.1) over D. Since

$$\iint\limits_{D} u_{\xi\eta}\, d\xi\, d\eta = u(R) - u(P) + u(S) - u(Q),$$

we obtain

$$(4.41) \quad u(R) = u(P) - u(S) + u(Q) + \iint\limits_{D} f(\xi, \eta, u, u_\xi, u_\eta) \, d\xi \, d\eta.$$

This is the desired integrodifferential equation because the first three terms on the right are given by the data of our problem.

Formula (4.41) shows that the value of u at R depends only on data along the arc of (4.7) between the origin and P and on data along the x-axis between the origin and Q. When f does not depend on u, p and q, (4.41) actually furnishes an explicit solution of the Goursat problem. In the general case we can define a sequence of successive approximations by the rule

$$(4.42) \quad u_{n+1}(R) = u(P) - u(S) + u(Q) + \iint\limits_{D} f(\xi, \eta, u_n, p_n, q_n) \, d\xi \, d\eta,$$

with

$$u_0(R) = u(P) - u(S) + u(Q).$$

The convergence of u_n to a solution u of the Goursat problem can be established in a sufficiently small neighborhood of the origin under the hypothesis that the Lipschitz condition (4.24) holds. Under these circumstances, uniqueness of the solution and its continuous dependence on the data follow, too. Therefore the Goursat problem is well posed. Finally, the solution is valid in the large when the partial differential equation (4.1) is linear, just as for Cauchy's problem.

EXERCISES

1. Work out the details of the proof of existence, uniqueness, and continuous dependence on data for the solution of the Goursat problem.

2. Use a Cauchy problem followed by a sequence of Goursat problems and characteristic initial value problems to solve the mixed initial and boundary value problem

$$u_{ss} = \frac{1}{c^2} u_{tt},$$

$$u(s, 0) = g(s), \qquad u_t(s, 0) = h(s),$$

$$u(0, t) = u(l, t) = 0$$

for a vibrating string of length l whose ends are pinned down. Show that this problem can still be solved even if the places where the string is pinned down are moved about at a rate slower than the speed of sound c.

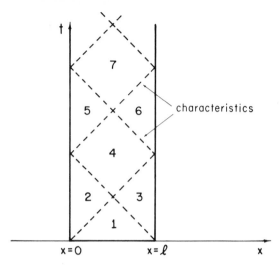

FIG. 8. Problem of the vibrating string with ends pinned.

3. Show that the successive approximations u_n defined by (4.42) for the specific characteristic initial value problem

(4.43) $$u_{xy} + \lambda u = 0,$$

$$u(x, 0) = u(0, y) = 1$$

are of the form

$$u_n(x, y) = 1 - \lambda xy + \frac{\lambda^2 x^2 y^2}{(2!)^2} - \cdots + \frac{(-1)^n \lambda^n x^n y^n}{(n!)^2}.$$

Here λ denotes a constant. Identify the solution u of this problem as the Bessel function

(4.44) $$u = J_0(2\sqrt{\lambda xy}).$$

4. Give an alternate treatment of the problem of Exercise 3 by asking for those solutions of (4.43) that depend on the product xy alone.

5. Show that the solution of the nonlinear Cauchy problem

$$u_{xy} = u^2,$$

$$u(x, -x) = 6, \qquad p(x, -x) = q(x, -x) = 12$$

becomes infinite on the line

$$x + y = 1$$

and hence does not exist in the large.

6. Use the method of successive approximations to establish the existence and uniqueness for small x of a solution of the Volterra integral equation

$$u(x) = \int_0^x F(x, y, u(y))\, dy$$

when the integrand F satisfies a Lipschitz condition as a function of its third argument u.

7. Let U stand for a closed set of elements $u, v, w, \ldots$ in some complete space with the norm $\|u\|$, and let T be any operator which maps U into itself and has the property

$$\|T(w) - T(v)\| \leq \epsilon \|w - v\|$$

on that set, where $\epsilon < 1$. Prove by means of Picard's method of successive approximations that there is in U a fixed point u of the contraction mapping T such that

$$u = T(u).$$

8. Use (4.37) to establish that for a linear equation (4.1) the solution of Cauchy's problem has at most exponential growth in its dependence on the square of the distance from the initial curve.

3. HYPERBOLIC SYSTEMS

We are interested in establishing that the procedure of successive approximations can be used to solve Cauchy's problem for a hyperbolic system, linear or nonlinear, of partial differential equations of the first order in two independent variables. Those who wish to skip this topic may proceed directly to Section 4.

The results of Section 3.5, and, in particular, of Exercise 3.5.7, show that it suffices to consider a canonical system

$$(4.45) \qquad \frac{\partial U_k}{\partial t} + \lambda_k \frac{\partial U_k}{\partial x} + C_k = 0, \qquad k = 1, \ldots, n,$$

where λ_k and C_k are given functions of t, x and

$$U = (U_1, \ldots, U_n).$$

We suppose that the initial values of the unknowns U_k are prescribed on a segment of the x-axis. Our first step will be to transform the latter Cauchy problem into a system of integral equations which lends itself to an iterative method of solution. For this purpose we shall integrate the equations (4.45) along the corresponding characteristic curves[2]

(4.46) $$\frac{dx}{dt} = \lambda_k, \qquad k = 1, \ldots, n.$$

It is convenient to represent the solutions of the ordinary differential equations (4.46) passing through the point (τ, ξ) of the (t,x)-plane in the form

$$x = X_k(t; \tau, \xi), \qquad k = 1, \ldots, n.$$

Let us integrate each equation (4.45) along the arc of the corresponding characteristic through (τ, ξ) which connects that point with the x-axis. Using t as a parameter along the characteristics (4.46) and taking into account the relation

$$\frac{dU_k}{dt} = \frac{\partial U_k}{\partial t} + \frac{\partial U_k}{\partial x}\frac{dx}{dt} = -C_k,$$

we thus derive

(4.47) $\quad U_k(\tau, \xi) = U_k(0, X_k(0; \tau, \xi))$

$$-\int_0^\tau C_k(\sigma, X_k(\sigma; \tau, \xi), U[\sigma, X_k(\sigma; \tau, \xi)]) \, d\sigma.$$

Since the unknown functions X_k appear on the right, we need to find integral equations for them as well. This can be done by integrating the ordinary differential equations (4.46) to obtain

(4.48) $\quad X_k(t; \tau, \xi) = \xi + \int_\tau^t \lambda_k(\sigma, X_k(\sigma; \tau, \xi), U[\sigma, X_k(\sigma; \tau, \xi)]) \, d\sigma.$

The formulas (4.47) and (4.48) comprise a system of simultaneous integral equations for the determination of the unknowns U_k and X_k which is equivalent to Cauchy's problem for the canonical system of partial differential equations (4.45).

Before solving (4.47) and (4.48) by successive approximations, we point out some qualitative properties of the solution of the Cauchy problem for a hyperbolic system which follow from its formulation as a set of integral equations. We observe that the domain of dependence of the solution U at a point R with the coordinates τ and ξ is a curvilinear triangle bounded by the initial line $x = 0$ and by two of the n characteristics through (τ, ξ). The characteristics in question are those through (τ, ξ) which intersect the initial line at the highest and lowest points. Intermediate characteristics, if such occur, do not influence the domain of dependence unless, by

[2] Cf. Courant-Lax 1.

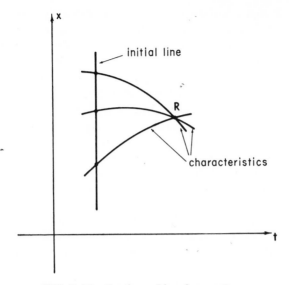

FIG. 9. The Cauchy problem for a system.

chance, they bulge outside the ones we have described as highest and lowest. The situation here is quite similar to that encountered for a single hyperbolic equation of the second order (cf. Section 1) and is only more complicated because of the larger number of families of characteristics that must be considered. The notion of a range of influence can be defined in the usual way, too, but it should be emphasized that the characteristics depend in general on the solution for a nonlinear system, so that the boundary of the range of influence cannot always be located without a knowledge of the relevant solution.

We define a sequence of successive approximations U_k^m, X_k^m for the solution of the set of integral equations (4.47), (4.48) by means of the formulas

(4.49) $U_k^0(\tau, \xi) = U_k(0, \xi), \qquad X_k^0(t; \tau, \xi) = \xi,$

(4.50) $U_k^{m+1}(\tau, \xi) = U_k(0, X_k^m(0; \tau, \xi))$

$$- \int_0^\tau C_k(\sigma, X_k^m(\sigma; \tau, \xi), U^m[\sigma, X_k^m(\sigma; \tau, \xi)]) \, d\sigma,$$

(4.51) $X_k^{m+1}(t; \tau, \xi) = \xi + \int_\tau^t \lambda_k(\sigma, X_k^m(\sigma; \tau, \xi), U^m[\sigma, X_k^m(\sigma; \tau, \xi)]) \, d\sigma.$

It is essential to specify a fixed region in which these definitions will make sense. Let K denote an upper bound on the absolute values of the coefficients C_k and λ_k and their first partial derivatives, and on the absolute values of the partial derivatives with respect to x of the functions $U_k(0, x)$ that are prescribed along the initial line $t = 0$. We shall confine our attention to a segment $-\beta \leq x \leq \beta$ of the initial line, and we shall denote by G the hexagonal neighborhood of this segment bounded by the two vertical lines $t = \pm\alpha$ and the four steeply sloped lines $x \pm \beta = \pm Kt$. The curves $x = X_k^m(t; \tau, \xi)$ defined by (4.51) have slopes dominated by K. Consequently, when its starting point (τ, ξ) lies in G, the arc of such a curve corresponding to choices of t between 0 and τ cannot exit from G. Now take α so small that for (τ, ξ) in G the formulas (4.50) and (4.51) have a meaning because the range of values of U_k^m and X_k^m is sufficiently restricted to make the bound K valid.

In order to verify the convergence of the successive approximations U_k^m and X_k^m, it will be necessary for us to establish that all the U_k^m satisfy a Lipschitz condition as functions of their second argument, for X_k^m appears as this argument under the integral signs in (4.50) and (4.51). We can differentiate (4.50) and (4.51) with respect to ξ to obtain

$$\frac{\partial U_k^{m+1}}{\partial \xi} = \frac{\partial U_k}{\partial x}\frac{\partial X_k^m}{\partial \xi} - \int_0^\tau \left[\frac{\partial C_k}{\partial x}\frac{\partial X_k^m}{\partial \xi} + \sum_{i=1}^n \frac{\partial C_k}{\partial U_i^m}\frac{\partial U_i^m}{\partial x}\frac{\partial X_k^m}{\partial \xi}\right] d\sigma,$$

$$\frac{\partial X_k^{m+1}}{\partial \xi} = 1 + \int_\tau^t \left[\frac{\partial \lambda_k}{\partial x}\frac{\partial X_k^m}{\partial \xi} + \sum_{i=1}^n \frac{\partial \lambda_k}{\partial U_i^m}\frac{\partial U_i^m}{\partial x}\frac{\partial X_k^m}{\partial \xi}\right] d\sigma.$$

If we set

$$\delta_m = \max \left|\frac{\partial U_k^m}{\partial \xi}\right|, \qquad \epsilon_m = \max \left|\frac{\partial X_k^m}{\partial \xi}\right|$$

for (τ, ξ) in G and for $k = 1, \ldots, n$, it follows that

$$\delta_{m+1} \leq K\epsilon_m + (K\epsilon_m + nK\delta_m\epsilon_m)\alpha,$$

$$\epsilon_{m+1} \leq 1 + (K\epsilon_m + nK\delta_m\epsilon_m)\alpha.$$

According to (4.49) we have $\delta_0 \leq K$ and $\epsilon_0 \leq 1$. Hence by induction we can pick α so small that

$$(4.52) \qquad \delta_m \leq 3K, \qquad \epsilon_m \leq 2.$$

The usefullness of the inequalities (4.52) lies in the fact that due to the mean value theorem δ_m is a Lipschitz constant for the functions U_k^m in their dependence on ξ. Thus we may write

$$|U_k^m(\tau, \xi'') - U_k^m(\tau, \xi')| \leq 3K |\xi'' - \xi'|.$$

The mean value theorem shows that K is a Lipschitz constant for the coefficients C_k and λ_k considered as functions of any of their arguments. We can use this information to deduce from (4.50) the estimate

(4.53)

$$
|U_k^{m+1} - U_k^m| \le K\,|X_k^m - X_k^{m-1}| + K \int_0^\tau [|X_k^m - X_k^{m-1}|
$$

$$
+ \sum_{i=1}^n |U_i^m(\sigma, X_k^m(\sigma; \tau, \xi)) - U_i^{m-1}(\sigma, X_k^{m-1}(\sigma; \tau, \xi))|]\,|d\sigma|
$$

$$
\le K\,|X_k^m - X_k^{m-1}| + K \int_0^\tau [(1 + 3nK)\,|X_k^m - X_k^{m-1}|
$$

$$
+ \sum_{i=1}^n |U_i^m(\sigma, X_k^m(\sigma; \tau, \xi)) - U_i^{m-1}(\sigma, X_k^m(\sigma; \tau, \xi))|]\,|d\sigma|.
$$

Similarly, (4.51) yields

(4.54) $$|X_k^{m+1} - X_k^m| \le K \int_\tau^t [(1 + 3nK)\,|X_k^m - X_k^{m-1}|$$

$$
+ \sum_{i=1}^n |U_i^m(\sigma, X_k^m(\sigma; \tau, \xi)) - U_i^{m-1}(\sigma, X_k^m(\sigma; \tau, \xi))|]\,|d\sigma|.
$$

We put

$$
\mu_m = \max |U_k^m - U_k^{m-1}|, \qquad \nu_m = \max |X_k^m - X_k^{m-1}|
$$

for (τ, ξ) ranging over G and for $k = 1, \ldots, n$. Formulas (4.53) and (4.54) show that

$$
\mu_{m+1} \le K\nu_m + \alpha K[(1 + 3nK)\nu_m + n\mu_m],
$$

$$
\nu_{m+1} \le \alpha K[(1 + 3nK)\nu_m + n\mu_m].
$$

Since obviously $\mu_1 \le \alpha K$ and $\nu_1 \le \alpha K$, we conclude that

$$
\mu_m \le (2nK\sqrt{\alpha})^m, \qquad \nu_m \le (2nK\sqrt{\alpha})^m \sqrt{\alpha}
$$

for small enough choices of α. Hence the geometric series

$$
\frac{2nK\sqrt{\alpha}}{1 - 2nK\sqrt{\alpha}} = \sum_{m=1}^\infty (2nK\sqrt{\alpha})^m
$$

majorizes the expressions

(4.55) $$U_k = \lim_{m \to \infty} U_k^m = U_k^0 + \sum_{m=1}^\infty (U_k^m - U_k^{m-1}),$$

(4.56) $$X_k = \lim_{m \to \infty} X_k^m = X_k^0 + \sum_{m=1}^\infty (X_k^m - X_k^{m-1})$$

for a solution of the set of integral equations (4.47) and (4.48). These infinite series therefore represent a convergent solution of the Cauchy problem for the nonlinear hyperbolic system (4.45), provided that α is sufficiently small.

It should be pointed out that in order to establish the convergence of our iterative scheme for solving (4.47) and (4.48), we had to assume that the Cauchy data $U_k(0, x)$ obeyed a Lipschitz condition with respect to the variable x. For a linear system no such restriction is necessary and, in fact, the whole procedure simplifies considerably because the characteristics are independent of the solution. Finally, our formulation of the Cauchy problem as a set of integral equations leads in the standard way (cf. Section 2) to a proof of the uniqueness of the solution and to a proof of its continuous dependence on data. Thus we can assert that the Cauchy problem is well posed in the sense of Hadamard for a linear or nonlinear hyperbolic system of partial differential equations in two independent variables. We shall not go into further detail about these matters here because such a discussion adds very little to what we have already said (cf. Exercises 1 and 2 below).

EXERCISES

1. Prove in detail the uniqueness and continuous dependence on data of the solution of Cauchy's problem for a nonlinear hyperbolic system in two independent variables, provided that the characteristics are distinct.

2. When the coefficients λ_k are independent of U, show that the second set of integral equations (4.48) can be eliminated from the scheme of successive approximations for the solution of the Cauchy problem. Establish that the solution is valid in the large when the λ_k are constants and the C_k depend linearly on U.

3. Precisely how small does α have to be taken to justify our proof that the infinite series (4.55) and (4.56) converge?

4. When two solutions of the hyperbolic system (4.45) are fitted together along a characteristic curve (4.46), what kind of discontinuities can appear in the derivatives of the result?

5. Show that the Cauchy problem is well posed for a quasi-linear hyperbolic equation (3.88) in two independent variables by discussing the same question for the characteristic system (3.93) to (3.97) to which it can be reduced. In what way does the introduction of characteristic coordinates make the analysis here simpler than our treatment of the general nonlinear canonical system (4.45)?

6. Solve the Cauchy problem for (3.88) by proving the convergence of a sequence of successive approximations defined by the formula

$$a(x, y, u_n, p_n, q_n)r_{n+1} + 2b(x, y, u_n, p_n, q_n)s_{n+1}$$
$$+ c(x, y, u_n, p_n, q_n)t_{n+1} + d(x, y, u_n, p_n, q_n) = 0,$$

which is to be interpreted as a linear hyperbolic equation for the determination of u_{n+1} as the solution of a Cauchy problem with the same Cauchy data as the original problem. What hypotheses about a, b, c, d and the data are needed here?

7. Show that the solution of the nonlinear initial value problem

$$u_t + u_x - u + w - 1 = 0,$$

$$v_t + 2u_x - w_x - 2u + w - v^2 + 2vw - w^2 = 0,$$

$$w_t + 2u_x - w_x - 2u + w = 0,$$

$$u(0, x) = 1, \quad v(0, x) = 3, \quad w(0, x) = 2$$

has a singularity on the line $t = 1$.

8. Show that the solution of the linear hyperbolic Cauchy problem

$$u_t - x^2 u_x + v + tx = 0,$$

$$v_t + v_x + (1 - tx)u + t = 0,$$

$$u(0, x) = x, \quad v(0, x) = 0$$

becomes singular on the hyperbola $tx = 1$. Explain why this happens.

9. Reduce any quasi-linear system of the first order whose characteristics are all identical to a system of ordinary differential equations. In particular, use this result to work out the theory of a single nonlinear equation of the first order that was presented in Section 2.2.

10. For any linear hyperbolic system in the canonical form (4.45), solve a mixed initial and boundary value problem over the first quadrant in which all the functions $U_1, \ldots, U_n$ are assigned at the initial time $t = 0$ but in which only those U_k corresponding to positive coefficients λ_k, that is, corresponding to characteristics that enter the first quadrant sloping forward, are assigned at the boundary $x = 0$.

4. THE RIEMANN FUNCTION

We shall attempt to represent the solution of Cauchy's problem for a linear equation

$$(4.57) \qquad u_{xy} + a(x, y)u_x + b(x, y)u_y + c(x, y)u = f(x, y)$$

in terms of definite integrals of the Cauchy data multiplied by appropriate factors. The reason for seeking a formula of this kind is that in the homogeneous case $f = 0$ any linear combination $\alpha_1 u_1 + \alpha_2 u_2$ of a pair of solutions u_1 and u_2 of (4.57) is again a solution. Hence the solution of Cauchy's problem defines a linear operation on the Cauchy data and the inhomogeneous term f. Such a linear operation might be expected to result in an expression like (4.10) for the solution involving integrals of the data. It will turn out, in particular, that when the coefficients a, b and c are constants we can exhibit the desired expression explicitly.

The principal tool we shall use to derive an integral formula for the solution of the Cauchy problem for (4.57) is *Green's theorem*[3]

$$(4.58) \qquad \iint\limits_D (U_x + V_y)\, dx\, dy = \int_C (U\, dy - V\, dx).$$

Here the line integral on the right is evaluated in the counterclockwise direction over the closed contour C bounding the region of integration D. The integrand on the left is recognized to be the divergence of the vector (U, V). Our first objective is to set up such a divergence expression $U_x + V_y$ involving the linear differential operator

$$(4.59) \qquad L[u] = u_{xy} + a u_x + b u_y + cu$$

that appears in the partial differential equation (4.57).

For the purpose at hand we introduce an *adjoint operator* $M[v]$ on a new unknown v. The operator $M[v]$ is defined so that the combination of terms $vL[u] - uM[v]$ becomes a divergence

$$(4.60) \qquad vL[u] - uM[v] = U_x + V_y.$$

To find $M[v]$, we set ourselves the task of integrating the product $vL[u]$ by parts so as to remove differentiations from the function u. Examining the terms in $L[u]$, we are led to write

$$vu_{xy} = (vu_x)_y - v_y u_x = (vu_x)_y - (v_y u)_x + v_{yx} u,$$

$$vau_x = (avu)_x - (av)_x u,$$

$$vbu_y = (bvu)_y - (bv)_y u,$$

where, of course, all the subscripts indicate partial derivatives. It follows that the adjoint operator $M[v]$ must have the form

$$(4.61) \qquad M[v] = v_{xy} - (av)_x - (bv)_y + cv$$

$$= v_{xy} - av_x - bv_y + (c - a_x - b_y)v,$$

and (4.60) becomes, more specifically,

$$(4.62)\quad vL[u] - uM[v] = (avu - v_y u)_x + (bvu + vu_x)_y$$

$$= (avu + \tfrac{1}{2}vu_y - \tfrac{1}{2}v_y u)_x + (bvu + \tfrac{1}{2}vu_x - \tfrac{1}{2}v_x u)_y.$$

[3] Cf. Buck 1.

Observe that the vector (U, V) whose divergence appears on the right is not uniquely determined and that the adjoint of M is the original operator L.

Solutions of the *adjoint equation*

$$(4.63) \qquad\qquad M[v] = 0$$

play an important role in the theory of the linear partial differential equation (4.57). Of special interest is the situation where the operators L and M are identical, in which case we call them *self-adjoint*. Comparison of (4.59) and (4.61) shows that L is self-adjoint if and only if

$$a \equiv b \equiv 0.$$

Our efforts will now be directed toward expressing the general solution of (4.57) in terms of particular solutions of the adjoint equation (4.63). We can insert (4.60) into Green's theorem (4.58) to establish the fundamental formula

$$(4.64) \qquad \iint_D (vL[u] - uM[v])\, dx\, dy = \int_C [(avu + \tfrac{1}{2}vu_y - \tfrac{1}{2}v_y u)\, dy$$
$$- (bvu + \tfrac{1}{2}vu_x - \tfrac{1}{2}v_x u)\, dx].$$

From Section 1 we recall the geometry of the solution of the Cauchy problem for (4.57) with data prescribed along an initial curve (4.7), which we shall assume here to be monotonically decreasing. We choose as the region of integration D in (4.64) the domain of dependence of the solution of Cauchy's problem with respect to a point R above the initial curve. Thus D is bounded by a contour C that consists of a vertical segment C_1 joining a point Q on (4.7) to R, plus a horizontal segment C_2 joining R to a point P on (4.7), plus the arc C_3 of (4.7) between P and Q.

We integrate the right-hand side of (4.64) by parts along the characteristic segments C_1 and C_2 to eliminate partial derivatives of u there, obtaining

$$(4.65) \qquad \iint_D (vL[u] - uM[v])\, dx\, dy = \int_{C_1} (av - v_y)u\, dy - \int_{C_2} (bv - v_x)u\, dx$$
$$+ v(R)u(R) - \tfrac{1}{2}v(P)u(P) - \tfrac{1}{2}v(Q)u(Q) + \int_{C_3} B[u, v],$$

where

$$(4.66) \qquad B[u, v] = (avu + \tfrac{1}{2}vu_y - \tfrac{1}{2}v_y u)\, dy - (bvu + \tfrac{1}{2}vu_x - \tfrac{1}{2}v_x u)\, dx.$$

Our next aim is to select v so that (4.65) will become a representation for the solution $u(R)$ of Cauchy's problem in terms of the Cauchy data along C_3 and the inhomogeneous term f on the right in (4.57).

Since we want to eliminate u from the integral over D in (4.65), we must choose v to be a solution of the adjoint equation

$$M[v] = 0.$$

If u is also not to occur in the integrals over the characteristic segments C_1 and C_2, we must require that

(4.67) $$v_y = av$$

on C_1 and that

(4.68) $$v_x = bv$$

on C_2. These are actually ordinary differential equations for v along C_1 and C_2. In addition we ask that

(4.69) $$v(R) = 1.$$

Consequently we obtain for (4.67) and (4.68) the explicit solutions

(4.70) $$v(\xi, y) = \exp\left[\int_\eta^y a(\xi, \sigma)\, d\sigma\right],$$

(4.71) $$v(x, \eta) = \exp\left[\int_\xi^x b(\sigma, \eta)\, d\sigma\right],$$

where ξ and η now denote the coordinates of R.

The conditions (4.70) and (4.71) describe a characteristic initial value problem for the adjoint equation (4.63). We conclude that they possess a uniquely determined solution v. Since v depends on the parameters ξ and η as well as on the variables x and y, we introduce the more detailed notation

(4.72) $$A(x, y; \xi, \eta) = v(x, y).$$

The quantity $A(x, y; \xi, \eta)$ is known as the *Riemann function*[4] associated with the linear partial differential equation (4.57). The identity (4.65) reduces to the representation

(4.73) $$u(\xi, \eta) = \tfrac{1}{2}A(P; \xi, \eta)u(P) + \tfrac{1}{2}A(Q; \xi, \eta)u(Q)$$

$$-\int_P^Q B[u(x, y), A(x, y; \xi, \eta)]$$

$$+ \iint_D f(x, y)A(x, y; \xi, \eta)\, dx\, dy$$

---

[4] Cf. Riemann 1.

for the solution u of Cauchy's problem in terms of this Riemann function, where P and Q stand for the pairs of coordinates $(x(\eta), \eta)$ and $(\xi, y(\xi))$, respectively, when they occur as arguments of A. The line integral between P and Q in (4.73) must be evaluated along the initial curve (4.7) and depends only on the Cauchy data prescribed there.

Various properties of the solution of Cauchy's problem for the linear equation (4.57), such as its uniqueness and its continuous dependence on the Cauchy data, are exhibited by the explicit representation (4.73). When the Riemann function A is known, this formula can also be regarded as an expression for the general solution of (4.57), since the Cauchy data involve precisely two arbitrary functions of one variable. Finally, notice that the solution (4.10) of the Cauchy problem for the wave equation (4.9) is a special case of (4.73), since (4.63), (4.70) and (4.71) characterize the Riemann function of (4.9) as the constant

$$A(x, y; \xi, \eta) \equiv 1.$$

Let us allow the path of integration C_3 from P to Q in (4.73) to degenerate to a segment of the vertical characteristic through P and a segment of the horizontal characteristic through Q, which intersect at a point to be denoted by S. If we perform preliminary integrations by parts to avoid integrals of u_x and u_y over these segments, we obtain the representation

(4.74) $u(R) = A(P; R)u(P) + A(Q; R)u(Q) - A(S; R)u(S)$

$$+ \int_S^P (aA - A_y)u \, dy + \int_S^Q (bA - A_x)u \, dx + \iint_D fA \, dx \, dy$$

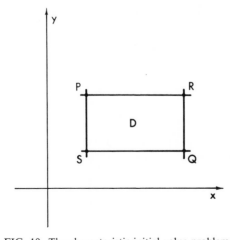

FIG. 10. The characteristic initial value problem.

for the solution of the characteristic initial value problem for (4.57) in terms of the Riemann function A. Of course D now stands for the rectangle with corners at the points R, P, S and Q. Observe that (4.74) has features in common with the integrodifferential equation (4.41). We shall see presently that the alternate formulation

$$(4.75) \qquad u(R) = A(S; R)u(S) + \int_S^P (au + u_y)A \, dy$$

$$+ \int_S^Q (bu + u_x)A \, dx + \iint_D fA \, dx \, dy$$

of (4.74) found by integrating the line integrals along the characteristics through S by parts again is quite useful, too.

At this juncture we emphasize that the Riemann function A has been defined as a solution of the adjoint equation (4.63). Thus we attach to it the subscript M and write

$$(4.76) \qquad A(x, y; \xi, \eta) = A_M(x, y; \xi, \eta).$$

Next we introduce another Riemann function $A_L = A_L(x, y; \alpha, \beta)$ which is to have in its dependence on x and y the properties

$$(4.77) \qquad L[A_L(x, y; \alpha, \beta)] = 0,$$

$$(4.78) \qquad A_L(\alpha, y; \alpha, \beta) = \exp \left[-\int_\beta^y a(\alpha, \sigma) \, d\sigma \right],$$

$$(4.79) \qquad A_L(x, \beta; \alpha, \beta) = \exp \left[-\int_\alpha^x b(\sigma, \beta) \, d\sigma \right]$$

associated with a characteristic initial value problem for (4.57) instead of (4.63).

When we put $u = A_L$ and take α and β as the coordinates of the point S, formula (4.75) yields an expression for A_L in terms of A_M; for then $f = 0$, and the line integrals from S to P and from S to Q drop out because of the requirements (4.78) and (4.79) on A_L. We are left with the important rule

$$(4.80) \qquad A_L(\xi, \eta; \alpha, \beta) = A_M(\alpha, \beta; \xi, \eta)$$

for interchanging the arguments and the parameters of the Riemann function. Note that our proof of (4.80) amounts to integrating the identity

$$(4.81) \qquad \iint_D (A_M L[A_L] - A_L M[A_M]) \, dx \, dy = 0$$

by parts in a suitable fashion.

If a differential operator L is self-adjoint, the Riemann functions A_L and A_M do not differ from one another and (4.80) becomes a rule of symmetry

$$(4.82) \qquad A(\xi, \eta; \alpha, \beta) = A(\alpha, \beta; \xi, \eta).$$

In the general case, (4.80) shows that as a function of its parameter point the Riemann function satisfies the adjoint of the linear partial differential equation used to define it as a function of its principal arguments. Consequently a direct proof can be given that the right-hand side of (4.73) solves Cauchy's problem (cf. Exercise 2 below).

When the Riemann function A can be found in closed form, (4.73) and (4.74) furnish explicit integral formulas for the solution of the Cauchy problem and the characteristic initial value problem. This solution of the Cauchy problem is valid for an arbitrary initial curve, although certain signs in front of the integrals must be reversed when the initial curve is monotonically increasing instead of monotonically decreasing. Thus it becomes important to evaluate the Riemann function for as many special examples as possible. We shall do so here for the case where the differential operator L has constant coefficients a, b and c.

The results of Section 3.2 show that it is sufficient to consider the canonical form

$$(4.83) \qquad L[u] = u_{xy} + \lambda u$$

of $L[u]$. We have already seen (cf. Exercise 2.3) that the relevant characteristic initial value problem for the corresponding self-adjoint equation (4.43) can be solved in terms of a Bessel function. Since solutions of (4.43) are obviously invariant under translation of the coordinates x and y, and since the initial conditions (4.70) and (4.71) imposed on the Riemann function are precisely those satisfied by (4.44) when $\xi = \eta = 0$, we conclude that the desired Riemann function is given by

$$(4.84) \qquad A(x, y; \xi, \eta) = J_0(2\sqrt{\lambda(x - \xi)(y - \eta)}).$$

The symmetry rule (4.82) is evident for the specific Riemann function (4.84). Notice that (4.84) is valid even if the argument $\lambda(x - \xi)(y - \eta)$ under the radical sign is negative, since the Bessel function J_0 is even and therefore no imaginary quantities are actually introduced in such circumstances.

It is of interest to work out the representation (4.75) of the solution of the characteristic initial value problem for the telegraph equation

$$u_{xy} + \lambda u = 0$$

in all detail, because the result provides a quite elegant expression for the general solution u in terms of the two arbitrary functions

$$u(x, \beta) = \phi(x), \qquad u(\alpha, y) = \psi(y).$$

The formula in question is

$$(4.85) \quad u(x, y) = \frac{\phi(0) + \psi(0)}{2} J_0(2\sqrt{\lambda x y}) + \int_0^x \phi'(\xi) J_0(2\sqrt{\lambda y(x - \xi)}) \, d\xi$$

$$+ \int_0^y \psi'(\eta) J_0(2\sqrt{\lambda x(y - \eta)}) \, d\eta,$$

where we have interchanged the roles of (x, y) and (ξ, η) in (4.75) and have put $\alpha = \beta = 0$ for the sake of simplicity. Observe that when $\lambda = 0$ this reduces, except for an additive constant, to the expression (4.6) for the general solution of the wave equation (4.5).

EXERCISES

1. Show that

$$M[v] = av_{xx} + 2bv_{xy} + cv_{yy} - (d - 2a_x - 2b_y)v_x$$

$$- (e - 2c_y - 2b_x)v_y + (f - d_x - e_y + a_{xx} + 2b_{xy} + c_{yy})v$$

is the adjoint of the general linear differential operator

$$L[u] = au_{xx} + 2bu_{xy} + cu_{yy} + du_x + eu_y + fu$$

of the second order in two independent variables.

2. Show that the integral formula (4.73) solves the Cauchy problem by verifying directly that the terms on the right satisfy the relevant partial differential equation and initial conditions.

3. Derive from (4.84) an explicit expression for the Riemann function of the general equation (4.57) with constant coefficients a, b and c.

4. Use the complex substitutions

$$z = x + iy, \quad z^* = x - iy, \quad \zeta = \xi + i\eta, \quad \zeta^* = \xi - i\eta$$

to transform (4.85) into the representation

$$(4.86) \quad u(x, y) = \text{Re} \left\{ \phi(0) J_0(k\sqrt{x^2 + y^2}) + 2 \int_0^z \phi'(\zeta) J_0(k\sqrt{z^*(z - \zeta)}) \, d\zeta \right\}$$

for the general solution of the elliptic equation

$$u_{xx} + u_{yy} + k^2 u = 0$$

in terms of an arbitrary analytic function $\phi(z)$ of the complex variable z, where k denotes a constant. The analogous formula for the more general equation

$$u_{xx} + u_{yy} + au_x + bu_y + cu = 0$$

with analytic coefficients a, b and c also generates solutions from analytic functions of a complex variable. It is related to the so-called *Bergman integral operator*.[5]

5. Find the Riemann function for the equation

(4.87)
$$u_{xx} - u_{tt} + \frac{\lambda}{x} u_x = 0,$$

where λ is a constant. First use the method of separation of variables to determine the solution which satisfies the mixed boundary and initial conditions

(4.88) $u(l_1, t) = u(l_2, t) = 0, \quad 0 < l_1 < l_2,$

(4.89) $u(x, 0) = g(x), \quad u_t(x, 0) = h(x), \quad l_1 \le x \le l_2,$

and then compare the result with (4.73) in the characteristic triangle $t \ge 0$, $t \le x - l_1, t \le l_2 - x$, where it depends exclusively on the initial data (4.89).

6. Find the adjoint equation for (4.87). Discuss the effect of the substitution

(4.90)
$$w = x^n u$$

on the form of (4.87) and show that w satisfies the self-adjoint equation

(4.91)
$$w_{xx} - w_{tt} + \frac{\lambda(2 - \lambda)}{4x^2} w = 0$$

when $n = \lambda/2$.

[5] Cf. Bergman 2.

5

The Fundamental Solution

1. TWO INDEPENDENT VARIABLES

Let L indicate the linear elliptic differential operator

$$L[u] = u_{xx} + u_{yy} + au_x + bu_y + cu,$$

which appears in canonical form to start with (cf. Section 3.1). We shall attempt to construct special solutions of the homogeneous partial differential equation

$$(5.1) \qquad L[u] = 0$$

in the case where the coefficients $a = a(x, y)$, $b = b(x, y)$ and $c = c(x, y)$ are real analytic functions of their arguments, i.e., have convergent double Taylor series expansions locally. In particular, we shall look for a generalization for (5.1) of the logarithmic solution

$$(5.2) \qquad u = \log \frac{1}{r}$$

of Laplace's equation

$$(5.3) \qquad \Delta u = 0,$$

where

$$\Delta u = u_{xx} + u_{yy}, \qquad r = \sqrt{(x - \xi)^2 + (y - \eta)^2}.$$

A *fundamental solution* S of (5.1) is defined to be a solution of the form

$$(5.4) \qquad S = S(x, y; \xi, \eta) = A(x, y; \xi, \eta) \log \frac{1}{r} + B(x, y; \xi, \eta),$$

136

depending on parameters ξ and η, such that the coefficients A and B are regular in some neighborhood of the parameter point (ξ, η) and such that

(5.5) $A(\xi, \eta; \xi, \eta) = 1.$

Thus the fundamental solution (which is also referred to occasionally in the literature as the *elementary solution*) becomes logarithmically infinite when $x \to \xi$ and $y \to \eta$. It is the simplest conceivable singular solution.

The importance of the fundamental solution is that it leads to an integral representation of other solutions which is analogous to the Cauchy formula for analytic functions of a complex variable, whose real parts, of course, satisfy the Laplace equation (5.3). This representation is related to the expression (4.73) for the solution of Cauchy's problem in terms of the Riemann function. Indeed, it turns out that the factor A in our definition (5.4) of the fundamental solution S is essentially a Riemann function. Also, we shall eventually solve boundary value problems by means of specialized fundamental solutions known as *Green's functions*. These, however, will not be discussed until Section 7.2.

Clearly, the logarithm (5.2) itself is a fundamental solution of Laplace's equation (5.3). For the more general equation (5.1) our hypothesis that the coefficients a, b and c are analytic has the advantage that it permits us to find a fundamental solution S by assuming the corresponding functions A and B to be analytic in their dependence on all four of the arguments x, y, ξ and η. Thus we are now motivated to perform an analytic continuation of S to complex values of the variables x, y, ξ and η and to introduce the complex substitution

(5.6) $z = x + iy, \quad z^* = x - iy, \quad \zeta = \xi + i\eta, \quad \zeta^* = \xi - i\eta.$

Here z^* is the complex conjugate of z only when x and y are both real, and similarly for ζ^*.

The complex variables z and z^* are actually *characteristic coordinates* (cf. Section 3.1) for the elliptic partial differential equation (5.1). Direct calculation shows that

$$\frac{\partial}{\partial x} = \frac{\partial}{\partial z} + \frac{\partial}{\partial z^*}, \qquad \frac{\partial}{\partial y} = i\frac{\partial}{\partial z} - i\frac{\partial}{\partial z^*},$$

$$\frac{\partial}{\partial z} = \frac{1}{2}\left(\frac{\partial}{\partial x} - i\frac{\partial}{\partial y}\right), \qquad \frac{\partial}{\partial z^*} = \frac{1}{2}\left(\frac{\partial}{\partial x} + i\frac{\partial}{\partial y}\right),$$

and, furthermore,

$$\Delta = \left(\frac{\partial}{\partial x} - i\frac{\partial}{\partial y}\right)\left(\frac{\partial}{\partial x} + i\frac{\partial}{\partial y}\right) = 4\frac{\partial^2}{\partial z\partial z^*}.$$

Consequently we have

$$L[u] = 4 \frac{\partial^2 u}{\partial z\, \partial z^*} + (a + bi)\frac{\partial u}{\partial z} + (a - bi)\frac{\partial u}{\partial z^*} + cu$$

in terms of the characteristic coordinates. Thus we can rewrite (5.1) in the complex form

(5.7)
$$\frac{\partial^2 u}{\partial z\, \partial z^*} + \alpha \frac{\partial u}{\partial z} + \beta \frac{\partial u}{\partial z^*} + \gamma u = 0,$$

where

$$\alpha = \frac{a + bi}{4}, \quad \beta = \frac{a - bi}{4}, \quad \gamma = \frac{c}{4}.$$

If we insert the expression (5.4) for a fundamental solution S into (5.7), we obtain the identity

(5.8) $$L[A] \log \frac{1}{r} - 2\frac{\partial A/\partial z + \beta A}{z^* - \zeta^*} - 2\frac{\partial A/\partial z^* + \alpha A}{z - \zeta} + L[B] = 0,$$

since
$$r^2 = (z - \zeta)(z^* - \zeta^*).$$

Suppose that S is an analytic function and that its definition has been extended to complex values

$$x = x_1 + ix_2, \qquad y = y + iy_2$$

of the arguments x and y. According to the principles of analytic continuation, (5.8) must remain valid in a four-dimensional region of the complex domain in which the variables z and z^* can vary independently. We proceed to examine the singularities of the left-hand side of (5.8) in such a region.

Since a logarithmic singularity is multiple-valued and cannot be canceled by contributions involving only poles, the logarithmic term ought to drop out of (5.8) completely. Therefore we have to require that

(5.9) $$L[A] = 0.$$

The two middle terms in (5.8) will become infinite when $z^* = \zeta^*$ or when $z = \zeta$ unless their numerators include, respectively, the factors $(z^* - \zeta^*)$ and $(z - \zeta)$. This observation leads us to impose on A the auxiliary conditions

(5.10) $$\left[\frac{\partial}{\partial z} + \beta(z, \zeta^*)\right] A(z, \zeta^*; \zeta, \zeta^*) = 0,$$

(5.11) $$\left[\frac{\partial}{\partial z^*} + \alpha(\zeta, z^*)\right] A(\zeta, z^*; \zeta, \zeta^*) = 0,$$

which assure us that the desired factors will be present. Our use of the abridged notation

$$A(z, z^*; \zeta, \zeta^*) = A(x, y; \xi, \eta)$$

here should not cause any confusion, although it does not indicate with strict accuracy the change in the functional form of A that is implied by the substitution of variables (5.6). When (5.9), (5.10) and (5.11) are fulfilled, the singularities of the first three terms on the left in (5.8) are annihilated, and hence it becomes feasible to cancel them by appropriate choice of the regular function B.

Viewing (5.10) and (5.11) as linear ordinary differential equations, we integrate them and take note of the initial condition (5.5) to obtain

$$(5.12) \qquad A(z, \zeta^*; \zeta, \zeta^*) = \exp\left[-\int_{\zeta}^{z} \beta(\sigma, \zeta^*) \, d\sigma\right],$$

$$(5.13) \qquad A(\zeta, z^*; \zeta, \zeta^*) = \exp\left[-\int_{\zeta^*}^{z^*} \alpha(\zeta, \sigma) \, d\sigma\right].$$

By Cauchy's theorem the definite integrals on the right are independent of the choice of the path of integration between prescribed limits in the complex σ-plane.

A comparison of the requirements (5.9), (5.12) and (5.13) with formulas (4.77), (4.78) and (4.79) from Section 4.4 shows in a formal way that the factor A in the expression (5.4) for a fundamental solution S of (5.1) should be identified as the *Riemann function* A_L associated with the partial differential equation (5.7) in the complex domain. Indeed, (5.9), (5.12) and (5.13) represent the formulation in analytical terms of the characteristic initial value problem which defines the Riemann function A_L. It is only the fact that z and z^* must be interpreted as independent complex variables that now complicates our construction of A as the Riemann function. As matters stand, it will be necessary to review the details of our proof of the existence of a solution of the characteristic initial value problem in order to check its validity in the complex domain for the special example given analytically by (5.9), (5.12) and (5.13).

We recall our derivation in Section 4.2 of the integrodifferential equation (4.41) for the solution of characteristic initial value problems. A quite analogous result may be formulated in the complex domain for the particular problem defining the complex analytic Riemann function A. Indeed, (5.9), (5.12) and (5.13) are equivalent to the single identity

(5.14)

$$A(z, z^*; \zeta, \zeta^*) = \exp\left[-\int_{\zeta}^{z} \beta(\sigma, \zeta^*) \, d\sigma\right] + \exp\left[-\int_{\zeta^*}^{z^*} \alpha(\zeta, \tau) \, d\tau\right] - 1$$

$$- \int_{\zeta^*}^{z^*} \int_{\zeta}^{z} \left[\alpha(\sigma, \tau)\frac{\partial}{\partial\sigma} A(\sigma, \tau; \zeta, \zeta^*) + \beta(\sigma, \tau)\frac{\partial}{\partial\tau} A(\sigma, \tau; \zeta, \zeta^*)\right.$$

$$\left. + \gamma(\sigma, \tau)A(\sigma, \tau; \zeta, \zeta^*)\right] d\sigma \, d\tau.$$

Here A is supposed to be analytic, so that by Cauchy's theorem the integrals from ζ to z and from ζ^* to z^* in the complex σ-plane and in the complex τ-plane are independent of path. Thus it is easy to verify by direct differentiation and substitution that a solution A of the integrodifferential equation (5.14) has to satisfy (5.9), (5.12) and (5.13).

We define a sequence of successive approximations A_n to the solution of (5.14) by the rules

$$A_0(z, z^*; \zeta, \zeta^*) = \exp\left[-\int_\zeta^z \beta(\sigma, \zeta^*)\, d\sigma\right] + \exp\left[-\int_{\zeta^*}^{z^*} \alpha(\zeta, \tau)\, d\tau\right] - 1$$

and

$$A_{n+1}(z, z^*; \zeta, \zeta^*) = \exp\left[-\int_\zeta^z \beta(\sigma, \zeta^*)\, d\sigma\right] + \exp\left[-\int_{\zeta^*}^{z^*} \alpha(\zeta, \tau)\, d\tau\right] - 1$$

$$-\int_{\zeta^*}^{z^*}\int_\zeta^z \left[\alpha(\sigma, \tau)\frac{\partial}{\partial\sigma} A_n(\sigma, \tau; \zeta, \zeta^*) + \beta(\sigma, \tau)\frac{\partial}{\partial\tau} A_n(\sigma, \tau; \zeta, \zeta^*)\right.$$

$$\left. + \gamma(\sigma, \tau)A_n(\sigma, \tau; \zeta, \zeta^*)\right] d\sigma\, d\tau, \qquad n \geq 0.$$

All the integrals are independent of path because each individual approximation A_n is by construction an analytic function of the two complex variables z and z^*. A proof of the uniform convergence of the sequence of analytic functions A_n to a limit

$$(5.15) \qquad A = \lim_{n\to\infty} A_n$$

can be modeled on the same result for the infinite series (4.22) when the paths of integration are selected in a reasonable fashion, for example, as straight lines. The key estimate here is found by straightforward calculation (cf. Exercise 1 below) to be of the form

$$(5.16)$$

$$|A_n - A_{n-1}| + \left|\frac{\partial A_n}{\partial z} - \frac{\partial A_{n-1}}{\partial z}\right| + \left|\frac{\partial A_n}{\partial z^*} - \frac{\partial A_{n-1}}{\partial z^*}\right| \leq \frac{K^{n+1}(2 + l)^n l^n}{n!},$$

where K denotes an upper bound on $|\alpha|$, $|\beta|$, $|\gamma|$ and

$$|A_0| + \left|\frac{\partial A_0}{\partial z}\right| + \left|\frac{\partial A_0}{\partial z^*}\right|,$$

and where l denotes an upper bound on the *sum* of the lengths of the paths of integration from ζ to z and from ζ^* to z^*. Thus when α, β and γ are entire, the convergence holds in the large, a property which stems from the linearity of the integrodifferential equation (5.14).

Either from the uniformity of the convergence of the approximations A_n in each bounded region, or, more directly, from the observation that the integrals involved in our procedure are independent of path, we deduce that the limit A is an analytic function of its arguments. The conclusion applies not just to z and z^* but also to ζ and ζ^*. In particular, it follows that A is a solution of the characteristic initial value problem (5.9), (5.12), (5.13). As such it should be viewed as the Riemann function for the partial differential equation (5.7) in the complex domain, although the order in which we have written its primary arguments z, z^* and its parameters ζ, ζ^* is the reverse of the order adopted earlier for the Riemann function (cf. Section 4.4).

We see from the analysis above that the factor A in the representation (5.4) for a fundamental solution S of (5.1) is uniquely determined. Therefore it must be real for real values of x, y, ξ and η, since its real part alone has all the properties required of it in the real domain. The remaining regular term B in (5.4) is to be found from the inhomogeneous equation

$$L[B(z, z^*; \zeta, \zeta^*)] = f(z, z^*),$$

where

$$f(z, z^*) = L[A(z, z^*; \zeta, \zeta^*) \log r].$$

We stress that the function f has no singularities in the vicinity of the parameter point (ζ, ζ^*). Evidently B is only determined up to an additive term solving the homogeneous equation (5.1).

We can use the integral formula (4.74) for the solution of characteristic initial value problems in terms of the Riemann function to derive the expression

$$(5.17) \qquad B(z, z^*; \zeta, \zeta^*) = \frac{1}{4} \int_{\zeta^*}^{z^*} \int_{\zeta}^{z} f(\sigma, \tau) A(z, z^*; \sigma, \tau) \, d\sigma \, d\tau$$

defining one possible choice of B whose initial data vanish. However, this particular representation for B has the disadvantage that when it is substituted into (5.4) it does not in general provide a fundamental solution S that satisfies the adjoint equation as a function of the parameters ξ and η, an additional property for S which happens to be useful in the applications (cf. Exercise 4 below). The explicit formula (5.17) does establish, on the other hand, that it is feasible to construct S in the large for a partial differential equation with entire coefficients that is written in the canonical form (5.7).

One of the principal applications of the fundamental solution consists in an integral representation for arbitrary solutions that is analogous to the Cauchy formula for analytic functions of a complex variable. The representation exhibits each solution in terms of the fundamental solution in such a way as to make evident certain of its more significant properties. For example, convergent Taylor series expansions of the solution may be developed in this fashion.

The integral representation with which we are concerned hinges on a consideration of the *adjoint partial differential equation*

$$(5.18) \qquad M[v] = v_{xx} + v_{yy} - av_x - bv_y + (c - a_x - b_y)v = 0.$$

From Section 4.4 we recall that the differential operator M is constructed so that the expression

$$(5.19) \qquad vL[u] - uM[v] = (vu_x - uv_x + auv)_x + (vu_y - uv_y + buv)_y$$

becomes a divergence. Hence Green's theorem yields the indentity

$$(5.20) \qquad \iint_D (vL[u] - uM[v]) \, dx \, dy = \int_C [(vu_x - uv_x + auv) \, dy$$
$$- (vu_y - uv_y + buv) \, dx],$$

where D stands for a region with smooth boundary curves C, and where the line integral on the right is evaluated in a direction such that the region lies on the left.

Observe that if ν stands for the inner normal and s for the arc length along C, then (5.20) can be presented in the frequently more useful form

$$\iint_D (vL[u] - uM[v]) \, dx \, dy + \int_C \left[v \frac{\partial u}{\partial \nu} - u \frac{\partial v}{\partial \nu} + \left(a \frac{\partial x}{\partial \nu} + b \frac{\partial y}{\partial \nu} \right) uv \right] ds = 0.$$

This reduces in the special case

$$a = b = c = 0$$

of Laplace's equation to the basic integral theorem

$$\iint_D (v \, \Delta u - u \, \Delta v) \, dx \, dy + \int_C \left(v \frac{\partial u}{\partial \nu} - u \frac{\partial v}{\partial \nu} \right) ds = 0$$

of potential theory. For the general equation (5.1) it is desirable to introduce the notation

$$(5.21) \qquad H[u, v] = -(vu_x - uv_x + auv) \, dy + (vu_y - uv_y + buv) \, dx$$
$$= \left\{ v \frac{\partial u}{\partial \nu} - u \frac{\partial v}{\partial \nu} + \left(a \frac{\partial x}{\partial \nu} + b \frac{\partial y}{\partial \nu} \right) uv \right\} ds$$

in order to arrive at the more concise version

(5.22) $$\iint_{D} (vL[u] - uM[v])\, dx\, dy + \int_{C} H[u, v] = 0$$

of (5.20).

We shall take for v in (5.22) any solution of the adjoint equation

$$M[v] = 0$$

which is defined and continuous together with its first and second partial derivatives in D and on C. We intend to obtain an expression for v by substituting for u a fundamental solution S of equation (5.1).

For a procedure of this kind to succeed, we must at first assume that the interior of a small circle E of radius ϵ about the singularity (ξ, η) of the fundamental solution S has been cut out of the region D. Thus E becomes an additional component of the boundary of the region to which (5.22) is actually applied. Since

$$L[S] = M[v] = 0,$$

we obtain the relation

(5.23) $$\int_{C} H[S, v] = \int_{E} H[S, v].$$

Motivated by the calculus of residues, we shall attempt to evaluate the term on the right by passing to the limit as $\epsilon \rightarrow 0$.

Let r and θ denote polar coordinates with respect to an origin located at the point (ξ, η). On E we find that

$$H[S, v] = -\left\{ v\,\frac{\partial S}{\partial r} - S\,\frac{\partial v}{\partial r} + (a\cos\theta + b\sin\theta)Sv \right\}\epsilon\, d\theta.$$

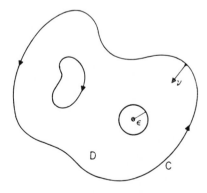

FIG. 11. Contours for the integral representation formula.

Because of the form

$$S = A \log \frac{1}{r} + B$$

of the fundamental solution we have

$$\lim_{\epsilon \to 0} \epsilon S = 0, \qquad \lim_{\epsilon \to 0} \epsilon \frac{\partial S}{\partial r} = -1.$$

It follows that

$$\lim_{\epsilon \to 0} \int_E H[S, v] = \lim_{\epsilon \to 0} \int_0^{2\pi} v \, d\theta = 2\pi v(\xi, \eta),$$

since v and its first derivatives are continuous. Therefore (5.23) reduces in the limit as $\epsilon \to 0$ to the important integral representation

$$(5.24) \qquad v(\xi, \eta) = \frac{1}{2\pi} \int_C H[S(x, y; \xi, \eta), v(x, y)]$$

for solutions v of (5.18) over the region D in terms of their Cauchy data on C. A more specific expression for the final result is

$$(5.25) \quad v(\xi, \eta) = \frac{1}{2\pi} \int_C \left[v \frac{\partial S}{\partial \nu} - S \frac{\partial v}{\partial \nu} + \left(a \frac{\partial x}{\partial \nu} + b \frac{\partial y}{\partial \nu} \right) Sv \right] ds.$$

Notice that when we replace S in our proof of the integral formula (5.24) by the singular term $-A \log r$ alone, we obtain the equally useful representation

$$(5.26) \quad v(\xi, \eta) = -\frac{1}{2\pi} \int_C H[A \log r, v] - \frac{1}{2\pi} \iint_D vL[A \log r] \, dx \, dy$$

for solutions v of the adjoint equation. Moreover, it is also true that solutions u of (5.1) fulfill

$$(5.27) \quad u(x, y) = \frac{1}{2\pi} \int_C H[u, A \log r] - \frac{1}{2\pi} \iint_D uM[A \log r] \, d\xi \, d\eta,$$

where we now integrate over D and C with respect to the point (ξ, η). This follows because the integral identity (5.22) remains valid in the complex domain and can be applied as in Section 4.4 to establish the rule (4.80) for interchanging the arguments z, z^* and the parameters ζ, ζ^* of the complex Riemann function A. Thus A is seen to be the Riemann function of the adjoint equation

$$\frac{\partial^2 A}{\partial \zeta \, \partial \zeta^*} - \alpha \frac{\partial A}{\partial \zeta} - \beta \frac{\partial A}{\partial \zeta^*} + \left(\gamma - \frac{\partial \alpha}{\partial \zeta} - \frac{\partial \beta}{\partial \zeta^*} \right) A = 0$$

in its dependence on ζ and ζ^*. We are therefore justified in switching the roles of L and M in our derivation of (5.26) to obtain (5.27) instead.

Finally, let

$$S_M(\xi, \eta; x, y) = A(x, y; \xi, \eta) \log \frac{1}{r} + B_M(\xi, \eta; x, y)$$

stand for an analytic fundamental solution of the adjoint equation (5.18), with B_M found from an appropriate modification of (5.17). It is a corollary of (5.24), or, for that matter, of (5.27), that

$$(5.28) \qquad u(x, y) = -\frac{1}{2\pi} \int_C H[u(\xi, \eta), S_M(\xi, \eta; x, y)]$$

holds for every twice continuously differentiable solution u of (5.1) defined in D and on C.

An important consequence of the integral representation (5.28) is that u must itself be a real analytic function of the variables x and y throughout D. This remarkable conclusion is based on the analyticity of S_M, which ensures that the integral on the right in (5.28) is analytic whenever $r \neq 0$. A similar investigation of (5.24) shows that every smooth solution v of the adjoint equation (5.18) has to be analytic, too. For the proof we have only assumed that (5.1) and (5.18) are of the elliptic type and that the coefficients a, b and c are analytic.

The regularity of solutions u of the elliptic equation (5.1) can alternately be deduced from formula (5.27), which involves the Riemann function A but no correction term B. Indeed, the kernel

$$M[A \log r] = 2 \frac{\partial A/\partial \zeta - \beta A}{\zeta^* - z^*} + 2 \frac{\partial A/\partial \zeta^* - \alpha A}{\zeta - z}$$

occurring in the double integral over D on the right in (5.27) is regular because of the characteristic initial conditions

$$\left[\frac{\partial}{\partial \zeta} - \beta(\zeta, z^*) \right] A(z, z^*; \zeta, z^*) = 0,$$

$$\left[\frac{\partial}{\partial \zeta^*} - \alpha(z, \zeta^*) \right] A(z, z^*; z, \zeta^*) = 0$$

satisfied by the Riemann function A in its dependence on ζ and ζ^*. Hence the integral in question is an analytic function of x and y. Similarly, the integral over C in (5.27) defines a real analytic function of x and y unless r vanishes. Since a situation where $r = 0$ can arise only for choices of

(x, y) on the boundary curve C itself, we conclude that the right-hand side of (5.27) is analytic in D.

The result about analyticity of solutions of a linear elliptic partial differential equation with analytic coefficients that we have just presented may be viewed as a generalization of the theorem that a continuously differentiable function of a complex variable, or, in other words, a pair of real functions satisfying the Cauchy-Riemann equations, can be expanded in convergent Taylor series. Furthermore, our proof has features in common with the demonstration of that classic theorem which is based on Cauchy's formula. It is significant that the representation (5.28) not only establishes that u is analytic but also provides a means of continuing it to complex values of the arguments x and y by direct substitution when the imaginary parts of these variables are so restricted that r does not vanish. Now let us think of u as a solution of (5.7) in the complex domain of independent values of $z = x + iy$ and $z^* = x - iy$. An intuitive explanation for the regularity of u in the real domain can be built around the remark that singularities of a solution of (5.7) should propagate exclusively along the characteristics, which are, however, imaginary in the elliptic case, so that each of them meets the real domain only in an isolated point.

An illuminating illustration of our theory of the fundamental solution S is provided by the special example of the self-adjoint equation

$$(5.29) \qquad \Delta u + k^2 u = 0$$

with a constant coefficient k^2. A closed expression for S can be found in this case by the method of separation of variables. In fact, the solutions u of (5.29) which depend on the distance

$$r = \sqrt{zz^*}$$

alone (cf. Exercise 4.2.4) must satisfy the ordinary differential equation

$$(5.30) \qquad r\frac{d^2u}{dr^2} + \frac{du}{dr} + k^2 ru = 0.$$

They are therefore given by linear combinations of the Bessel functions

$$J_0(kr) = \sum_{n=0}^{\infty} \frac{(-1)^n (kr)^{2n}}{2^{2n}(n!)^2},$$

$$Y^{(0)}(kr) = J_0(kr) \log kr - \sum_{n=1}^{\infty} \left(1 + \frac{1}{2} + \cdots + \frac{1}{n}\right) \frac{(-1)^n (kr)^{2n}}{2^{2n}(n!)^2}.$$

Obviously

$$S(x, y; \xi, \eta) = -Y^{(0)}(kr)$$

is a fundamental solution of (5.29). If we identify the coefficient of the logarithm here as the Riemann function A, we are led to conclude that

$$(5.31) \qquad A(x, y; \xi, \eta) = J_0(kr).$$

It is interesting to observe that the complex substitution (5.6) puts the reduced wave equation (5.29) into the form

$$\frac{\partial^2 u}{\partial z \, \partial z^*} + \frac{k^2}{4} u = 0$$

of the telegraph equation and reduces (5.31) to our earlier result (4.84) for the Riemann function associated with the hyperbolic differential operator

$$L[u] = u_{xy} + \lambda u.$$

Thus in certain cases where a fundamental solution S is easier to determine than the Riemann function A, it is convenient to deduce an expression for the latter by exploiting the relation

$$S = A \log \frac{1}{r} + B$$

between them. Incidentally, the appearance of a logarithmic term in the expansion of $Y^{(0)}$ is a consequence of the fact that zero is a multiple root of the indicial equation associated with solving the linear ordinary differential equation (5.30) by means of power series. It comes as no surprise, therefore, that the coefficient of the logarithm is another solution of (5.30).

Our discussion of the example (5.29) indicates that the notion of a fundamental solution has meaning for hyperbolic as well as for elliptic equations. Suffice it to say in this connection that a fundamental solution of the linear hyperbolic equation

$$u_{xy} + au_x + bu_y + cu = 0$$

should be required to have the form

$$(5.32) \quad S(x, y; \xi, \eta) = A(x, y; \xi, \eta) \log (x - \xi)(y - \eta) + B(x, y; \xi, \eta),$$

where A is once more the relevant Riemann function and where B is supposed to be regular along the pair of characteristics $x = \xi$ and $y = \eta$. It is clear how one can verify that such a solution exists by using the methods we have presented in the elliptic case.

In Section 3 we shall establish the local existence of fundamental solutions for a linear partial differential equation even when the coefficients are not analytic. For the case of two independent variables, an especially simple construction based on successive approximations in the real domain is suggested in Exercise 3.4. Therefore we do not go into such matters here.

EXERCISES

1. Derive the estimate (5.16), taking particular care to achieve the appropriate denominator $n!$ on the right. Complete the remaining details of the proof of convergence of the expression (5.15) for the Riemann function A of (5.7) in the complex domain.

2. Give a detailed derivation of the rule for interchanging the arguments and parameters of the complex analytic Riemann function associated with the elliptic partial differential equation (5.1).

3. Derive Cauchy's integral formula from (5.25).

4. What boundary conditions ought to be imposed on S in order to convert (5.28) into a formula giving the solution of (5.1) in D that assumes prescribed values on C?

5. Let u_1 and u_2 be twice continuously differentiable solutions of (5.1) that are specified over disjoint domains D_1 and D_2 whose boundaries have a smooth arc C in common. If u_1 and u_2 have identical Cauchy data along C, prove by means of (5.28) that they fit together to define a single analytic solution u of (5.1) throughout the union of D_1, D_2 and C.

6. Interpret

$$(5.33) \qquad u_{zz} + u_{rr} + \frac{1}{r} u_r = 0$$

as Laplace's equation in cylindrical coordinates for a harmonic function u in three-dimensional space that possesses axial symmetry. Verify that the complete elliptic integral

$$(5.34) \qquad S(z, r; \zeta, \rho) = \int_0^\pi \frac{\rho \, d\theta}{\sqrt{(z - \zeta)^2 + (r - \rho)^2 + 4r\rho \sin^2 \theta}},$$

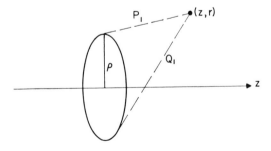

FIG. 12. Gravitational potential of a circular wire.

which represents the gravitational potential of a homogeneous circular wire of radius ρ, is a fundamental solution of (5.33). Set

$$P_1 = \sqrt{(z - \zeta)^2 + (r - \rho)^2}, \qquad Q_1 = \sqrt{(z - \zeta)^2 + (r + \rho)^2},$$

$$P_{n+1} = \sqrt{P_n Q_n}, \qquad Q_{n+1} = \frac{P_n + Q_n}{2}, \qquad n \geq 1,$$

and show that[1]

$$P_n < P_{n+1} < Q_{n+1} < Q_n,$$

$$\int_0^\pi \frac{d\theta}{\sqrt{P_{n+1}^2 \cos^2 \theta + Q_{n+1}^2 \sin^2 \theta}} = \int_0^\pi \frac{d\theta}{\sqrt{P_n^2 \cos^2 \theta + Q_n^2 \sin^2 \theta}},$$

$$S(z, r; \zeta, \rho) = \lim_{n \to \infty} \frac{\pi \rho}{P_n} = \lim_{n \to \infty} \frac{\pi \rho}{Q_n}.$$

7. Recast (5.34) in the form

$$S(z, r; \zeta, \rho) = \frac{\rho}{Q_1} \int_{-\infty}^\infty \frac{dt}{\sqrt{(1 + t^2)(k^2 + t^2)}},$$

where

$$k = \frac{P_1}{Q_1} = \sqrt{\frac{(z - \zeta)^2 + (r - \rho)^2}{(z - \zeta)^2 + (r + \rho)^2}}.$$

Establish that the Riemann function $A = A(z, r; \zeta, \rho)$ for (5.33) can be identified, except for a factor of $-2\pi i$, as the period of the term on the right corresponding to an increment of 2π in the argument of the parameter k. Show in this way that

$$(5.35) \quad A(z, r; \zeta, \rho) = \frac{4\rho}{\pi Q_1} \int_0^1 \frac{dt}{\sqrt{(1 - t^2)(1 - k^2 t^2)}}$$

$$= \frac{2\rho}{\sqrt{(z - \zeta)^2 + (r + \rho)^2}} F\left(\frac{1}{2}, \frac{1}{2}; 1; \frac{(z - \zeta)^2 + (r - \rho)^2}{(z - \zeta)^2 + (r + \rho)^2}\right),$$

where

$$F(a, b; c; \sigma) = 1 + \frac{ab}{c} \frac{\sigma}{1!} + \frac{a(a + 1)b(b + 1)}{c(c + 1)} \frac{\sigma^2}{2!} + \cdots$$

is the hypergeometric series defined by the ordinary differential equation[2]

$$\sigma(1 - \sigma) \frac{d^2 F}{d\sigma^2} + [c - (a + b + 1)\sigma] \frac{dF}{d\sigma} - abF = 0.$$

8. Apply a complex substitution to the result (5.35) in order to determine the Riemann function of the hyperbolic equation

$$u_{xx} - u_{tt} + \frac{1}{x} u_x = 0.$$

[1] Cf. Kellogg 1.
[2] Cf. Whittaker-Watson 1.

9. Establish that the Riemann function for equation (4.87) has the form

$$A(\xi, \tau; x, t) = \frac{(2\xi)^\lambda}{[(\xi + x)^2 - (\tau - t)^2]^{\lambda/2}} F\left(\frac{\lambda}{2}, \frac{\lambda}{2}; 1; \frac{(\xi - x)^2 - (\tau - t)^2}{(\xi + x)^2 - (\tau - t)^2}\right)$$

by seeking solutions of the related self-adjoint equation (4.91) which depend only on the quantity

$$\sigma = \frac{(\xi - x)^2 - (\tau - t)^2}{(\xi + x)^2 - (\tau - t)^2}.$$

Show in a similar way that the equation

(5.36) $$u_{\xi\eta} + \frac{\lambda}{2(\xi + \eta)}(u_\xi + u_\eta) = 0$$

has the Riemann function

(5.37) $$A(x, y; \xi, \eta) = \frac{(x + y)^\lambda}{(x + \eta)^{\lambda/2}(y + \xi)^{\lambda/2}} F\left(\frac{\lambda}{2}, \frac{\lambda}{2}; 1; \frac{(x - \xi)(y - \eta)}{(x + \eta)(y + \xi)}\right).$$

10. Find a fundamental solution of the elliptic equation

$$u_{zz} + u_{rr} + \frac{\lambda}{r} u_r = 0.$$

Discuss the limiting case in which λ and r both become infinite while their ratio remains bounded.

11. If the transformation of coordinates

$$\xi = \xi(x, y), \qquad \eta = \eta(x, y)$$

brings a linear elliptic equation

(5.38) $$au_{xx} + 2bu_{xy} + cu_{yy} + du_x + eu_y + fu = 0$$

with analytic coefficients a, b, c, d, e and f into canonical form, prove that the quantity

$$\Gamma = (\xi - \xi_0)^2 + (\eta - \eta_0)^2$$

satisfies the first order partial differential equation

$$a\Gamma_x^2 + 2b\Gamma_x\Gamma_y + c\Gamma_y^2 = 4\sqrt{ac - b^2}\,\frac{\partial(\xi, \eta)}{\partial(x, y)}.$$

Show that fundamental solutions of (5.38) can be found in the form

$$S(x, y; x_0, y_0) = A(x, y; x_0, y_0) \log \Gamma + B(x, y; x_0, y_0)$$

with coefficients $A \neq 0$ and B that are regular at the point (x_0, y_0), where it is understood that

$$\xi_0 = \xi(x_0, y_0), \qquad \eta_0 = \eta(x_0, y_0).$$

Use this result to establish that all solutions of (5.38) are real analytic functions of the variables x and y.

12. Find a fundamental solution of the most general elliptic equation (5.38) with constant coefficients (cf. Exercise 3.2.7).

13. In the special case $a \equiv b \equiv 0$ of a self-adjoint equation (5.1), continue the identity (5.25) analytically into the complex domain

$$\xi = \xi_1 + i\xi_2, \qquad \eta = \eta_1 + i\eta_2$$

and draw the contour of integration C down on a slit from $x = \xi_1 + \eta_2$, $y = \eta_1 - \xi_2$ to $x = \xi_1 - \eta_2$, $y = \eta_1 + \xi_2$ in order to derive the formula

$$v(\xi_1 + i\xi_2, \eta_1 + i\eta_2) = \frac{1}{2} v(\xi_1 + \eta_2, \eta_1 - \xi_2) + \frac{1}{2} v(\xi_1 - \eta_2, \eta_1 + \xi_2)$$

$$+ \frac{1}{2i} \int_{\xi_1 + \eta_2, \eta_1 - \xi_2}^{\xi_1 - \eta_2, \eta_1 + \xi_2} \{v(x, y)[A_x(x, y; \xi, \eta)\, dy - A_y(x, y; \xi, \eta)\, dx]$$
$$+ A(x, y; \xi, \eta)[v_y(x, y)\, dx - v_x(x, y)\, dy]\},$$

related to (4.73), for the analytic extension of v into the complex domain (cf. Section 16.3).

14. Work out the first term of an asymptotic expansion for the Riemann function

$$A(\xi, \tau; x, t) = \exp\left(-\frac{t - \tau}{2\epsilon^2}\right) J_0\left(\frac{1}{2\epsilon^2} \sqrt{\epsilon^2(x - \xi)^2 - (t - \tau)^2}\right)$$

of the hyperbolic equation

$$(5.39) \qquad\qquad u_{xx} - \epsilon^2 u_{tt} - u_t = 0$$

in the limit as $\epsilon \to 0$. Derive from it the expression

$$T(x, t; \xi, \tau) = \lim_{\epsilon \to 0} \frac{1}{2\epsilon} A(\xi, \tau; x, t)$$

$$= \frac{1}{2\sqrt{\pi(t - \tau)}} \exp\left\{-\frac{(x - \xi)^2}{4(t - \tau)}\right\}$$

for the fundamental solution of the heat equation

$$(5.40) \qquad\qquad u_{xx} = u_t,$$

to which (5.39) reduces when $\epsilon = 0$. Use Green's theorem to show that the integral

$$u(x, t) = \int_{-\infty}^{\infty} f(\xi) T(x, t; \xi, 0)\, d\xi$$

represents for $t > 0$ the unique solution of the parabolic equation (5.40) assuming the initial values

$$u(x, 0) = f(x)$$

along the characteristic line $t = 0$. Interpret the characteristic initial value problem for (5.39) as a singular perturbation of this result.

2. SEVERAL INDEPENDENT VARIABLES

We shall discuss the determination of fundamental solutions of linear elliptic and hyperbolic partial differential equations of the second order in more than two independent variables. We confine our attention at first to the case of analytic coefficients in order to explore a procedure that is applicable in the complex domain. We are thus able to avoid for the moment the issue of whether the equation to be treated is of the elliptic or the hyperbolic type. Our analysis will involve a rather complicated construction,[3] and the reader who prefers to skip this material and pass directly to the study in Chapter 7 of boundary value problems appropriate for an elliptic equation will find it is possible to do so without difficulty; what is to follow here is not actually a prerequisite for understanding that later topic.

It is important to establish the existence of a fundamental solution in space of arbitrary dimension because it occurs in integral formulas for other solutions analogous to the ones described in Section 1 for the plane case. Moreover, the fundamental solution will be needed in Chapter 6 to develop an explicit representation of the solution of Cauchy's problem for linear hyperbolic equations in more than two independent variables. In this context it plays a role similar to that of the Riemann function in two independent variables (cf. Section 4.4). The construction of the fundamental solution that will be presented here is based on a power series expansion related to those used in Section 1.2 to prove the Cauchy-Kowalewski theorem.

We set

$$L[u] = \sum_{i,j=1}^{n} a_{ij} \frac{\partial^2 u}{\partial x_i \, \partial x_j} + \sum_{i=1}^{n} b_i \frac{\partial u}{\partial x_i} + cu,$$

and we consider the linear partial differential equation

(5.41) $$L[u] = 0,$$

where the coefficients a_{ij}, b_i and c are supposed to be real analytic functions of the real variables $x_1, \ldots, x_n$, and where the matrix of elements a_{ij} is assumed to be symmetric. In order to describe what a fundamental solution S of (5.41) should look like, we must refer back to Section 2.6 and recall our introduction there of a quantity Γ defined to be the square of the *geodesic distance* I between two points with respect to the metric given by (2.85). We shall take the two points in question to have the coordinates $\xi_1, \ldots, \xi_n$ and $x_1, \ldots, x_n$ so that we can rewrite the first

[3] Cf. Hadamard 3.

order partial differential equation (2.94) for Γ in the form

$$(5.42) \qquad \sum_{i,j=1}^{n} a_{ij} \frac{\partial \Gamma}{\partial x_i} \frac{\partial \Gamma}{\partial x_j} = 4\Gamma,$$

where the coefficients a_{ij} are the same as those appearing in the definition of L. It is important to emphasize that Γ is a symmetric function of $x = (x_1, \ldots, x_n)$ and $\xi = (\xi_1, \ldots, \xi_n)$.

A *fundamental solution* $S = S(x, \xi)$ of (5.41) is defined to be a solution of that equation in its dependence on x possessing at the parameter point ξ a singularity characterized by the representation

$$(5.43) \qquad S = \frac{U}{\Gamma^m} + V \log \Gamma + W,$$

where U, V and W are supposed to be regular functions of x in a neighborhood of ξ, with $U \neq 0$ at ξ, and where the exponent m should have the specific value

$$m = \frac{n-2}{2}.$$

The most elementary example is the fundamental solution

$$S = \frac{1}{r^{n-2}}, \qquad r = \sqrt{(x_1 - \xi_1)^2 + \cdots + (x_n - \xi_n)^2},$$

of Laplace's equation

$$\Delta u = \frac{\partial^2 u}{\partial x_1^2} + \cdots + \frac{\partial^2 u}{\partial x_n^2} = 0,$$

which is a function of the Euclidean distance r.

Further motivation for the hypothesis (5.43) about the nature of S is to be found in Exercise 3.2.7, which concerned the case of an equation (5.41) with constant coefficients. As we shall see shortly, when n is even the term W is redundant, whereas, on the other hand, the coefficient V of the logarithm in (5.43) vanishes for n odd. Distinctions of this kind between space of an even number of dimensions and space of an odd number of dimensions play a significant role in the theory of the fundamental solution.

We begin with the odd choices of n. Let us put $V = W = 0$ and seek U as a convergent series

$$U = U_0 + U_1 \Gamma + U_2 \Gamma^2 + \cdots$$

that advances in integral powers of Γ, with regular coefficients U_l. A direct calculation using the symmetry of a_{ij} shows that

$$(5.44) \quad L[U_l\Gamma^{l-m}] = (l-m)(l-m-1)U_l\Gamma^{l-m-2}\sum_{i,j=1}^{n}a_{ij}\frac{\partial\Gamma}{\partial x_i}\frac{\partial\Gamma}{\partial x_j}$$

$$+ (l-m)\left\{2\sum_{i,j=1}^{n}a_{ij}\frac{\partial U_l}{\partial x_i}\frac{\partial\Gamma}{\partial x_j} + 4CU_l\right\}\Gamma^{l-m-1} + L[U_l]\Gamma^{l-m},$$

where

$$(5.45) \qquad C = \frac{1}{4}\sum_{i,j=1}^{n}a_{ij}\frac{\partial^2\Gamma}{\partial x_i\,\partial x_j} + \frac{1}{4}\sum_{i=1}^{n}b_i\frac{\partial\Gamma}{\partial x_i}.$$

The possibility of eliminating the lowest power of Γ on the right in (5.44) by means of the first order partial differential equation (5.42) now explains why the fundamental solution S ought to be expanded in terms of this particular function.

It is convenient to introduce a parameter s which indicates the direction of, and which is measured along, the geodesics (2.95) that served to generate Γ. We can then use the relation of Exercise 2.6.3 in order to establish that

$$\sum_{i,j=1}^{n}a_{ij}\frac{\partial U_l}{\partial x_i}\frac{\partial\Gamma}{\partial x_j} = 2s\sum_{i=1}^{n}\frac{\partial U_l}{\partial x_i}\frac{dx_i}{ds} = 2s\frac{dU_l}{ds}.$$

These simplifications help us to transform (5.44) into

$$(5.46)$$
$$L[U_l\Gamma^{l-m}] = 4(l-m)\left\{s\frac{dU_l}{ds} + (C+l-m-1)U_l\right\}\Gamma^{l-m-1} + L[U_l]\Gamma^{l-m}.$$

Thus if we work out formally the coefficients of the various powers of Γ in the identity

$$L[S] = L\left[\sum_{l=0}^{\infty}U_l\Gamma^{l-m}\right] = 0$$

and set each of them equal to zero, we obtain the sequence of recursion formulas

$$(5.47) \qquad s\frac{dU_0}{ds} + (C-m-1)U_0 = 0,$$

$$(5.48) \quad s\frac{dU_l}{ds} + (C+l-m-1)U_l = -\frac{1}{4(l-m)}L[U_{l-1}], \qquad l\geq 1,$$

for the determination of U_0, U_1, U_2, ... For odd n the division by $(l-m)$ is always feasible.

Before we go on to compute the functions U_l, a few words are in order about the significance of our procedure. When (5.41) is of the hyperbolic type, the fundamental solution (5.43) becomes infinite along a conoid $\Gamma = 0$ of two sheets separating n-dimensional space into three parts. This conoid is in fact a *characteristic surface* for the second order equation (5.41), since (5.42) reduces on the level surface $\Gamma = 0$ to the first order partial differential equation (3.71) for such a characteristic. The basic principle involved here is that any locus of singularities of a solution of a linear hyperbolic equation can be expected to form a characteristic surface.

The geodesics (2.95) that lie on the conoid $\Gamma = 0$, which is a special case of the Monge cone described in Section 2.2, are themselves characteristics of the first order equation (5.42), or (3.71). We therefore refer to them as the *bicharacteristics* of the original equation (5.41). We see that along the characteristic conoid $\Gamma = 0$ the ordinary differential operators occurring on the left in the transport equations (5.47) and (5.48) apply in the directions of the bicharacteristics. This phenomenon is related to the observation that within any of its characteristic surfaces (5.41) reduces to an ordinary differential equation imposed on the Cauchy data along each bicharacteristic. Indeed, if we put $m = 0$ and define U_0 and U_1 so that they represent at each point of the characteristic $\Gamma = 0$ the value and the normal derivative, respectively, of a regular solution of (5.41), then (5.48) becomes, for $l = 1$, the ordinary differential equation in question.

Our remarks apply to the elliptic case, too, when we use the analyticity of the coefficients a_{ij}, b_i and c to extend our construction of a fundamental solution to complex values of independent variables $x_1, \ldots, x_n$. The first order equation (5.42) and the associated geodesics have a quite definite meaning in the complex domain, provided that all the differentiations we have had occasion to use are interpreted in the sense of the theory of functions of a complex variable. Thus it is permissible to operate with a fundamental solution throughout the complex as well as the real domain of regularity of the coefficients. It should be mentioned in this connection that not all the geodesics needed to generate the function Γ are bicharacteristics of (5.41) but only those which actually lie on a characteristic surface.

Returning to the calculation of the functions $U_0, U_1, U_2, \ldots$, we find it convenient to work in a new space with coordinates $\theta_1, \ldots, \theta_n$ defined by (2.97) rather than to persist in using the independent variables $x_1, \ldots, x_n$. We are able to do this in a sufficiently small neighborhood of the parameter point $\xi = (\xi_1, \ldots, \xi_n)$ because the relevant Jacobian does not vanish. The advantage of the new space is that the geodesics (2.95) become rays there emanating from the origin, while the parameter s can be chosen to coincide with the distance from the origin along each such ray, as is suggested by (2.97). With these ideas in mind, we shall write each coefficient U_l from the expansion of U in the form of an infinite series

$$(5.49) \qquad\qquad U_l = \sum_{j=0}^{\infty} P_{lj}$$

of polynomials P_{lj} homogeneous in the coordinates $\theta_1, \ldots, \theta_n$ of a degree equal to the index j.

Observe that the differential operator $s(d/ds)$ appearing on the left in (5.47) and (5.48) will not alter the degree of any of the polynomials P_{lj}, with the single exception that it reduces a polynomial of degree zero, or, in other words, a constant, to zero. It follows that unless the coefficient $C - m - 1$ vanishes for $\theta_1 = \cdots = \theta_n = 0$ there does not exist a solution U_0 of (5.47) satisfying the requirement $P_{00} \neq 0$, which must be imposed to obtain a true fundamental solution. However, our choice of the exponent $m = (n - 2)/2$ has been made precisely so that this will be the case, as can be verified by using the Taylor series (2.98) for Γ and the definition (5.45) of C to show that $C = n/2$ at the parameter point $x = \xi$. Thus we can integrate (5.47) explicitly to obtain

$$(5.50) \qquad\qquad U_0 = P_{00} \exp \left\{ -\int_0^s (C - m - 1)\frac{ds}{s} \right\},$$

where P_{00} is a constant that might conceivably vary with ξ [cf. formula (5.71)].

A similar integration of (5.48) based on an integrating factor leads to the recursion formula

$$(5.51) \qquad U_l = -\frac{U_0}{4(l - m)s^l} \int_0^s \frac{L[U_{l-1}]s^{l-1}}{U_0} \, ds, \qquad l \geq 1.$$

The linear operator on the right transforms any convergent series U_{l-1} of the type (5.49) into another series of the same kind for U_l. We have still to show that when these results are substituted back into the expansion of U in powers of Γ, it converges uniformly for small enough choices of s. That can be accomplished by a *method of majorants* analogous to the Cauchy-Kowalewski technique of Section 1.2. However, it will suffice for the purpose of the convergence proof to treat only the special case $U_0 \equiv$ const., since the substitution $u_1 = u/U_0$, with U_0 given by (5.50), reduces (5.41) to a new partial differential equation defined by

$$L_1[u_1] = L[U_0 u_1] = 0$$

for which such an assumption is valid.

We choose positive numbers K and ϵ such that the geometric series

$$\frac{K\epsilon}{\epsilon - |\theta_1| - \cdots - |\theta_n|} = \sum_{j=0}^{\infty} \frac{K}{\epsilon^j}(|\theta_1| + \cdots + |\theta_n|)^j$$

is a majorant for the Taylor expansions in powers of $\theta_1, \ldots, \theta_n$ of all the coefficients of L, which is now understood to have been expressed as a differential operator with respect to these coordinates. It follows that if

$$(5.52) \qquad M\{U_l\} = \frac{M_l}{\left(1 - \dfrac{|\theta_1| + \cdots + |\theta_n|}{\epsilon}\right)^{2l}}$$

denotes a majorant for U_l, with M_l taken as a suitably large constant, then

$$(5.53) \qquad M\{L[U_l]\} = \frac{2l(2l+1)[1 + n/\epsilon + n^2/\epsilon^2]KM_l}{\left(1 - \dfrac{|\theta_1| + \cdots + |\theta_n|}{\epsilon}\right)^{2l+3}}$$

majorizes $L[U_l]$. Here we use the notation $M\{U\}$ to indicate a majorant of the function U. We proceed to apply the recursion formula (5.51) to (5.53) in order to establish that with l replaced by $l + 1$ and with

$$(5.54) \qquad M_{l+1} = \frac{l(2l+1)}{2(l+1)(l-m+1)}\left[1 + \frac{n}{\epsilon} + \frac{n^2}{\epsilon^2}\right]KM_l$$

the rule (5.52) also defines a majorant for U_{l+1}.

Since we have been able to confine our attention to the case $U_0 \equiv \mathrm{const.}$, the proof of the result just stated comes down to a verification that

$$M\left\{s^{-l-1}\int_0^s \frac{s^l \, ds}{(1 - \gamma s)^{2l+3}}\right\} = \frac{1}{l+1}\frac{1}{(1 - \gamma s)^{2l+2}}$$

is a majorant for the integral indicated inside the braces on the left. This can be deduced by using the convenient choice

$$M\left\{\frac{s^l}{(1 - \gamma s)^{2l+3}}\right\} = [1 + \gamma s]\frac{s^l}{(1 - \gamma s)^{2l+3}}$$

$$= \frac{1}{l+1}\frac{d}{ds}\frac{s^{l+1}}{(1 - \gamma s)^{2l+2}}$$

of a majorant for the integrand. Here γ is specified by the requirement

$$|\theta_1| + \cdots + |\theta_n| = \epsilon\gamma s.$$

The principle of mathematical induction permits us to conclude that the majorants (5.52) are valid for all $l \geq 1$, provided that M_1 is sufficiently large and that $M_2, M_3, \ldots$ are given by (5.54). Therefore the series for U in powers of Γ converges throughout a neighborhood

$$(5.55) \qquad |\Gamma| < \frac{\left(1 - \dfrac{|\theta_1| + \cdots + |\theta_n|}{\epsilon}\right)^2}{\left[1 + \dfrac{n}{\epsilon} + \dfrac{n^2}{\epsilon^2}\right] K}$$

of the parameter point $\xi = (\xi_1, \ldots, \xi_n)$.

Locally we now obtain a fundamental solution S of (5.41) in the special form

$$(5.56) \qquad S = \frac{U}{\Gamma^m} = \sum_{l=0}^{\infty} U_l \Gamma^{l-m}$$

when the number n of independent variables is *odd*. The addition of a regular term W is optional. If n is *even*, on the other hand, so that the exponent $m = (n - 2)/2$ is a positive integer, the construction of U that we have presented breaks down, for the recursion formula (5.51) for the coefficient U_m involves a division by zero and therefore has no meaning. Thus only the first m functions $U_0, U_1, \ldots, U_{m-1}$ can be defined by means of (5.50) and (5.51). It is for this reason that a logarithmic term is needed in the representation (5.43) for the fundamental solution in space of an even number of dimensions. We are led to put

$$(5.57) \qquad S = \sum_{l=0}^{m-1} U_l \Gamma^{l-m} + V \log \Gamma + W$$

for even values of n and to seek conditions determining the factor V. Insertion of (5.57) into the partial differential equation (5.41) gives

$$\sum_{l=0}^{m-1} L[U_l \Gamma^{l-m}] + L[V \log \Gamma] + L[W] = L[U_{m-1}] \frac{1}{\Gamma} - \frac{V}{\Gamma^2} \sum_{i,j=1}^{n} a_{ij} \frac{\partial \Gamma}{\partial x_i} \frac{\partial \Gamma}{\partial x_j}$$

$$+ \left\{ 2 \sum_{i,j=1}^{n} a_{ij} \frac{\partial V}{\partial x_i} \frac{\partial \Gamma}{\partial x_j} + 4CV \right\} \frac{1}{\Gamma} + L[V] \log \Gamma + L[W] = 0,$$

by virtue of (5.46), (5.47) and (5.48). Hence

$$(5.58)$$

$$\left\{ L[U_{m-1}] + 4\left[s \frac{dV}{ds} + (C - 1)V \right] \right\} \frac{1}{\Gamma} + L[V] \log \Gamma + L[W] = 0,$$

according to (5.42). We shall prove that the identity (5.58) determines V uniquely but that W can be selected in a variety of ways to meet the requirements imposed on it.

Since the logarithmic term must cancel out altogether in (5.58), we have to ask that

$$(5.59) \qquad L[V] = 0,$$

just as in our discussion of (5.8). Once this has been granted, we see that the coefficient of $1/\Gamma$ on the left in (5.58) ought to vanish all along the characteristic conoid $\Gamma = 0$, since the remaining regular term $L[W]$ cannot become infinite there. If follows that V should fulfill the ordinary differential equation

$$(5.60) \qquad s\frac{dV}{ds} + (C - 1)V = -\frac{1}{4}L[U_{m-1}]$$

on each bicharacteristic generating the conoid.

Our previous treatment of (5.47) and (5.48) shows that (5.60) determines V uniquely on the characteristic surface $\Gamma = 0$ and that V must in fact coincide there with the function V_0 defined in a full neighborhood of the parameter point ξ by the integral

$$(5.61) \qquad V_0 = -\frac{U_0}{4s^m}\int_0^s \frac{L[U_{m-1}]s^{m-1}}{U_0}\,ds.$$

Thus we have a *characteristic initial value problem* for the partial differential equation (5.59) in which the unknown V is prescribed on the conoid $\Gamma = 0$. This interpretation of the situation is reminiscent of the conditions (5.9), (5.12) and (5.13) determining the Riemann function A in the complex domain.

From another point of view, we can think of (5.61) as a substitute for the recursion formula (5.51) in the degenerate case $l = m$. This suggests that we attempt to find V as a convergent power series

$$(5.62) \qquad V = \sum_{l=0}^{\infty} V_l\Gamma^l.$$

We insert (5.62) into (5.59), pick out and equate to zero the coefficient of each power of Γ in the result, and integrate the ordinary differential equations analogous to (5.48) so obtained to establish the recursion formula

$$(5.63) \qquad V_l = -\frac{U_0}{4ls^{l+m}}\int_0^s \frac{L[V_{l-1}]s^{l+m-1}}{U_0}\,ds,$$

which is valid for $l \geq 1$. Here the first term V_0 is still to be found from (5.61). The method of majorants can be used to deduce estimates like (5.52) for the functions V_l given by (5.63). Therefore the series (5.62) for V converges uniformly in a region (5.55) surrounding the parameter point ξ. The details of the proof of these remarks follow so closely our treatment of the representation (5.56) of the fundamental solution in space of odd dimension that we do not go into the matter further in the present case.

We have now shown how V can be constructed in such a way that (5.58) becomes a partial differential equation for W with an inhomogeneous term that is regular in the neighborhood of ξ. This determines W only up to the addition of an arbitrary solution of the homogeneous equation (5.41). A special choice for W that is in the spirit of the procedure we have adopted thus far is obtained by writing

$$W = \sum_{l=1}^{\infty} W_l \Gamma^l$$

and exploiting a recursion formula like (5.63) to find the coefficients $W_1, W_2, \ldots$. By insisting that the series for W does not include a term W_0 corresponding to the value $l = 0$ of the index of summation, we achieve a unique determination of the fundamental solution (5.57) such that the factors $U_0, \ldots, U_{m-1}$, V and W are all regular in their dependence on the parameter point ξ. It is to be understood in this connection that the value of U_0 at ξ should be normalized in a natural way, for example, according to a rule (5.71) that we shall describe momentarily in connection with some applications.

Before discussing how the fundamental solution can be used in practice, we should emphasize that the method by which we have constructed it is valid in the domain of complex values of the independent variables $x_1, \ldots, x_n$ as well as in the strictly real domain. If we confine ourselves to the real case, then for a hyperbolic equation the bicharacteristics and the characteristic conoid $\Gamma = 0$ along which S becomes infinite appear to play a central role, whereas for an elliptic equation the geodesics seem a more natural concept and the interpretation of Γ in terms of geodesic distance from an isolated singularity at ξ is more relevant. In both situations a significant limitation of our approach is that it only yields a fundamental solution in some sufficiently small neighborhood of the prescribed singularity at ξ. In fact, for variable coefficients a_{ij} even the basic function Γ satisfying the nonlinear first order equation (5.42) is in general defined only in the small.

For a linear equation of the elliptic type it turns out that fundamental solutions can actually be found in the large, as will become apparent from our treatment in Chapters 8 and 9 of deeper existence theorems connected with boundary value problems. It can be seen from Exercise 1.11 that this question is related to the problem raised in Section 3.1 of reducing an elliptic equation in two independent variables to canonical form in the large. Furthermore, when such an equation presents itself in canonical form to begin with, our procedure does furnish a global fundamental solution. Finally, we call the reader's attention to the elegant method of spherical means developed by John (cf. John 1) which defines fundamental solutions for a wider class of linear elliptic equations with analytic coefficients.

We wish next to generalize the important integral formula (5.25) representing arbitrary solutions in terms of a fundamental solution to the case of an elliptic equation of the second order in any number of independent variables. For this purpose we have to associate with the linear equation (5.41) its *adjoint*

$$(5.64) \qquad M[v] = \sum_{i,j=1}^{n} \frac{\partial^2(a_{ij}v)}{\partial x_i \partial x_j} - \sum_{i=1}^{n} \frac{\partial(b_i v)}{\partial x_i} + cv$$

$$= \sum_{i,j=1}^{n} a_{ij} \frac{\partial^2 v}{\partial x_i \partial x_j} - \sum_{i=1}^{n} \left[b_i - 2 \sum_{j=1}^{n} \frac{\partial a_{ij}}{\partial x_j} \right] \frac{\partial v}{\partial x_i}$$

$$+ \left[c - \sum_{i=1}^{n} \frac{\partial b_i}{\partial x_i} + \sum_{i,j=1}^{n} \frac{\partial^2 a_{ij}}{\partial x_i \partial x_j} \right] v = 0.$$

The expression

$$vL[u] - uM[v] = \sum_{i=1}^{n} \frac{\partial}{\partial x_i} \left\{ \left[b_i - \sum_{j=1}^{n} \frac{\partial a_{ij}}{\partial x_j} \right] uv + \sum_{j=1}^{n} a_{ij} \left[v \frac{\partial u}{\partial x_j} - u \frac{\partial v}{\partial x_j} \right] \right\}$$

is, of course, a divergence. Therefore the divergence theorem[4]

$$\int_D \left[\sum_{i=1}^{n} \frac{\partial F_i}{\partial x_i} \right] dx_1 \cdots dx_n = \int_{\partial D} \left[\sum_{i=1}^{n} (-1)^{i-1} F_i \, dx_1 \cdots \widehat{dx_i} \cdots dx_n \right]$$

yields the fundamental identity

$$(5.65) \qquad \int_D (vL[u] - uM[v]) \, dx + \int_{\partial D} B[u, v] = 0,$$

where $dx = dx_1 \cdots dx_n$, where

$$(5.66) \qquad B[u, v] = \sum_{i=1}^{n} (-1)^i \left\{ \left[b_i - \sum_{j=1}^{n} \frac{\partial a_{ij}}{\partial x_j} \right] uv \right.$$

$$\left. + \sum_{j=1}^{n} a_{ij} \left[v \frac{\partial u}{\partial x_j} - u \frac{\partial v}{\partial x_j} \right] \right\} dx_1 \cdots \widehat{dx_i} \cdots dx_n,$$

[4] Cf. Buck 1.

and where D is any region of n-dimensional space with a sufficiently smooth set of boundary surfaces denoted by ∂D.

In specifying the signs of the terms in the surface integrals here we have adopted the conventions of the calculus of exterior differential forms (cf. Buck 1). Thus odd permutations of the factors in a product $dx_j \cdots dx_k$ of differentials dx_i are understood to alter the sign in front of the surface element involved, whereas even permutations leave that sign invariant. Furthermore, $dx_1 \cdots \widehat{dx_i} \cdots dx_n$ indicates the product $dx_1 \cdots dx_n$ with the factor dx_i deleted. We shall find that these simple notations prove to be quite convenient for applications of the divergence theorem or of Stokes' theorem to the theory of partial differential equations.

We are interested in using (5.65) to develop an integral representation for solutions v of an equation (5.64) of the elliptic type in terms of a fundamental solution S of the adjoint equation (5.41). For this purpose we suppose that D is a region in which S is regular everywhere except for its prescribed singularity at ξ, and we remove from D a small neighborhood of the point ξ defined by the inequality $\Gamma \leq \epsilon^2$. If D is replaced by such a truncated domain in (5.65) and S is substituted for u, while v is taken to be a solution of (5.64), then the volume integral on the left vanishes and we obtain the result

$$(5.67) \qquad \int_{\Gamma=\epsilon^2} B[S(x, \xi), v(x)] = \int_{\partial D} B[S(x, \xi), v(x)]$$

comparable to (5.23). We shall establish that when the value at ξ of the coefficient U_0 of the leading term in our expansion (5.56) or (5.57) of the fundamental solution S is normalized appropriately, we have

$$(5.68) \qquad \lim_{\epsilon \to 0} \int_{\Gamma=\epsilon^2} B[S(x, \xi), v(x)] = v(\xi).$$

The desired representation

$$(5.69) \qquad v(\xi) = \int_{\partial D} B[S(x, \xi), v(x)]$$

for solutions v of (5.64) follows from (5.67) and the residue theorem (5.68).

It is a consequence of the Taylor expansion (2.98) that the $(n-1)$-dimensional area of the surface $\Gamma = \epsilon^2$ over which the integral (5.68) is evaluated approaches zero like ϵ^{n-1}. Therefore we may neglect terms in the integrand that become infinite more slowly than ϵ^{-n+1}. Thus in estimating $B[S, v]$ we need consider only those contributions that involve differentiating the denominator of the principal term U_0/Γ^m in the expansion of S. Hence we can write

$$(5.70) \quad \lim_{\epsilon \to 0} \int_{\Gamma = \epsilon^2} B[S, v]$$

$$= \lim_{\epsilon \to 0} \int_{\Gamma = \epsilon^2} \left[\sum_{i,j=1}^{n} (-1)^{i+1} m a_{ij} \frac{v U_0}{\Gamma^{m+1}} \frac{\partial \Gamma}{\partial x_j} dx_1 \cdots \widehat{dx_i} \cdots dx_n \right].$$

Furthermore, we can replace Γ by the quadratic terms

$$\Gamma_2 = \sum_{j,k=1}^{n} A_{jk}(x_j - \xi_j)(x_k - \xi_k)$$

in its Taylor series representation (2.98) and we can substitute for the variable argument of a_{ij}, v and U_0 the fixed limiting value ξ without invalidating (5.70). Therefore

$$\lim_{\epsilon \to 0} \int_{\Gamma = \epsilon^2} B[S, v]$$

$$= \lim_{\epsilon \to 0} \frac{2 m v(\xi) U_0(\xi)}{\epsilon^{2m+2}} \int_{\Gamma_2 = \epsilon^2} \left[\sum_{i,j,k=1}^{n} (-1)^{i+1} a_{ij} A_{jk}(x_k - \xi_k) dx_1 \cdots \widehat{dx_i} \cdots dx_n \right]$$

$$= \lim_{\epsilon \to 0} \frac{2 m v(\xi) U_0(\xi)}{\epsilon^{2m+2}} \int_{\Gamma_2 = \epsilon^2} \left[\sum_{i=1}^{n} (-1)^{i-1} (x_i - \xi_i) dx_1 \cdots \widehat{dx_i} \cdots dx_n \right],$$

since we have

$$\sum_{j=1}^{n} a_{ij} A_{jk} = \delta_{ik}.$$

The divergence theorem and a linear change of variables diagonalizing the quadratic form Γ_2 give

$$\int_{\Gamma_2 = \epsilon^2} \left[\sum_{i=1}^{n} (-1)^{i-1} (x_i - \xi_i) dx_1 \cdots \widehat{dx_i} \cdots dx_n \right]$$

$$= \int_{\Gamma_2 < \epsilon^2} n \, dx_1 \cdots dx_n = n \tau_n \epsilon^n \sqrt{\Delta},$$

where τ_n denotes the volume of the unit sphere in n-dimensional space and where

$$\Delta = \begin{vmatrix} a_{11} & \cdots & a_{1n} \\ a_{21} & \cdots & a_{2n} \\ \cdots\cdots\cdots\cdots \\ a_{n1} & \cdots & a_{nn} \end{vmatrix}.$$

Since $n = 2m + 2$, we conclude that

$$v(\xi) = \frac{1}{n(n-2)\tau_n\sqrt{\Delta}U_0(\xi)} \lim_{\epsilon \to 0} \int_{\Gamma=\epsilon^2} B[S, v].$$

Thus the desired result (5.68) follows if we impose the normalization

(5.71) $$U_0(\xi) = \frac{1}{n(n-2)\tau_n\sqrt{\Delta}}$$

on the fundamental solution S. It will be convenient to require that (5.71) be fulfilled, although it is sometimes preferable to omit the factor $\sigma_{n-1} = n\tau_n$, which represents the area of the $(n-1)$-dimensional surface of an n-dimensional unit sphere.

Observe that when U_0 is determined by (5.71) all the terms U_l in the power series expansion of U become regular in their dependence on the parameter point ξ because the coordinates θ_j are regular (cf. Section 2.6). Consequently U, V and W are analytic functions of ξ, too. Hence the fundamental solution S we have constructed is regular in its dependence on ξ for $\xi \neq x$.

A first consequence of the integral formula (5.69) in space of arbitrary dimension is the theorem that any solution v of a linear elliptic equation (5.64) with analytic coefficients must itself be an analytic function of its arguments. This follows from the regularity of the integrand $B[S, v]$ for choices of ξ inside the region D and of x on the boundary ∂D. Because our construction of the fundamental solution S is valid in the complex domain, formula (5.69) also provides a means for continuing solutions of (5.64) analytically into regions of complex values of the independent variables.

A modified version of (5.69) may be used to examine the relationship between a fundamental solution $S_L(x, \xi)$ of (5.41) and a fundamental solution $S_M(x, \eta)$ of the adjoint equation (5.64). Let us put $u = S_L$ and $v = S_M$ in (5.65) after first removing from D small geodesic spheres $\Gamma(x, \xi) \leq \epsilon^2$ and $\Gamma(x, \eta) \leq \epsilon^2$ around the singular points ξ and η, and let us then allow ϵ to approach zero. In analogy with (5.69) we obtain the identity

(5.72) $$S_M(\xi, \eta) - S_L(\eta, \xi) = \int_{\partial D} B[S_L(x, \xi), S_M(x, \eta)].$$

Consider a situation in which ξ and η are close, and recall that

$$\Gamma(\xi, \eta) = \Gamma(\eta, \xi)$$

by virtue of the interpretation of this quantity as the square of the geodesic distance between ξ and η. Since the integral on the right in (5.72) is regular for all ξ and η in D, we see that the singular parts of $S_M(\xi, \eta)$ and $S_L(\eta, \xi)$ must be identical. In particular, for even values of n the coefficient of the logarithmic term in the expansion (5.57) satisfies the rule

$$(5.73) \qquad\qquad V_L(\eta, \xi) = V_M(\xi, \eta)$$

analogous to (4.80) for interchanging arguments and parameters.

A stronger conclusion can be drawn from (5.72) for odd n because in that case the regular terms that might be added to a fundamental solution can be chosen in a natural way to fix it uniquely. We have in mind here taking the fundamental solutions S_L and S_M to have the special form (5.56), with every term involving the radical $\sqrt{\Gamma}$ in an essential way. Observe that the right-hand side of (5.72) is a regular function of ξ and of η in a portion of the complex domain containing the region D, whereas the left-hand side can be expressed as a regular function of the same kind multiplied by $\sqrt{\Gamma}$. We maintain that the latter function must vanish, for otherwise we could represent the square root $\sqrt{\Gamma}$ as a quotient of two single-valued functions in a complex domain including loci $\Gamma = 0$ around which it changes sign. Our conclusion is that the integral on the right in (5.72) is zero, too, and that the specific fundamental solution (5.56) has the symmetry property

$$(5.74) \qquad\qquad S_L(\eta, \xi) = S_M(\xi, \eta).$$

This exhibits it as a solution of the adjoint equation in its dependence on the parameter point. The derivation of both of the results (5.73) and (5.74) relies heavily on our normalization requirement (5.71).

In closing we stress that the representation (5.69) expresses the values of v inside the region D in terms of the fundamental solution S and the Cauchy data of v on the boundary surface ∂D. An analogous integral formula that serves to define solutions u of the inhomogeneous equation

$$L[u] = \rho$$

will be developed in Exercise 3.2.

EXERCISES

1. Use Exercise 3.2.7 to determine a fundamental solution of (5.41) explicitly for the case of constant coefficients a_{ij}, b_i and c.

2. Use infinite series of the type (5.62) to establish the existence of fundamental solutions (5.4) of the equation (5.1) in two independent variables.

3. Complete all details of the proof of convergence of the series (5.62) outlined in the text. Use such a series to establish the existence in the small of a solution of the *characteristic initial value problem* which asks for a solution V of (5.59) that coincides on a characteristic conoid $\Gamma = 0$ with the values of an arbitrary analytic function V_0 of the real variables $x_1, \ldots, x_n$.

4. Prove that solutions of (5.41) with singularities of higher order than that described by (5.43) with $2m = n - 2$ can be found by differentiating the fundamental solution S with respect to the parameters $\xi_1, \ldots, \xi_n$. For an elliptic equation, show that any singularity of an order lower than this must be *removable* if the solution is single-valued in the real domain.

5. Consider two solutions v_1 and v_2 of a linear elliptic equation (5.64) that are defined in disjoint domains whose boundaries possess a common section. Assume that on this section v_1 and v_2 have identical Cauchy data, that is, assume that their values and first derivatives are identical there. Use (5.69) to establish that in such a situation v_1 and v_2 combine to form a single solution of (5.64) which is regular throughout the union of the original domains of definition of v_1 and v_2 and, more specifically, on the common portion of their boundaries. Discuss the relationship between this result and the corresponding theorem about *unique continuation* of analytic functions of a complex variable.

6. Let ϕ be an analytic solution of the first order partial differential equation

$$(5.75) \qquad \sum_{i,j=1}^{n} a_{ij} \frac{\partial \phi}{\partial x_i} \frac{\partial \phi}{\partial x_j} = 0$$

defining the characteristics of (5.41). Consider a mixed problem for (5.41) in which the values of the solution u are assigned analytically and consistently on the characteristic surface $\phi = 0$ and on some other analytic surface crossing it. Show how to represent the answer as a power series[5]

$$(5.76) \qquad u = \sum_{l=0}^{\infty} U_l \phi^l$$

that converges near the intersection of the two surfaces.

7. Show that the bicharacteristics of the wave equation may be interpreted as light rays and determine them explicitly. Explain why the envelope of a family of bicharacteristics may be viewed as a focal curve, or caustic, in the sense of Chapter 2.

8. For large values of the frequency parameter k consider an asymptotic solution

$$(5.77) \qquad u = e^{ik\phi} \sum_{l=0}^{\infty} \frac{l! \, U_l}{(ik)^l}$$

of the wave equation, where ϕ stands for a function satisfying the corresponding eiconal equation (5.75) of geometrical optics (cf. Section 2.4). Show that for (5.77) to be a valid representation of u asymptotically as $k \to \infty$, the amplitudes

[5] Cf. Hadamard 3.

U_l must fulfill a sequence of transport equations[6] similar to (5.48) along the bicharacteristics generating ϕ.

9. Develop an analogy between the power series expansion (5.76) and the asymptotic expansion (5.77) of solutions of the wave equation that is based on the formal identity

$$\frac{1}{2\pi i} \oint e^{ik\phi} \sum_{l=0}^{\infty} \frac{l! \, U_l}{(ik)^l} \frac{dk}{k} = \sum_{l=0}^{\infty} U_l \phi^l,$$

where the contour integration is performed over a large circle in the complex k-plane.

10. Represent the bicharacteristics of the hyperbolic equation

$$(5.78) \qquad (1 + f_x^2 + f_y^2)u_{tt} = (1 + f_y^2)u_{xx} - 2f_x f_y u_{xy} + (1 + f_x^2)u_{yy}$$

in terms of geodesics on the surface $z = f(x, y)$, and determine the function f so that some pair of these bicharacteristics has more than one point of intersection. Discuss the global behavior of fundamental solutions of the equation (5.78) corresponding to such a choice of f.

11. Verify that our construction of a fundamental solution of (5.41) is applicable in the large when the a_{ij} are constants and the b_i and c are entire.

12. Suppose that L is an elliptic differential operator with constant coefficients, and write

$$L(\xi) = \sum_{j,k=1}^{n} a_{jk}\xi_j\xi_k + \sum_{j=1}^{n} b_j\xi_j + c,$$

$$x\xi = \sum_{j=1}^{n} x_j\xi_j, \qquad d\xi = d\xi_1 \ldots d\xi_n.$$

Use the theory of Fourier transforms to prove[7] that for large enough m the formula

$$S(x, y) = \frac{(-1)^{m+1}}{(2\pi)^n} \Delta^m \int \frac{e^{i(x-y)\xi} \, d\xi}{\xi^{2m}L(i\xi)}$$

defines a fundamental solution S of equation (5.41) when the manifold of integration is selected appropriately. More specifically, integrate first with respect to ξ_1 along a parallel to the real axis depending in a suitable way on $\xi_2, \ldots, \xi_n$, and then integrate with respect to the remaining variables along the real axis from $-\infty$ to $+\infty$. Generalize the result to partial differential equations of higher order.

13. Use Stokes' theorem and the Cauchy-Riemann equations to show that the fundamental identity (5.65) is valid in the complex domain of real dimension $2n$ generated by the real and imaginary parts of n complex variables $x_j = x_j' + ix_j''$, where D can be chosen as any n-dimensional bounded manifold and where it is

---

[6] Cf. Keller-Lewis-Seckler 1.
[7] Cf. Nirenberg 1.

understood that $dx_j = dx'_j + idx''_j$ must be interpreted in the sense of the calculus of exterior differential forms.

14. Discuss the fundamental solution of the wave equation in any number of independent variables.

3. THE PARAMETRIX

The fundamental solution is a useful tool for the construction and representation of more general solutions of a linear partial differential equation. However, some of the problems in which it plays a role can be handled just as easily, or perhaps better, by means of functions possessing a singularity that is not annihilated but merely smoothed out by the differential operator in question. The smoothing might even be so weak that the singularity actually becomes augmented but acquires a less rapid growth than was to be anticipated from the order of the operator. Such a function is called a *parametrix* associated with the relevant partial differential equation. In particular, any fundamental solution is automatically a parametrix. The notion of a parametrix also concerns functions that nearly fulfill a given boundary condition (cf. Section 8.3). The principal application of the parametrix occurs in developing integral equations[8] for the solution of initial or boundary value problems (cf. Sections 6.4 and 9.3).

To be precise, for an equation

$$L[u] = \sum_{i,j=1}^{n} a_{ij} \frac{\partial^2 u}{\partial x_i \, \partial x_j} + \sum_{i=1}^{n} b_i \frac{\partial u}{\partial x_i} + cu = 0$$

of the elliptic type we shall say that a function $P = P(x, \xi)$ of two points x and ξ is a parametrix if it has at $x = \xi$ a pole whose leading term is equivalent to that prescribed for the fundamental solution and becomes infinite like $r^{-(n-2)}$, yet the operator L applied to P with respect to the variable x gives a result $L[P]$ that grows not faster than $r^{-(n-1)}$, where r is the distance between x and ξ. The simplest example of such a parametrix is the expression

$$(5.79) \qquad P(x, \xi) = \frac{1}{\sigma'_{n-1}\sqrt{\Delta}\left[\displaystyle\sum_{j,k=1}^{n} A_{jk}(x_j - \xi_j)(x_k - \xi_k)\right]^{(n-2)/2}},$$

where $\sigma'_{n-1} = n(n-2)\tau_n$ and where the inverse matrix of coefficients A_{jk} and the determinant Δ are understood to be evaluated at ξ. Formula

---

[8] Cf. Hadamard 3, Hilbert 1, Levi 1.

(5.79) defines a quantity with the desired properties because the quadratic form Γ_2 inside the square brackets satisfies the condition

$$\lim_{x \to \xi} \frac{1}{\Gamma_2} \sum_{i,j=1}^{n} a_{ij} \frac{\partial \Gamma_2}{\partial x_i} \frac{\partial \Gamma_2}{\partial x_j} = 4$$

ensuring that the operator L increases the order of the pole of P by at most one (cf. Section 2). We have intentionally normalized P here in accordance with (5.71). Finally, for the case $n = 2$ of a plane problem we must replace (5.79) by

$$(5.80) \qquad P(x, \xi) = \frac{1}{2\pi\sqrt{\Delta}} \log \frac{1}{\sqrt{\sum_{j,k=1}^{2} A_{jk}(x_j - \xi_j)(x_k - \xi_k)}}.$$

One advantage of formula (5.79) is that the parametrix it represents has meaning in the large. On the other hand, any partial sum

$$(5.81) \qquad P(x, \xi) = \sum_{l=0}^{\nu} U_l \Gamma^{l-m} + \sum_{l=0}^{\mu} V_l \Gamma^l \log \Gamma$$

of terms from either of the expansions (5.56) or (5.57) for a fundamental solution can be used equally well as a parametrix in any appropriately small region where the squared geodesic distance Γ is single-valued.

By choosing the integers ν and μ in (5.81) large enough, we can achieve any desired degree of smoothness for $L[P]$ and, in particular, we can arrange that it have as many bounded derivatives as we please in some neighborhood of ξ. A parametrix of this kind is better than (5.79) in the sense that it approximates more closely an actual solution of the partial differential equation (5.41). Furthermore, the function (5.81) can be used as a parametrix even when (5.41) is of the hyperbolic type, since it possesses a suitable singularity all along the characteristic conoid $\Gamma = 0$, a property not shared in general by the more elementary expression (5.79). For most applications to hyperbolic equations it turns out that we must take $\nu \geq m - 1$ in order to obtain a workable parametrix. If, more specifically, we put $\nu = m + \frac{1}{2}$ for odd n and $\nu = m - 1$, $\mu = 1$ for even n, we find that $L[P]$ remains finite as the point x approaches the conoid $\Gamma = 0$.

Since the definition (5.81) of a parametrix does not involve infinite series, it remains valid in situations where the coefficients a_{ij}, b_i and c of (5.41) are no longer analytic. It is sufficient, indeed, for these coefficients to possess continuous partial derivatives of an order high enough to ensure the success of our construction of the factors Γ, U_l and V_l that actually appear in (5.81). This construction was based on solving ordinary differential equations, a problem that can be treated by the Picard method of successive approximations (cf. Section 4.2). Our principal objective in introducing the notion of a parametrix is to exploit it to determine solutions of a partial differential equation whose coefficients are not analytic. We shall proceed to derive an alternate representation for the fundamental solution in the elliptic case which is based on an integral equation involving a parametrix and which therefore applies without the hypothesis of analyticity . In preparation for such an analysis we must develop several basic identities involving the parametrix.

We consider (5.81) as a function of the parameter point ξ and insert it to play the role of v in formula (5.65). More precisely, we must first delete from the domain of integration D a small geodesic sphere $\Gamma(\xi, x) \leq \epsilon^2$ about the singular point x, but by letting $\epsilon \to 0$ afterward we derive

$$(5.82) \qquad u(x) = - \int_{\partial D} B[u, P] - \int_D (PL[u] - uM[P]) \, d\xi$$

in a fashion similar to our proof of (5.69). The improper integral on the right is seen to be absolutely convergent because the volume element $d\xi = d\xi_1 \ldots d\xi_n$, when expressed in a spherical coordinate system whose origin lies at x, includes the factor r^{n-1}. If ν and μ are large enough so that $L[P]$ does not grow as rapidly as P itself when $\xi \to x$, we obtain

$$(5.83) \qquad 0 = \int_{\partial D} B[u, L[P]] + \int_D (L[P]L[u] - uM[L[P]]) \, d\xi$$

by the same procedure that led to (5.82). From here on L is intended to operate on P with respect to x, whereas M is applied with respect to ξ.

Given a sufficiently differentiable function ρ, choose u so that $L[u] - \rho$ and u both vanish at the point x, together with their first partial derivatives. Because P is a parametrix for the adjoint operator M in its dependence on ξ, the product $uM[P]$ becomes infinite at x no faster than r^{-n+3}. Hence we are permitted to differentiate under the sign of integration to derive

$$(5.84) \qquad L\left[\int_D uM[P] \, d\xi \right] = \int_D uL[M[P]] \, d\xi.$$

Consequently application of the operator L to (5.82) yields

$$(5.85) \quad L[u] = - \int_{\partial D} B[u, L[P]] + \int_D uM[L[P]] \, d\xi - L\left[\int_D PL[u] \, d\xi \right].$$

By adding (5.83) to (5.85) we deduce that

$$(5.86) \qquad L[u] = \int_D L[P]L[u] \, d\xi - L\left[\int_D PL[u] \, d\xi \right].$$

On the other hand, the argument used to establish (5.84) also shows that

$$L\left[\int_D (L[u] - \rho)P \, d\xi\right] = \int_D (L[u] - \rho)L[P] \, d\xi.$$

This combines with (5.86) to prove the *generalized Poisson equation*

$$(5.87) \qquad L\left[\int_D \rho P \, d\xi\right] = -\rho(x) + \int_D \rho L[P] \, d\xi,$$

since $L[u] = \rho$ at the point x.

Equation (5.87) has been proved under the assumption that the density factor ρ and the coefficients a_{ij}, b_i and c of the elliptic partial differential equation (5.41) are sufficiently differentiable, though not necessarily analytic. Incidentally, our justification of (5.84) also shows that we can substitute the elementary parametrix (5.79) into (5.87) instead of the more sophisticated version (5.81), since their difference has a singularity that behaves no worse than r^{-n+3}.

Our aim will be to use the Poisson equation (5.87) in order to determine ρ so that the expression

$$(5.88) \qquad S(x, \eta) = P(x, \eta) + \int_D \rho(\xi)P(x, \xi) \, d\xi$$

defines a fundamental solution S of (5.41) even in the non-analytic case. Since the appropriate pole appears on the right, we have only to require that the term occurring there be completely annihilated by the operator L. For this purpose ρ must satisfy the linear integral equation

$$(5.89) \qquad \rho(x) - \int_D L[P(x, \xi)]\rho(\xi) \, d\xi = L[P(x, \eta)],$$

according to (5.87), where L operates with respect to x. We shall see that we can treat (5.89) by means of a natural iteration scheme when the domain of integration D is small enough.

We suppose for the sake of simplicity that ν and μ have been taken so large in (5.81) that $L[P]$ is continuously differentiable and is bounded by a fixed positive number λ. Moreover, we assume D to be so small that its n-dimensional volume τ satisfies

$$(5.90) \qquad \lambda\tau < 1.$$

After these preliminaries, we introduce the sequence of successive approximations

$$\rho_l(x) = L[P(x, \eta)] + \int_D L[P(x, \xi)]\rho_{l-1}(\xi) \, d\xi, \qquad l = 1, 2, \ldots,$$

to the solution of (5.89), with $\rho_0 = 0$. Such an iteration procedure leads to the *Neumann series* (cf. Section 10.1)

(5.91) $$\rho(x) = \sum_{l=1}^{\infty} K_l(x, \eta)$$

for ρ, where

$$K_1(x, \eta) = L[P(x, \eta)],$$

$$K_{l+1}(x, \eta) = \int_D L[P(x, \xi)]K_l(\xi, \eta)\, d\xi, \qquad l = 1, 2, \ldots.$$

The series converges geometrically because of the elementary estimate

$$|K_{l+1}| \leq \int_D |L[P]K_l|\, d\xi \leq \lambda\tau \sup_D |K_l|$$

and the hypothesis (5.90). Since $L[P]$ is continuously differentiable, (5.91) defines a smooth enough solution of the integral equation (5.89) to justify an application of the Poisson equation (5.87). It follows that (5.88) and (5.91) together represent a fundamental solution S of the elliptic equation (5.41) in the (small) region D.

As an illustration of the need for a formula like (5.88) that can be used to generate solutions of a linear elliptic equation whose coefficients are not analytic, we recall the problem of reducing a second order equation (3.1) in two independent variables to the canonical form (3.5). This led to the self-adjoint *Beltrami equation* (3.43), which we are now in a position to solve, at least in the small, even when the functions a, b and c involved are not analytic but merely possess continuous derivatives of a sufficiently high order. Thus the construction of a fundamental solution of the Beltrami equation permits us to transform an arbitrary linear elliptic equation in the plane to canonical form locally. The same analysis shows, according to Exercise 3.1.8, that any sufficiently small portion of a curved surface can be mapped conformally onto a plane section. A verification of the validity of these results in the large must still be postponed, however, until the more difficult existence theorems of Chapter 8 become available to us.

Further applications of the concept of a parametrix, which we shall have occasion to discuss later, are concerned with integral equations derived for the purpose of constructing solutions of specific initial or boundary value problems. The question of how the solution depends on the data is especially amenable to an approach of this kind (cf. Section 15.1). Finally, such methods are also useful for certain nonlinear problems (cf. Exercise 16.3.9).

EXERCISES

1. Show that the function $-A \log r$ occurring in (5.27) is a parametrix.

2. Establish the Poisson equation

(5.92) $$L\left[\int_D \rho(\xi) S(x, \xi)\, d\xi\right] = -\rho(x),$$

where S is a fundamental solution of (5.41) in D. In particular, show that

(5.93) $$\left(\frac{\partial^2}{\partial x^2} + \frac{\partial^2}{\partial y^2}\right) \iint_D \rho(\xi, \eta) \log \frac{1}{r}\, d\xi\, d\eta = -2\pi \rho(x, y),$$

where $r = \sqrt{(x - \xi)^2 + (y - \eta)^2}$.

3. Prove[9] that (5.93) holds whenever the density ρ fulfills a *Hölder condition*

(5.94) $$|\rho(x, y) - \rho(\xi, \eta)| \leq \beta r^\alpha$$

of order α, where $0 < \alpha < 1$, $\beta > 0$. Under the same mild hypothesis, show that all the second derivatives of the integral on the left in (5.93), which represents a *logarithmic potential*, exist and are *Hölder continuous* in the sense that they satisfy a requirement of the form (5.94).

4. Use a Neumann series

$$S = \frac{1}{2\pi} \log \frac{1}{r} + \frac{1}{4\pi^2} \iint c \log \frac{1}{r_1} \log \frac{1}{r_2}\, dx\, dy$$

$$+ \frac{1}{8\pi^3} \iint \iint c_1 c_2 \log \frac{1}{r_1} \log \frac{1}{r_2} \log \frac{1}{r_3}\, dx\, dy\, d\xi\, d\eta + \cdots$$

like (5.91) to establish the existence in the small of a fundamental solution of the equation

(5.95) $$u_{xx} + u_{yy} + cu = 0$$

under the assumption that the coefficient $c = c(x, y)$ is merely Hölder continuous.

5. Show that

$$P(x, y; \xi, \eta) = -\frac{1}{2\pi} \log |f(z) - f(\zeta)|$$

is a parametrix for (5.95) provided that f is an analytic function of the complex variable $z = x + iy$ with a derivative f' which does not vanish.

6. Under the hypothesis (5.90), prove the uniqueness of the solution ρ of the integral equation (5.89).

7. Show that the elementary parametrix (5.79) can be used in the representation (5.88) of a fundamental solution of (5.41).

[9] Cf. Kellogg 1.

174 THE FUNDAMENTAL SOLUTION

8. Use an analogue of (5.72) to establish that (5.81) defines a function P which is simultaneously a parametrix for (5.41) in its dependence on x and a parametrix for the adjoint equation (5.64) in its dependence on ξ.

9. Use (5.82) to prove that any solution of an equation (5.41) of the elliptic type must be differentiable infinitely often if the coefficients a_{ij}, b_i and c are differentiable infinitely often.

10. For even values of n, precisely how many derivatives must a_{ij}, b_i and c possess in order to ensure that (5.81) have a meaning with $\nu = m - 1$ and $\mu = 1$?

11. Define the notions of a fundamental solution and of a parametrix for the linear ordinary differential equation

$$y'' + ay' + by = 0.$$

12. Show in detail how fundamental solutions of the Beltrami equation (3.43) in two independent variables can be used to bring a linear elliptic equation of the second order into canonical form.

6

Cauchy's Problem in Space
of Higher Dimension

1. CHARACTERISTICS; UNIQUENESS

It is considerably harder to treat the Cauchy problem in space of
dimension larger than two than it is to discuss the same problem in the
plane. One explanation for the difficulties encountered in the case of
more than two independent variables lies in our inability to introduce
·characteristic coordinates which reduce an arbitrary linear equation of
the second order to a canonical form such that the coefficients of the
highest derivatives are constant, a goal that can be achieved in two-
dimensional space. However, the results of Chapter 4, together with our
knowledge of mathematical physics, do suggest that it is primarily for
equations of the hyperbolic type that Cauchy's problem makes sense.
Thus we confine our attention here to equations that can be expressed in
the canonical form

(6.1) $$\frac{\partial^2 u}{\partial t^2} = \sum_{j,k=1}^{n} a_{jk} \frac{\partial^2 u}{\partial x_j \, \partial x_k} + \cdots,$$

which was introduced in Exercise 3.2.3. The dots stand, as usual, for
terms involving derivatives of the unknown u of at most the first order, and
the n-dimensional differential operator of the second order on the right is
supposed to be elliptic. The coefficients a_{jk} can be functions of t as well as
of $x_1, \ldots, x_n$, but for each choice of these arguments they must generate a
positive-definite quadratic form.

Cauchy's problem asks for a solution of (6.1) such that both u and its
normal derivative $\partial u/\partial v$ assume prescribed values on a given surface, or
section of surface, T. We are interested in finding out under what circumstances this question is correctly set in the sense that the answer exists, is
unique, and depends continuously on the data. From the discussions of

Chapter 3 we are already aware that for our present purposes the surface T should not be chosen as a characteristic of the partial differential equation (6.1), for then the second normal derivative of u is not uniquely determined on T by (6.1), which becomes a restriction on the Cauchy data u and $\partial u/\partial v$. More remarkable is the observation that in space of dimension greater than two it is not sufficient to take T distinct from a characteristic in order to achieve a well posed problem.

To prove the latter assertion we have only to recall from Section 4.1 the Hadamard example

$$u = \frac{1}{m^2}\,\text{sh}\ mx \sin my$$

of a solution of Laplace's equation

$$u_{xx} + u_{yy} = 0.$$

Let it now be considered as a solution of the wave equation

(6.2) $$u_{tt} = u_{xx} + u_{yy}$$

assuming the Cauchy data

$$u = 0, \qquad \frac{\partial u}{\partial v} = \frac{1}{m}\sin my$$

on the plane $x = 0$ in three-dimensional space. Hadamard's example does not approach the trivial solution zero as the Cauchy data do when the parameter m becomes infinite. Therefore not only is the original elliptic Cauchy problem for which the example was devised unreasonable, but even Cauchy's problem for the hyperbolic equation (6.2) is not correctly set when data are assigned on the plane $x = 0$. A noteworthy feature of our argument is that the above function u simply does not depend on the time t.

In order to deduce from Hadamard's analysis a requirement on the initial surface T ensuring that Cauchy's problem is well posed, we consider at first an equation

(6.3) $$\frac{\partial^2 u}{\partial t^2} - \sum_{j,k=1}^{n} a_{jk}\frac{\partial^2 u}{\partial x_j\,\partial x_k} = 0$$

with constant coefficients $a_{jk} = a_{kj}$. Let T be the hyperplane

$$\lambda_0 t + \sum_{j=1}^{n} \lambda_j x_j = 0.$$

If the Cauchy problem is to be correctly set with data assigned on T, there should exist no exponential solutions

$$u = \exp\left\{i\left(\mu_0 t + \sum_{j=1}^{n} \mu_j x_j\right) + \lambda_0 t + \sum_{j=1}^{n} \lambda_j x_j\right\}$$

of (6.3) with the parameters μ_j real; otherwise we could construct a counterexample like Hadamard's. Substituting into (6.3), we conclude that the quadratic equation

$$(6.4) \qquad (\lambda_0 + i\mu_0)^2 - \sum_{j,k=1}^{n} a_{jk}(\lambda_j + i\mu_j)(\lambda_k + i\mu_k) = 0$$

ought to have no real roots μ_j.

The real and imaginary parts of (6.4) are

$$\mu_0^2 - \sum_{j,k=1}^{n} a_{jk}\mu_j\mu_k = \lambda_0^2 - \sum_{j,k=1}^{n} a_{jk}\lambda_j\lambda_k,$$

$$2\lambda_0\mu_0 - 2\sum_{j,k=1}^{n} a_{jk}\lambda_j\mu_k = 0.$$

Eliminating μ_0, we find that

$$\frac{1}{\lambda_0^2}\left(\sum_{j,k=1}^{n} a_{jk}\lambda_j\mu_k\right)^2 - \sum_{j,k=1}^{n} a_{jk}\mu_j\mu_k = \lambda_0^2 - \sum_{j,k=1}^{n} a_{jk}\lambda_j\lambda_k.$$

Whenever the expression on the left has the same sign as the one on the right we can solve by adjusting the length of the vector $(\mu_0, \ldots, \mu_n)$ appropriately. To determine the sign on the left we appeal to Schwarz's inequality

$$\left(\sum_{j,k=1}^{n} a_{jk}\lambda_j\mu_k\right)^2 \leq \sum_{j,k=1}^{n} a_{jk}\lambda_j\lambda_k \sum_{j,k=1}^{n} a_{jk}\mu_j\mu_k,$$

which holds because the matrix of coefficients a_{jk} is supposed to be positive-definite. Thus any conceivable roots of (6.4) must satisfy the estimate

$$\frac{1}{\lambda_0^2}\left(\sum_{j,k=1}^{n} a_{jk}\mu_j\mu_k\right)\left(\sum_{j,k=1}^{n} a_{jk}\lambda_j\lambda_k - \lambda_0^2\right) \geq \lambda_0^2 - \sum_{j,k=1}^{n} a_{jk}\lambda_j\lambda_k,$$

which prevents the quantity on the right from being positive. Therefore the desired condition that (6.4) have no real roots μ_j is simply

$$(6.5) \qquad \lambda_0^2 - \sum_{j,k=1}^{n} a_{jk}\lambda_j\lambda_k > 0.$$

Since $\lambda_0, \lambda_1, \ldots, \lambda_n$ may be interpreted as the components $\partial t/\partial\nu$, $\partial x_1/\partial\nu, \ldots, \partial x_n/\partial\nu$ of the normal to the hyperplane T, our requirement (6.5) on T for the Cauchy problem to be well posed is equivalent to the inequality

$$(6.6) \qquad \left(\frac{\partial t}{\partial\nu}\right)^2 > \sum_{j,k=1}^{n} a_{jk}\frac{\partial x_j}{\partial\nu}\frac{\partial x_k}{\partial\nu}.$$

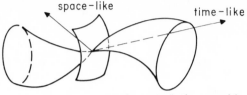

characteristic conoid
FIG. 13.   Space-like and time-like directions.

In this form we might expect that it even applies to curved initial surfaces T and to the general hyperbolic equation (6.1) with variable coefficients a_{jk}. Therefore it is natural to describe any surface element satisfying (6.6) as *space-like* with respect to the partial differential equation (6.1). Clearly (6.1) is hyperbolic if and only if space-like surface elements exist.

Next recall that in Section 3.2 we defined the *characteristics* of (6.1) to be level surfaces

$$\phi(t, x_1, \ldots, x_n) = \text{const.}$$

whose normals $(\phi_t, \phi_{x_1}, \ldots, \phi_{x_n})$ fulfill the first order equation

$$(6.7) \qquad \phi_t^2 = \sum_{j,k=1}^{n} a_{jk} \phi_{x_j} \phi_{x_k}.$$

The above stability analysis now leads to the characteristics as a limiting case of space-like surfaces for which the inequality (6.6) degenerates into an equality (6.7).

Of special interest in the present context are the *characteristic conoids* $\Gamma = 0$ introduced in Section 5.2, which are conoidal solutions (cf. Exercise 2.2.1) of (6.7). Because (6.1) is of the hyperbolic type, each characteristic conoid consists of two sheets dividing a neighborhood of its vertex $(t, x_1, \ldots, x_n)$ into three disjoint regions. Surface elements at the vertex are space-like if they lie in the region bounded by both sheets, so that (6.6) is satisfied. On the other hand, directions pointing into either of the remaining two regions, each of which is bounded by a single sheet of the conoid, are called *time-like*. Intermediate directions tangent to the generators of the conoid have been described in Section 5.2 as *bicharacteristic*. We have encountered space-like and time-like directions previously in connection with our treatment of Lorentz transformations (cf. Section 3.3).

It is instructive to study the example of the wave equation (6.2) in two space variables x and y. The relation (6.7) for the characteristics of (6.2) is simply

$$(6.8) \qquad \phi_t^2 = \phi_x^2 + \phi_y^2,$$

which may be reduced to the equation of geometrical optics investigated in Section 2.4. The special solution

$$\phi = (t - t_0)^2 - (x - x_0)^2 - (y - y_0)^2 = 0$$

of (6.8) obviously represents a characteristic cone whose vertex is located at the point (x_0, y_0, t_0). Any space-like surface element through (x_0, y_0, t_0) has to fulfill the requirement

$$\left(\frac{\partial t}{\partial \nu}\right)^2 > \left(\frac{\partial x}{\partial \nu}\right)^2 + \left(\frac{\partial y}{\partial \nu}\right)^2.$$

Consequently its normal has a t-component exceeding those of the characteristic cone, which are given by

$$\left(\frac{\partial t}{\partial \nu}\right)^2 = \left(\frac{\partial x}{\partial \nu}\right)^2 + \left(\frac{\partial y}{\partial \nu}\right)^2.$$

It is for this reason that the space-like surface must lie between the two sheets of the cone. Physically speaking, the characteristic cone represents circular wave fronts propagating from a source at (x_0, y_0). The generators of the cone, which are bicharacteristics, may be interpreted as light rays perpendicular to the wave fronts (cf. Section 2.4).

We have now explained why Cauchy's problem for the hyperbolic equation (6.1) will be well posed only if the data are assigned on a space-like surface T. In the particular case of the wave equation (6.2) with Cauchy data prescribed at $t = 0$ the space-like requirement is fulfilled. Thus it fits in logically with the natural formulation of a physically significant initial value problem in which the data describe a situation observed at a specified time. On the other hand, we have seen that a solution of (6.2) does not have to depend continuously on Cauchy data given along the plane $x = 0$. We should ascribe the latter phenomenon to the fact that the locus $x = 0$ contains time-like directions; it includes, indeed, the t-axis itself.

The greater portion of this chapter will be devoted to a verification that the Cauchy problem is correctly set for a hyperbolic equation in space of arbitrary dimension when the initial data are prescribed on a space-like surface. Since data of physical interest need not always be analytic, but might conceivably be associated with an impulsive motion, we shall want to develop constructions of the solution of Cauchy's problem that are valid for initial values which merely possess continuous derivatives of a suitable order. It is typical of well posed problems that such an objective can be achieved.

We begin by asking just how much of the Cauchy data is needed to determine uniquely the value of the solution of Cauchy's problem at a given point. The answer will be expressed in terms of the characteristic conoid whose vertex is placed at that point. The conoid cuts out of the initial surface T precisely the section of data which we are seeking, or at least does so when its vertex is close enough to T. In analogy with the remarks in Section 4.1, we describe this cell of T as the *domain of dependence* of the solution at the original point, or vertex. It follows that either of the time-like regions which are separated from T by a second characteristic conoid whose vertex is located at some point of T may be called a *range of influence* for the initial point in question. We shall attempt to justify these statements by proving a uniqueness theorem for the wave equation (6.2) which shows that the Cauchy data over a domain of dependence do actually suffice to fix the solution at the vertex of the associated characteristic cone. The corresponding result for more general equations will be taken up afterward in the exercises.

The wave equation

$$L[u] = u_{tt} - u_{xx} - u_{yy} = 0$$

serves to describe the small vibrations of a thin membrane. To settle the uniqueness question for it we shall estimate the variation with time of the combined kinetic and potential energy contained in a conveniently specified portion of the membrane. Let D be a three-dimensional region bounded by a sheath ∂D, and let T be the intersection of D with a horizontal plane located t units above the (x,y)-plane. We are interested in the rate of change

$$(6.9) \quad \frac{\partial}{\partial t} \iint_T \frac{1}{2} [u_t^2 + u_x^2 + u_y^2] \, dx \, dy = \iint_T [u_t u_{tt} + u_x u_{xt} + u_y u_{yt}] \, dx \, dy$$

$$+ \frac{1}{2} \int_{\partial T} [u_t^2 + u_x^2 + u_y^2] \frac{t_\nu \, ds}{\sqrt{x_\nu^2 + y_\nu^2}}$$

of the energy contained in T, where ν stands for the inner normal on ∂D and s denotes arc length along the boundary ∂T of T.

According to the divergence theorem we have, in the notation of the calculus of exterior differential forms,

$$\iint_T [u_t u_{tt} + u_x u_{xt} + u_y u_{yt}] \, dx \, dy$$

$$= \iint_T u_t [u_{tt} - u_{xx} - u_{yy}] \, dx \, dy + \int_{\partial T} (u_t u_x \, dy - u_t u_y \, dx).$$

Consequently (6.9) becomes

(6.10) $\dfrac{\partial}{\partial t} \displaystyle\iint_T \dfrac{1}{2} [u_t^2 + u_x^2 + u_y^2] \, dx \, dy = \iint_T u_t L[u] \, dx \, dy$

$+ \dfrac{1}{2} \displaystyle\int_{\partial T} \left\{ 2u_t u_x \, dy - 2u_t u_y \, dx + \dfrac{u_t^2 + u_x^2 + u_y^2}{\sqrt{x_v^2 + y_v^2}} \, t_v \, ds \right\}.$

For a solution u of the wave equation, the double integral on the right vanishes. Moreover, it is evident that

$$2u_t u_x y_s - 2u_t u_y x_s \leq 2\, |u_t|\, \sqrt{u_x^2 + u_y^2} \leq u_t^2 + u_x^2 + u_y^2.$$

Thus if T shrinks as the time t increases and if the sheath ∂D is space-like, so that

$$t_v < -\sqrt{x_v^2 + y_v^2},$$

then the line integral on the right in (6.10) is non-positive and

(6.11) $\dfrac{\partial}{\partial t} \displaystyle\iint_T \dfrac{1}{2} [u_t^2 + u_x^2 + u_y^2] \, dx \, dy \leq 0.$

The energy inequality (6.11) shows that when the Cauchy data u and u_t are zero on T at, say, the initial time $t = 0$, we must have

$$0 \leq \iint_T [u_t^2 + u_x^2 + u_y^2] \, dx \, dy \leq 0$$

for all $t > 0$, provided only that ∂D is space-like. Observe that the result remains valid in the limiting case where ∂D is a characteristic surface, which could be taken as the lower sheet of the characteristic cone

$$(t - t_0)^2 = (x - x_0)^2 + (y - y_0)^2, \qquad t_0 > 0.$$

Consider now the difference

$$u = u_2 - u_1$$

between two solutions of Cauchy's problem assuming the same initial data on the disc which the characteristic cone cuts out of the plane $t = 0$. In order not to violate the above energy inequality, u must vanish identically throughout the region D enclosed by the cone and the disc. In particular, $u_1 = u_2$ at the vertex of the cone, which means that the solution of Cauchy's problem is uniquely determined there by the initial data specified on the disc.

For the purpose of generalizing the above uniqueness proof to curved initial surfaces T, we prefer to consider directly a triple integral of the expression $u_t L[u]$ over the domain D. An application of the three-dimensional version of the divergence theorem yields the identity

$$(6.12) \quad \iiint_D u_t[u_{tt} - u_{xx} - u_{yy}] \, dx \, dy \, dt$$

$$= \iint_{\partial D} \left[\frac{1}{2} (u_t^2 + u_x^2 + u_y^2) \, dx \, dy - u_t u_x \, dy \, dt + u_t u_y \, dx \, dt \right],$$

provided that the function u has continuous second derivatives. Note that (6.12) also follows from an integration of (6.10) with respect to t, because

$$- \frac{t_v \, ds \, dt}{\sqrt{x_v^2 + y_v^2}} = dx \, dy.$$

If we substitute for u a solution of the wave equation, the triple integral drops out of (6.12) and we are led to the relation

$$(6.13) \quad \iint_{\partial D} [t_v(u_t^2 + u_x^2 + u_y^2) - 2x_v u_t u_x - 2y_v u_t u_y] \, d\sigma = 0,$$

where $d\sigma$ stands for the area element on the boundary ∂D of D.

We choose ∂D to consist of a sheath C that is capped at both ends by space-like surfaces T and T'. We shall suppose that T is situated underneath T' and that C fulfills the requirement

$$t_v \leq -\sqrt{x_v^2 + y_v^2}$$

defining surface elements which are either space-like or characteristic. The integrand

$$(6.14) \quad t_v(u_t^2 + u_x^2 + u_y^2) - 2x_v u_t u_x - 2y_v u_t u_y$$

$$= \frac{1}{t_v} [(t_v^2 - x_v^2 - y_v^2)u_t^2 + (t_v u_x - x_v u_t)^2 + (t_v u_y - y_v u_t)^2]$$

in (6.13) is a positive-definite quadratic form on T because

$$t_v > \sqrt{x_v^2 + y_v^2} \geq 0$$

there. Similarly, it is negative-definite on T' and non-positive on C. We are primarily interested in the special case where C is a characteristic.

Then (6.13) reduces to the more revealing identity

$$(6.15) \quad \iint\limits_{T} [(t_v^2 - x_v^2 - y_v^2)u_t^2 + (t_v u_x - x_v u_t)^2 + (t_v u_y - y_v u_t)^2] \frac{d\sigma}{t_v}$$

$$= \iint\limits_{C} [(t_v u_x - x_v u_t)^2 + (t_v u_y - y_v u_t)^2] \frac{d\sigma}{(-t_v)}$$

$$+ \iint\limits_{T'} [(t_v^2 - x_v^2 - y_v^2)u_t^2 + (t_v u_x - x_v u_t)^2 + (t_v u_y - y_v u_t)^2] \frac{d\sigma}{(-t_v)}$$

in which all the integrands that appear are non-negative.

Because of the sign of the first term on the right in (6.15), we see that the integral over T is at least as large as the one over T'. Since the two integrands involved are positive-definite quadratic forms in u_x, u_y and u_t that can be estimated both from above and from below, for some $\epsilon > 0$ this conclusion yields the possibly weaker, but also more transparent, *energy inequality*

$$(6.16) \quad \iint\limits_{T} [u_t^2 + u_x^2 + u_y^2] \, d\sigma \geq \epsilon \iint\limits_{T'} [u_t^2 + u_x^2 + u_y^2] \, d\sigma \geq 0.$$

In particular, we can set $\epsilon = 1$ in the simplest case where T and T' are plane sections perpendicular to the t-axis.

If we assume that the Cauchy data vanish on T, then the integral on the left in (6.16), and therefore the one on the right, too, must be zero. It follows that the integrand vanishes or, in other words, that

$$u_t = u_x = u_y = 0$$

on T'. Now let C be composed of the lower sheet of a characteristic cone. By varying the position of T' we deduce that $u \equiv 0$ in the entire pyramid bounded by T and C.

The basic uniqueness theorem for Cauchy's problem is a direct consequence of the analysis just presented. If u_1 and u_2 are twice continuously differentiable solutions of the wave equation (6.2) with identical Cauchy data on the space-like section of surface T, then their difference

$$u = u_2 - u_1$$

satisfies (6.16) and has first derivatives that are zero on T. Hence it vanishes at the vertex (x_0, y_0, t_0) of every characteristic cone cutting a domain of dependence from T, and we must have $u_1 = u_2$ there. The

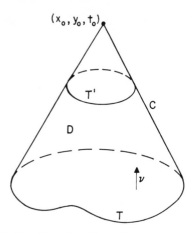

FIG. 14.   Domain of dependence in space.

significant feature about the conclusion is that it has been deduced merely from the hypothesis that the Cauchy data for u_1 and u_2 coincide throughout the domain of dependence on T associated with the point (x_0, y_0, t_0). No information was required concerning the behavior of the two solutions beyond that domain.

If we think of the wave equation as governing the motion of a vibrating membrane, the identity (6.15) shows that the energy corresponding to T exceeds the energy associated in a like fashion with T'. In this context the integral over C represents the energy lost because T' is receding too fast for inward flow of energy to occur. Their interpretation as energy integrals explains why the individual terms in (6.15) turn out to be non-negative.

It is easy to extend our uniqueness theorem to the *characteristic initial value problem*, which asks for a solution of (6.2) assuming given values on one sheet of a characteristic cone (cf. Section 5.2). We can treat this as a limiting case of Cauchy's problem such that the space-like initial surface T degenerates into some portion of the upper sheet of a characteristic cone surrounding the vertex. Since $t_\nu^2 - x_\nu^2 - y_\nu^2$ is now zero on T, formula (6.15) becomes

$$(6.17) \quad \iint\limits_{T} [(t_\nu u_x - x_\nu u_t)^2 + (t_\nu u_y - y_\nu u_t)^2] \frac{d\sigma}{t_\nu}$$

$$= \iint\limits_{C} [(t_\nu u_x - x_\nu u_t)^2 + (t_\nu u_y - y_\nu u_t)^2] \frac{d\sigma}{(-t_\nu)},$$

where C denotes the lower sheet of a second cone that combines with T to bound a finite domain D. Since the vectors $(t_\nu, 0, -x_\nu)$ and $(0, t_\nu, -y_\nu)$

are both orthogonal to the normal (x_ν, y_ν, t_ν), the squared terms appearing in the integrals (6.17) represent independent tangential derivatives of u. Hence if u vanishes on T, it must also vanish on C. The uniqueness of the solution of the characteristic initial value problem follows from these remarks.

We have used the energy inequality (6.16) to establish the uniqueness of the solution of Cauchy's problem for the wave equation in three independent variables. Not only can the proof be extended to cover more general equations of the normal hyperbolic type in any number of variables, like (6.1), but it can even serve as the basis for a corresponding existence theorem. Although it is not our intention to go into such matters until Section 12.2, we should observe that techniques of this kind have certain advantages over the more explicit construction of the solution of the Cauchy problem in terms of a fundamental solution that we shall present in Section 4. The reason is that for a single equation of order higher than two or for a system of many equations the characteristic conoid along which the fundamental solution becomes infinite may consist of a multitude of different sheets, so that its geometry becomes quite complicated.

Before leaving the topic of uniqueness we wish to discuss a theorem due to Holmgren which, although it should not be associated with the notion of a correctly set problem that has preoccupied us here, does have a wide scope of application and ought therefore to be added to the list of techniques available to us. Holmgren's theorem asserts the uniqueness of the solution of Cauchy's problem for linear analytic partial differential equations of arbitrary type when the data, which need not themselves be analytic, are assigned on a surface that is not characteristic. The proof hinges on an application of the Cauchy-Kowalewski theorem to the adjoint equation. We shall develop it only in the special case of a second order equation

$$(6.18) \qquad L[u] = \sum_{j,k=1}^{n} a_{jk} \frac{\partial^2 u}{\partial x_j \, \partial x_k} + \sum_{j=1}^{n} b_j \frac{\partial u}{\partial x_j} + cu = 0,$$

for which we already became familiar with the adjoint operator

$$M[v] = \sum_{j,k=1}^{n} \frac{\partial^2 (a_{jk}v)}{\partial x_j \, \partial x_k} - \sum_{j=1}^{n} \frac{\partial (b_j v)}{\partial x_j} + cv$$

in Section 5.2.

Let D stand for an n-dimensional domain with a smooth boundary ∂D. The significance of the adjoint operator M stems from the role that it plays in the important identity

$$(6.19) \qquad \int_D (vL[u] - uM[v]) \, dx_1 \cdots dx_n + \int_{\partial D} B[u, v] = 0,$$

which is a consequence of the divergence theorem. The differential form

$$B[u, v] = \sum_{j=1}^{n}(-1)^j\left\{\left[b_j - \sum_{k=1}^{n}\frac{\partial a_{jk}}{\partial x_k}\right]uv\right.$$

$$\left. + \sum_{k=1}^{n}a_{jk}\left[v\frac{\partial u}{\partial x_k} - u\frac{\partial v}{\partial x_k}\right]\right\} dx_1 \cdots dx_{j-1}\, dx_{j+1} \cdots dx_n$$

occurring in the boundary integral here was defined in Section 5.2. Observe that it only depends on Cauchy data for the two functions u and v over the surface ∂D.

Since our aim is to show that the difference between any two solutions of (6.18) with identical Cauchy data vanishes, it will be enough to consider in the first place a solution u with the data zero. The essential difficulty is that we cannot assume analyticity of the solution, and, indeed, the surface T along which the data are assigned may not even be analytic. However, we may suppose without loss of generality that an analytic substitution of variables has been made so that T has become tangent to the hyperplane $x_1 = 0$ at the origin and is convex there. Thus if the equation of T is written by expressing x_1 as a function of the remaining coordinates $x_2, \ldots, x_n$, then the second order terms of the Taylor series

$$x_1 = \sum_{j,k=2}^{n} f_{jk}x_j x_k + \cdots$$

for this function comprise a positive-definite quadratic form. Moreover, because T is not a characteristic of (6.18) we must have $a_{11} \neq 0$ at the origin.

Under the preceding circumstances there exists a small positive value of the parameter X such that the hyperplane

$$x_1 = X$$

bounds together with T a thin crescent domain D near the origin to which the integral identity (6.19) can be applied. It is to be understood that u is defined in D but might have no meaning on the opposite side of T from D.

Since the Cauchy data for u vanish on T, it suffices to choose v to be a function with zero Cauchy data on the hyperplane $x_1 = X$ in order to arrange for the differential form $B[u, v]$ to vanish on the entire boundary of D. When this is done the surface integral drops out of (6.19) and we are left with the relation

(6.20) $$\int_D uM[v]\, dx_1 \cdots dx_n = 0,$$

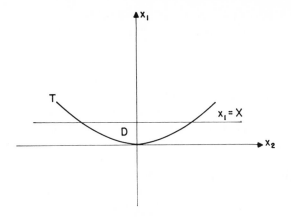

FIG. 15.   Domain of integration for Holmgren's theorem.

because of (6.18). We want to conclude from (6.20) that $u \equiv 0$ in D, which would be feasible if we knew that $M[v]$ could be replaced by an arbitrary polynomial $p(x_1, \ldots, x_n)$. Then the polynomial could be taken as close as we pleased to u, according to the Weierstrass approximation theorem.[1] Hence (6.20) would imply that

$$\int_D u^2 \, dx_1 \cdots dx_n = 0,$$

and the desired result would follow. Thus we turn our attention to establishing that for $X > 0$ small enough there is in D, for every choice of the polynomial p, a solution v of the analytic partial differential equation

(6.21) $$M[v] = p$$

which assumes the Cauchy data zero on the hyperplane $x_1 = X$.

For the purpose at hand we have only to recall the Cauchy-Kowalewski theorem and, more specifically, Exercise 1.2.3. We take X so small that $a_{11} \neq 0$ on the part of the hyperplane $x_1 = X$ bounding D. Therefore (6.21) can be solved for the partial derivative $\partial^2 u/\partial x_1^2$, and a power series solution about the point $x_1 = X$, $x_2 = \cdots = x_n = 0$ can be found possessing zero Cauchy data. The region of convergence of this solution includes a domain independent of the polynomial p because the Taylor series expansion of p converges everywhere and because (6.21) is linear.

[1] Cf. Courant-Hilbert 2.

Thus v will be defined in particular throughout D when the positive parameter X is taken small enough. The latter remark suffices to show that $u \equiv 0$ in D, as we have already observed.

The argument just presented provides a demonstration of Holmgren's uniqueness theorem in some neighborhood of T. The analysis can be repeated for any other smooth initial surface in that neighborhood which is not a characteristic. Therefore the region in which u vanishes must actually contain all points of its domain of definition that cannot be separated from T by a characteristic surface. It is of interest to compare such a result about the unique continuation of solutions of a linear analytic partial differential equation of arbitrary type with the sharper statements concerning a domain of dependence that were deduced from the energy inequality (6.16) in the hyperbolic case (cf. Exercise 14 below). However, there is a significant difference between our geometrical requirements on the region of integration for the two identities (6.15) and (6.20) involved here, so the comparison has its limitations. Finally, note that our application of the integral formula (6.19) to Holmgren's theorem shows that the existence of enough solutions of the adjoint equation (6.21) implies the uniqueness of solutions of the original linear equation (6.18).

EXERCISES

1. Use the method of separation of variables to show that Cauchy's problem is not well posed for the wave equation (6.2) when the data are assigned on the circular cylinder $x^2 + y^2 = 1$.

2. If D stands for a fixed region of the (x,y)-plane and u is a solution of the wave equation (6.2) which vanishes on the boundary of D for all choices of the time t, show that

$$\frac{\partial}{\partial t} \iint_D [u_t^2 + u_x^2 + u_y^2] \, dx \, dy = 0.$$

Use this result to establish a uniqueness theorem for the mixed initial and boundary value problem that asks for a solution of (6.2) possessing prescribed initial data u and u_t on D at $t = 0$ and assuming prescribed boundary values u on the perimeter of D for all t.

3. Use an energy inequality to establish that the solution of the mixed initial and boundary value problem for the wave equation (6.2) is unique when the boundary values are assigned on a sheath satisfying the requirement $t_\nu^2 < x_\nu^2 + y_\nu^2$ that each of its tangent planes include time-like directions. Generalize this result to the case of the inhomogeneous wave equation in any number of independent variables.

4. The mixed initial and boundary value problem for the heat equation

$$u_{xx} + u_{yy} = u_t$$

asks for a solution u assuming prescribed initial values inside a given plane domain D at the time $t = 0$ and assuming prescribed boundary values on ∂D for all $t > 0$. Show that the solution of this problem is unique by deriving either of the two energy inequalities

$$\frac{\partial}{\partial t} \iint\limits_{D} (u_x^2 + u_y^2)\, dx\, dy = -2 \iint\limits_{D} u_t^2\, dx\, dy \leq 0,$$

$$\frac{\partial}{\partial t} \iint\limits_{D} u^2\, dx\, dy = -2 \iint\limits_{D} (u_x^2 + u_y^2)\, dx\, dy \leq 0$$

in the case where the data vanish.

5. Prove the energy conservation theorem

$$\frac{\partial}{\partial t} \iiint\limits_{D} (E^2 + B^2)\, dx\, dy\, dz = 0$$

for every solution E, B of Maxwell's equations

$$\dot{B} + \nabla \times E = 0, \qquad \dot{E} - \nabla \times B = 0$$

fulfilling the homogeneous boundary condition $\nu \times E = 0$ on the surface ∂D. Establish uniqueness of the solution of a mixed initial and boundary value problem for Maxwell's equations that is suggested by this theorem.[2]

6. Let u denote a vector of m unknown functions, and let $A_1, \ldots, A_n$ and B denote m-by-m matrix functions of the variables $t, x_1, \ldots, x_n$. We say that the system of m linear partial differential equations

(6.22)
$$\frac{\partial u}{\partial t} = \sum_{j=1}^{n} A_j \frac{\partial u}{\partial x_j} + Bu$$

of the first order is *symmetric hyperbolic* when the matrices $A_1, \ldots, A_n$ are all symmetric (cf. Section 3.5). Under a sufficient hypothesis about the differentiability of the coefficients involved, reduce any single second order equation (6.1) of the hyperbolic type to a symmetric hyperbolic system (6.22) by introducing suitable linear combinations of the first derivatives of the unknown as additional new functions to be determined.

[2] Cf. Rubinowicz 1.

7. Use the divergence theorem to derive the energy formula

$$(6.23) \qquad \int_T u^2 \, d\sigma = \int_{T'} u^2 \, d\sigma + \int_C u\left(-\frac{\partial t}{\partial \nu} I + \sum_{j=1}^{n} \frac{\partial x_j}{\partial \nu} A_j\right) u \, d\sigma$$

$$- \int_D \left\{2uL[u] - \sum_{j=1}^{n} u \frac{\partial A_j}{\partial x_j} u + 2uBu\right\} dt \, dx_1 \dots dx_n$$

analogous to (6.15), where I is the identity matrix and

$$L = I \frac{\partial}{\partial t} - \sum_{j=1}^{n} A_j \frac{\partial}{\partial x_j} - B$$

is a symmetric hyperbolic differential operator (cf. Section 12.2). Here D is a domain bounded by sections T and T' of two hyperplanes $t = \tau$ and $t = \tau'$, with $\tau < \tau'$, and by an intermediate sheath C.

8. The characteristics of (6.22) are surfaces on which the normal derivatives of $t, x_1, \dots, x_n$ satisfy the determinant condition

$$(6.24) \qquad \left| -\frac{\partial t}{\partial \nu} I + \sum_{j=1}^{n} \frac{\partial x_j}{\partial \nu} A_j \right| = 0.$$

Suppose that the sheath C of Exercise 7 is less steeply inclined than any such characteristic or, in other words, is *space-like* in the sense that the matrix occurring in (6.24) generates a positive-definite quadratic form when evaluated on C. Then deduce from (6.23) the *energy inequality*

$$(6.25) \qquad \frac{\partial}{\partial \tau} \int_T u^2 \, d\sigma \leq K \int_T u^2 \, d\sigma$$

for solutions u of the symmetric hyperbolic system (6.22), where K is a sufficiently large constant. Use (6.25) to establish the uniqueness of the solution of the initial value problem for (6.22) with u prescribed at $t = 0$.

9. Show how the results of Exercises 6, 7 and 8 explain our concept of a domain of dependence for any hyperbolic equation (6.1) of the second order. State and prove the relevant uniqueness theorem.

10. Prove the theorem of Exercise 9 directly, without reference to symmetric hyperbolic systems.

11. Prove Holmgren's uniqueness theorem for the most general analytic system of linear partial differential equations that can be expressed in the canonical form (1.24) introduced in connection with our demonstration of the Cauchy-Kowalewski theorem.

12. Apply the Holmgren theorem to show that any solution of the wave equation (6.2) assuming Cauchy data on the plane $x = 0$ which do not depend on the time t must be a harmonic function of x and y.

13. Use Holmgren's theorem and the Cauchy-Kowalewski theorem to establish that when the coefficients and the initial data for (6.1) are analytic, then the solution is analytic, too, at least near the initial surface.

14. Arrive at the conical domains of dependence for the solution of Cauchy's problem in the case of the wave equation by means of the Holmgren uniqueness theorem.

2. THE WAVE EQUATION

It is natural to seek a generalization for higher dimensional space of the representation (4.73) of the solution of Cauchy's problem in terms of the Riemann function which we developed in Section 4.4 for linear hyperbolic equations in two independent variables. There is actually an integral formula of the desired kind for linear hyperbolic partial differential equations of the second order in any number of independent variables. It provides an expression for the solution of the Cauchy problem in terms of the fundamental solution described in Section 5.2. We shall derive it now for the wave equation, postponing until Section 4 our discussion of more complicated equations with variable coefficients. In both cases the derivation will be based on applying the method of contour integration to an analytic extension of the Green's identity (6.19) into the complex domain (cf. Exercise 5.2.13).

For a geometrical visualization of our procedure it is convenient to keep in mind the example

$$u_{tt} = u_{xx} + u_{yy}$$

involving only three independent variables. However, to avoid excessive repetition we shall proceed directly to the wave equation

$$(6.26) \qquad \frac{\partial^2 u}{\partial t^2} = \frac{\partial^2 u}{\partial x_1^2} + \cdots + \frac{\partial^2 u}{\partial x_n^2}$$

in n space variables, with initial data

$$(6.27) \qquad u(0, x_1, \ldots, x_n) = f(x_1, \ldots, x_n),$$

$$u_t(0, x_1, \ldots, x_n) = g(x_1, \ldots, x_n)$$

prescribed on the hyperplane $t = 0$. We shall see that this special Cauchy problem is by no means trivial and suffices to exhibit most of the interesting and important phenomena concerning hyperbolic equations in space.

Our point of departure is the representation formula (5.69) for solutions of a linear elliptic equation in terms of the fundamental solution of the adjoint equation. More specifically, we consider the Laplace equation

$$(6.28) \qquad \frac{\partial^2 v}{\partial x_0^2} + \frac{\partial^2 v}{\partial x_1^2} + \cdots + \frac{\partial^2 v}{\partial x_n^2} = 0$$

in $n + 1$ independent variables, which is self-adjoint.

Let τ_{n+1} stand for the volume of the unit ball E_{n+1} of dimension $n + 1$ and let σ_n stand for the surface area of the n-dimensional sphere ∂E_{n+1} bounding E_{n+1}. We may write

$$\sigma_n = (n + 1)\tau_{n+1}, \qquad \sigma_n' = (n - 1)\sigma_n.$$

For $n > 1$ the expression

$$S(\mathbf{x}, \boldsymbol{\xi}) = \frac{1}{\sigma_n' r^{n-1}}$$

is known (cf. Section 5.2) to provide a fundamental solution of Laplace's equation, where

$$r = \sqrt{(x_0 - \xi_0)^2 + \cdots + (x_n - \xi_n)^2}$$

is the distance between the points $\mathbf{x} = (x_0, \ldots, x_n)$ and $\boldsymbol{\xi} = (\xi_0, \ldots, \xi_n)$. Here (5.69) takes the elementary form

$$(6.29) \qquad v(x_0, \ldots, x_n) = \frac{1}{\sigma_n'} \int_{\partial D} \left[v \frac{\partial}{\partial \nu} \frac{1}{r^{n-1}} - \frac{1}{r^{n-1}} \frac{\partial v}{\partial \nu} \right] d\sigma,$$

where ∂D is any closed surface bounding a domain D that contains the point $\mathbf{x}$.

Suppose that the hyperplane $x_0 = 0$ intersects D. For a fixed choice of $x_1, \ldots, x_n$ on that intersection we proceed to exploit the integral (6.29) as a means for performing an analytic continuation of the harmonic function v to complex values of the argument x_0. Our aim in so doing is to convert v into a solution of the wave equation (6.26).[3]

The explicit nature of the representation (6.29) shows that extension of v into the complex domain is feasible by direct substitution of complex values of the variables $x_0, \ldots, x_n$ in the integrand. It is only necessary to forbid their imaginary parts to be taken so large that r could vanish for some position of the variables of integration $\xi_0, \ldots, \xi_n$ on ∂D.

We are interested in the special case where $x_1, \ldots, x_n$ remain real, while

$$(6.30) \qquad x_0 = it$$

[3] Cf. Garabedian 8.

becomes pure imaginary. Since v is an analytic function of x_0, it satisfies the Cauchy-Riemann equations in the complex x_0-plane. Its derivative there can be computed by taking, in particular, pure imaginary increments of the argument x_0. Therefore we can write

$$\frac{\partial^2 v}{\partial x_0^2} = \frac{1}{i^2}\frac{\partial^2 v}{\partial t^2} = -\frac{\partial^2 v}{\partial t^2}.$$

Comparison with (6.28) establishes that the substitution (6.30) transforms v into a possibly complex-valued solution of the wave equation

$$(6.31) \qquad -\frac{\partial^2 v}{\partial t^2} + \frac{\partial^2 v}{\partial x_1^2} + \cdots + \frac{\partial^2 v}{\partial x_n^2} = 0$$

when $t, x_1, \ldots, x_n$ are interpreted as the independent variables.

We examine next the region of validity of the integral (6.29) as a solution of (6.31). If $t > 0$, for example, then the surface of integration ∂D must be kept large enough to enclose the locus of roots ξ of the equation

$$r^2 = (it - \xi_0)^2 + (x_1 - \xi_1)^2 + \cdots + (x_n - \xi_n)^2 = 0,$$

where the integrand becomes singular. When broken down into its real and imaginary parts, this equation defines an $(n-1)$-dimensional sphere

$$(6.32) \qquad \xi_0^2 + (\xi_1 - x_1)^2 + \cdots + (\xi_n - x_n)^2 = t^2, \qquad 2t\xi_0 = 0$$

of radius t in the n-dimensional hyperplane $\xi_0 = 0$. Since the integral (6.29) is independent of the surface of integration ∂D, there is nothing to prevent us from drawing ∂D down around a thin generalized torus that contains the sphere (6.32).

It is perhaps easier to visualize the geometry here in a cylindrical coordinate system $\xi, \rho, \theta_1, \ldots, \theta_{n-1}$ with

$$\xi = \xi_0, \qquad \rho = \sqrt{(\xi_1 - x_1)^2 + \cdots + (\xi_n - x_n)^2}.$$

We are free to pick ∂D as an axially symmetric surface corresponding to some curve in the meridian (ξ,ρ)-plane. Let the curve be a loop that rises from the origin along the ρ-axis until it reaches and circumscribes a small circle

$$(6.33) \qquad \xi^2 + (\rho - t)^2 = \epsilon^2$$

whose center is the image of the sphere (6.32), whereupon it returns to the origin on what might be conceived of as the opposite side of a slit along

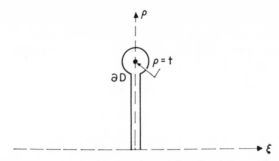

FIG. 16.   Surface of integration solving Cauchy's problem.

the ρ-axis from $\rho = 0$ to $\rho = t$. In particular, for three independent variables our choice of the surface of integration ∂D consists of a torus cut around its inner rim and joined there to a pair of circular discs.

The portion of the integral (6.29) extended over the pair of solid spheres, or discs, comprising the sections of ∂D that correspond in the meridian plane to the two edges of a slit $0 \leq \rho \leq t - \epsilon$ along the ρ-axis simplifies. Since these discs actually coincide, the integrals over them can only differ because of multiple-valuedness of the integrand. Observe that

$$ r = \sqrt{(\xi - it)^2 + \rho^2} = \sqrt{\xi - i\rho - it} \, \sqrt{\xi + i\rho - it} $$

changes its sign when the complex variable $\xi + i\rho$ is allowed to traverse the small circle (6.33) around the branch point located at $\xi = 0$, $\rho = t$ in the meridian plane. Under the same circumstances, however, the direction of the inner normal v on the section of ∂D projecting onto the ρ-axis becomes reversed. Since v is supposed to be single-valued, we conclude that a circuit around (6.33) does or does not alter the sign of the integrand in (6.29) depending on whether the exponent $n - 1$ to which r must be raised is even or odd. Thus the opposite orientations of the normals on the two edges of the ρ-axis cancel out the contributions to the integral (6.29) there for odd n, since only the single-valued square of r is involved. However, for even n the effects of the simultaneous reversals in sign of v and of r compensate for each other, and the integrations over the two edges of the ρ-axis combine to form a single non-trivial integral.

If we proceed to reduce the formula (6.29) for a solution v of the wave equation (6.31) still further by allowing the radius ϵ of the circle (6.33) to approach zero, we find from the above analysis that the results for odd n are quite distinct from those for even n, which will be discussed later. In particular, when the number n of space variables is odd, we have only to deal with an integral over the circle itself. The calculus of residues suggests that in the limit this should become an average of v and of its derivatives of order not exceeding $(n - 1)/2$ over the spherical surface of dimension $n - 1$ defined by (6.32).

To be more precise, denote by

$$w(\xi, \rho) = \int v(\xi_0, \ldots, \xi_n)\, d\Omega$$

the mean value, multiplied by σ_{n-1}, of v over the $(n - 1)$-dimensional sphere corresponding to the point (ξ, ρ) in the meridian plane. Here $d\Omega$ indicates the surface element on an associated unit sphere. Now let n be odd. In order to achieve appropriate derivatives of the spherical mean w in the final version of (6.29), it turns out to be advisable to shift our path of integration from the circle (6.33) to an equivalent rectangle R with sides described by the lines

$$\xi = \pm \epsilon, \qquad \rho = t \pm l.$$

These preliminaries enable us to bring (6.29) into the more tangible form

$$(6.34) \quad v(it, x_1, \ldots, x_n) = \frac{1}{\sigma_n'} \int_R \left[\left(\frac{1}{r^{n-1}} \frac{\partial w}{\partial \xi} - w \frac{\partial}{\partial \xi} \frac{1}{r^{n-1}} \right) \rho^{n-1}\, d\rho \right.$$

$$\left. - \left(\frac{1}{r^{n-1}} \frac{\partial w}{\partial \rho} - w \frac{\partial}{\partial \rho} \frac{1}{r^{n-1}} \right) \rho^{n-1}\, d\xi \right]$$

by putting $d\sigma = \rho^{n-1}\, ds\, d\Omega$, $ds^2 = d\xi^2 + d\rho^2$, and then carrying out the integration with respect to the surface element $d\Omega$. We shall evaluate the right-hand side of (6.34) by letting $\epsilon \to 0$ with $l > 0$ held fixed.

In the limit process just indicated, the integrals over the two short sides of the rectangle R approach zero because $1/r$ remains bounded there. Furthermore, when we come presently to perform $(n - 3)/2$ integrations by parts with respect to ρ over the long sides of R, the integrated terms that will be obtained must approach zero for the same reason. Thus, denoting by $O(\epsilon)$ an expression that tends toward zero with ϵ, we can write

$$v(it, x_1, \ldots, x_n) = \frac{1}{\sigma_n'} \int_R \left[\frac{\rho}{r^{n-1}} \frac{\partial w}{\partial \xi} - w \frac{\partial}{\partial \xi} \frac{\rho}{r^{n-1}} \right] \rho^{n-2}\, d\rho + O(\epsilon)$$

$$= \frac{1}{(n - 3)\sigma_n'} \int_R \left[\frac{1}{r^{n-3}} \frac{\partial}{\partial \rho} \left(\rho^{n-2} \frac{\partial w}{\partial \xi} \right) - \frac{\partial}{\partial \rho} (\rho^{n-2} w) \frac{\partial}{\partial \xi} \frac{1}{r^{n-3}} \right] d\rho + O(\epsilon),$$

since

$$\frac{\rho}{r^{n-1}} = - \frac{1}{n - 3} \frac{\partial}{\partial \rho} \frac{1}{[(\xi - it)^2 + \rho^2]^{(n-3)/2}} = - \frac{1}{n - 3} \frac{\partial}{\partial \rho} \frac{1}{r^{n-3}}.$$

Integrating by parts this way $(n - 5)/2$ more times, we derive the formula

$$(6.35) \quad v(it, x_1, \ldots, x_n) = \frac{1}{\alpha_n} \int_R \left[\frac{\rho}{r^2} \left(\frac{\partial}{\rho \, \partial \rho} \right)^{(n-3)/2} \left(\rho^{n-2} \frac{\partial w}{\partial \xi} \right) \right.$$

$$\left. - \left(\frac{\partial}{\rho \, \partial \rho} \right)^{(n-3)/2} (\rho^{n-2} w) \frac{\partial}{\partial \xi} \frac{\rho}{r^2} \right] d\rho + O(\epsilon)$$

where

$$\alpha_n = 2 \cdot 4 \cdots (n - 3)\sigma'_n = 2 \cdot 4 \cdots (n + 1)\tau_{n+1}.$$

The calculus of residues shows that for any function $F = F(\xi, \rho)$ with bounded first partial derivatives we have

$$\lim_{\epsilon \to 0} \int_R \frac{F}{r^2} d\rho = \frac{\pi i F(0, t)}{t},$$

since the value of the limit is not altered if $F(\xi, \rho)$ is replaced by $F(0, t)$ and if $d\rho$ is replaced by $d(\xi + i\rho)/i$. Furthermore, because r is a function of $\xi - it$ we can substitute $i\partial/\partial t$ for $\partial/\partial \xi$ in the second term on the right in (6.35) and then bring this differential operator outside the sign of integration. Therefore in the limit as $\epsilon \to 0$, formula (6.35) reduces to the result

$$(6.36) \quad v(it, x_1, \ldots, x_n) = \frac{\pi}{\alpha_n} \left[\frac{\partial}{\partial t} \left(\frac{\partial}{t \, \partial t} \right)^{(n-3)/2} t^{n-2} w(0, t) \right.$$

$$\left. + \left(\frac{\partial}{t \, \partial t} \right)^{(n-3)/2} t^{n-2} i w_\xi(0, t) \right].$$

Our aim is to interpret (6.36) as an explicit representation of the solution of Cauchy's problem for the wave equation (6.26) in any odd number n of space variables $x_1, \ldots, x_n$. For this purpose we substitute an actual solution

$$(6.37) \qquad u(t, x_1, \ldots, x_n) = v(it, x_1, \ldots, x_n)$$

of the wave equation (6.26) for the harmonic function v of an imaginary argument it.

Let

$$(6.38) \qquad \omega[u; x, t] = \frac{1}{\sigma_{n-1}} \int u(0, \xi_1, \ldots, \xi_n) \, d\Omega$$

be the average of the function u over the $(n - 1)$-dimensional sphere

$$(\xi_1 - x_1)^2 + \cdots + (\xi_n - x_n)^2 = t^2$$

of radius t about the point $x = (x_1, \ldots, x_n)$ in the initial hyperplane $t = 0$. For this spherical mean we have

$$w(0, t) = \sigma_{n-1}\omega[u;x, t], \qquad iw_s(0, t) = \sigma_{n-1}\omega[u_t;x, t].$$

It follows that (6.36) can be expressed in the more easily understood form

$$(6.39) \quad u(t, x_1, \ldots, x_n) = \frac{1}{1 \cdot 3 \cdots (n - 2)} \left\{ \frac{\partial}{\partial t}\left(\frac{\partial}{t\, \partial t}\right)^{(n-3)/2} t^{n-2}\omega[u\,;\,x,\,t] \right.$$

$$\left. + \left(\frac{\partial}{t\, \partial t}\right)^{(n-3)/2} t^{n-2}\omega[u_t\,;\,x,\,t]\right\},$$

since[4] for odd n

$$\frac{n\tau_{n+1}}{\sigma_{n-1}} = \frac{\tau_{n+1}}{\tau_n} = \frac{1 \cdot 3 \cdots n}{2 \cdot 4 \cdots (n + 1)}\pi.$$

Formula (6.39) exhibits the dependence of u on Cauchy data prescribed at $t = 0$. It is a remarkable fact that for an evaluation of the terms on the right we only need data over the $(n - 1)$-dimensional boundary of the n-dimensional sphere of radius t cut out of the initial hyperplane $\tau = 0$ by the characteristic cone

$$(\tau - t)^2 = (\xi_1 - x_1)^2 + \cdots + (\xi_n - x_n)^2$$

for (6.26). The theory of the characteristics of a hyperbolic partial differential equation already indicated (cf. Section 1) that the solid sphere itself would constitute a domain of dependence for u at the vertex $(t, x_1, \ldots, x_n)$ of the cone. However, for the wave equation in an *odd* number $n \geq 3$ of space variables, the more specific representation (6.39) by means of the averages ω shows that the solution actually depends on the data and their derivatives exclusively over the *perimeter* of the domain of dependence.

The phenomenon just described is known as *Huygens' principle*. We shall return to a more detailed study of its implications later (cf. Section 3). We merely observe now that it does not apply to the wave equation when the number of space variables is not odd. This conclusion can be drawn without a more explicit calculation of the answer, for it follows directly from our earlier remark that for even n the integrand in (6.29) does

[4] Cf. Courant 1.

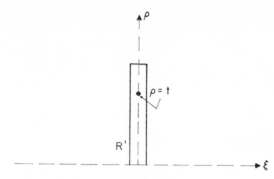

FIG. 17.   Contour of integration for even n.

not cancel out over the section of the initial hyperplane which projects onto a slit $0 \leq \rho \leq t$ along the ρ-axis in the meridian plane. One of the important advantages of solving Cauchy's problem by means of the identity (6.29) for harmonic functions is the explanation thus obtained, without actual recourse to the more specific computation leading to (6.39), of the reasons why Huygens' principle should hold for the wave equation in an odd, but not an even, number of space variables.

If we attempt to transform (6.29) into a formula for the solution of the wave equation when n is *even*, we notice that the contribution from integrals over the circle (6.33) has an expansion in exclusively odd powers of $\sqrt{\epsilon}$. Since $v(x_0, \ldots, x_n)$ does not really depend on ϵ, we expect all such terms to drop out in the limit as $\epsilon \to 0$, despite the fact that some of the exponents involved are negative. What is left ought to be expressible in terms of an integral over the segment $0 \leq \rho \leq t$ of the ρ-axis in the meridian plane. To verify this we find it convenient to shift the surface of integration ∂D for (6.29), which has only to enclose the sphere (6.32) where $r = 0$, so that it corresponds to the three sides $\xi = \pm\epsilon$, $0 \leq \rho \leq t + l$ and $\rho = t + l$, $|\xi| \leq \epsilon$ of an incomplete rectangle R' in the (ξ,ρ)-plane. Imitating our derivation of (6.39), we shall hold $l > 0$ fixed while allowing $\epsilon \to 0$. Our intention is to integrate by parts appropriately with respect to ρ in order to ascertain what becomes of (6.29) in these circumstances.

Using the spherical mean $w(\xi, \rho)$ again, we can put (6.29) into a form identical with (6.34), except that R must be replaced by R' as the path of integration. We integrate the result by parts with respect to ρ over the longer legs of R' in a fashion similar to that used to derive (6.35). If this process is repeated only $(n-2)/2$ times, all the integrated terms include the factor ρ and hence vanish at $\rho = 0$. On the other hand, the contributions evaluated on the short side of R' approach zero with ϵ and can therefore be denoted by the symbol $O(\epsilon)$. Thus we obtain for even n the formula

$$(6.40) \quad v(it, x_1, \ldots, x_n) = \frac{1}{\beta_n} \int_{R'} \left[\frac{\rho}{r} \left(\frac{\partial}{\rho \, \partial \rho} \right)^{(n-2)/2} \left(\rho^{n-2} \frac{\partial w}{\partial \xi} \right) \right.$$

$$\left. - \left(\frac{\partial}{\rho \, \partial \rho} \right)^{(n-2)/2} (\rho^{n-2} w) \frac{\partial}{\partial \xi} \frac{\rho}{r} \right] d\rho + O(\epsilon)$$

analogous to (6.35), where

$$\beta_n = 1 \cdot 3 \cdots (n-3)\sigma_n' = 1 \cdot 3 \cdots (n+1)\tau_{n+1}.$$

In the second term on the right in (6.40) we are permitted once more to replace the operator $\partial/\partial \xi$ by $i\partial/\partial t$ and then to bring it outside the sign of integration. Also, we see that as $\epsilon \to 0$ the integrals over those sections of R' which lie in the range $t < \rho \leq t + l$ beyond the branch point of r located at $\xi = 0$, $\rho = t$ cancel out. However, for $\rho \leq t$ the same cannot be said. There the sign of r changes so that the terms coming from the two opposite sides of the ρ-axis combine to yield the final expression

$$(6.41) \quad v(it, x_1, \ldots, x_n) = \frac{2}{\beta_n} \left\{ \frac{\partial}{\partial t} \int_0^t \left(\frac{\partial}{\rho \, \partial \rho} \right)^{(n-2)/2} [\rho^{n-2} w(0, \rho)] \frac{\rho \, d\rho}{\sqrt{t^2 - \rho^2}} \right.$$

$$\left. + i \int_0^t \left(\frac{\partial}{\rho \, \partial \rho} \right)^{(n-2)/2} [\rho^{n-2} w_\xi(0, \rho)] \frac{\rho \, d\rho}{\sqrt{t^2 - \rho^2}} \right\}$$

for v involving improper integrals that are convergent.

We may perform the substitutions (6.37) and (6.38) in order to reduce (6.41) to a representation of the solution u of Cauchy's problem for the wave equation in an even number n of space variables. In addition, we are able to replace the differentiations with respect to ρ under the integral sign by differentiations with respect to t through $(n-2)/2$ applications of the rule

$$\int_0^t \frac{\partial F(\rho)}{\rho \, \partial \rho} \frac{\rho \, d\rho}{\sqrt{t^2 - \rho^2}} = \frac{\partial}{t \, \partial t} \int_0^t \frac{\partial F(\rho)}{\partial \rho} \sqrt{t^2 - \rho^2} \, d\rho$$

$$= \frac{\partial}{t \, \partial t} \int_0^t F(\rho) \frac{\rho \, d\rho}{\sqrt{t^2 - \rho^2}},$$

which is established, under the hypothesis $F(0) = 0$, by means of an integration by parts. Taking into account the relation

$$\frac{n \tau_{n+1}}{\sigma_{n-1}} = \frac{\tau_{n+1}}{\tau_n} = 2 \frac{2 \cdot 4 \cdots n}{1 \cdot 3 \cdots (n+1)}$$

valid[5] for even n, we derive in this manner the fundamental result

(6.42)

$$
u(t, x_1, \ldots, x_n) = \frac{1}{2 \cdot 4 \cdots (n-2)} \left\{ \frac{\partial}{\partial t} \left(\frac{\partial}{t \, \partial t} \right)^{(n-2)/2} \int_0^t \omega[u; x, \rho] \frac{\rho^{n-1} \, d\rho}{\sqrt{t^2 - \rho^2}} \right.
$$

$$
\left. + \left(\frac{\partial}{t \, \partial t} \right)^{(n-2)/2} \int_0^t \omega[u_t; x, \rho] \frac{\rho^{n-1} \, d\rho}{\sqrt{t^2 - \rho^2}} \right\},
$$

where the coefficient in front of the braces should be interpreted as unity in the special case $n = 2$.

An inspection of (6.42) suffices to show that Cauchy data over a full domain of dependence in the initial hyperplane $t = 0$ are required to determine u, as was predicted for even n by our earlier analysis. Observe, however, that the infinity of the integrand at $\rho = t$ still tends to weight more heavily the part of the initial data located near the perimeter of the domain of dependence.

We have arrived at the expressions (6.39) and (6.42) for the solution u of the initial value problem for the wave equation under the rather stringent assumption that it can be represented as a harmonic function v of an imaginary argument. It is essential to verify that the final results of our computation hold even when the data (6.27) are not analytic and that they define a solution of the problem provided only that the initial functions f and g possess derivatives of a high enough order to ensure appropriate differentiability of the spherical means ω. We shall present a proof of these facts here only for odd n, since they follow afterward for even n by an application of the method of descent to be described in the next section. Note, incidentally, that our formulas for u really do involve derivatives of f and g at the edge of the domain of dependence.

Motivated by (6.39), we set

$$
(6.43) \quad u(t, x_1, \ldots, x_n) = \frac{1}{1 \cdot 3 \cdots (n-2)} \left\{ \frac{\partial}{\partial t} \left(\frac{\partial}{t \, \partial t} \right)^{(n-3)/2} t^{n-2} \omega[f; x, t] \right.
$$

$$
\left. + \left(\frac{\partial}{t \, \partial t} \right)^{(n-3)/2} t^{n-2} \omega[g; x, t] \right\}
$$

and proceed to establish that the function u so defined fulfills the requirements (6.26) and (6.27) of Cauchy's problem under the hypothesis that the partial derivatives of f and g of order $(n + 3)/2$ exist and are continuous.

[5] Cf. Courant 1.

For this purpose we write

$$\omega[g; x, t] = \frac{1}{\sigma_{n-1}} \int_{\partial E} g(x_1 + \eta_1, \ldots, x_n + \eta_n) \, d\Omega,$$

where E stands for the solid sphere

$$\eta_1^2 + \cdots + \eta_n^2 \leq t^2$$

and ∂E is its surface. We note that

(6.44) $$\Delta\omega[g; x, t] = \frac{1}{\sigma_{n-1}} \int_{\partial E} \Delta g \, d\Omega = \frac{1}{\sigma_{n-1} t^{n-1}} \int_{\partial E} \Delta g \, d\sigma$$

$$= \frac{1}{\sigma_{n-1} t^{n-1}} \frac{\partial}{\partial t} \int_E \Delta g \, d\eta_1 \cdots d\eta_n,$$

with

$$\Delta = \frac{\partial^2}{\partial x_1^2} + \cdots + \frac{\partial^2}{\partial x_n^2}.$$

In view of the divergence theorem and the relation

$$-\frac{\partial g}{\partial \nu} = \frac{\partial g}{\partial \eta_1} \frac{\partial \eta_1}{\partial t} + \cdots + \frac{\partial g}{\partial \eta_n} \frac{\partial \eta_n}{\partial t} = \frac{\partial g}{\partial t},$$

we obtain

$$\Delta\omega[g; x, t] = -\frac{1}{\sigma_{n-1} t^{n-1}} \frac{\partial}{\partial t} \int_{\partial E} \frac{\partial g}{\partial \nu} \, d\sigma$$

$$= \frac{1}{\sigma_{n-1} t^{n-1}} \frac{\partial}{\partial t} t^{n-1} \frac{\partial}{\partial t} \int_{\partial E} g \, d\Omega = \frac{1}{t^{n-1}} \frac{\partial}{\partial t} \left(t^{n-1} \frac{\partial}{\partial t} \omega[g; x, t] \right)$$

from (6.44). It is now possible to deduce by setting $F = \omega$ in the quite general identity

(6.45) $$\frac{\partial^2}{\partial t^2} \left(\frac{\partial}{t \, \partial t} \right)^{(n-3)/2} t^{n-2} F(t) = \left(\frac{\partial}{t \, \partial t} \right)^{(n-1)/2} t^{n-1} \frac{\partial}{\partial t} F(t)$$

that the second term on the right in (6.43) must satisfy the wave equation (6.26). The same is true of the first term and therefore of u, too, because only an additional partial differentiation with respect to t is involved and that transforms one solution of (6.26) into another. Incidentally, to prove (6.45) it suffices to carry out the verification for arbitrary powers $F = t^k$ only, in view of Taylor's theorem.

The initial conditions (6.27) follow most easily if we consider first the case $f \equiv 0$. Then the relevant contributions in (6.43) are those resulting from applications of the operator $\partial/t\partial t$ exclusively to the factor t^{n-2} and not to $\omega[g;x, t]$. Such a process cancels the product of integers $1 \cdot 3 \cdots (n-2)$ in the denominator and yields the leading term

$$t\omega[g;x, 0] = tg(x_1, \ldots, x_n)$$

for the Taylor expansion of u about $t = 0$. This shows that (6.27) is satisfied when $f \equiv 0$ because it turns out that $u = 0$ and $u_t = g$ at $t = 0$. Furthermore, we have $u_{tt} = \Delta u = 0$ there also. Thus if we now interchange the hypotheses about f and g, taking $g = 0$, and notice that u has the same form it had before except for an additional partial differentiation with respect to t, we are led to the conclusion that in the new situation $u = f$ and $u_t = 0$ at $t = 0$. Hence (6.27) is again verified. Since the initial conditions (6.27) are linear, they must also be satisfied by the solution (6.43) of the general problem, for it is expressed merely as the sum of the two special wave functions just described.

What we have succeeded in establishing is the existence of a solution of the initial value problem (6.26), (6.27) when n is odd. The question of uniqueness has already been settled in the affirmative in Section 1. It remains to discuss the issue of continuous dependence of the solution u on the initial data f and g. The explicit representation (6.43) provides an answer. It shows that u changes continuously when the variations in f and g are small, together with all their partial derivatives of orders less than or equal to $(n - 1)/2$. We must view this result as satisfactory evidence that the Cauchy problem is well posed for the wave equation. However, the large number of independent variables here does entail a new feature in as much as the data must be chosen to lie in a class of functions possessing continuous derivatives of an appropriately high order if our analysis of existence and of continuous dependence on initial values is to meet with success. It should be observed that the dependence of the solution on data over only a spherical surface of dimension $n - 1$ rather than over its n-dimensional interior, which we have described as the Huygens phenomenon, is connected with the requirement of extra differentiability.

As indicated earlier, the method of descent to be developed in the next section can be applied to establish results of the above kind for the wave equation in any even number n of space variables. The same thing could also be accomplished directly, of course, on the basis of formula (6.42). Finally, for the case of a curved initial surface the reader is referred to Section 4.

EXERCISES

1. Show that the solution of the Cauchy problem (6.26), (6.27) is an odd function of t when $f \equiv 0$ and is an even function of t when $g \equiv 0$.

2. Give a geometrical discussion of the simplest cases $n = 2$ and $n = 3$ of the procedure described in the text. In particular, make a sketch of the surface of integration ∂D when $n = 2$. What becomes of the method if $n = 1$?

3. Use Fourier analysis[6] to derive formulas (6.39) and (6.42).

4. Show that the *retarded potential*[7]

(6.46)
$$u(t, x, y, z) = \frac{1}{4\pi} \iiint_{\rho \leq t} \frac{F(t - \rho, \xi, \eta, \zeta)}{\rho} \, d\xi \, d\eta \, d\zeta,$$

in which

$$\rho = \sqrt{(x - \xi)^2 + (y - \eta)^2 + (z - \zeta)^2},$$

satisfies the inhomogeneous wave equation

$$u_{tt} - u_{xx} - u_{yy} - u_{zz} = F(t, x, y, z),$$

together with the homogeneous initial conditions

$$u(0, x, y, z) = u_t(0, x, y, z) = 0.$$

Work out the analogous solution of the inhomogeneous wave equation in any number of independent variables (cf. also Exercise 3.8 below). Combine these results with (6.42) or (6.43) to solve the initial value problem for the wave equation explicitly in the inhomogeneous case, with data assigned at $t = 0$.

5. Use a Lorentz transformation (cf. Section 3.3) to derive from (6.39) or (6.42) an expression for the solution of the Cauchy problem for the wave equation with data prescribed on an arbitrary space-like hyperplane.

6. If λ is an arbitrary constant, show that Cauchy's problem for the equation

$$\frac{\partial^2 u}{\partial t^2} = \frac{\partial^2 u}{\partial x_1^2} + \frac{\partial^2 u}{\partial x_2^2} + \frac{\partial^2 u}{\partial x_3^2} + \lambda u$$

in three space variables, with initial data of the type (6.27) prescribed at $t = 0$, has a solution of the form[8]

$$u(t, x_1, x_2, x_3) = \frac{\partial}{\partial t} \left\{ t\omega[f; x, t] + \lambda \int_0^t \rho^2 \omega[f; x, \rho] I(\lambda t^2 - \lambda \rho^2) \, d\rho \right\}$$

$$+ t\omega[g; x, t] + \lambda \int_0^t \rho^2 \omega[g; x, \rho] I(\lambda t^2 - \lambda \rho^2) \, d\rho,$$

where

$$I(\theta) = \frac{I_0'(\sqrt{\theta})}{\sqrt{\theta}} = \sum_{k=0}^{\infty} \frac{\theta^k}{2^{2k+1} k! \, (k + 1)!}.$$

[6] Cf. Courant-Hilbert 1.
[7] Cf. Sommerfeld 2.
[8] Cf. Hadamard 3.

7. With the help of Exercise 5.2.1, solve Cauchy's problem for the most general linear hyperbolic equation (6.1) with constant coefficients and with initial data given by (6.27).

8. By applying Stokes' theorem directly to the surface integral in formula (6.44), with the Laplacian there expressed in terms of spherical coordinates, avoid integration over the solid sphere E in the verification that (6.43) solves the initial value problem for the wave equation in the case of three space variables.

9. For initial data f and g that are not analytic, give an alternate proof that (6.43) represents a solution of the Cauchy problem basing the argument on the Cauchy-Kowalewski theorem, on an approximation to f and g by polynomials, and on the continuous dependence of the representation (6.39) on derivatives of u of a sufficiently high order.

10. Explain in terms of a characteristic initial value problem why the formula[9]

$$v(x, y, z) = \int_0^{2\pi} F\left(\frac{x^2 + y^2 + z^2}{z + ix \cos \theta + iy \sin \theta}, \theta \right) \frac{d\theta}{\sqrt{x^2 + y^2 + z^2}},$$

where F is analytic, should furnish the general solution of Laplace's equation

$$v_{xx} + v_{yy} + v_{zz} = 0.$$

11. Derive (6.36) from (6.29) starting with $\xi_0 - x_0$ real and taking $x_0 = it$ fixed, but introducing an analytic continuation to general complex values of the variable of integration ξ_0. The result is to be achieved by a shift of the surface of integration ∂D through the complex domain until it reaches a final limiting position in the real hyperplane $\xi_0 = 0$. The latter step must be justified by means of Stokes' theorem and the Cauchy-Riemann equations.

12. Show at least for $n = 5$ that the result obtained by using a circular path of integration in (6.34) instead of a rectangular one reduces to (6.39) because u satisfies the wave equation at $t = 0$.

3. THE METHOD OF DESCENT

It is interesting to compare the various formulas for the solution of Cauchy's problem in different numbers of independent variables. Observe that because a solution of the wave equation in n space variables is automatically a solution in one more space variable, too, we should be able to deduce the formula for the solution of the Cauchy problem in n space variables from that derived for $n + 1$ space variables. This simple idea for solving a problem with a specified number of independent variables by recourse to a known answer in space of higher dimension is called the *method of descent*. It is particularly significant in the case of

[9] Cf. Whittaker 1.

Cauchy's problem because of the change in the nature of the solution when we pass from an even to an odd number of independent variables. Here we shall make a detailed investigation of the implications of the method of descent for the wave equation in one, two and three space variables.

We start by eliciting from (6.43) the classical *Kirchhoff solution*

$$(6.47) \quad t\omega[g; x, t] = \frac{t}{4\pi} \int_0^{2\pi} \int_0^\pi g(x_1 + t \sin \theta \cos \phi, x_2 + t \sin \theta \sin \phi, x_3$$
$$+ t \cos \theta) \sin \theta \, d\theta \, d\phi$$

of the wave equation $u_{tt} = \Delta u$ in four independent variables x_1, x_2, x_3 and t. It corresponds, of course, to *Poisson's solution*

$$(6.48) \qquad u(t, x) = \frac{\partial}{\partial t} t\omega[f; x, t] + t\omega[g; x, t]$$

of the initial value problem $u(0, x) = f$, $u_t(0, x) = g$ with $n = 3$. We shall proceed to discuss the analogous problem for $n = 2$ by descent.

Suppose that g depends only on its first two arguments. If we think now in terms of two space variables x_1 and x_2, we should write

$$\omega[g; x, t] = \frac{1}{2\pi} \int_0^{2\pi} g(x_1 + t \cos \phi, x_2 + t \sin \phi) \, d\phi.$$

Thus the substitution $\rho = t \sin \theta$ leads to a modified version

$$\frac{t}{4\pi} \int_0^{2\pi} \int_0^\pi g(x_1 + t \sin \theta \cos \phi, x_2 + t \sin \theta \sin \phi) \sin \theta \, d\theta \, d\phi$$
$$= \int_0^t \omega[g; x, \rho] \frac{\rho \, d\rho}{\sqrt{t^2 - \rho^2}}$$

of the integral (6.47). Together with an analogous expression in f, this yields the solution of the initial value problem in the case $n = 2$. If we interpret ρ and ϕ as polar coordinates in the (ξ_1, ξ_2)-plane, the final result can be stated in the form

$$(6.49) \qquad u(t, x) = \frac{\partial}{\partial t} \frac{1}{2\pi} \iint_{\xi_1^2 + \xi_2^2 \leq t^2} \frac{f(x_1 + \xi_1, x_2 + \xi_2) \, d\xi_1 \, d\xi_2}{\sqrt{t^2 - \xi_1^2 - \xi_2^2}}$$
$$+ \frac{1}{2\pi} \iint_{\xi_1^2 + \xi_2^2 \leq t^2} \frac{g(x_1 + \xi_1, x_2 + \xi_2) \, d\xi_1 \, d\xi_2}{\sqrt{t^2 - \xi_1^2 - \xi_2^2}}$$

associated with the name of *Parseval*.

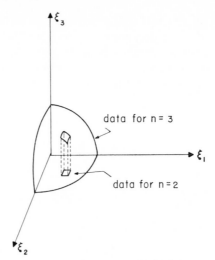

FIG. 18.   Projection of initial data.

Notice that the radical appearing in (6.49) serves to convert the area element $d\xi_1 d\xi_2$ over the domain of dependence $\xi_1^2 + \xi_2^2 \leq t^2$ for the wave equation in three variables into the spherical surface element $t \sin \theta \, d\theta \, d\phi$ that features in the averages used to generate the solution (6.48) of Cauchy's problem in four variables. The latter obeys Huygens' principle, but its initial data over a sphere project onto data for (6.49) over a circular disc.

We can descend again to derive the solution of the initial value problem for $n = 1$. Indeed, if g depends only on its first argument, the second term on the right in (6.49) becomes

$$\frac{1}{2\pi} \int_{-t}^{t} g(x_1 + \xi_1) \int_{-\sqrt{t^2 - \xi_1^2}}^{\sqrt{t^2 - \xi_1^2}} \frac{d\xi_2}{\sqrt{t^2 - \xi_1^2 - \xi_2^2}} \, d\xi_1 = \frac{1}{2} \int_{-t}^{t} g(x_1 + \xi_1) \, d\xi_1.$$

A similar reduction of the other integral in (6.49) leads to *d'Alembert's formula*

$$(6.50) \quad u(t, x) = \frac{1}{2} [f(x + t) + f(x - t)] + \frac{1}{2} \int_{-t}^{t} g(x + \xi) \, d\xi,$$

which we encountered earlier in the form (4.17).

This solution of the wave equation in two variables depends exclusively on initial data at the ends $x - t$ and $x + t$ of the interval of dependence only when $g \equiv 0$. The failure of Huygens' principle to hold more generally for one space variable can be traced to the occurrence of a logarithm in the analogue of (6.29) for the degenerate case $n = 1$. Indeed, for harmonic functions v of two variables the representation corresponding to (6.29) is

$$(6.51) \qquad v(x_0, x_1) = \frac{1}{2\pi} \int_{\partial D} \left[v \frac{\partial}{\partial v} \log \frac{1}{r} - \frac{\partial v}{\partial v} \log \frac{1}{r} \right] ds.$$

If we perform the complex substitution $x_0 = it$ in (6.51) and then draw the contour of integration ∂D down on a slit joining the two roots of the equation $r = 0$, we are led to the result (6.50). According to such an analysis it is the period of πi in $\log r$ around the branch points at the ends of the above slit that contributes the non-trivial integral of g in (6.50) over the interior of the domain of dependence. This situation is to be contrasted with the single-valuedness of the fundamental solution of Laplace's equation when $n + 1$ is an even integer exceeding two, which gives rise to Huygens' principle.

It is instructive to examine Huygens' principle in more detail. A glance at the explicit solution

$$u(t, x, y, z) = \frac{1}{4\pi} \iiint_{\rho \leq t} \frac{F(t - \rho, \xi, \eta, \zeta)}{\rho} \, d\xi \, d\eta \, d\zeta$$

of the inhomogeneous wave equation in three space variables (cf. Exercise 2.4) shows that it, too, depends only on the values of the prescribed function F over a characteristic cone $\tau = t - \rho$ whose vertex lies at the point where u is to be evaluated. This example of the Huygens phenomenon can be expressed quite differently if it is formulated in terms of *radiation* emanating from a given source. Let F be the product of a function of $t - \rho$ alone multiplied by an auxiliary factor which depends on the space variables in such a way that it has a fixed average but differs from zero only in some neighborhood of a single point (ξ, η, ζ). We can generate a solution of the homogeneous wave equation of the form

$$(6.52) \qquad u(t, x, y, z) = \frac{F(t - \rho)}{\rho}$$

by allowing the diameter of the neighborhood in question to shrink toward zero.

When F vanishes outside an infinitesimal interval about some value τ of its argument, formula (6.52) represents a spherical wave that radiates to infinity in three-dimensional space from a source at (ξ, η, ζ). The fact that the influence of such a wave can be felt at a specific time t only *on* the surface of the sphere

$$(6.53) \qquad \rho = t - \tau$$

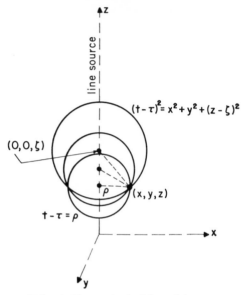

FIG. 19. Huygens' principle and descent.

or radius $t - \tau$ about the source, where $F \neq 0$, and *not inside* that sphere, where $F = 0$, is synonymous with Huygens' principle. In this context the equation (6.53) should be interpreted as describing a characteristic cone that bounds the range of influence for our radiation problem.

When Huygens' principle is stated in terms of a radiating spherical wave of the type (6.52), an explanation for its failure in the case of two space variables can be given on physical grounds that involve the idea of descent. We have seen that for an appropriate choice of F the spherical wave (6.52) corresponds to a signal that is heard in a specified place at just one single instant and leaves behind it there no residual effect. The analogue for two variables x and y of (6.52) is a cylindrical wave generated by what amounts in three-dimensional space to a uniform distribution of sources along a line $x = \xi$, $y = \eta$ parallel to the z-axis. An observer located at a distance ρ from that line will receive signals from the source only after an interval of time $t - \tau = \rho$ has elapsed from the moment τ when they were emitted. Thereafter signals coming to the same location from positions farther and farther away on the line source will, however, continue to be heard indefinitely. It is for this reason that Huygens' principle is not valid in two-dimensional space.

The method of descent can be combined with the technique of separation of variables to produce from (6.43) solutions of Cauchy's problem for partial differential equations other than the wave equation. In fact, most of the cases for which an explicit answer has been found are amenable to this approach. As an illustration, we shall solve the initial value problem for the equation

$$(6.54) \qquad \frac{\partial^2 u}{\partial t^2} = \frac{\partial^2 u}{\partial x_1^2} + \frac{\partial^2 u}{\partial x_2^2} + \lambda u$$

with data assigned at $t = 0$, where the coefficient $\lambda = k^2$ is a constant.

Our point of departure is the Kirchhoff solution (6.47) of the wave equation with an integrand g of the special separated form

$$g(x_1, x_2, x_3) = g(x_1, x_2)e^{kx_3}.$$

In such circumstances (6.47) represents a solution $u = u(t, x_1, x_2)$ of (6.54) multiplied by the factor e^{kx_3}. Dividing out this exponential factor, we obtain for u itself the expression

$$(6.55) \quad u(t, x_1, x_2) = \frac{t}{4\pi} \int_0^{2\pi} \int_0^{\pi} g(x_1 + t \sin \theta \cos \phi, \, x_2 + t \sin \theta \sin \phi)$$
$$\cdot e^{kt \cos \theta} \sin \theta \, d\theta \, d\phi.$$

The substitution $\rho = t \sin \theta$ used earlier, followed by the transformation

$$\xi_1 = \rho \cos \phi, \qquad \xi_2 = \rho \sin \phi$$

from polar to rectangular coordinates, brings (6.55) into a form similar to the second term on the right in (6.49). From that we pass directly to the representation

$$(6.56) \quad u(t, x_1, x_2)$$
$$= \frac{\partial}{\partial t} \frac{1}{2\pi} \iint\limits_{\xi_1^2 + \xi_2^2 \le t^2} \frac{f(x_1 + \xi_1, x_2 + \xi_2) \operatorname{ch}(k\sqrt{t^2 - \xi_1^2 - \xi_2^2}) \, d\xi_1 \, d\xi_2}{\sqrt{t^2 - \xi_1^2 - \xi_2^2}}$$
$$+ \frac{1}{2\pi} \iint\limits_{\xi_1^2 + \xi_2^2 \le t^2} \frac{g(x_1 + \xi_1, x_2 + \xi_2) \operatorname{ch}(k\sqrt{t^2 - \xi_1^2 - \xi_2^2}) \, d\xi_1 \, d\xi_2}{\sqrt{t^2 - \xi_1^2 - \xi_2^2}},$$

based on (6.48), of the solution of the initial value problem for equation (6.54).

To close our discussion of the method of descent, we note that it also applies to situations involving neither initial nor boundary conditions. For example, the fundamental solution of a partial differential equation can be investigated by such a procedure, which enables us in particular to avoid a separate treatment of the case of an even number of independent variables by simply observing that it is a consequence of our results for an odd number. For the purpose at hand it is merely necessary to add an extra term involving a new independent variable to the original equation under study. We thus ascend to a space of higher dimension for the sole purpose of descending once more. Exercise 5.1.6 may be interpreted as an illustration of this technique.

EXERCISES

1. Prove that Kirchhoff's solution (6.47) of the wave equation is an odd function of the time t.

2. Derive (6.56) from the result of Exercise 2.6, noting that in neither of the two cases does Huygens' principle hold. What becomes of these results if λ is negative? Generalize them to the case of an arbitrary number of space variables.

3. Apply the method of descent to formula (6.49) in order to find the Riemann function of equation (4.87) when $\lambda = 1$.

4. Use (6.43) and the method of descent to establish the existence of the solution of the initial value problem (6.26), (6.27) for even n. Deduce (6.42) as a corollary.

5. Derive (6.39) from (6.42) and thus show in a different way that Huygens' principle is valid for odd $n \geq 3$.

6. Give details of the derivation of (6.50) from (6.51) by means of a complex substitution.

7. Prove the following version of *Duhamel's principle*. Let $v(x, t; \tau)$ solve the linear homogeneous system

$$L[v] = \frac{\partial v}{\partial t} - \sum_{j=1}^{n} A_j \frac{\partial v}{\partial x_j} - Bv = 0,$$

with inhomogeneous initial data

$$v(x, \tau; \tau) = f(x, \tau)$$

assigned at $t = \tau$. Then the *Duhamel integral*

(6.57) $$u(x, t) = \int_0^t v(x, t; \tau)\, d\tau$$

satisfies the inhomogeneous equation

$$L[u] = f,$$

together with the homogeneous initial condition

$$u(x, 0) = 0.$$

8. Formulate Duhamel's principle for the wave equation. Use it to derive the retarded potential (6.46) from Kirchhoff's solution (6.47).

9. If $u(x; y) = u(x_1, \ldots, x_n, y_1, \ldots, y_n)$ stands for a solution of the ultra-hyperbolic equation

$$(6.58) \qquad \frac{\partial^2 u}{\partial x_1^2} + \cdots + \frac{\partial^2 u}{\partial x_n^2} = \frac{\partial^2 u}{\partial y_1^2} + \cdots + \frac{\partial^2 u}{\partial y_n^2},$$

prove *Asgeirsson's mean value theorem*[10]

$$(6.59) \qquad \omega[u(x; y); x, t] = \omega[u(x; y); y, t],$$

where the $(n-1)$-dimensional spherical means ω are defined as in Section 2, with y interpreted as a parameter on the left and x interpreted as a parameter on the right.

10. Use the method of descent to generalize Asgeirsson's theorem to a situation where the number of independent variables occurring on one side of (6.58) is less than that occurring on the other side. Apply the result to obtain an alternate solution of the initial value problem (6.26), (6.27).

11. By performing Lorentz transformations (cf. Section 3.3) on the identity

$$(6.60) \qquad u(2t, x) = \frac{\partial}{\partial t} t\omega[u(t, x); x, t],$$

valid for any solution of the wave equation (6.26) with $n = 3$, solve the characteristic initial value problem for that equation in closed form. Determine precisely what data are needed to calculate the solution, and relate the answer to Huygens' principle. Show that (6.60) is a consequence of (6.48), or of (6.59), taking care to observe that the average involved is not evaluated over the usual hyperplane $t = 0$.

12. By considering suitable odd functions of x_n, show how to solve a mixed initial and boundary value problem for the wave equation (6.26) with initial data (6.27) prescribed only over the half-space $x_n > 0$, but with the additional boundary condition $u = 0$ imposed at $x_n = 0$, $t \geq 0$. The answer is to be determined for positive values of x_n and t.

4. EQUATIONS WITH ANALYTIC COEFFICIENTS

It will be our aim in what follows to extend the procedure used in Section 2 for the solution of the special Cauchy problem (6.26), (6.27) to the case of a hyperbolic equation with variable coefficients, under the hypothesis that the data are assigned on a surface T which is duly inclined in the sense of the space-like requirement (6.6). Using a fundamental solution, we intend to derive integral formulas like (6.39) and (6.42) that express the solution of Cauchy's problem in terms of the initial data. The distinguishing feature of our treatment is an analytic continuation

[10] Cf. Courant-Hilbert 1.

into the complex domain that enables us to apply contour integration techniques and the calculus of residues.[11] Therefore we place the main emphasis on equations whose coefficients are real analytic functions of the independent variables. However, we shall indicate later on how such an objectionable restriction can be set aside both with regard to the coefficients and with regard to the initial data.

The Cauchy problem for equations with coefficients that are not analytic will be reduced to an integral equation based on the parametrix of Section 5.3. The integral equation may be solved by successive approximations in a fashion reminiscent of the method of Section 4.2. Underlying the existence proof so obtained are certain *a priori estimates* which follow from the integral equation. The energy inequalities of Section 1 actually furnish a more elementary set of estimates that can be exploited to the same end. This will ultimately be achieved in Section 12.2. However, we note that the resulting construction of the solution of Cauchy's problem is correspondingly more existential than the one to be described here.

The Cauchy-Kowalewski theorem of Section 1.2 assures us of the existence of an analytic solution of Cauchy's problem for an analytic equation

$$(6.61) \qquad \frac{\partial^2 u}{\partial t^2} = \sum_{j,k=1}^{n} a_{jk} \frac{\partial^2 u}{\partial x_j\, \partial x_k} + b\,\frac{\partial u}{\partial t} + \sum_{j=1}^{n} b_j\,\frac{\partial u}{\partial x_j} + cu$$

of the hyperbolic type with data prescribed analytically on a space-like section of surface T. We shall be concerned with the explicit representation of this solution by means of a fundamental solution S of the adjoint equation. The desired formula will be derived from a suitable modification of the basic identity

$$(6.62) \qquad v(\xi) = \int_{\partial D} B[S(x, \xi), v(x)]$$

for solutions v of a linear elliptic partial differential equation (cf. Section 5.2).

The manipulations needed to carry out the relevant transformation of (6.62) for variable coefficients are more devious than those we used in the case of the wave equation, since it is not possible in general to find an elliptic equation equivalent through a simple complex substitution to the given hyperbolic one. We shall be able to attain our objective, however, by converting (6.61) into a requirement on an operator which is elliptic only at points where $t, x_1, \ldots, x_n$ actually remain real. Indeed, analytic extension to complex values of the independent variable t will permit us to write down an analogue of (6.62) at such points, with the integral that

---

[11] Cf. Garabedian 8.

appears evaluated over a manifold in the complex domain. The essential difference between the elliptic and hyperbolic cases lies in the choice of this manifold. Our representation of the solution of Cauchy's problem will be obtained by deforming the manifold of integration in a fashion similar to that developed for the wave equation until it finally collapses onto the domain of dependence where the data are assigned in the space of the real coordinates $t, x_1, \ldots, x_n$.

We emphasize that the basic integral identity

$$(6.19) \qquad \int_D (vL[u] - uM[v]) \, dx_1 \cdots dx_n + \int_{\partial D} B[u, v] = 0,$$

listed as formula (5.65) in Section 5.2, is valid in the complex domain whenever u and v are analytic functions of their arguments. We wish to apply it here to the differential operator

$$L[u] = \frac{\partial^2 u}{\partial t^2} - \sum_{j,k=1}^{n} a_{jk} \frac{\partial^2 u}{\partial x_j \, \partial x_k} - b \frac{\partial u}{\partial t} - \sum_{j=1}^{n} b_j \frac{\partial u}{\partial x_j} - cu$$

occurring in equation (6.61). Hence we must now deal with $n + 1$ independent variables $t, x_1, \ldots, x_n$, each of which is allowed to assume complex values. Therefore the region of integration D must be conceived of as an $(n + 1)$-dimensional cell (or manifold) in the complex domain, and its boundary ∂D is an n-dimensional cycle there. Integrals over D and over ∂D are to be interpreted in the sense of the calculus of exterior differential forms,[12] which attaches a meaning to them even when some of the differentials involved are complex. In this context identities of the type (6.19) follow from Stokes' theorem and from the Cauchy-Riemann equations, which make their formal derivation step by step the same as that familiar to us in the real case (cf. Exercise 5.2.13).

Our choice of u in (6.19) is to be the analytic solution of an analytic Cauchy problem for (6.61), whereas we take for v a fundamental solution S of the *adjoint equation*

$$M[S] = 0$$

whose pole lies at a point $(t, x_1, \ldots, x_n)$ in the real domain. We denote by $\tau, \xi_1, \ldots, \xi_n$ the variables of integration; it will suffice for the moment to take all of them real except τ. Thus we pick D to be a small $(n + 1)$-dimensional sphere

$$(6.63) \qquad \text{Im } \{\tau\}^2 + (\xi_1 - x_1)^2 + \cdots + (\xi_n - x_n)^2 \leq \epsilon^2,$$

$$(6.64) \qquad \text{Re } \{\tau - t\} = \text{Im } \{\xi_1\} = \cdots = \text{Im } \{\xi_n\} = 0$$

[12] Cf. Buck 1.

of radius ϵ in the complex domain, centered at the real point $(t, x_1, \ldots, x_n)$.

To generalize (6.62), we apply the complex analogue of formula (6.19) to D with a neighborhood

$$|\Gamma| \leq \delta^2$$

of the singularity of S deleted, where Γ stands for the square of the geodesic distance associated with the operator L. When δ is small enough relative to ϵ, this yields the identity

$$(6.65) \qquad \int_{|\Gamma|=\delta^2} B[u, S] = \int_{\partial D} B[u, S].$$

Note that the differential form $B[u, S]$ is now given by

$$B[u, S] = \left(buS - \left[S \frac{\partial u}{\partial \tau} - u \frac{\partial S}{\partial \tau} \right] \right) d\xi_1 \cdots d\xi_n$$

$$+ \sum_{j=1}^{n} (-1)^j \left(\left[b_j - \sum_{k=1}^{n} \frac{\partial a_{jk}}{\partial \xi_k} \right] uS \right.$$

$$+ \sum_{k=1}^{n} a_{jk} \left[S \frac{\partial u}{\partial \xi_k} - u \frac{\partial S}{\partial \xi_k} \right] \right) d\tau \, d\xi_1 \cdots \widehat{d\xi_j} \cdots d\xi_n,$$

in view of our definition of L.

The limit as $\delta \to 0$ of the integral on the left in (6.65), which is, of course, evaluated within the hyperplane (6.64), is to be calculated by an analysis similar to our derivation of (5.68) in Section 5.2. The experience gained there shows that the answer is not altered if we neglect contributions involving $b, b_1, \ldots, b_n$, and if we replace the coefficients a_{jk} by constants equal to their values at $(t, x_1, \ldots, x_n)$. Thus it turns out that in calculating residues within the hyperplane (6.64) we may confine our attention to a partial differential equation that can be approximated by an elliptic operator with constant coefficients, which we could even reduce to a Laplacian by performing a suitable affine transformation.

About its pole in the above hyperplane we can write the leading term in the expansion of S in the form (cf. Chapter 5)

$$(6.66) \quad \frac{1}{(-1)^{n/2} \sigma_n' \sqrt{\Delta} \, \Gamma_2^{(n-1)/2}}$$

$$= \frac{\left[-\operatorname{Im} \{\tau\}^2 - \sum_{j,k=1}^{n} A_{jk} (\xi_j - x_j)(\xi_k - x_k) \right]^{-(n-1)/2}}{(-1)^{n/2} \sigma_n' \sqrt{\Delta}},$$

where the matrix of elements A_{jk} is the inverse of that generated by the coefficients a_{jk} at $(t, x_1, \ldots, x_n)$, and where

$$\Delta = \begin{vmatrix} a_{11} \cdots a_{1n} \\ a_{21} \cdots a_{2n} \\ \cdots \cdots \\ a_{n1} \cdots a_{nn} \end{vmatrix}.$$

Because it is permissible to replace S on the left in (6.65) by its principal part (6.66) as $\delta \to 0$, we find in the limit that

$$(6.67) \qquad u(t, x_1, \ldots, x_n) = \int_{\partial D} B[u, S],$$

just as in our derivation of formula (5.69). To avoid an awkward factor of $(-1)^{n-1}$ in front of the integral here we make a convenient determination of the branch of S to be used in the complex domain. Namely, we ask that $(-\Gamma_2)^{(n-1)/2} > 0$ over the sphere (6.63), (6.64). Notice that in this matter a significant role is played by the factor $(-1)^{n/2}$ in the normalization (6.66) of S for the hyperbolic equation (6.61). Unfortunately the normalization makes S imaginary in the real domain when n is odd.

Our intention is to convert (6.67) into an explicit representation of the solution of Cauchy's problem by deforming the manifold, or cycle, of integration ∂D through the complex domain until it can be shrunk down around the appropriate domain of dependence in the initial surface T. That the integral concerned is independent of the surface of integration follows from our generalization of the identity (6.19) to a space with complex coordinates. This step can be viewed as a combined application of Cauchy's integral theorem and of the divergence theorem to analytic solutions u and S of an adjoint pair of partial differential equations. The only precaution that must be taken is to avoid intersecting the cycle of integration ∂D with the locus of singularities of the fundamental solution S, which consists of the characteristic conoid $\Gamma = 0$ whose vertex is located at the point $(t, x_1, \ldots, x_n)$. The restriction thus imposed on our deformation of ∂D will be what eventually leads to the emergence of an appropriate domain of dependence within T as the final manifold of integration.

It is helpful to execute the deformation just described in a coordinate system that will make evident its connection with the technique of Section 2. We refer to the coordinates θ_j defined by (2.97), which are now $n + 1$ in number, of course. In view of Exercise 2.6.4, an additional affine transformation of the variables θ_j can be introduced so that the character-istic surface $\Gamma = 0$ becomes a genuine cone like the one for the wave equation. Furthermore, there is no loss of generality in assuming that the surface T on which the initial data are assigned is a hyperplane perpendic-ular to the axis of the cone in question, for we can write the equation of T in terms of spherical coordinates and then substitute, if necessary, a new expression for the radius to achieve such a simplification. Thus it suffices, at least locally, to consider only the case of the wave equation with initial data of the type (6.27) in specifying our deformation of the cycle of inte-gration ∂D from the surface of the small sphere (6.63), (6.64) around the point $(t, x_1, \ldots, x_n)$ into a closed surface degenerating to the associated domain of dependence on T as a limit.

Confining our geometrical discussion to the wave equation with Cauchy data prescribed at $\tau = 0$, we use the cylindrical coordinate

$$\rho = \sqrt{(\xi_1 - x_1)^2 + \cdots + (\xi_n - x_n)^2}$$

to define an appropriate shift of the manifold of integration ∂D. While ρ is to remain real, together with the various angular coordinates associated with it, we allow $\tau = \tau_1 + i\tau_2$ to be complex. Thus the characteristic conoid $\Gamma = 0$ has a realization as two straight lines

$$(6.68) \qquad\qquad \tau_1 = t \pm \rho, \qquad \tau_2 = 0$$

in the three-dimensional space spanned by ρ, τ_1 and τ_2, which are to be interpreted now as rectangular coordinates. Our original choice for ∂D as the surface of the sphere (6.63), (6.64) appears in that space as a circle

$$(6.69) \qquad\qquad \tau_2^2 + \rho^2 = \epsilon^2, \qquad \tau_1 = t$$

of radius ϵ, and the relevant domain of dependence corresponds to the segment

$$(6.70) \qquad\qquad \tau_1 = \tau_2 = 0, \qquad -t \leq \rho \leq t.$$

In terms of ρ, τ_1 and τ_2, our problem is to draw the circle (6.69) down onto the segment (6.70) without permitting it to cross either of the lines (6.68). This can be achieved by specifying a curve into which the circle should be moved in each replica of the family of planes $\tau_1 = \text{const.}$, where it ought to correspond to some convenient loop around the two points of intersection of the characteristic lines (6.68) with that plane (cf. Figure 20). The desired deformation of ∂D is defined by translating the plane $\tau_1 = \text{const.}$ that supports the loop with which it is associated. The starting position is $\tau_1 = t$ and the final one is $\tau_1 = 0$. There the loop

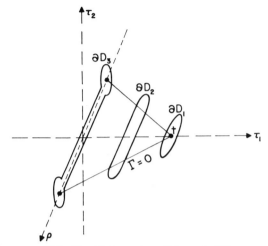

FIG. 20.   Deformation of the cycle ∂D.

is made to shrink down afterward onto the segment (6.70) joining the
pair of points it has been required to enclose.

As the path of integration thus approaches its limit in the plane $\tau_1 = 0$,
formula (6.67) becomes a representation[13]

$$(6.71) \qquad u(t, x) = \lim_{\partial D \to T} \int_{\partial D} B[u(\tau, \xi), S(\tau, \xi; t, x)]$$

of the solution of Cauchy's problem for (6.61) at the point $(t, x_1, \ldots, x_n)$
in terms of initial data over the corresponding domain of dependence
within T. Here we have set

$$x = (x_1, \ldots, x_n), \qquad \xi = (\xi_1, \ldots, \xi_n),$$

and we have resorted to the somewhat symbolic notation $\partial D \to T$ to
indicate the above process for drawing the cycle ∂D down around the
domain of dependence on T. Notice that actually a quite complicated
calculation of residues is involved in (6.71) due to the singularity of the
fundamental solution S on the characteristic conoid $\Gamma = 0$.

It is apparent that we are allowed considerable freedom in our selection
of the deformation of ∂D and that the use of special coordinates suggested
by the example of the wave equation, while helpful, is not essential.
However, our derivation of the basic integral formula (6.71) is in general
valid only in some neighborhood of the point (t, x), since the geometry of
the characteristic conoid $\Gamma = 0$ is usually not simple in the large when the

[13] Cf. Hadamard 3.

coefficients a_{jk} of (6.61) are variable (cf. Exercise 5.2.10). Furthermore, the precise fashion in which ∂D approaches its limit affects the computation of a residue from the integral (6.71) at the edge of the domain of dependence on T, where $\Gamma = 0$ and S becomes infinite. On the other hand, if the partial differential equation (6.61) satisfied by u is used to express the final result of such a computation exclusively in terms of tangential derivatives of the Cauchy data through elimination of high order normal derivatives of u, a step which is made feasible by the Cauchy-Kowalewski construction, the answer obtained must be unique (cf. Exercise 2.12). In referring to (6.71) we shall assume in the future that the reduction just described has been performed.

We are not so much interested in a more explicit evaluation of the representation (6.71) as we are in deducing from its present formulation the most important properties of the solution of Cauchy's problem for a hyperbolic equation. As a first example, we cite the fact that u depends continuously on the partial derivatives of the Cauchy data of order $(n - 1)/2$ or less when the number n of space-like variables is odd. Indeed, the expansion

$$(6.72) \qquad S = \frac{1}{\Gamma^m} \sum_{l=0}^{\infty} U_l \Gamma^l + V \log \Gamma, \qquad m = \frac{n-1}{2},$$

of the fundamental solution (cf. Section 5.2) shows that it becomes infinite at the edge of the domain of dependence like the reciprocal of the distance to that boundary raised to the power $(n - 1)/2$. Hence the residue there in the integral formula (6.71) involves derivatives of u of order $(n - 1)/2$ or less, multiplied by factors depending on the elementary coefficients $U_1, \ldots, U_{m-1}$. From this observation, which is related to the Huygens phenomenon, we can also conclude that the second partial derivatives of u vary continuously with the derivatives of the Cauchy data of order $(n + 3)/2$.

For even values of n a similar analysis can be made on the basis of the expansion

$$(6.73) \qquad S = \frac{1}{\Gamma^m} \sum_{l=0}^{\infty} U_l \Gamma^l, \qquad m = \frac{n-1}{2},$$

of S. However, the theorem in question is proved more readily in that case by an application of the method of descent. It asserts that u, together with its first and second derivatives, depends continuously on the derivatives of the data of order $(n + 4)/2$.

The preceding remarks have the additional consequence that a solution of the Cauchy problem for an analytic equation (6.61) of the hyperbolic type, with initial data prescribed on a space-like surface T, exists provided only that the data possess continuous partial derivatives of order $(n + 3)/2$ for odd n and of order $(n + 4)/2$ for even n. The verification is made by approximating the data, together with their derivatives of the order mentioned, by polynomials, solving the approximate problem by the Cauchy-Kowalewski method, and then using (6.71) to justify a passage to the limit. Since we have just shown that in such circumstances the second derivatives of u must converge, we obtain in this way a solution of (6.61) and, indeed, one assuming the prescribed initial data. It is important to recognize that the Cauchy-Kowalewski solutions on which the proof is based are defined in a fixed region because of the linearity of (6.61), which permits us to reduce the scale of the polynomials featuring as data until their coefficients become as small as we like (cf. Exercise 1.2.3).

Our derivation of the solution of Cauchy's problem in the case of data that are not necessarily analytic shows that it can be represented by means of (6.71), provided that the infinities of the integrand have been removed by either calculating residues or integrating by parts, so that only definite integrals of the Cauchy data and their derivatives over the domain of dependence and over its boundary remain. Furthermore, substitution of arbitrary initial data possessing suitable higher derivatives into a refinement of formula (6.71) of the kind described must always generate a solution of the Cauchy problem for (6.61) simply because the solution associated with that data satisfies (6.71). Incidentally, the solution in question is unique according to theorems from Section 1, and we have already seen that it depends continuously on a specified set of derivatives of the data. Consequently the initial value problem with data assigned on a space-like surface is correctly set in the sense of Hadamard for a hyperbolic equation (6.61) with analytic coefficients, and it can even be solved in the more or less explicit form (6.71).

The validity of the argument developed thus far is confined to a neighborhood of the initial surface T because of difficulties with the geometry of the characteristic conoid $\Gamma = 0$ when the coefficients a_{jk} are variable. However, the linearity of (6.61), combined with its special normal form ensuring that the hyperplanes $t = \text{const.}$ are space-like everywhere, permits us to avoid such a restriction. The device for extension of the solution into the large consists in applying (6.71) repeatedly to initial data computed in succession at various levels of t that are spaced sufficiently close together. Since the range of validity of (6.71) depends only on the geometry of the characteristic conoids, and therefore only on the coefficients of (6.61) and not on the Cauchy data, the neighborhood of the initial surface where (6.71) can be used has a width in the direction of increasing t which is bounded away from zero. It is, in fact, determined by the size of the coefficients a_{jk} alone. Thus we can advance a prescribed

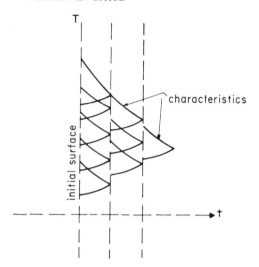

FIG. 21.   Step-by-step solution of Cauchy's problem in the large.

distance in the desired direction by subdividing $(n + 1)$-dimensional space into a finite number of layers perpendicular to the t-axis and using (6.71) to solve the Cauchy problem separately in each such layer. When the original initial data are given in the large, this iterated construction shows that the solution exists in the large, too (but cf. Exercise 4.1.5). However, differentiability of the data of an order proportional to the number of iterations employed must be assumed.

Just as in the case of the wave equation, there are significant changes in the nature of the solution of Cauchy's problem for an equation with variable coefficients as we pass from an even to an odd number n of space variables. When n is even it is convenient to omit regular terms in the expansion of the fundamental solution S and to take it in the special form (6.73) involving only fractional powers of Γ. Under these circumstances the sign of S becomes reversed if its arguments $\tau, \xi_1, \ldots, \xi_n$ perform the equivalent in the complex domain of a circuit around either of the two characteristic lines (6.68). On the other hand, the cycle of integration ∂D has opposite orientations on the two sides of the initial surface T at points where it meets the domain of dependence. The simultaneous changes of sign caused by these effects offset each other in the evaluation of the differential form $B[u, S]$. Hence there are nontrivial contributions in the representation (6.71) of the solution of Cauchy's problem from all interior points of the domain of dependence.

We conclude from the above analysis that Huygens' principle cannot hold for any second order partial differential equation (6.61) in an even number n of space variables. More specifically, suppose that initial data are given at $t = 0$, and suppose that $u = 0$ there, while u_τ has an integral equal to $\frac{1}{2}$ but differs from zero only in some neighborhood σ of a single point $(0, \xi_1, \ldots, \xi_n)$. Far enough inside the range of influence associated with that point we can determine the limit in (6.71) without reference to the singularity of S, since, according to the uniqueness theorem of Section 1, u vanishes identically near points of T outside σ. Examination of the way B is defined shows that in such a situation we have

$$B[u, S] = -S \frac{\partial u}{\partial \tau} d\xi_1 \cdots d\xi_n$$

on T. Therefore when the diameter of σ is allowed to approach zero we obtain for u the explicit result

(6.74) $$u(t, x) = -S(0, \xi; t, x).$$

From (6.74) we see that as a function of its parameter point S turns out to be a solution of (6.61) representing a disturbance that radiates from a source at its·argument point. Furthermore, the effect of the radiation is felt throughout the interior of the range of influence of the source. Since, on the other hand, nothing is observed outside that region, it is customary in many applications of physical interest to define the fundamental solution S to be zero in the exterior of the range of influence of the parameter point, whereas our previous conventions would have assigned to it imaginary values there.

The theory is quite different for odd n. Here the expansion

$$S = \frac{U}{\Gamma^m} + V \log \Gamma$$

is relevant, and $m = (n - 1)/2$ is an integer, so that only the term $V \log \Gamma$ is multiple-valued in the complex domain. Consequently all contributions to the integral on the right in (6.71) which do not involve V cancel out over the interior of the domain of dependence, because ∂D approaches T from its two sides with opposite orientations. The determinations of $V \log \Gamma$ on the two sides of T differ by $-2\pi i V$. Therefore when u vanishes near the boundary of the domain of dependence, and no residues are left there, formula (6.71) takes on the simplified form

(6.75) $$u(t, x) = -2\pi i \int B[u(\tau, \xi), V(\tau, \xi; t, x)].$$

In particular, an initial value problem with data of the degenerate kind described in the previous example has for odd n the solution

$$u(t, x) = \pi i V(0, \xi; t, x)$$

inside the range of influence associated with the point $(0, \xi)$.

The quantity $\pi i V$, which is actually real according to our normalization (6.66) of S, is seen to be a solution of (6.61) in its dependence on the parameter point (t, x). However, only in the interior of the range of influence does it describe the effect of a disturbance emanating from a source located at the argument point. Either from this particular radiation problem or from the contribution (6.75) to the solution of a more general Cauchy problem, we can conclude that Huygens' principle is valid for the partial differential equation (6.61) in an odd number of space variables only when the coefficient V of the logarithmic term in the expansion of the fundamental solution vanishes identically. We know that $V \equiv 0$ for the wave equation, but it is a more difficult question to decide whether or not this is the case for a given equation with variable coefficients. On the other hand, it is remarkable that we have been able to establish the connection between Huygens' principle and the disappearance of the coefficient V by a qualitative discussion of the integral formula (6.71) for the solution of the Cauchy problem that avoids detailed evaluation of the residues at the edge of the domain of dependence.

Our use of contour integration in the complex domain to derive the basic formula (6.71) causes difficulties in solving Cauchy's problem by the same method for linear hyperbolic equations (6.61) whose coefficients are not analytic. However, if the coefficients possess partial derivatives of a sufficiently high order we can find the desired solution by substituting for S in our analysis a parametrix P of the type (5.81). We shall outline here the steps required to execute such a procedure, but to simplify our treatment we shall assume that the coefficients and the initial data are differentiable infinitely often. In this way we can present the essentials of the argument without going into an elaborate discussion about the precise number of derivatives that are needed.

First we observe that for the inhomogeneous equation

$$(6.76) \qquad\qquad L[u] = f$$

with analytic coefficients and right-hand side we can write down a direct analogue

$$(6.77) \quad u(t, x) = \lim_{\partial D \to T} \left\{ \int_{\partial D} B[u, P] + \int_D (Pf - u M[P]) \, d\tau \, d\xi_1 \cdots d\xi_n \right\}$$

of (6.71) based on (5.82) and on an analytic parametrix P defined by (5.81). As $\partial D \to T$ here, the manifold of integration D is supposed to approach the real domain in such a way that it folds around the characteristic conoid $\Gamma = 0$ without intersecting it. This means that the image of D in the three-dimensional space of the meridian variables ρ, τ_1 and τ_2 should collapse on the triangle bounded by the initial segment (6.70) and the characteristic lines (6.68).

Either by calculating residues at the conoid $\Gamma = 0$ or integrating by parts appropriately, we can evaluate (6.77) in terms of convergent definite integrals involving only a finite number of partial derivatives of u and of P, and therefore also of the coefficients of the partial differential equation (6.76). The integrals in question are extended over the real domain of dependence bounded by T and by the characteristic conoid $\Gamma = 0$, or else over certain subsets of that domain having lower dimension. From now on we shall interpret (6.77) as a symbolic notation for these real integrals. Except on T, they do not contain derivatives of u when the integers ν and μ in the representation (5.81) of P are taken so large that $M[P]$ remains finite on the characteristic conoid.

Formula (6.77) defines a *Volterra integral equation* (cf. Section 10.1) for the solution u of Cauchy's problem[14]. It can be solved, in turn, by the method of successive approximations described in Section 4.2. In fact, we may consider the integrodifferential equation (4.18) introduced there as a special two-dimensional case of (6.77) corresponding to the parametrix

$$P = \frac{1}{4\pi i \sqrt{\Delta}} \log \frac{1}{\Gamma}.$$

It follows that u varies continuously with the derivatives of the coefficients of equation (6.76) which feature in the real formulation of (6.77) referred to above. Similarly, the second partial derivatives of u depend continuously on the derivatives of the coefficients of a high enough order. Hence when they are no longer analytic we may replace these coefficients by polynomials approximating an appropriate set of their derivatives in order to establish the validity of (6.77) in the general case by passage to the limit. It is important to observe in this connection that the integral equation (6.77) has a meaning in the real domain even when the partial differential equation (6.76) is not analytic, since the construction of the parametrix P and of Γ in Chapter 5 only requires differentiability of the coefficients of a sufficient order.

To be more precise, for coefficients that possess partial derivatives of all orders we introduce a polynomial approximation that includes enough of those derivatives to ensure that the solution of the corresponding

---

[14] Cf. Hadamard 3, Hilbert 1, Levi 1.

224 CAUCHY'S PROBLEM IN SPACE

approximate equation (6.77) converges together with its second derivatives. Consequently the limit has to be a solution of the Cauchy problem associated with the more general coefficients and must itself satisfy (6.77), too. Thus there is a Volterra integral equation defined by means of (6.77) which provides a method to treat Cauchy's problem for arbitrary second order linear hyperbolic equations (6.76) that are not necessarily analytic. That the solution found in such a way is the only one to exist can be concluded from the uniqueness theorem developed in Exercises 6 to 9 at the end of Section 1. It follows that the initial value problem is well posed for arbitrary linear equations of the hyperbolic type in several independent variables when the data are assigned on a space-like surface.

Note that for the purpose of the demonstration given here it was not necessary to work out a more detailed formulation of the underlying integral equation (6.77), although to achieve this simplification we were obliged to employ a somewhat undesirable approximation argument. Finally, to extend the solution into the large we have only to subdivide space again into layers perpendicular to the t-axis and construct it in each layer successively.

To show that Cauchy's problem is correctly set, at least in the small, for nonlinear hyperbolic equations, too, it is feasible to apply an iterative scheme based on (6.77). For example, if the equation in question is quasi-linear in the sense that it is of the form (6.76) with coefficients depending on the unknown and its first derivatives, then we may define successive approximations by solving at each stage a linear problem obtained by substituting into those coefficients the values and the first derivatives of the previous approximation. However, it should be emphasized that in order to establish the convergence of such a procedure in space of arbitrary dimension it is necessary to incorporate into the method a system of integral equations of the type (6.77) for suitably many higher derivatives of the solution. Furthermore, the parametrix has to be modified at each step. We prefer to relegate detailed analysis of these more complicated matters to the exercises below, since they entail an unwieldy set of estimates.

Other treatments of the problem we have discussed in this section are available in the literature. Significant simplifications can be achieved by expressing the answer as a convergent integral to which a differential operator of appropriate order is applied.[15] Integrals of fractional order are also useful,[16] and the method of taking residues in the complex domain has generalizations to the case of equations of higher order.[17]

[15] Cf. Courant-Hilbert 3, John 1, Volterra 1.
[16] Cf. Duff 1, Riesz 1.
[17] Cf. Gårding-Leray 1.

EXERCISES

1. Show that the representation (4.73) for the solution of Cauchy's problem in terms of the Riemann function is really a special case of (6.75).

2. Use (6.71) to derive a formula for the solution of the Cauchy problem for the wave equation in four independent variables when the initial surface T is curved. Obtain the expression (6.60) for the solution of the characteristic initial value problem as a limiting case of this result.

3. Motivate the introduction of the retarded potential (6.46) by means of (6.77) with P replaced by S.

4. Carry out details of the proof based on (6.77) of the existence of the solution of Cauchy's problem for a linear hyperbolic equation whose coefficients are not analytic. In terms of the number n of space variables involved, find a precise order of differentiability of the coefficients that will suffice for this purpose. Include the case where the initial surface T is not analytic, either, and show that under appropriate hypotheses the solution is valid in the large.

5. Show that the solution of Cauchy's problem is analytic in the large for a linear hyperbolic equation with analytic coefficients and analytic initial data. Prove that the theorem remains true when the data are merely analytic at the edge of the domain of dependence.

6. Show that the Cauchy problem is well posed for a quasi-linear equation

$$au_{xx} + 2bu_{xy} + cu_{yy} + d = 0$$

of the hyperbolic type in two independent variables without using a preliminary reduction to canonical form. More specifically, prove the convergence of a sequence of successive approximations defined by three linear equations of the form

$$a^{(m)}u_{xx}^{(m+1)} + 2b^{(m)}u_{xy}^{(m+1)} + c^{(m)}u_{yy}^{(m+1)} + d^{(m)} = 0,$$

$$a^{(m)}p_{xx}^{(m+1)} + 2b^{(m)}p_{xy}^{(m+1)} + c^{(m)}p_{yy}^{(m+1)} + \cdots = 0,$$

$$a^{(m)}q_{xx}^{(m+1)} + 2b^{(m)}q_{xy}^{(m+1)} + c^{(m)}q_{yy}^{(m+1)} + \cdots = 0,$$

where $a^{(m)} = a(x, y, u^{(m)}, p^{(m)}, q^{(m)})$, and where $b^{(m)}$, $c^{(m)}$ and $d^{(m)}$ have a similar meaning.

7. Prove[18] in detail, at least for $n = 3$, that the solution of Cauchy's problem for any nonlinear hyperbolic partial differential equation of the second order exists in the small, is unique, and depends continuously on the data. To achieve an adequate estimate of the convergence of a sequence of successive approximations to the solution, how many of its derivatives have to be treated as additional unknowns?

8. Describe the range of influence of a disturbance satisfying the wave equation which radiates from sources spread over a region of arbitrarily prescribed shape.

[18] Cf. Fourès-Bruhat 1.

7

The Dirichlet
and Neumann Problems

1. UNIQUENESS

The boundary value problems of potential theory are suggested by physical phenomena from such varied fields as electrostatics, steady heat conduction and incompressible fluid flow. They stand in marked contrast to the initial value problems that have occupied our attention thus far. In particular, the investigations of Chapters 4 and 6 showed that while initial value problems are reasonable for partial differential equations of the hyperbolic type, they are not well posed for equations of the elliptic type. Here we shall formulate boundary value problems that are appropriate in the elliptic case.

Laplace's equation

$$u_{xx} + u_{yy} = 0$$

is the most elementary example of an elliptic partial differential equation. Any solution u may be interpreted as describing a steady distribution of temperature throughout its domain of definition D. From a physical point of view one might naturally expect that u could be determined inside D from measurements of its values at the boundary ∂D of that domain. The boundary value problem so specified differs significantly from what we have had occasion to deal with before and is appreciably more difficult to handle. Indeed, because boundary data rather than initial data serve to fix properly the solution of an elliptic equation, it is usually necessary to find an answer in the large. This is harder to achieve than is the local treatment of initial value problems, which often have physical interest even in the small, say over a short time interval. The need for global constructions makes it especially awkward to master nonlinear elliptic equations. Thus many of our results will be restricted in an essential way to linear problems.

Consider the quite general linear equation

$$(7.1) \qquad L[u] = \sum_{j,k=1}^{n} a_{jk} \frac{\partial^2 u}{\partial x_j \partial x_k} + \sum_{j=1}^{n} b_j \frac{\partial u}{\partial x_j} + cu = 0,$$

where the coefficients a_{jk}, b_j and c are suitably differentiable functions of the independent variables $x_1, \ldots, x_n$, and where it may be assumed that $a_{jk} = a_{kj}$. We shall suppose that (7.1) is of the *elliptic type* in some n-dimensional region D, which means that the quadratic form

$$Q = \sum_{j,k=1}^{n} a_{jk} \lambda_j \lambda_k$$

is positive-definite there (provided the signs are adjusted to make $a_{11} > 0$).

The *first boundary value problem*, also called the *Dirichlet problem*, asks for a solution u of (7.1) in D which takes on prescribed values

$$u = f$$

at the boundary ∂D of that region. Under various hypotheses about the geometry of ∂D and the behavior of the coefficients a_{jk}, b_j and c, the most essential of which states that $c \leq 0$, it is possible to establish the existence, uniqueness and continuous dependence on boundary data of the solution of Dirichlet's problem. (See Exercise 1 below for an example with $c > 0$ for which the uniqueness theorem fails.) The question of existence is by all means the hardest to settle, and we therefore postpone discussing it until Chapters 8 and 9, where a variety of methods of proof will be described. Here we develop formal procedures for the construction of the solution u, with an emphasis on special cases. We also intend to show in a quite elementary way why u is unique and depends continuously on the data.

Even for the Laplace equation

$$\Delta u = \frac{\partial^2 u}{\partial x_1^2} + \cdots + \frac{\partial^2 u}{\partial x_n^2} = 0,$$

which has constant coefficients, it is not an easy matter to solve the Dirichlet problem in a region D of arbitrary shape. The solution can be obtained in closed form only for special choices of D, such as a sphere or a cube. Most examples for which an explicit answer is known yield to the technique of separation of variables and Fourier analysis. Thus it turns out that the treatment of Laplace's equation for more general regions is as elaborate as that required for the study of a large class of elliptic equations (7.1) with variable coefficients. For this reason we shall be

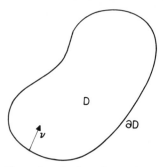

FIG. 22.   Geometry of a boundary value problem.

content to present many aspects of the theory for the typical self-adjoint equation

(7.2) $\Delta u = Pu$

with just one variable coefficient

$$P = P(x_1, \ldots, x_n) \geq 0.$$

Such an approach permits many simplifications which facilitate an understanding of the fundamental principles without obscuring their more essential features.

Since the formulation of reasonable boundary conditions involving the derivatives of u is rather complicated for an equation (7.1) with arbitrary coefficients a_{jk}, we restrict our statement about such matters to the case of equation (7.2). In particular, the *second boundary value problem* for (7.2), more often referred to as the *Neumann problem*, consists in the determination in some region D of a solution u that possesses prescribed normal derivatives

$$\frac{\partial u}{\partial \nu} = f$$

on the surface ∂D bounding D. To be specific, we take ν in this context to stand for the unit normal directed into the interior of D, and we assume, of course, that ∂D is smooth enough to have a meaningful normal. Finally, a mixed boundary condition that imposes given values

$$\frac{\partial u}{\partial \nu} + \alpha u = f$$

on a linear combination of u and $\partial u/\partial \nu$ along ∂D, rather than on either of them separately, defines what is known as the *third boundary value problem* for solutions of (7.2). We shall see that the latter question is well posed only for suitably restricted choices of the coefficient α, for example, for $\alpha \leq 0$.

We commence our analysis of the first, second and third boundary value problems for the elliptic equation (7.2) by establishing a uniqueness theorem. Our point of departure is the *Green's identity*

$$(7.3) \quad \int_D \left[\frac{\partial u}{\partial x_1}\frac{\partial v}{\partial x_1} + \cdots + \frac{\partial u}{\partial x_n}\frac{\partial v}{\partial x_n} + u\,\Delta v\right] dx_1 \cdots dx_n$$
$$= \int_{\partial D}\left[\sum_{j=1}^{n}(-1)^{j-1}u\,\frac{\partial v}{\partial x_j}\,dx_1\cdots \widehat{dx_j}\cdots dx_n\right],$$

which follows from the divergence theorem (cf. Section 5.2) and the relations

$$\frac{\partial u}{\partial x_j}\frac{\partial v}{\partial x_j} + u\,\frac{\partial^2 v}{\partial x_j^2} = \frac{\partial}{\partial x_j}\left(u\,\frac{\partial v}{\partial x_j}\right), \quad j = 1,\ldots,n.$$

Here the notation $\widehat{dx_j}$ indicates that the differential dx_j is to be omitted from the product $dx = dx_1 \ldots dx_n$. Introducing the operator $\nabla = (\partial/\partial x_1, \ldots, \partial/\partial x_n)$ and letting $d\sigma$ stand for the area element on the surface ∂D, we may express (7.3) in the more convenient form

$$\int_D [\nabla u \cdot \nabla v + u\,\Delta v]\,dx + \int_{\partial D} u\,\frac{\partial v}{\partial \nu}\,d\sigma = 0.$$

If v is a solution of (7.2) and $u = v$, then (7.3) reduces to

$$(7.4) \quad \int_D\left[\left(\frac{\partial u}{\partial x_1}\right)^2 + \cdots + \left(\frac{\partial u}{\partial x_n}\right)^2 + Pu^2\right] dx = -\int_{\partial D} u\,\frac{\partial u}{\partial \nu}\,d\sigma.$$

We call the term on the left the *Dirichlet integral* associated with (7.2), and we introduce for it the notation

$$\|u\|^2 = \int_D\left[\left(\frac{\partial u}{\partial x_1}\right)^2 + \cdots + \left(\frac{\partial u}{\partial x_n}\right)^2 + Pu^2\right] dx.$$

The importance of the norm $\|u\|$ for uniqueness proofs lies in the fact that it is positive when u is not a constant, or even when u differs from zero and $P \not\equiv 0$, in view of the assumption $P \geq 0$. It is comparable to the energy integral studied in Section 6.1.

In a region D consider two solutions u_1 and u_2 of the Dirichlet problem for (7.2) which possess the same boundary values $u_1 = u_2 = f$ on ∂D. Suppose that their first partial derivatives behave so well at ∂D that formula (7.4) can be applied to the difference

$$u = u_2 - u_1.$$

Since u satisfies (7.2) in D and is zero on ∂D, we have

$$\|u\|^2 = -\int_{\partial D} u \frac{\partial u}{\partial v} \, d\sigma = 0.$$

Hence u is a constant because of the positive-definite structure of $\|u\|$, and, indeed, it must vanish everywhere in D because it does so on ∂D. This proves the uniqueness $u_1 = u_2$ of the solution of Dirichlet's problem.

When $P > 0$ a similar argument shows that the solution of the Neumann problem is unique. However, in the case $P \equiv 0$ of Laplace's equation we are only able to conclude that two solutions of the same Neumann problem differ at most by a constant. Furthermore, such a constant can be added arbitrarily to the answer, since it is a harmonic function with zero normal derivatives.

The difference u between any pair of solutions of the third boundary value problem possessing identical data f fulfills the homogeneous condition

$$\frac{\partial u}{\partial v} + \alpha u = 0$$

on ∂D. In these circumstances (7.4) can be transformed into

$$(7.5) \qquad \|u\|^2 - \int_{\partial D} \alpha u^2 \, d\sigma = 0.$$

Under the hypothesis that $\alpha < 0$ we can deduce from (7.5) not only that $\|u\| = 0$ but also that u vanishes on ∂D, for otherwise the expression on the left would be positive. Thus uniqueness is assured for the third boundary value problem provided that the coefficient α is negative.

We emphasize at this point that the Neumann problem cannot be solved for Laplace's equation unless the prescribed values f of the normal derivative $\partial u / \partial v$ satisfy the *compatibility condition*

$$(7.6) \qquad \int_{\partial D} f \, d\sigma = 0.$$

Indeed, if we set one of the functions occurring in (7.3) equal to unity, we are led to the relation

$$\int_{\partial D} \frac{\partial u}{\partial v} \, d\sigma = -\int_D \Delta u \, dx = 0.$$

By viewing u as the velocity potential of a flow of incompressible fluid (cf. Section 14.1) we can interpret (7.6) as a formulation of the *law of conservation of mass*, for it states that the rate at which fluid pours into D just balances the rate of flow outward from D. It should be observed in this connection that the compatibility condition (7.6) does not apply when D is an infinite region in, say, three-dimensional space, since in that case

$$\frac{1}{r} = \frac{1}{\sqrt{x_1^2 + x_2^2 + x_3^2}}$$

is considered to be a regular harmonic function at infinity, yet it represents flow issuing from a source located there.

An objection to proving the uniqueness theorem for Dirichlet's problem by means of the identity (7.4) can be raised because of the assumption that is needed about the behavior of the first derivatives of the solution at the boundary. A more satisfactory proof avoiding this difficulty can be based on the *maximum principle*, which states that the solutions of certain partial differential equations of the elliptic type never achieve a strong relative maximum or minimum in the interior of their domain of definition.

Perhaps the simplest example of the phenomenon we have in mind is furnished by the equation (7.2) when the coefficient P exceeds zero. We maintain that no solution u of such an equation could possibly assume a positive maximum at an interior point $x = (x_1, \ldots, x_n)$ of the region D where it is defined. If it did, the usual necessary conditions

$$\frac{\partial u}{\partial x_j} = 0, \quad \frac{\partial^2 u}{\partial x_j^2} \leq 0, \quad j = 1, \ldots, n,$$

for a maximum would be fulfilled at that point. Hence we would have

(7.7) $$Pu = \Delta u \leq 0$$

there; but, on the other hand, both factors P and u on the left are positive by hypothesis. This contradiction establishes our result. Furthermore, substitution of $-u$ for u in the proof shows that no solution of (7.2) could have a negative minimum inside its domain of regularity, either, when $P > 0$.

To deduce a uniqueness theorem for Dirichlet's problem from the preceding remarks it suffices to show that any solution u of (7.2) specified in a region D where $P > 0$ must be identically zero if it vanishes continuously on the boundary ∂D. The demonstration consists in observing that if u differed from zero inside D, it would have to have either a positive maximum or a negative minimum there, which we have seen to be impossible. A similar argument can even be used to establish the continuous dependence of the solution of the Dirichlet problem on the boundary values that are assigned. Indeed, let u_1 and u_2 be two solutions of (7.2) in D satisfying

$$(7.8) \qquad\qquad |u_2 - u_1| \leq \epsilon$$

on ∂D. Then according to what has been said, their difference $u_2 - u_1$ cannot be larger than ϵ or less than $-\epsilon$ in the interior of D, either, for otherwise it would achieve a positive maximum or a negative minimum there. It follows that the inequality (7.8) is valid inside D, too, which provides us with an eminently satisfactory statement about continuous dependence on boundary values.

We turn our attention to a more usual formulation of the maximum principle which applies to a quite general class of linear elliptic partial differential equations of the second order and which leads for them, too, directly to the uniqueness and continuous dependence on data of the solution of Dirichlet's problem. To be precise, we consider the elliptic equation (7.1) with

$$(7.9) \qquad\qquad c \equiv 0,$$

and we assume that the remaining coefficients are continuous. It is clear that in this case any constant represents a solution of the equation. The maximum principle in its most general form states that the constants are the only solutions which can assume a maximum or a minimum value in the interior of their domain of definition. In other words, no solution can possess a strong relative maximum or a strong relative minimum at an interior point of its domain of existence. An alternate formulation of the principle which is useful in the applications asserts that when (7.9) is fulfilled, every solution of (7.1) achieves its maximum and minimum values on the boundary ∂D of any region D where it is known to be (uniformly) continuous.

We shall develop the proof[1] of the maximum principle only for an equation

$$(7.10) \qquad L[u] = au_{xx} + 2bu_{xy} + cu_{yy} + du_x + eu_y = 0$$

in two independent variables x and y. The argument is equally valid for arbitrarily many independent variables (cf. Exercise 5 below), but it is more easily understood and visualized within the framework of plane geometry. Furthermore, since the minima of u correspond to the maxima

[1] Cf. Hopf 1.

of $-u$, it suffices to consider only maximum points in our analysis. We suppose that u achieves a maximum in the interior of some region D where it satisfies (7.10), and we proceed to show by contradiction that it must be constant there. The conclusion hinges on several constructions which will bring us to a situation analogous to that encountered in connection with the inequality (7.7).

We may assume without loss of generality that the maximum value of u in D is zero, since otherwise subtraction of an appropriate constant would provide a new solution with that property. If u is not identically zero, it is possible to find a circle E_1 within D which includes on its perimeter an interior point (ξ, η) of D where the maximum value

$$u(\xi, \eta) = 0$$

is attained, but such that

(7.11) $u(x, y) < 0$

inside E_1. Let E denote a smaller circle whose closure lies entirely in the interior of E_1, except that it touches the boundary of E_1 at (ξ, η). We observe that (7.11) must be satisfied at every point of E and of its circumference ∂E, with the exception of (ξ, η).

Let R stand for the radius of E, and for the sake of simplicity locate the origin at the center of E. We define E_2 to be a circle about (ξ, η) of radius less than R, and we even choose it so small that its closure is confined

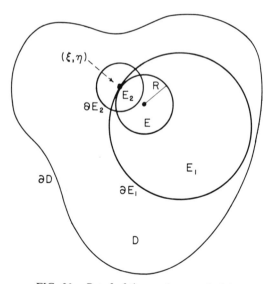

FIG. 23.   Proof of the maximum principle.

to the interior of D. Then we divide the boundary of E_2 into two components consisting of the closed arc σ_1 where it intersects E and the complementary open arc, to be denoted by σ_2. It follows from our construction of E that there exists a small positive number ϵ such that

(7.12) $u \leq -\epsilon < 0$

on σ_1, since this is a closed set where (7.11) holds.

We introduce the *comparison function*

$$U = e^{-Nr^2} - e^{-NR^2},$$

where

$$r = \sqrt{x^2 + y^2}.$$

Clearly, for positive values of the parameter N we have $U > 0$ inside E, while $U = 0$ on ∂E, and $U < 0$ outside E. Direct computation shows that

$$L[U] = \{4N^2(ax^2 + 2bxy + cy^2) - 2N(a + c + dx + ey)\}\, e^{-Nr^2}.$$

Since

$$Q = ax^2 + 2bxy + cy^2$$

is a positive-definite quadratic form with continuous coefficients a, b and c, it has a positive lower bound on E_2, which was selected to be a circle not enclosing the origin. Therefore we can pick N so large that

(7.13) $L[U] > 0$

throughout E_2. Finally, in view of (7.12) it is possible to find a small number $\theta > 0$ such that

(7.14) $u + \theta U < 0$

on both of the arcs σ_1 and σ_2 of ∂E_2, whereas, of course,

(7.15) $L[u + \theta U] > 0$

throughout E_2 by virtue of (7.10) and (7.13).

Since u and U both vanish at (ξ, η), which is a point inside E_2, the inequality (7.14) shows that the auxiliary function $u + \theta U$ must achieve a non-negative maximum at some interior point (ξ', η') of E_2. We can perform an affine transformation of variables which reduces (7.10) to the canonical form

$$\Delta u + \cdots = 0$$

at such a point (cf. Chapter 3). Hence no generality is lost if we assume that

$$b = 0, \qquad a = c = 1$$

there. In these circumstances (7.15) implies that

$$(7.16) \qquad u_{xx} + \theta U_{xx} + u_{yy} + \theta U_{yy} > 0$$

at (ξ', η'), where the necessary conditions

$$(7.17) \qquad u_x + \theta U_x = u_y + \theta U_y = 0,$$

$$(7.18) \qquad u_{xx} + \theta U_{xx} \leq 0, \qquad u_{yy} + \theta U_{yy} \leq 0$$

for a maximum hold, too. Because the inequalities (7.16) and (7.18) are contradictory, we conclude that u must be identically zero after all. This completes our proof of the maximum principle, for the original supposition that u takes on a maximum value at some interior point of its domain of existence has served to establish that it cannot be anything but a constant.

We have already explained how the maximum principle is used to show that the solution of Dirichlet's problem is unique and depends continuously on its boundary values. These results are now seen to apply to the elliptic equation (7.10), or even more generally to (7.1) when the hypothesis (7.9) is fulfilled.

It is interesting to discuss such matters in the context of a physical interpretation of the maximum principle for the special case of Laplace's equation

$$(7.19) \qquad u_{xx} + u_{yy} = 0.$$

Once again we think of u as describing a steady distribution of temperature over some plane region D. To be more precise, observe that (7.19) is the form that the heat equation

$$u_t = u_{xx} + u_{yy}$$

takes when the temperature u is independent of the time t. In this terminology the maximum principle merely asserts that in a steady state the largest and the smallest values of the temperature occur on the boundary ∂D of the material occupying the domain D. Furthermore, the only way in which a maximum or a minimum of the distribution could appear inside D is for the temperature to be constant everywhere. These facts are altogether plausible physically, and they contribute to a more thorough understanding of the implications of the maximum principle. Incidentally, for Laplace's equation an alternate and quite elementary proof of the maximum principle will be given in Section 2.

Except for the matter of existence, which will be settled in Chapters 8 and 9, we have now succeeded in justifying our contention that Dirichlet's problem is correctly set for partial differential equations of the elliptic type. In order to indicate that such a conclusion would not be evident without a study of the kind we have undertaken, we proceed to show that the Dirichlet problem is not well posed in the hyperbolic case. A counterexample is easily developed for the wave equation in the normal form

$$(7.20) \qquad\qquad u_{xy} = 0.$$

We choose our region D to be the unit square $0 < x < 1, 0 < y < 1$. If the values of u are assigned smoothly on two adjacent sides of D, say if

$$u(x, 0) = f(x), \qquad 0 \le x \le 1,$$
$$u(0, y) = g(y), \qquad 0 \le y \le 1,$$

we can interpret them as data for a characteristic initial value problem which fixes the solution of (7.20) uniquely throughout the square (cf. Section 4.2). In fact, we have

$$u(x, y) = f(x) + g(y)$$

when $f(0) = g(0) = 0$. Hence the boundary values of u cannot be prescribed arbitrarily on the remaining two sides of D. Thus a solution does not in general exist when data are imposed on the complete perimeter of the characteristic square, which means that the Dirichlet problem for (7.20) is *overdetermined* in that region.

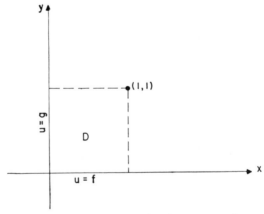

FIG. 24.   Counterexample for the wave equation.

Finally, observe that it is not clear how to count the number of arbitrary functions that feature in Dirichlet's problem for an equation of the second order. In the foregoing case of a square we might consider the boundary data to consist of four real functions, each depending on one variable, whereas for a circle it is more natural to express the boundary values in terms of a single arbitrary function. Thus it is false to state that the general solution of an elliptic equation of the second order involves two arbitary functions in the real sense. The concept of a general solution is therefore not very appropriate for equations of the elliptic type and is unsuited to the geometry of Dirichlet's problem. However, it does have a meaning in the complex domain for analytic elliptic equations, as was indicated by our treatment of the Cauchy-Kowalewski theorem in Chapter 1. The general solution may also play an important role in specific examples, as we shall see from the close relationship between analytic functions of a complex variable and the theory of Laplace's equation in the plane (cf. Section 3).

EXERCISES

1. To show that a condition of the type $P \geq 0$ is necessary in order that the solution of Dirichlet's problem for (7.2) be unique, verify that the eigenfunction (cf. Chapter 11)

$$u(x, y) = \sin x \sin y$$

solves the elliptic equation

$$u_{xx} + u_{yy} + 2u = 0,$$

but vanishes on the boundary of the square $0 < x < \pi, 0 < y < \pi$.

2. Find a solution different from zero of the homogeneous mixed boundary value problem defined by

$$u_{xx} + u_{yy} = 0, \qquad x^2 + y^2 < 1,$$

$$\frac{\partial u}{\partial v} + u = 0, \qquad x^2 + y^2 = 1.$$

3. Use the method of Fourier series to verify that the Neumann problem for Laplace's equation is solvable in the unit circle whenever the compatibility condition (7.6) is satisfied.

4. Formulate and solve Dirichlet's problem for the ordinary differential equation

$$u_{xx} - u = 0$$

on the interval $0 < x < 1$.

5. Under the hypothesis (7.9), prove the maximum principle for the general elliptic equation (7.1) in more than two independent variables.

6. If (7.10) is quasi-linear in the sense that the coefficients a, b, c, d and e depend on x, y, u, u_x and u_y, show that the maximum principle is still valid for any solution fulfilling the ellipticity requirement

$$b^2 - ac < 0.$$

7. Let D be a plane region bounded by a smooth curve ∂D, and suppose that u is a non-negative harmonic function in D possessing continuous first partial derivatives on ∂D. If u vanishes at a point of ∂D, show that

(7.21) $$\frac{\partial u}{\partial v} \geq 0$$

there. Furthermore, prove (and this is where the difficulty lies)·that the equality sign can hold in (7.21) only when u vanishes identically. *Hint:* Introduce a suitable comparison function.[2]

8. Use the result of Exercise 7 to discuss uniqueness of the solution of the Neumann problem for Laplace's equation (7.19).

9. Let u be a solution of the mixed initial and boundary value problem

$$u_t = u_{xx}, \qquad 0 < x < l, 0 < t < T,$$

$$u(x, 0) = f(x), \qquad 0 \leq x \leq l,$$

$$u(0, t) = u(l, t) = 0, \qquad 0 \leq t \leq T,$$

for the distribution of heat in a rod of length l. Show that it achieves its maximum and minimum values at $t = 0$; use this result to establish a uniqueness theorem for the problem.

10. By means of the identity

(7.22) $$\int_D \sum_{j,k=1}^{n} \left[a_{jk} \frac{\partial u}{\partial x_j} \frac{\partial u}{\partial x_k} + u \frac{\partial}{\partial x_j} \left(a_{jk} \frac{\partial u}{\partial x_k} \right) \right] dx_1 \cdots dx_n$$

$$= \int_{\partial D} \left[\sum_{j,k=1}^{n} (-1)^{j-1} u a_{jk} \frac{\partial u}{\partial x_k} dx_1 \cdots \widehat{dx_j} \cdots dx_n \right]$$

analogous to (7.3), prove the uniqueness of the solution of Dirichlet's problem for the general *self-adjoint* elliptic equation

(7.23) $$\sum_{j,k=1}^{n} \frac{\partial}{\partial x_j} \left(a_{jk} \frac{\partial u}{\partial x_k} \right) = Pu$$

with $P \geq 0$ and $a_{11} > 0$.

11. Under the hypothesis $P > 0$, use the integral identity (7.22) to establish the uniqueness of the solution u of the Neumann problem for equation (7.23), which is supposed to fulfill a boundary condition of the form

$$\sum_{j,k=1}^{n} a_{jk} \frac{\partial x_j}{\partial v} \frac{\partial u}{\partial x_k} = f.$$

[2] Cf. Lavrentiev 1.

12. Suppose that u solves the inhomogeneous biharmonic equation

$$\Delta \Delta u = u_{xxxx} + 2u_{xxyy} + u_{yyyy} = f$$

in a plane region D and satisfies the edge conditions

$$u = \frac{\partial u}{\partial \nu} = 0$$

at the boundary ∂D. Thus it represents the infinitesimal deflection of a *clamped plate*[3] supporting the prescribed load f. Show that the solution of this higher order boundary value problem is unique through an investigation of the identity

$$\iint_D (\Delta u)^2 \, dx \, dy = \iint_D uf \, dx \, dy.$$

13. Denote by u a possibly complex-valued solution of the reduced wave equation

(7.24) $$u_{xx} + u_{yy} + k^2 u = 0$$

in some region D containing the point at infinity, and assume that it satisfies the *radiation condition*

(7.25) $$\lim_{r \to \infty} \sqrt{r}\left(\frac{\partial u}{\partial r} - iku\right) = 0, \qquad \overline{\lim_{r \to \infty}} \sqrt{r}\,|u| < \infty.$$

By representing u in terms of a specific fundamental solution of (7.24) that fulfills (7.25), prove[4] that it vanishes identically whenever

$$\lim_{R \to \infty} \int_{r=R} |u|^2 r \, d\theta = 0,$$

where r and θ are polar coordinates (cf. Exercises 2.20 and 2.21).

14. The *exterior Dirichlet and Neumann problems* for (7.24), subject to the radiation condition (7.25), consist in prescribing the values of u or of $\partial u/\partial \nu$, respectively, on the boundary ∂D of the infinite region D where u is to be determined. Establish the uniqueness of the solution u of either of these problems by proving and then using the identity

$$\mathrm{Im}\left\{\int_{\partial D} \bar{u}\,\frac{\partial u}{\partial \nu}\, ds\right\} = \lim_{R \to \infty} k \int_{r=R} |u|^2 r \, d\theta,$$

where s stands for the arc length along ∂D and $\bar{u}$ stands for the complex conjugate of u.

[3] Cf. Love 1.
[4] Cf. Rellich 1.

15. Establish the uniqueness of the solution of the Dirichlet problem for the quasi-linear equation

$$(1 + u_y^2)u_{xx} - 2u_x u_y u_{xy} + (1 + u_x^2)u_{yy} = 0,$$

which characterizes *minimal surfaces* (cf. Section 15.4) over a convex region D, by showing (cf. Section 11.3) that if u_1 and u_2 are two solutions with the same boundary values and if

$$u = \lambda u_1 + (1 - \lambda)u_2,$$

then the area integral

$$I(\lambda) = \iint_D \sqrt{1 + u_x^2 + u_y^2}\, dx\, dy$$

is a convex function of the parameter λ in the interval $0 \le \lambda \le 1$, yet

$$I'(0) = I'(1) = 0.$$

16. Prove uniqueness of the solution u of a mixed boundary value problem, or *Robin problem*, for equation (7.2) in which the normal derivative $\partial u/\partial \nu$ is prescribed on one portion of ∂D, but u itself is given on the remainder of that surface.

2. THE GREEN'S AND NEUMANN'S FUNCTIONS

In this section we develop in a formal way explicit representations for the solution of the Dirichlet problem and of the Neumann problem which have a significance similar to that of the integral formula (4.73) expressing the solution of Cauchy's problem in terms of the Riemann function. The analogue of the Riemann function for an elliptic boundary value problem is a fundamental solution which fulfills an appropriate condition at the boundary of the region where the problem is posed. The Green's function and the Neumann's function are examples of fundamental solutions of this kind. However, they turn out to be far more difficult to determine than the Riemann function, which is merely a factor in the expansion of the fundamental solution (cf. Section 5.1). The boundary conditions involved in their definition make them depend not only on the coefficients of the partial differential equation at hand but also on a specified region. Therefore we postpone until Chapters 8 and 9 the analysis of their existence and confine our attention here to a description of their properties and of the principal uses to which they can be put.

For the sake of simplicity we shall discuss in detail only the case of the equation

(7.2) $\Delta u = Pu,$

and special emphasis will be placed on examples from potential theory. Our basic tool is the integral representation (5.69) for solutions of a linear elliptic equation in terms of a fundamental solution of the adjoint equation. In order to achieve a complete treatment without making unnecessary references to previous material of a less elementary nature (cf. Chapter 5) we rederive that result from (7.3) for the self-adjoint equation (7.2). To start with, we interchange the roles of u and v in (7.3) and proceed to subtract the new formula so obtained from the old one to establish the Green's identity

$$(7.26) \qquad \int_D [u\,\Delta v - v\,\Delta u]\,dx + \int_{\partial D}\left[u\,\frac{\partial v}{\partial \nu} - v\,\frac{\partial u}{\partial \nu}\right]\,d\sigma = 0,$$

which plays an important role in potential theory. If u and v are both solutions of (7.2) throughout D, the multiple integral over that region drops out of (7.26) and we have

$$(7.27) \qquad \int_{\partial D}\left[u\,\frac{\partial v}{\partial \nu} - v\,\frac{\partial u}{\partial \nu}\right]\,d\sigma = 0.$$

It will be our aim to choose v so that (7.27) becomes an expression for u exclusively in terms of either its boundary values or its normal derivatives.

For the equation (7.2) in $n > 2$ independent variables, a fundamental solution $S = S(x, \xi)$ has the form

$$(7.28) \qquad S = \frac{1}{\sigma'_{n-1}}\frac{1}{r^{n-2}} + W,$$

where

$$r = \sqrt{(x_1 - \xi_1)^2 + \cdots + (x_n - \xi_n)^2},$$

and where $W = W(x, \xi)$ grows less rapidly than $r^{-(n-2)}$ as the argument point $x = (x_1, \ldots, x_n)$ approaches the parameter point $\xi = (\xi_1, \ldots, \xi_n)$. Again we have put

$$\sigma'_{n-1} = (n - 2)\sigma_{n-1},$$

with σ_{n-1} standing for the $(n - 1)$-dimensional surface area of a unit sphere. When $n = 2$, the singularity of S becomes

$$(7.29) \qquad S = \frac{1}{2\pi}\log\frac{1}{r} + W,$$

where W and its first derivatives remain bounded at ξ. Although it is not general practice to divide by σ'_{n-1} and 2π as we have in (7.28) and (7.29), we prefer these definitions because they serve to eliminate corresponding factors from our final representation formula. Incidentally, for the special case $P \equiv 0$ of Laplace's equation we can simply take $W = 0$.

We propose to set $v = S$ in (7.27) with the singular point ξ located inside D. Hence it is necessary to delete from that region a small sphere of radius ϵ about ξ. Applying (7.27) to the punctured domain, we find

$$(7.30) \qquad \int_{r=\epsilon} \left[u \frac{\partial S}{\partial \nu} - S \frac{\partial u}{\partial \nu} \right] d\sigma = \int_{\partial D} \left[u \frac{\partial S}{\partial \nu} - S \frac{\partial u}{\partial \nu} \right] d\sigma.$$

This relation holds for all sufficiently small values of $\epsilon > 0$; we intend to transform it into a formula for u by allowing ϵ to approach zero.

Our specifications about W are such that

$$(7.31) \qquad \lim_{r \to 0} r^{n-1} \frac{\partial S}{\partial r} = - \frac{1}{\sigma_{n-1}}, \qquad \lim_{r \to 0} r^{n-1} S = 0.$$

Since the area of the surface of an n-dimensional sphere of radius ϵ around ξ is

$$\int_{r=\epsilon} d\sigma = \sigma_{n-1} \epsilon^{n-1},$$

we can establish the equality

$$\lim_{\epsilon \to 0} \int_{r=\epsilon} \left[u \frac{\partial S}{\partial \nu} - S \frac{\partial u}{\partial \nu} \right] d\sigma = - \lim_{\epsilon \to 0} \int_{r=\epsilon} u \frac{\partial S}{\partial r} d\sigma = u(\xi)$$

by substituting the results (7.31) into the integrand on the left. Letting $\epsilon \to 0$ in (7.30), we obtain in this way the basic representation

$$(7.32) \qquad u(\xi) = \int_{\partial D} \left[u(x) \frac{\partial S(x, \xi)}{\partial \nu} - S(x, \xi) \frac{\partial u(x)}{\partial \nu} \right] d\sigma,$$

analogous to (5.69), for solutions u of (7.2) in terms of an arbitrary fundamental solution S defined in the region D.

Unfortunately, the integral on the right in (7.32) involves both u and its normal derivative $\partial u / \partial \nu$ along ∂D. Therefore it does not immediately yield a formula for solving boundary value problems. We shall endeavor to convert (7.32) into such an integral formula, however, by imposing suitable boundary conditions on S.

We define the *Green's function* $G = G(x, \xi)$ of equation (7.2) with respect to the region D to be a fundamental solution of (7.2) there which satisfies the homogeneous boundary condition

$$(7.33) \qquad G(x, \xi) = 0$$

when the argument point x is located on ∂D. Thus the Green's function has the form

$$G = S - u_S,$$

where u_S is a regular solution of (7.2) in D that reduces to the particular fundamental solution S on the boundary ∂D. The existence and uniqueness of such a term u_S are consequences of the solvability of Dirichlet's problem in D for the linear elliptic equation (7.2), which is feasible when $P \geq 0$.

In the case of Laplace's equation the Green's function has an electrostatic interpretation[5] that lends credence to its existence. Let us think of the domain D as a vacuum bounded by a perfectly conducting surface ∂D. The presence of a positive charge at the point ξ inside D induces an equilibrium distribution of negative charge over the surface ∂D. When normalized appropriately, the potential of the electrostatic field created by these charges has precisely the properties ascribed to the Green's function.

The importance of the Green's function G with regard to Dirichlet's problem is that when we insert it into (7.32) to play the role of S, the contribution from the normal derivative $\partial u/\partial \nu$ disappears, by virtue of (7.33). Thus we obtain the specific representation

$$(7.34) \qquad u(\xi) = \int_{\partial D} u(x) \frac{\partial G(x, \xi)}{\partial \nu} \, d\sigma$$

for the solution u of the Dirichlet problem as a definite integral of the prescribed boundary values, multiplied by the kernel $\partial G/\partial \nu$. Incidentally, since $G \to +\infty$ as $x \to \xi$, and since the maximum principle shows that G cannot have a negative minimum in the interior of D, its minimum must occur on the boundary ∂D. Since (7.33) holds on ∂D, we therefore have

$$G(x, \xi) > 0$$

inside D. If follows that the kernel $\partial G/\partial \nu$ featuring in the integral formula (7.34) is non-negative.

For $P > 0$ we can introduce the *Neumann's function*, or *Green's function of the second kind*, for equation (7.2) with respect to D, which is a fundamental solution $N = N(x, \xi)$ fulfilling the requirement

$$(7.35) \qquad \frac{\partial N}{\partial \nu} = 0$$

on ∂D. It can be expressed as the difference

$$N = S - v_S$$

[5] Cf. Jackson 1.

between an arbitrary fundamental solution S in D and a regular function v_S which solves the Neumann problem for (7.2) there defined by the boundary condition

$$\frac{\partial v_S}{\partial \nu} = \frac{\partial S}{\partial \nu}.$$

Substitution of N for S in (7.32) eliminates the term involving boundary values of u and provides the integral representation

$$(7.36) \qquad u(\xi) = -\int_{\partial D} N(x, \xi) \frac{\partial u(x)}{\partial \nu} \, d\sigma$$

for the solution of the general Neumann problem. Formulas (7.34) and (7.36) furnish a useful and quite basic tool for the study of the Dirichlet and Neumann problems. However, it is not feasible in most applications to determine in closed form the kernels $\partial G/\partial \nu$ and N that are needed for a more explicit evaluation of the results.

Since (7.34) and (7.36) exhibit u as a function of the parameter point ξ where the Green's and Neumann's functions become infinite, we might expect that G and N are actually solutions of the partial differential equation (7.2) in their dependence on ξ for each fixed position of x. The truth of this conjecture is apparent from the symmetry property

$$(7.37) \qquad G(x, \xi) = G(\xi, x),$$

$$(7.38) \qquad N(x, \xi) = N(\xi, x)$$

of the Green's and Neumann's functions, which we shall now proceed to deduce from (7.27). Note the analogy between the relations (7.37), (7.38) and the rule (4.82) for interchanging the arguments and parameters of the Riemann function.

Let y stand for the variable of integration and set

$$(7.39) \qquad u = G(y, \xi), \qquad v = G(y, x)$$

in (7.27), after first deleting from D a pair of small spheres of radius ϵ about the singular points ξ and x. Residue calculations quite like our derivation of (7.32) show that the resulting identity reduces to the relation

$$(7.40) \quad G(x, \xi) - G(\xi, x) = \int_{\partial D} \left[G(y, \xi) \frac{\partial G(y, x)}{\partial \nu} - G(y, x) \frac{\partial G(y, \xi)}{\partial \nu} \right] d\sigma,$$

comparable to (5.72), in the limit as $\epsilon \to 0$. The boundary conditions $G(y, \xi) = G(y, x) = 0$ which hold when y lies on ∂D imply that the integral on the right in (7.40) is zero, whence (7.37) follows. A similar argument based on the substitution

$$u = N(y, \xi), \qquad v = N(y, x)$$

instead of (7.39) serves to verify the second symmetry rule (7.38.)

There are difficulties in introducing a Neumann's function for the Laplace equation

(7.41) $$\Delta u \doteq 0$$

because of the compatibility condition (7.6) required for the solvability of the Neumann problem in that case. One device to avoid encountering an incompatible problem is to specify two singular points ξ_1 and ξ_2 in D. We then ask that the Neumann's function N have the structure

$$N = N(x; \xi_1, \xi_2) = S(x, \xi_1) - S(x, \xi_2) - v_S(x),$$

where S is any fundamental solution of (7.41) in D and v_S stands for a regular harmonic function there such that

(7.42) $$\frac{\partial v_S}{\partial \nu} = \frac{\partial S(x, \xi_1)}{\partial \nu} - \frac{\partial S(x, \xi_2)}{\partial \nu}$$

when x lies on ∂D, which makes $\partial N/\partial \nu = 0$. To see that the boundary condition (7.42) is consistent with the requirement (7.6), it suffices to put $u = 1$ in the representation (7.32) for arbitrary harmonic functions and then to compute the difference of the results obtained at $\xi = \xi_1$ and at $\xi = \xi_2$. This gives

(7.43) $$\int_{\partial D} \left[\frac{\partial S(x, \xi_1)}{\partial \nu} - \frac{\partial S(x, \xi_2)}{\partial \nu} \right] d\sigma = 1 - 1 = 0.$$

Observe that the Neumann's function of Laplace's equation has a physical interpretation as the velocity potential of an incompressible fluid flow in D from a source at ξ_2 into a sink at ξ_1. The relation (7.43) merely states that the source has the same strength as the sink, so that no liquid has to enter or exit through ∂D. That makes it feasible to impose the boundary condition $\partial N/\partial \nu = 0$. Taking into account the two singularities of N, we are now able to write down an analogue of (7.36) for Laplace's equation. In fact, the integral formula

(7.44) $$u(\xi_2) - u(\xi_1) = \int_{\partial D} N(x; \xi_1, \xi_2) \frac{\partial u(x)}{\partial \nu} d\sigma$$

provides a solution of the Neumann problem when one exists, but, of course, the answer is only determined up to an additive constant that cancels out of the difference on the left.

Having developed the general theory, we turn our attention to the explicit calculation of the Green's function of Laplace's equation for special domains. Consider first a half-space $x_n > 0$. We might expect that G is odd in its dependence on x_n in this case because it must vanish when $x_n = 0$ and because substitution of $-x_n$ for x_n does not alter the form of the Laplace equation (7.41). Thus, corresponding to its singularity at $\xi = (\xi_1, \ldots, \xi_{n-1}, \xi_n)$ in the upper half-space $x_n > 0$, the Green's function $G = G(x, \xi)$ ought to posses an opposite singularity in the lower half-space $x_n < 0$ at a point $\tilde{\xi} = (\xi_1, \ldots, \xi_{n-1}, -\xi_n)$ which is the *reflected image* of ξ in the hyperplane $x_n = 0$. These remarks suggest that G can be expressed as the difference

$$(7.45) \qquad G = \frac{1}{\sigma'_{n-1}}\left[\frac{1}{r^{n-2}} - \frac{1}{\tilde{r}^{n-2}}\right]$$

between two fundamental solutions of (7.41) with singularities located at ξ and $\tilde{\xi}$, where

$$\tilde{r} = \sqrt{(x_1 - \xi_1)^2 + \cdots + (x_{n-1} - \xi_{n-1})^2 + (x_n + \xi_n)^2}.$$

That (7.45) does, indeed, represent the Green's function of the half-space $x_n > 0$ follows directly from the observation that both terms on the right are harmonic and that they just cancel each other out on the boundary $x_n = 0$, since $\tilde{r} = r$ there.

By way of an application, we insert (7.45) into (7.34) to derive an explicit integral formula for the solution of Dirichlet's problem for the Laplace equation in a half-space. On the hyperplane $x_n = 0$ we find that

$$\frac{\partial G}{\partial \nu} = \frac{\partial G}{\partial x_n} = \frac{\xi_n}{\sigma_{n-1}r^n} + \frac{\xi_n}{\sigma_{n-1}\tilde{r}^n} = \frac{2\xi_n}{\sigma_{n-1}r^n}.$$

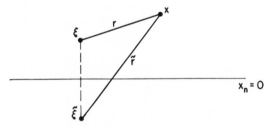

FIG. 25. The singularities of G.

Hence (7.34) reduces to the simple representation

$$(7.46) \quad u(\xi_1, \ldots, \xi_n)$$

$$= \frac{2}{\sigma_{n-1}} \int_{-\infty}^{\infty} \cdots \int_{-\infty}^{\infty} \frac{f(x_1, \ldots, x_{n-1}) \xi_n \, dx_1 \cdots dx_{n-1}}{[(x_1 - \xi_1)^2 + \cdots + (x_{n-1} - \xi_{n-1})^2 + \xi_n^2]^{n/2}}$$

for bounded harmonic functions in the half-space $\xi_n > 0$, where

$$f(x_1, \ldots, x_{n-1}) = u(x_1, \ldots, x_{n-1}, 0)$$

stands for the boundary values of u. No unexpected contribution to (7.34) results from the fact that the half-space is an infinite domain, as can be shown by performing an intermediate integration over a large hemisphere (cf. Exercise 5 below). Note that (7.46) holds for any number of independent variables $n \geq 2$, although the validity of (7.45) is restricted to the cases with $n > 2$.

We take up next the problem of determining the harmonic Green's function for the interior of a circle

$$(7.47) \qquad\qquad x^2 + y^2 < R^2$$

in the plane. It is convenient to use polar coordinates ρ and ϕ to describe the position

$$(7.48) \qquad\qquad \xi = \rho \cos \phi, \qquad \eta = \rho \sin \phi$$

of the logarithmic singularity of $G = G(x, y; \xi, \eta)$. The *method of images* suggests placing a reflected singularity at the point *inverse* to (7.48) in the circle (7.47), which has R^2/ρ and ϕ at its polar coordinates (cf. Exercise 3.3.1). Denoting by r and $\tilde{r}$ the distances from (x, y) to the point (ξ, η) and to its inverse, respectively, we proceed to establish the identity

$$(7.49) \qquad\qquad \frac{r}{\rho} = \frac{\tilde{r}}{R}$$

when (x, y) lies on the circumference of (7.47). Consider two triangles which have the radial segment joining the origin to (x, y) as a common side and have either (ξ, η) or its inverse as the opposite vertex. The obvious relationship

$$\frac{\rho}{R} = \frac{R}{R^2/\rho}$$

between the ratios of corresponding sides of these triangles shows that they are similar. Formula (7.49) is merely a restatement of the similarity in terms of another pair of sides.

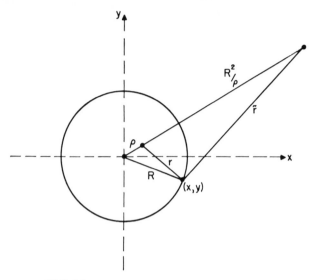

FIG. 26. Inverse points with respect to a circle.

We deduce from (7.49) that the Green's function of the circle (7.47) is given by the elementary formula

(7.50)
$$G(x, y; \xi, \eta) = \frac{1}{2\pi} \log \frac{\rho \tilde{r}}{Rr}.$$

To verify this, observe that the logarithm on the right provides an appropriate fundamental solution of Laplace's equation in the circle (7.47) because the distance $\tilde{r}$ from (x, y) to the inverse of the singular point (ξ, η) does not vanish there. On the other hand, the boundary condition $G = 0$ required of the Green's function along the circumference of (7.47) is fulfilled by the expression (7.50), too, in view of (7.49). One of the most important consequences of (7.50) is an explicit representation of the solution of Dirichlet's problem for a circle which merits a detailed analysis here.

It is an easy matter to calculate the normal derivative $\partial G/\partial \nu$ of (7.50) along the boundary of the circle (7.47). For the general boundary point

$$x = R \cos \theta, \qquad y = R \sin \theta$$

we find that

(7.51)
$$\frac{\partial G}{\partial \nu} = \frac{1}{4\pi} \left[\frac{1}{\tilde{r}^2} \frac{\partial \tilde{r}^2}{\partial \nu} - \frac{1}{r^2} \frac{\partial r^2}{\partial \nu} \right]$$

$$= \frac{1}{2\pi R} \frac{R^2 - \rho^2}{R^2 - 2R\rho \cos (\theta - \phi) + \rho^2},$$

since $\partial/\partial v = -\partial/\partial R$ and

$$r^2 = R^2 - 2R\rho \cos(\theta - \phi) + \rho^2.$$

After a multiplication by R, the expression on the right in (7.51) is called the *Poisson kernel*. We substitute it into (7.34) to obtain the *Poisson integral formula*

$$(7.52) \quad u(\rho \cos \phi, \rho \sin \phi) = \frac{1}{2\pi} \int_0^{2\pi} \frac{(R^2 - \rho^2)f(\theta)\,d\theta}{R^2 - 2R\rho \cos(\theta - \phi) + \rho^2}$$

for the solution of Dirichlet's problem in a circle of radius R about the origin, where

$$f(\theta) = u(R \cos \theta, R \sin \theta)$$

denotes the prescribed boundary values of the harmonic function u.

In order to derive the Poisson integral formula we have assumed that u is a regular solution of Laplace's equation in the closure of the circle (7.47). We shall now proceed to show, conversely, that the formula actually defines a solution of the Dirichlet problem in the circle of radius R whenever f is a continuous function of θ with the period 2π. We might think of this result as a rather special existence theorem.

The symmetry property (7.37) of the Green's function implies that the Poisson kernel is harmonic in its dependence on the parameter point (ξ, η). Hence differentiation under the integral sign serves to establish that formula (7.52) always defines a harmonic function u of the two independent variables ξ and η inside the circle of radius R around the origin. What we have to prove is that this harmonic function takes on the specified boundary values or, in other words, that

$$(7.53) \qquad \lim_{\rho \to R} u(\rho \cos \phi, \rho \sin \phi) = f(\phi).$$

Three basic properites of the Poisson kernel (7.51) are required for the demonstration. First, it is positive when $\rho < R$ because $r^2 > 0$; and second, its average has the value

$$(7.54) \qquad \frac{1}{2\pi} \int_0^{2\pi} \frac{(R^2 - \rho^2)\,d\theta}{R^2 - 2R\rho \cos(\theta - \phi) + \rho^2} = 1$$

because we can put $u \equiv f \equiv 1$ in (7.52). Finally, we have

$$(7.55) \qquad \lim_{\rho \to R} \frac{1}{2\pi} \frac{R^2 - \rho^2}{R^2 - 2R\rho \cos(\theta - \phi) + \rho^2} = 0$$

uniformly for $0 < \delta \leq |\theta - \phi| \leq 2\pi - \delta$, since the denominator r^2 is subject to the inequality

$$r^2 \geq R^2 - 2R\rho \cos \delta + \rho^2 = (R - \rho)^2 + 4R\rho \sin^2 (\delta/2)$$

in that range.

To pursue the analysis, let any number $\epsilon > 0$ be given, and choose $\delta > 0$ so small that

(7.56) $$|f(\theta) - f(\phi)| \leq \epsilon$$

for $|\theta - \phi| \leq \delta$. We can write

$$|u(\rho \cos \phi, \rho \sin \phi) - f(\phi)| = \left| \frac{1}{2\pi} \int_0^{2\pi} \frac{(R^2 - \rho^2)[f(\theta) - f(\phi)] \, d\theta}{R^2 - 2R\rho \cos (\theta - \phi) + \rho^2} \right|$$

$$\leq \frac{1}{2\pi} \int_0^{2\pi} \frac{(R^2 - \rho^2) \, |f(\theta) - f(\phi)| \, d\theta}{R^2 - 2R\rho \cos (\theta - \phi) + \rho^2}$$

because of (7.54) and the positivity of the Poisson kernel. Therefore

(7.57) $$\overline{\lim_{\rho \to R}} \, |u(\rho \cos \phi, \rho \sin \phi) - f(\phi)|$$

$$\leq \overline{\lim_{\rho \to R}} \frac{1}{2\pi} \int_{\phi-\delta}^{\phi+\delta} \frac{(R^2 - \rho^2)\epsilon \, d\theta}{R^2 - 2R\rho \cos (\theta - \phi) + \rho^2} \leq \epsilon$$

by virtue of (7.55), (7.56), and (7.54) again. Since $\epsilon > 0$ is arbitrary, the boundary condition (7.53) follows from (7.57). This completes our proof that the Poisson integral formula (7.52) furnishes a solution u of the Dirichlet problem for Laplace's equation in a circle when the boundary values f are prescribed continuously. It is a straightforward exercise to work out the analogous result for spheres of higher dimension.

An important special case of the representation (7.52) occurs when the point (ξ, η) is placed at the origin $\rho = 0$. We obtain by such a simplification the *mean value theorem*

(7.58) $$u(0, 0) = \frac{1}{2\pi} \int_0^{2\pi} u(R \cos \theta, R \sin \theta) \, d\theta,$$

which states that the average of any harmonic function of two variables over a circle is equal to its value at the center (cf. Exercises 6.3.9 and 6.3.10).

It is interesting to note that the maximum principle (cf. Section 1) can be derived as a consequence of (7.58). In fact, suppose that the harmonic function u has a maximum in the interior of its domain of definition. We can locate the origin at the maximum point without any loss of generality.

Letting M stand for the maximum value of u, we conclude from (7.58) that

$$(7.59) \qquad M = \frac{1}{2\pi} \int_0^{2\pi} u(R \cos \theta, R \sin \theta) \, d\theta \leq \frac{1}{2\pi} \int_0^{2\pi} M \, d\theta = M$$

for small enough choices of the radius R. Since the equality sign has to hold in the estimate (7.59), u must be identical with M on every circle about the origin. Therefore u is a constant. This deduction is, of course, synonymous with one formulation of the maximum principle for harmonic functions.

Poisson's integral formula leads to several more subtle inequalities within the class of positive harmonic functions

$$u > 0.$$

It is evident from the interval $-1 \leq \cos(\theta - \phi) \leq 1$ to which the values of the cosine are confined that

$$(R - \rho)^2 \leq R^2 - 2R\rho \cos(\theta - \phi) + \rho^2 \leq (R + \rho)^2.$$

Therefore the Poisson kernel lies between the bounds

$$(7.60) \qquad \frac{1}{2\pi} \frac{R - \rho}{R + \rho} \leq \frac{1}{2\pi} \frac{R^2 - \rho^2}{R^2 - 2R\rho \cos(\theta - \phi) + \rho^2} \leq \frac{1}{2\pi} \frac{R + \rho}{R - \rho}.$$

For a harmonic function $u \geq 0$ we can use (7.60) to estimate the integrand in (7.52) without introducing any absolute values. Thus we have

$$\frac{1}{2\pi} \int_0^{2\pi} \frac{R - \rho}{R + \rho} u(R \cos \theta, R \sin \theta) \, d\theta \leq u(\rho \cos \phi, \rho \sin \phi)$$

$$\leq \frac{1}{2\pi} \int_0^{2\pi} \frac{R + \rho}{R - \rho} u(R \cos \theta, R \sin \theta) \, d\theta.$$

Exploiting the mean value theorem (7.58) to evaluate the integrals on the left and on the right, we obtain the *Harnack inequalities*

$$(7.61) \qquad \frac{R - \rho}{R + \rho} u(0, 0) \leq u(\rho \cos \phi, \rho \sin \phi) \leq \frac{R + \rho}{R - \rho} u(0, 0),$$

which serve to describe the growth of a positive harmonic function in any circle of radius R. There is no real need to locate the center of the circle in question at the origin, of course, since the class of harmonic functions is invariant under a translation of coordinates.

A principal application of the Harnack inequalities arises in the demonstration of the *Harnack convergence theorem*. This theorem states that any monotonically increasing sequence of harmonic functions

$$u_1 \leq u_2 \leq u_3 \leq \cdots \leq u_m \leq \cdots$$

in, say, the circle of radius R about the origin must either approach infinity everywhere or else converge to a limit function that is again harmonic.

Insertion of the non-negative differences $u_{m+1} - u_m$ into the Harnack inequalities (7.61) establishes that the distinction of cases above depends only on the convergence at the origin, for $u_{m+1} - u_m$ is seen to have the same order of magnitude there that it has at any point inside the circle $\rho < R$. If the limit of the increasing sequence of numbers $u_m(0, 0)$ is finite, then the estimate

$$0 \leq u_{m+1}(\rho \cos \phi, \rho \sin \phi) - u_m(\rho \cos \phi, \rho \sin \phi)$$

$$\leq \frac{R + \rho}{R - \rho} [u_{m+1}(0, 0) - u_m(0, 0)]$$

based on (7.61) shows that

$$u(\rho \cos \phi, \rho \sin \phi) = \lim_{m \to \infty} u_m(\rho \cos \phi, \rho \sin \phi)$$

exists uniformly in every closed circle of radius $R_0 < R$ about the origin. Therefore as $m \to \infty$ we can pass to the limit under the integral sign in the Poisson representation

$$(7.62) \quad u_m(\rho \cos \phi, \rho \sin \phi) = \frac{1}{2\pi} \int_0^{2\pi} \frac{(R_0^2 - \rho^2)u_m(R_0 \cos \theta, R_0 \sin \theta) \, d\theta}{R_0^2 - 2R_0\rho \cos(\theta - \phi) + \rho^2}$$

for u_m in each circle of that kind; hence the limit function u is given by the same formula. It follows that u must be harmonic, which is what we set out to prove.

Passage to the limit under the sign of integration in (7.62) shows more generally that the limit of any uniformly convergent sequence of harmonic functions u_m must itself be harmonic. This statement is equivalent to the *Weierstrass convergence theorem* for analytic functions of a complex variable.[6] It is valid in regions of arbitrary shape, since each point of such a region can be covered by a circle so small that (7.62) applies to it, and we can appeal to the Heine-Borel theorem.[7] Furthermore, the convergence

---

[6] Cf. Ahlfors 1.
[7] Cf. Buck 1.

is uniform in closed subdomains not only for the function concerned but also for its partial derivatives of all orders. Finally, the result remains true for the solutions u of any linear elliptic partial differential equation possessing a regular Green's function, since it is actually a consequence of the integral representation (7.34) for u in terms of its boundary values.

EXERCISES

1. Introduce a Green's function for the ordinary differential equation $u_{xx} = 0$ on the interval $0 < x < 1$.

2. Find the harmonic function in the circle (7.47) which assumes boundary values that are 0 for $0 < \theta < \pi$ and that are 1 for $\pi < \theta < 2\pi$.

3. Show that at discontinuities of f of the first kind the boundary condition (7.53) takes the form

$$\lim_{\rho \to R} u(\rho \cos \phi, \rho \sin \phi) = \tfrac{1}{2}[f(\phi +) + f(\phi -)],$$

where

$$f(\phi +) = \lim_{\theta \to \phi, \theta > \phi} f(\theta), \qquad f(\phi -) = \lim_{\theta \to \phi, \theta < \phi} f(\theta).$$

4. Use the method of images to find the Green's function of the elliptic partial differential equation

$$u_{xx} + u_{yy} + u_{zz} = u$$

for the half-space $z > 0$.

5. Justify the integral formula (7.34) for Laplace's equation in a half-space by performing an intermediate integration over a large hemisphere and then passing to the limit as its radius becomes infinite.

6. Prove that formula (7.46) defines a solution of Dirichlet's problem for the Laplace equation in the upper half-space $\xi_n > 0$ whenever the assigned boundary values f are continuous and bounded.

7. By applying the maximum principle to a suitable comparison function, show that any bounded solution of Laplace's equation in a half-space whose value at each finite boundary point is zero must vanish identically.

8. Show by means of the Poisson integral formula (7.52) that the sequence of trigonometric polynomials $\cos n\theta$ and $\sin n\theta$ is complete on the interval $0 \leq \theta < 2\pi$.

9. Use formula (7.46) to obtain polynomials that approximate a continuously differentiable function f together with its first partial derivatives simultaneously on any bounded set.

10. Conclude from (7.49) and Exercise 3.3.1 that the Green's function for Laplace's equation in an n-dimensional sphere of radius R is given, in an obvious notation, by the expression

$$G(x, \xi) = \frac{1}{\sigma'_{n-1}} \left[\frac{1}{r^{n-2}} - \frac{R^{n-2}}{\rho^{n-2} \bar{r}^{n-2}} \right].$$

Extend the Poisson integral formula to this case and use it to establish the existence of the solution of Dirichlet's problem for a sphere. Finally, show that the mean value of any harmonic function u of n variables over the $(n - 1)$-dimensional surface of a sphere is equal to its value at the center, i.e., prove that

$$(7.63) \qquad u(0) = \frac{1}{\sigma_{n-1} R^{n-1}} \int_{r=R} u \, d\sigma.$$

11. Show that any continuous function satisfying the mean value theorem (7.63) over spheres of arbitrary center and radius must be harmonic.

12. Use the method of images to find the Green's function of the first octant $x_1 > 0$, $x_2 > 0$, $x_3 > 0$ for Laplace's equation in three-dimensional space. Find the harmonic Green's function of the infinite strip $a < x < b$ in the (x,y)-plane in the same way.

13. In the case of Laplace's equation, work out in closed form the representation (7.34) for the solution of Dirichlet's problem in a hemisphere

$$x_1 > 0, \qquad x_1^2 + \cdots + x_n^2 < R^2.$$

Deduce from it a proof of the *Schwarz principle of reflection*, which asserts that if u is harmonic in a region D including a section of the hyperplane $x_1 = 0$ on its boundary, and if u takes on the boundary values zero continuously on that section, then the rule

$$(7.64) \qquad u(-x_1, x_2, \ldots, x_n) = -u(x_1, x_2, \ldots, x_n)$$

defines an analytic extension of u as a harmonic function throughout the interior of the set obtained by adding to D its reflection in the hyperplane $x_1 = 0$.

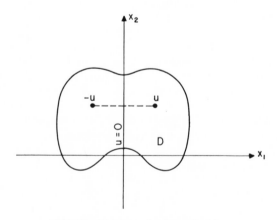

FIG. 27. Schwarz reflection principle.

14. Let u be any harmonic function in the region D of Exercise 13 which satisfies a boundary condition of the form

$$u(0, x_2, \ldots, x_n) = f(x_2, \ldots, x_n)$$

on the flat part of ∂D, where f is a real analytic function. Use the Cauchy-Kowalewski theorem (cf. Section 1.2) and the Schwarz principle of reflection to prove that u can be continued analytically across the hyperplane $x_1 = 0$. What does this result assert about the differentiability of the solution of Dirichlet's problem at the boundary?

15. Establish the formula

$$u(\xi, \eta) = \frac{1}{2\pi} \int_{-\infty}^{\infty} g(x) \log [(x - \xi)^2 + \eta^2] \, dx + \text{const.}$$

for the solution of the Neumann problem for Laplace's equation in the upper half-plane, where

$$g(x) = u_y(x, 0).$$

Generalize this result to the case of $n > 2$ independent variables and prove the associated existence theorem.

16. What becomes of the Schwarz principle of reflection if it is the normal derivative of u rather than u itself that vanishes on some section of a hyperplane?

17. Introduce the *Green's function of the third kind* for equation (7.2) in any region D as a fundamental solution $G_3 = G_3(x, \xi)$ there satisfying the mixed boundary condition

$$\frac{\partial G_3}{\partial \nu} + \alpha G_3 = 0$$

for x on ∂D. Show that a solution u of the third boundary value problem has the representation

$$u(\xi) = -\int_{\partial D} G_3(x, \xi) \left[\frac{\partial u(x)}{\partial \nu} + \alpha u(x) \right] d\sigma$$

in terms of this Green's function.

18. The Green's function of a linear elliptic equation (7.1) defined by a differential operator L that is not self-adjoint is a fundamental solution $G_L = G_L(x, \xi)$ in some specified region D which vanishes at the boundary ∂D. If G_M stands for the Green's function of the adjoint equation in the same region, establish the *interchange rule*

$$G_L(x, \xi) = G_M(\xi, x),$$

which shows that G_L satisfies the adjoint equation in its dependence on the parameter point ξ. Prove that the solution of Dirichlet's problem for (7.1) in D has an integral representation of the form

$$u(\xi) = -\int_{\partial D} B[u(x), G_M(x, \xi)],$$

analogous to (5.69), which involves only the boundary values of u along ∂D.

19. Prove the Harnack convergence theorem for an arbitrary plane region by covering it with circles. Is the result valid in space of higher dimension?

20. The first Hankel function[8]

$$H_0^{(1)}(kr) = \frac{1}{\pi} \int_{-\frac{\pi}{2}+i\infty}^{\frac{\pi}{2}-i\infty} \exp\left(ikr \cos z\right) dz$$

is sometimes referred to as the *free space Green's function* of the reduced wave equation (7.24) because it fulfills the radiation condition (7.25) at infinity. Establish the identity

$$u(\xi, \eta) = \frac{i}{4} \int_{\partial D} \left[u(x, y) \frac{\partial H_0^{(1)}(kr)}{\partial \nu} - H_0^{(1)}(kr) \frac{\partial u(x, y)}{\partial \nu} \right] ds$$

for arbitrary solutions u of (7.24) and (7.25) in the exterior of any simple closed curve ∂D.

21. Find the Green's function G for (7.24) and (7.25) in the exterior of a circle of radius R by the method of separation of variables.[9] Observe that the radiation condition is included among the requirements of the exterior Dirichlet problem for the reduced wave equation.

3. THE KERNEL FUNCTION; CONFORMAL MAPPING

It is instructive to introduce the difference

$$(7.65) \qquad K(x, \xi) = N(x, \xi) - G(x, \xi)$$

between the Neumann's function and the Green's function of the linear elliptic partial differential equation

$$(7.2) \qquad \Delta u = Pu$$

in a region D where the coefficient P is positive. Since the singularities of the two fundamental solutions N and G cancel each other out in the combination (7.65), the quantity K, which is known as the *kernel function* of the region D with respect to (7.2), is regular in its dependence on both the argument point x and the parameter point ξ throughout D. Because it has no physical interpretation, the kernel function is of more recent origin[10] than the Green's and Neumann's functions. However, in certain respects it provides a more effective tool for the solution of boundary value problems.

[8] Cf. Sommerfeld 1.
[9] Cf. Sommerfeld 1.
[10] Cf. Bergman-Schiffer 1.

The boundary conditions (7.33) and (7.35) that apply to the Green's and Neumann's functions show that

$$K(x, \xi) = N(x, \xi), \qquad \frac{\partial K(x, \xi)}{\partial \nu} = -\frac{\partial G(x, \xi)}{\partial \nu}$$

when x lies on ∂D. Consequently the integral formulas (7.34) and (7.36) for the solution of the Dirichlet and Neumann problems take the form

$$(7.66) \qquad u(\xi) = -\int_{\partial D} u(x) \frac{\partial K(x, \xi)}{\partial \nu} \, d\sigma,$$

$$(7.67) \qquad u(\xi) = -\int_{\partial D} K(x, \xi) \frac{\partial u(x)}{\partial \nu} \, d\sigma$$

when they are expressed in terms of the kernel function. Thus the kernel function is a regular solution of (7.2) in the region D whose normal derivative and boundary values along ∂D are precisely the factors required for the representation of solutions of both the first and second boundary value problems.

An illuminating formulation of (7.66) and (7.67) can be expressed in terms of the bilinear expression

$$(7.68) \qquad (u, v) = \int_D \left[\frac{\partial u}{\partial x_1} \frac{\partial v}{\partial x_1} + \cdots + \frac{\partial u}{\partial x_n} \frac{\partial v}{\partial x_n} + Puv \right] dx,$$

which we choose to interpret as the *scalar product* of two functions u and v defined over the region D. Observe that the Dirichlet integral is the natural *norm* (cf. Section 8.1) to associate with this scalar product, since it is equal to

$$\|u\|^2 = (u, u).$$

In particular, if u and v have a quotient that is not constant, the quadratic form

$$(7.69) \qquad \|\lambda u + \mu v\|^2 = (u, u)\lambda^2 + 2(u, v)\lambda\mu + (v, v)\mu^2$$

is positive-definite. It follows from an examination of the discriminant of (7.69) that the *Schwarz inequality*

$$(7.70) \qquad (u, v)^2 \leq (u, u)\,(v, v)$$

is valid for the scalar product (7.68), with equality holding only when u and v are proportional.

The significance of the scalar product (u, v) for equation (7.2) lies in the fact that if v is a solution, then

$$(u, v) = -\int_{\partial D} u \frac{\partial v}{\partial \nu} d\sigma$$

by virtue of (7.3). This identity shows that (7.66) and (7.67) are equivalent to the single relation

(7.71) $$u(\xi) = (u(x), K(x, \xi)),$$

since both u and K are solutions of (7.2). We often refer to (7.71) as the *reproducing property* of the kernel function.

Combined with the Schwarz inequality, the reproducing property (7.71) yields the estimate

(7.72) $$u(\xi)^2 = (u, K)^2 \leq (u, u)(K, K) = K(\xi, \xi) \|u\|^2$$

for solutions u of (7.2) possessing a finite Dirichlet integral over the region D. A sharp formulation of the same result states that for each fixed choice of the point ξ we can introduce $K = K(x, \xi)$ as an extremal function for the minimum problem

(7.73) $$\min \frac{\|u\|^2}{u(\xi)^2} = \frac{1}{K(\xi, \xi)},$$

where u is subject to the restriction that it must satisfy the partial differential equation (7.2). In particular, we conclude that

$$K(\xi, \xi) > 0.$$

Notice that the reproducing property (7.71) determines the kernel function uniquely, whereas the minimum problem (7.73) only serves, according to the Schwarz inequality, to fix it to within multiplication by a constant factor. Neither of these characterizations involves the Green's function or the Neumann's function, however. The minimum problem (7.73) will turn out to be important not only because it furnishes a construction of the kernel function but also because it leads to a proof of the existence of the Green's function (cf. Exercise 8.2.3).

We shall exploit (7.71) in order to develop K in a convergent series of solutions of (7.2) that are orthogonal with respect to the scalar product (7.68). Such a series is useful in that it suggests a method of calculating the kernels needed for the solution of the Dirichlet and Neumann problems by means of the integral formulas (7.66) and (7.67).

What we have in mind is to approximate an arbitrary solution u of (7.2) in D possessing a finite norm

$$\|u\|^2 < \infty$$

by linear combinations of a special sequence of solutions u_m with the same property. In this connection we say that the system of functions u_m is *complete* if, corresponding to any choice of u above and for any number $\epsilon > 0$, there exists an index μ and coefficients $a_1, \ldots, a_\mu$ such that

(7.74)
$$\left\| u - \sum_{m=1}^{\mu} a_m u_m \right\|^2 < \epsilon.$$

An application of the estimate (7.72) to (7.74) suffices to establish the inequality

(7.75)
$$\left[u(\xi) - \sum_{m=1}^{\mu} a_m u_m(\xi) \right]^2 < \epsilon K(\xi, \xi).$$

Therefore we see that the approximations to u in the mean by solutions drawn from a complete system can be made point-wise as accurate as we please over each closed subregion of D, where, of course, the continuous function $K(\xi, \xi)$ must remain bounded. Sometimes a complete system can be found by the method of separation of variables.

We say that the system of functions u_m is *orthonormal* if

(7.76)
$$(u_l, u_m) = \delta_{lm}, \quad l, m = 1, 2, \ldots,$$

where δ_{lm} is the Kronecker delta. From any complete system we can construct a complete orthonormal system by means of the *Gram-Schmidt orthogonalization process*. Briefly, this can be described as follows. Suppose that the first m functions $u_1, \ldots, u_m$ are already orthonormal, although the same may not be true after we adjoin to them the next function u_{m+1}. Consider the remainder

$$v_{m+1}(x) = u_{m+1}(x) - (u_1, u_{m+1})u_1(x) - \cdots - (u_m, u_{m+1})u_m(x)$$

of a truncated Fourier development for u_{m+1}. If v_{m+1} vanishes identically, then u_{m+1} is *linearly dependent* on $u_1, \ldots, u_m$ and we can throw it out of the system altogether. Otherwise we replace u_{m+1} by the normalized function $v_{m+1}/\|v_{m+1}\|$ and add it to the original set $u_1, \ldots, u_m$. The new set of $m + 1$ functions so obtained is orthonormal. By continuing in this way, we can build up the desired complete orthonormal system inductively.

It is a familiar fact[11] that for an orthonormal system u_m the norm on the

---

[11] Cf. Churchill 1.

left in (7.74) becomes a minimum when the parameters a_m are chosen to be the *Fourier coefficients*

$$(7.77) \qquad\qquad a_m = (u_m, u)$$

of u. This follows from the relation

$$(7.78) \quad \left\| u - \sum_{m=1}^{\mu} a_m u_m \right\|^2 = \|u\|^2 - 2\sum_{m=1}^{\mu} a_m (u_m, u) + \sum_{m=1}^{\mu} a_m^2$$

$$= \left\| u - \sum_{m=1}^{\mu} (u_m, u) u_m \right\|^2 + \sum_{m=1}^{\mu} [a_m - (u_m, u)]^2$$

based on the orthogonality condition (7.76). In particular, when (7.77) holds we derive the *Bessel inequality*

$$(7.79) \qquad\qquad \sum_{m=1}^{\mu} a_m^2 \leq \|u\|^2$$

from the observation that the norm on the left in (7.78) is non-negative. Furthermore, under the hypothesis that the orthonormal system u_m is complete we have

$$\lim_{\mu \to \infty} \left\| u - \sum_{m=1}^{\mu} a_m u_m \right\| = 0,$$

for ϵ can then be made arbitrarily small in (7.74). Therefore (7.79) becomes the *Parseval identity*

$$\|u\|^2 = \sum_{m=1}^{\infty} a_m^2$$

in the limit as $\mu \to \infty$, which can simply be viewed as another formulation of the completeness property.

In contrast with the standard theory of Fourier series, we find in the present investigation of the expansion

$$u(x) = \sum_{m=1}^{\infty} a_m u_m(x), \qquad a_m = (u_m, u),$$

of a solution u of the linear elliptic partial differential equation (7.2) in terms of a complete orthonormal system of special solutions u_m that the series converges uniformly to u in every closed subdomain of the region D. This unusually strong convergence theorem is a consequence of the fundamental estimate (7.75), since the number ϵ occurring there can be taken as small as we please because of the completeness of the system u_m. As an example, we expand the kernel function K itself in terms of any complete orthonormal system u_m. In this case the reproducing property (7.71) of K furnishes the formula

$$(u_m(x), K(x, \xi)) = u_m(\xi)$$

for the Fourier coefficients a_m. Thus we conclude that the kernel function is given by the remarkable series representation

$$(7.80) \qquad K(x, \xi) = \sum_{m=1}^{\infty} u_m(x)u_m(\xi).$$

We have already pointed out why the infinite series (7.80) must converge uniformly in every closed subregion of D. In this connection an analogy with the power series expansion of an analytic function of a complex variable is more appropriate than the more evident similarity to Fourier series. The special form of (7.80) explains our description of K as a kernel function, too. It is interesting that the development is quite independent of the individual choice of the complete orthonormal system involved. Note, incidentally, that (7.80) serves to establish the *symmetry property*

$$K(x, \xi) = K(\xi, x)$$

of the kernel function, which follows also, of course, from (7.37), (7.38) and the definition (7.65) of K in terms of the Green's and Neumann's functions. Finally, we emphasize that our analysis of the convergence of the series (7.80) is inconclusive on the boundary of the region D, where a special study of the representation would be required before it could be inserted into the integral formulas (7.66) and (7.67) for the solution of the Dirichlet and Neumann problems.

A modification of our theory of the kernel function is needed for Laplace's equation

$$\Delta u = 0,$$

since not every solution whose norm

$$\|u\|^2 = \int_D (\nabla u)^2 \, dx$$

vanishes is zero. We recall that because the Dirichlet integral of any constant is zero, the Neumann's function N of Laplace's equation must have two singularities. For the sake of simplicity we assume that the origin lies inside the region D, and we put $\xi_2 = 0$ in the specification

$$N(x; \xi_1, \xi_2) = S(x, \xi_1) - S(x, \xi_2) - v_S(x)$$

of N. Let us consider only those harmonic functions u which vanish at the origin. In short, let us impose the normalizations

$$\xi_2 = 0, \qquad u(0) = 0.$$

Then we may introduce a harmonic kernel function K by setting

$$K(x, \xi) = N(x; \xi, 0) - G(x, \xi) + G(x, 0) - \kappa(\xi),$$

where the additive constant κ is adjusted so that

$$K(0, \xi) = 0.$$

After the above preliminaries, our discussion of the kernel function K remains valid verbatim with (7.2) replaced by Laplace's equation and with the scalar product defined by

$$(u, v) = \int_D \left[\frac{\partial u}{\partial x_1} \frac{\partial v}{\partial x_1} + \cdots + \frac{\partial u}{\partial x_n} \frac{\partial v}{\partial x_n} \right] dx.$$

In particular, we find that (7.66), (7.67) and (7.71) hold as before, although the extremal problem (7.73) should only be posed for $\xi \neq 0$. Most significant of all, however, is the fact that the series expansion (7.80) is still correct under our hypothesis that all the harmonic functions of the complete orthonormal system u_m satisfy the side condition $u_m(0) = 0$.

We turn our attention to the explicit construction of the harmonic kernel function for the special case in the plane where the region D is a circle of radius R about the origin. As happens all too often in the applications, we find here that a specific complete orthonormal system of harmonic functions vanishing at the origin is most readily identified by the method of separation of variables. Indeed, the normalized *circular harmonics*

$$u_{2m-1} = \frac{\rho^m \cos m\phi}{\sqrt{\pi m} R^m}, \qquad u_{2m} = \frac{\rho^m \sin m\phi}{\sqrt{\pi m} R^m}$$

provide a system of the desired kind.

We introduce polar coordinates and use them to describe the two points

$$x = r \cos \theta, \qquad y = r \sin \theta$$

and

$$\xi = \rho \cos \phi, \qquad \eta = \rho \sin \phi$$

that feature as the argument and the parameter, respectively, for $K = K(r, \theta; \rho, \phi)$. In this notation, substitution of the complete orthonormal system of circular harmonics into the general expansion (7.80) yields the expression

$$K(r, \theta; \rho, \phi) = \frac{1}{\pi} \sum_{m=1}^{\infty} \frac{r^m \rho^m}{mR^{2m}} \cos m(\theta - \phi)$$

for the kernel function of the circle of radius R, since

$$\cos m\theta \cos m\phi + \sin m\theta \sin m\phi = \cos m(\theta - \phi).$$

Representing the cosine in terms of exponentials, we can sum the power series on the right to obtain the final closed formula

$$(7.81) \qquad K(r, \theta; \rho, \phi) = -\frac{1}{2\pi} \log \left[1 - 2 \frac{r\rho}{R^2} \cos(\theta - \phi) + \frac{r^2 \rho^2}{R^4} \right]$$

for K. Note the similarity between the argument of the logarithm on the right and the denominator of the Poisson kernel (7.51).

Both the method of images and the expansion of the kernel function in terms of a complete orthonormal system suggest procedures for the determination of the Green's function in space of arbitrary dimension. On the other hand, for Laplace's equation in the plane a more flexible and far-reaching tool is proffered by the technique of *conformal mapping*. In a certain sense conformal mapping provides the analogue for elliptic equations of the method of characteristics that we developed for hyperbolic equations (cf. Chapters 3 and 4). Like characteristic coordinates, however, it is only applicable in the case of two independent variables.

We have already seen in Section 3.1 that the homogeneous Laplace equation

$$u_{xx} + u_{yy} = 0$$

is invariant under any conformal transformation

$$w = \alpha + i\beta = F(x + iy) = F(z),$$

where F denotes an analytic function of the complex variable z. Indeed, we have

$$(7.82) \qquad u_{xx} + u_{yy} = |F'(z)|^2 (u_{\alpha\alpha} + u_{\beta\beta})$$

by virtue of the Cauchy-Riemann equations

$$\frac{\partial \alpha}{\partial x} = \frac{\partial \beta}{\partial y}, \qquad \frac{\partial \beta}{\partial x} = -\frac{\partial \alpha}{\partial y}.$$

The identity (7.82) is a direct statement of the invariance at issue. However, a more revealing explanation of the result is provided by the observation that, whereas each harmonic function of two variables is the real part of some analytic function, any analytic function of another analytic function is again analytic. Thus it is feasible to solve the Dirichlet problem in a simply connected plane region D, or, more specifically, in a plane region bounded by a single simple closed curve ∂D, by first mapping that region conformally onto a circle and then finding the solution of the transformed problem afterward by means of the Poisson integral formula (7.52). The difficulties are shifted in this fashion to the question of mapping a given region onto, say, the unit circle.

From a slightly different point of view we can attack the Dirichlet problem for Laplace's equation by representing the Green's function directly as the real part of a suitable mapping function. To be specific, let

$$w = F(z; \zeta)$$

stand for that analytic function of z which provides a conformal transformation of the simply connected region D onto the unit circle $|w| < 1$ so that

$$F(\zeta; \zeta) = 0, \qquad F'(\zeta; \zeta) > 0.$$

For values of z lying on the boundary curve ∂D we must have

$$|F(z; \zeta)| = 1.$$

That a function F with the above properties exists is the content of the *Riemann mapping theorem*[12] (cf. Sections 9.1 and 15.4). From this theorem it also follows, of course, that Dirichlet's problem for harmonic functions of two variables is solvable.

We proceed to verify that the formula

$$(7.83) \qquad G(z, \zeta) = -\frac{1}{2\pi} \log |F(z; \zeta)|$$

expresses the Green's function G in terms of the mapping function F. As the real part of an analytic function, the logarithm on the right must be harmonic in D except at $z = \zeta$. There it has the necessary form of a fundamental solution of the Laplace equation in view of our assumption that the point ζ corresponds to the origin. Furthermore, the boundary condition $G = 0$ required of the Green's function is a consequence of the relation $|F| = 1$ stating that ∂D maps onto the circumference of the unit

---

[12] Cf. Ahlfors 1.

circle. These remarks suffice to prove (7.83). Finally, we observe that the map function F has an inverse representation

$$F = e^{-2\pi(G+iH)}$$

in terms of the Green's function G and its conjugate harmonic function H.

An elementary application of the basic identity (7.83) can be made to determine in a new way the Green's function of the unit circle $|z| < 1$. For this purpose we recall[13] that all conformal mappings of the unit circle onto itself have the linear fractional form

$$w = e^{i\delta} \frac{z - \zeta}{1 - \bar{\zeta}z},$$

where $|\zeta| < 1$, where $\bar{\zeta}$ stands for the complex conjugate of the parameter ζ, and where δ denotes a real number. Therefore the desired Green's function is given by the expression

$$(7.84) \qquad G(z, \zeta) = -\frac{1}{2\pi} \log \left| \frac{z - \zeta}{1 - \bar{\zeta}z} \right|,$$

which the reader should compare with our previous formulation (7.50) of essentially the same result. Further examples of the calculation of the Green's function by techniques of conformal mapping will be taken up in the exercises.

It is suggestive of the generality of the rather explicit procedures we have outlined thus far for treating Dirichlet's problem for an equation of the second order to describe briefly what can be done in a similar fashion for the *biharmonic equation*

$$(7.85) \qquad \Delta\Delta u = 0$$

in two independent variables x and y. The *first boundary value problem* for (7.85) consists in finding a solution in a given region D which assumes prescribed values and normal derivatives

$$(7.86) \qquad u = f, \qquad \frac{\partial u}{\partial v} = g$$

along the boundary ∂D of that region (cf. Exercise 1.12). It is natural to impose two boundary conditions (7.86) because the elliptic equation (7.85) is of the fourth order.

---

[13] Cf. Nehari 1.

A fundamental solution of the biharmonic equation (7.85) is supposed to have the form

$$\Gamma(z, \zeta) = \frac{1}{8\pi} r^2 \log r + v,$$

where $r = |z - \zeta|$ is the distance between the two points z and ζ and where v stands for a regular biharmonic function of z with derivatives of all orders at $z = \zeta$. The *biharmonic Green's function* Γ of a plane region D is defined to be a fundamental solution there which fulfills the homogeneous boundary conditions

$$(7.87) \qquad\qquad \Gamma = \frac{\partial \Gamma}{\partial \nu} = 0$$

when the argument z lies on ∂D. We often refer to Γ as the Green's function of a *clamped plate* because it can be interpreted physically as the deflection at z of an infinitesimally thin elastic plate clamped at its edges and supporting a point load at ζ. The Green's function of the clamped plate can be determined by solving the first boundary value problem for v that results from the pair of boundary conditions (7.87).

Two applications of Green's identity (7.26) serve to establish the similar relation

$$(7.88) \quad \int_D [u \, \Delta \, \Delta v - v \, \Delta \, \Delta u] \, dx \, dy + \int_{\partial D} \left[u \, \frac{\partial \Delta v}{\partial \nu} \right.$$
$$\left. - \Delta v \, \frac{\partial u}{\partial \nu} + \Delta u \, \frac{\partial v}{\partial \nu} - v \, \frac{\partial \Delta u}{\partial \nu} \right] ds = 0,$$

which is important for a study of the biharmonic equation. In (7.88) suppose that u is biharmonic, and replace v by the Green's function $\Gamma(z, \zeta)$, after first deleting from D a small circle around ζ which may later be allowed to shrink down on that point. Then we obtain in the usual way a representation

$$(7.89) \qquad u(\zeta) = - \int_{\partial D} \left[u(z) \, \frac{\partial \Delta \Gamma(z, \zeta)}{\partial \nu} - \Delta \Gamma(z, \zeta) \, \frac{\partial u(z)}{\partial \nu} \right] ds$$

for the solution of the first boundary value problem for biharmonic functions in D. Our aim will now be to convert this basic integral formula into a reproducing property like (7.71) for a new kernel function[14] that is helpful in the calculation of Γ.

---

[14] Cf. Garabedian 2, Zaremba 1.

We begin by applying the Laplace operator to both sides of (7.89). To avoid ambiguity we find it convenient to write

$$\Delta_z = \frac{\partial^2}{\partial x^2} + \frac{\partial^2}{\partial y^2}, \qquad \Delta_\zeta = \frac{\partial^2}{\partial \xi^2} + \frac{\partial^2}{\partial \eta^2},$$

where z indicates the point (x, y) and ζ indicates the point (ξ, η). Thus we have

$$(7.90) \qquad \Delta_\zeta u = - \int_{\partial D} \left[u \frac{\partial}{\partial \nu} \Delta_z \Delta_\zeta \Gamma - \Delta_z \Delta_\zeta \Gamma \frac{\partial}{\partial \nu} u \right] ds.$$

This suggests that we attempt to interpret the quantity

$$k(z, \zeta) = -\Delta_z \Delta_\zeta \Gamma(z, \zeta)$$

as a *kernel function*.

Since Γ is biharmonic, we see that, at least as a function of its principal argument, k is harmonic. Furthermore, it is regular at $z = \zeta$. In fact, the double operator $\Delta_z \Delta_\zeta$ annihilates the singularity of the Green's function Γ because the distance r depends only on the differences $x - \xi$ and $y - \eta$, so that

$$\Delta_z \Delta_\zeta \, r^2 \log r = \Delta_z \Delta_z \, r^2 \log r = 0.$$

Due to the regularity of the harmonic function k, we can evaluate the integral on the right in (7.90) by means of Green's theorem to derive

$$(7.91) \qquad \Delta_\zeta u = - \int_D [u \, \Delta_z k - k \, \Delta_z u] \, dx \, dy = \int_D k \, \Delta_z u \, dx \, dy.$$

It will soon become apparent that (7.91) should be regarded as the *reproducing property* of the kernel function k.

With each solution u of the biharmonic equation we can associate the function

$$h = \Delta u,$$

which is obviously harmonic. In terms of h, the identity (7.91) assumes the more natural form

$$(7.92) \qquad h(\zeta) = \int_D h(z) k(z, \zeta) \, dx \, dy.$$

By introducing the new scalar product

$$[h, k] = \int_D hk \, dx \, dy$$

between any pair of functions h and k whose squares are integrable over D, we can transform (7.92) into the reproducing property

$$h(\zeta) = [h(z), k(z, \zeta)]$$

of the kernel function $k(z, \zeta)$. It follows that k enjoys with respect to $[h, k]$ all the properties we established earlier for K with respect to the quite different scalar product (u, K). In particular, $k(z, \zeta)$ minimizes the dimensionless quotient

$$\frac{[h, h]}{h(\zeta)^2} \geq \frac{1}{k(\zeta, \zeta)}$$

within the class of functions h that are harmonic in D. In this connection it should be observed that the norm

$$[h, h] = \int_D (\Delta u)^2 \, dx \, dy$$

plays the role of a Dirichlet integral for the biharmonic equation.

Relative to the norm $[h, h]$ we can speak of a complete system of harmonic functions h_m in the region D which are orthonormal in the sense

$$[h_l, h_m] = \delta_{lm}, \qquad l, m = 1, 2 \ldots .$$

The representation

(7.93) $$k(z, \zeta) = \sum_{m=1}^{\infty} h_m(z) h_m(\zeta)$$

of the kernel function k by means of such a complete orthonormal system is a consequence of the reproducing property, as we have seen in our derivation of (7.80). Despite the fact that it has an expansion in terms of harmonic functions, we prefer to describe k as the *biharmonic kernel function* because of its relationship to the clamped plate Green's function Γ. A principal point of interest here is the possibility (7.93) offers of solving the first boundary value problem for the biharmonic equation through recourse to a series of orthogonal harmonic functions. In practice the latter can sometimes be determined by orthogonalizing an appropriate set of polynomials.

An alternative approach to the biharmonic equation can be developed[15] around a formula for the *general solution* in terms of analytic functions of the complex variable $z = x + iy$. If we think of z and $\bar{z} = x - iy$ as

[15] Cf. Garabedian 4.

characteristic coordinates (cf. Section 3.1), it is natural to write the biharmonic equation (7.85) in the form

$$\frac{\partial^4 u}{\partial z^2 \, \partial \bar{z}^2} = 0.$$

Two integrations with respect to the independent variable $\bar{z}$ show that

$$\frac{\partial^2 u}{\partial z^2} = \bar{z}\Phi''(z) + \Psi''(z),$$

where Φ'' and Ψ'' are arbitrary analytic functions of z. Integrating two more times, now with respect to z instead of $\bar{z}$, we find that

$$u = \bar{z}\Phi(z) + \Psi(z) + z\Phi^*(\bar{z}) + \Psi^*(\bar{z}),$$

where Φ^* and Ψ^* are analytic functions of $\bar{z}$. If we require u to be real, we obtain in this way the representation

(7.94) $$u = 2\mathrm{Re}\,\{\bar{z}\Phi(z) + \Psi(z)\}$$

for the general solution of the biharmonic equation in terms of a pair of aribitrary analytic functions Φ and Ψ.

In order to apply (7.94) to the solution of boundary value problems in a simply connected region D, we first consider the equation of the boundary ∂D in the form

$$\tilde{F}(x, y) = 0.$$

Under the hypothesis that $\tilde{F}$ is analytic, we can write

$$\tilde{F}\!\left(\frac{z + \bar{z}}{2}, \frac{z - \bar{z}}{2i}\right) = 0$$

and solve for $\bar{z}$ in terms of z to obtain the less usual expression

(7.95) $$\bar{z} = F(z)$$

for the equation of the closed curve ∂D. We now make the important assumption that F is a single-valued analytic function of z throughout the region D, except at a finite number of points where it has poles.

To facilitate an understanding of this requirement we reformulate it differently. Let $\phi(w)$ denote a function which is meromorphic in the unit circle $|w| < 1$ and which is real on $|w| = 1$, and let $\psi(w)$ be another analytic function with the same poles as $\phi(w)$ in $|w| < 1$ but possessing boundary values on $|w| = 1$ that are pure imaginary. If the (rational) function

$$z = \phi(w) - \psi(w)$$

provides a conformal mapping of the unit circle $|w| < 1$ onto some schlicht region D, then the equation (7.95) of the corresponding curve ∂D is defined by the function

$$F(z) = \phi(w) + \psi(w).$$

The latter expression is seen to have the desired property of being meromorphic in its dependence on z.

Consider the class of biharmonic functions

(7.96) $$u = \mathrm{Re}\,\{[\bar{z} - F(z)]\,\Phi(z) + \Psi(z)\},$$

where Φ and Ψ are analytic or meromorphic functions of z such that u is regular throughout D. Since Φ is multiplied by $\bar{z}$, we must choose it to be analytic everywhere in D. On the other hand, at the poles of F we should ascribe to Ψ corresponding poles such that the combination $\Psi - F\Phi$ remains without singularities in D.

This representation suggests how to solve the first boundary value problem for the biharmonic equation when F is meromorphic. We have only to seek the solution in the form (7.96). Inserting it into the prescribed boundary conditions (7.86) and taking advantage of (7.95), we derive the equivalent requirements

(7.97) $$\mathrm{Re}\,\{\Psi\} = f,$$

(7.98) $$\mathrm{Re}\,\left\{\left[\frac{\partial \bar{z}}{\partial \nu} - \frac{\partial F}{\partial \nu}\right]\Phi + \frac{\partial \Psi}{\partial \nu}\right\} = g$$

on Φ and Ψ along ∂D. If s stands for the arc length along ∂D, measured to increase as we advance with the region D on our left, (7.95) gives

$$\frac{\partial \bar{z}}{\partial \nu} = -i\,\frac{\partial \bar{z}}{\partial s} = -i\,\frac{dF}{dz}\,\frac{\partial z}{\partial s} = -F'\,\frac{\partial z}{\partial \nu}\,.$$

Therefore (7.98) can be replaced by the relation

$$\mathrm{Re}\,\left\{-2F'\,\frac{\partial z}{\partial \nu}\,\Phi + \frac{\partial \Psi}{\partial \nu}\right\} = g,$$

which we prefer to write in the form

$$(7.99) \qquad \mathrm{Re} \left\{ \frac{\partial \Theta}{\partial \nu} \right\} = g,$$

where

$$\Theta = \Theta(z) = -2 \int F'(z) \Phi(z) \, dz + \Psi(z).$$

Our problem is to determine a pair of singular harmonic functions $\mathrm{Re}\{\Psi\}$ and $\mathrm{Re}\{\Theta\}$ which fulfill the boundary conditions (7.97) and (7.99) along ∂D and simultaneously generate a biharmonic function (7.96) that is regular in D. From its definition we see that the analytic function Θ might have both poles and logarithmic singularities in D at the poles of F. For the sake of simplicity suppose that F has only simple poles in D. Since both Φ and

$$\Psi - F\Phi = \tfrac{1}{2}[\Theta + \Psi] - \int F\Phi' \, dz$$

have to be regular, the poles of Ψ must have the same residues as those of $-\Theta$. Thus only the residues of Θ are at our disposal in adjusting Θ and Ψ so that

$$(7.100) \qquad \Phi(z) = \frac{\Psi'(z) - \Theta'(z)}{2F'(z)}$$

remains finite in D. Nevertheless, the existence of a solution of the first boundary value problem for the biharmonic equation is enough to convince us that these residues can be chosen in such a way that the numerator in (7.100) vanishes at the roots of the denominator. After the values of the residues are found, the final determination of the analytic functions Ψ and Θ may be carried out by solving Dirichlet and Neumann problems with the assistance of a conformal transformation of D onto the unit circle.

As an application of the method just described, we proceed to construct the biharmonic Green's function Γ of the unit circle. When solved for $\bar{z}$, the equation $z\bar{z} = 1$ of the circumference becomes

$$(7.101) \qquad \bar{z} = F(z) = \frac{1}{z}.$$

To avoid difficulties with the pole of F at the origin we simply replace Φ in (7.96) by $z\Phi$. Thus we attempt to find Γ in the form

$$\Gamma = \frac{1}{8\pi} |z - \zeta|^2 \log |z - \zeta| + \mathrm{Re}\left\{ [|z|^2 - 1]\Phi(z) + \Psi(z) \right\}.$$

The condition $\Gamma = 0$ along the boundary curve (7.101) reduces to

$$\text{Re}\,\{\Psi(z)\} = -\frac{1}{8\pi}(z - \zeta)(\bar{z} - \bar{\zeta})\log|z - \zeta|$$

$$= \text{Re}\,\left\{-\frac{1}{8\pi}(z - \zeta)\left(\frac{1}{z} - \bar{\zeta}\right)\log(1 - \bar{\zeta}z)\right\}$$

since $|z - \zeta| = |1 - \bar{\zeta}z|$ there. Because the function of z inside the braces on the right is analytic in the unit circle, it follows that

$$\Psi(z) = -\frac{1}{8\pi}(z - \zeta)\left(\frac{1}{z} - \bar{\zeta}\right)\log(1 - \bar{\zeta}z).$$

From the combined boundary conditions $\Gamma = \partial\Gamma/\partial\nu = 0$, which imply $\Gamma_x = \Gamma_y = 0$, we conclude that

$$\frac{\partial\Gamma}{\partial z} = \bar{z}\,\text{Re}\,\{\Phi(z)\} + \tfrac{1}{2}\Psi'(z) + \frac{1}{16\pi}(\bar{z} - \bar{\zeta}) + \frac{1}{16\pi}(\bar{z} - \bar{\zeta})\log(z - \zeta)(\bar{z} - \bar{\zeta}) = 0$$

along ∂D. Hence we have

$$\text{Re}\,\{\Phi(z)\} = \text{Re}\,\left\{\frac{|\zeta|^2 - 1}{16\pi} - \frac{1}{8\pi}\left(1 - \frac{\zeta}{z}\right)\log(1 - \bar{\zeta}z)\right\}$$

there. Thus

$$\Phi(z) = \frac{|\zeta|^2 - 1}{16\pi} - \frac{1}{8\pi}\left(1 - \frac{\zeta}{z}\right)\log(1 - \bar{\zeta}z)$$

for all values of z; therefore, finally,

$$(7.102)\quad \Gamma(z, \zeta) = \frac{1}{8\pi}|z - \zeta|^2\log\left|\frac{z - \zeta}{1 - \bar{\zeta}z}\right| + \frac{1}{16\pi}(|z|^2 - 1)(|\zeta|^2 - 1).$$

Observe that this closed expression for the biharmonic Green's function of the unit circle can be inserted into (7.89) to yield an explicit integral formula for the solution of the first boundary value problem (7.85), (7.86) for a circular plate.

EXERCISES

1. Use (7.66) and (7.81) to check the Poisson integral formula for harmonic functions that vanish at the origin. Similarly, use (7.67) and (7.81) to derive the representation

$$u(\rho\cos\phi, \rho\sin\phi) = \frac{1}{2\pi}\int_0^{2\pi} g(\theta)\log[1 - 2\rho\cos(\theta - \phi) + \rho^2]\,d\theta$$

for the solution of the Neumann problem in the unit circle, where

$$g(\theta) = -\lim_{\rho \to 1} \frac{\partial u(\rho \cos \theta, \, \rho \sin \theta)}{\partial \rho}.$$

2. Use (7.66) and a suitable complete orthonormal system of harmonic functions to find an explicit integral formula for the solution of the Dirichlet problem for Laplace's equation in a circular ring $r^2 < x^2 + y^2 < R^2$.

3. Show how the *conjugate*

$$v = \int (u_x \, dy - u_y \, dx)$$

of a harmonic function u of two independent variables x and y can be used to convert the Neumann problem for Laplace's equation in the plane into a Dirichlet problem.

4. Derive the estimate

(7.103) $$[u(\rho \cos \phi, \, \rho \sin \phi) - u(0, 0)]^2 \leq \frac{1}{\pi} \|u\|^2 \log \frac{R^2}{R^2 - \rho^2}$$

for harmonic functions u in a circle of radius R about the origin by substituting (7.81) into (7.72), which is subject to the restriction $u(0, 0) = 0$.

5. By performing an appropriate infinitesimal variation, show that the reproducing property (7.71) of the kernel function K is a consequence of the extremal property (7.73).

6. Show formally that the Green's function G is a Lagrange multiplier for the minimum problem (7.73) when it is treated as a *Bolza problem*[16] in the calculus of variations, with the partial differential equation (7.2) imposed as a side condition.

7. Use the Schwarz principle of reflection (cf. Exercise 2.13) to study the *Schwarz-Christoffel transformation*[17]

(7.104) $$z = \int \prod_{j=1}^{N} (w - w_j)^{-\theta_j} \, dw$$

of the upper half-plane Im $\{w\} > 0$ onto the interior of a polygon of N sides in the z-plane, where $w_1, \dots, w_N$ stand for the points on the real axis corresponding to the vertices of the polygon, and where the exponents θ_j add up to 2.

8. Show by means of (7.104) that the conformal mapping of the unit circle onto a rectangle can be found in terms of an elliptic integral. Combine this result with (7.83) to express the harmonic Green's function of a rectangle in terms of elliptic functions.

---

[16] Cf. Courant-Hilbert 2.
[17] Cf. Nehari 1.

9. When a solution of Laplace's equation in the plane is expressed as a function $u = u(z, \bar{z})$ of the two complex characteristic coordinates z and $\bar{z}$, prove that it obeys the reflection rule

(7.105) $$u(\overline{F(z)}, F(z)) = -u(z, \bar{z})$$

across any analytic curve (7.95) on which it vanishes (cf. Section 16.4). Compare this result with the Schwarz reflection principle (7.64), and use (7.96) to generalize it to the case of a biharmonic function u that satisfies the homogeneous boundary conditions $u = \partial u / \partial v = 0$ along an analytic arc.

10. Let G be the Green's function of Laplace's equation in a plane domain D bounded by a simple closed curve ∂D that is analytic. Use the Schwarz reflection principle (cf. Exercise 9) to show that G can be continued analytically across the boundary curve ∂D and is therefore defined in the closed domain $D + \partial D$. Imitating our discussion of the Poisson integral formula (7.52), apply these results to the representation formula (7.34) in order to establish that the existence of a solution of the Dirichlet problem in D follows from the existence of the Green's function.

11. Formulate the first boundary value problem for the biharmonic equation in the exterior of an ellipse. Solve it explicitly by means of the reduction (7.96) to a pair of boundary value problems (7.97) and (7.99) for Laplace's equation.

12. Establish the symmetry
$$\Gamma(z, \zeta) = \Gamma(\zeta, z)$$

of the Green's function of a clamped plate.

13. Discuss a boundary value problem for the biharmonic equation that consists in replacing (7.86) by
$$u = f, \qquad \Delta u = g.$$

14. Show that the system of harmonic polynomials

$$h_{2m-1} = 2\sqrt{\frac{1}{\pi} \frac{m}{\text{sh } 2m\lambda + m \text{ sh } 2\lambda}} \text{ Re} \left\{ \frac{\text{sh}(m \text{ ch}^{-1} z)}{\text{sh}(\text{ch}^{-1} z)} \right\}, \qquad m = 1, 2, \ldots,$$

$$h_{2m-2} = 2\sqrt{\frac{1}{\pi} \frac{m}{\text{sh } 2m\lambda - m \text{ sh } 2\lambda}} \text{ Im} \left\{ \frac{\text{sh}(m \text{ ch}^{-1} z)}{\text{sh}(\text{ch}^{-1} z)} \right\}, \qquad m = 2, 3, \ldots$$

is complete and orthonormal over the ellipse

$$\frac{x^2}{\text{ch}^2\lambda} + \frac{y^2}{\text{sh}^2\lambda} < 1$$

in the sense $[h_l, h_m] = \delta_{lm}$, where $z = x + iy$. Deduce that for this ellipse

(7.106)
$$\Delta_z \Delta_\zeta \Gamma = -\frac{4}{\pi} \sum_{m=1}^{\infty} \frac{m}{\text{sh } 2m\lambda + m \text{ sh } 2\lambda} \text{ Re} \left\{ \frac{\text{sh}(m \text{ ch}^{-1} z)}{\text{sh}(\text{ch}^{-1} z)} \right\} \text{ Re} \left\{ \frac{\text{sh}(m \text{ ch}^{-1} \zeta)}{\text{sh}(\text{ch}^{-1} \zeta)} \right\}$$
$$-\frac{4}{\pi} \sum_{m=2}^{\infty} \frac{m}{\text{sh } 2m\lambda - m \text{ sh } 2\lambda} \text{ Im} \left\{ \frac{\text{sh}(m \text{ ch}^{-1} z)}{\text{sh}(\text{ch}^{-1} z)} \right\} \text{ Im} \left\{ \frac{\text{sh}(m \text{ ch}^{-1} \zeta)}{\text{sh}(\text{ch}^{-1} \zeta)} \right\}.$$

15. Verify that the Green's function (7.102) is positive in the circle $|z| < 1$. Use (7.106) to establish that there exists an ellipse whose Green's function Γ assumes negative as well as positive values.[18]

16. In a plane region D bounded by a smooth closed curve ∂D, let f_m be a complete system of analytic functions which are orthonormal in the sense

$$\int_{\partial D} f_l(z)\,\overline{f_m(z)}\,ds = \delta_{lm}, \qquad l, m = 1, 2, \ldots.$$

Establish the relation[19]

$$-\frac{1}{\pi}\frac{\partial^2 G(z, \zeta)}{\partial z\,\partial \bar{\zeta}} = \left[\sum_{m=1}^{\infty} f_m(z)\,\overline{f_m(\zeta)}\right]^2,$$

where G stands for the harmonic Green's function of D. Moreover, prove that

$$\frac{\partial G(z, \zeta)}{\partial \nu} = \frac{\left|\displaystyle\sum_{m=1}^{\infty} f_m(z)\,\overline{f_m(\zeta)}\right|^2}{\displaystyle\sum_{m=1}^{\infty} |f_m(\zeta)|^2}$$

along ∂D if the series in the numerator on the right is convergent there. Examine the special case in which the functions f_m are polynomials orthogonal over the perimeter of the unit circle.

17. The *interior property* of a conformal transformation asserts that around the image of any point there is a circle which is the map of some neighborhood of the original point. Deduce the maximum principle for harmonic functions of two variables from this result.

18. If $S(x, \xi)$ is a fundamental solution of (7.2) in a region containing the closure of D, and if v is any function that vanishes on ∂D, show that

(7.107) $$v(\xi) = (v(x), S(x, \xi)),$$

which can be viewed as a modified reproducing property of S.

---

[18] Cf. Garabedian 2.
[19] Cf. Garabedian 1, Szegö 1.

8

Dirichlet's Principle

1. ORTHOGONAL PROJECTION

The task that now lies before us is to establish the existence of a solution of the Dirichlet problem for a linear elliptic partial differential equation of the second order. We shall discuss this matter in detail for the equation

$$(8.1) \qquad \Delta u - Pu = 0,$$

where the coefficient P is supposed to be a non-negative function of the independent variables $x_1, \ldots, x_n$ that possesses continuous partial derivatives of several orders. We intend to develop two different existence proofs based on the class of functions w that have a finite *Dirichlet integral*

$$\|w\|^2 = \int_D \left[\left(\frac{\partial w}{\partial x_1} \right)^2 + \cdots + \left(\frac{\partial w}{\partial x_n} \right)^2 + Pw^2 \right] dx$$

over the region D where the boundary value problem is to be solved.

Our analysis may be viewed as an example of the *direct method of the calculus of variations* for multiple integral extremum problems. The particular properties of the Dirichlet integral that we shall exploit are suggested on the one hand by the extremal characterization (7.73) of the kernel function and on the other hand by equilibrium situations in mathematical physics where the energy is a minimum. The variational approach we shall describe is important, too, because it leads to approximations to the solution of Dirichlet's problem that come under the heading of the *Rayleigh-Ritz procedure*.

In Chapter 7 we introduced the *scalar product*

$$(u, v) = \int_D \left[\frac{\partial u}{\partial x_1} \frac{\partial v}{\partial x_1} + \cdots + \frac{\partial u}{\partial x_n} \frac{\partial v}{\partial x_n} + Puv \right] dx$$

276

of any pair of functions u and v whose Dirichlet integrals over D are finite. An application of the formulation (7.3) of Green's theorem yields the alternate expression

$$(8.2) \qquad (u, v) = -\int_D [\Delta u - Pu]v \, dx - \int_{\partial D} v \frac{\partial u}{\partial v} \, d\sigma$$

for the scalar product when the partial derivatives involved exist and are continuous in D and on its boundary ∂D. An important consequence of (8.2) is that u and v are orthogonal in the sense

$$(8.3) \qquad\qquad (u, v) = 0$$

whenever u is a solution of the partial differential equation (8.1) in the region D and v satisfies the boundary condition

$$(8.4) \qquad\qquad v = 0$$

along ∂D.

The result (8.3) leads to certain procedures for studying the Dirichlet problem which will be more readily grasped if we assume for the moment that a solution can always be found. Given any function w that possesses a finite Dirichlet integral over D and takes on continuous boundary values on ∂D, let u be the corresponding solution of Dirichlet's problem which reduces to w along ∂D. Then w has a decomposition

$$(8.5) \qquad\qquad w = u + v,$$

where v satisfies the requirement (8.4) that it vanishes on ∂D. Since the two components u and v of the representation (8.5) for w are orthogonal in the sense (8.3), we have

$$(w, w) = (u, u) + 2(u, v) + (v, v) = (u, u) + (v, v),$$

a relation that we prefer to write in the form

$$(8.6) \qquad\qquad \|w\|^2 = \|u\|^2 + \|v\|^2$$

of the *Pythagorean theorem*. This suggests extremal characterizations of the two functions u and v which will exhibit them ultimately as *orthogonal projections* of the original function w.

To begin with, we maintain that among all functions defined in D which take on the same boundary values as w, the solution u of the partial differential equation (8.1) is the one that minimizes the Dirichlet integral. Indeed, any such function is expressible as a sum, like (8.5), of u and of a term v that vanishes on ∂D. Therefore its Dirichlet integral is subject to the estimate

$$\|w\| \geq \|u\|,$$

in view of (8.6), with the equality sign holding only when $v \equiv 0$, so that

$$\|v\| = 0.$$

The characterization of a solution u of the first boundary value problem for (8.1) as the extremal function for the problem in the calculus of variations that has just been described is known as *Dirichlet's principle*. If we think of v as an infinitesimal perturbation of u, the orthogonality condition $(u, v) = 0$ is equivalent to the statement that the first variation of the Dirichlet integral is zero. Thus it follows from (8.2) that (8.1) is the *Euler equation* associated with the minimum problem in question.[1]

An alternate formulation of Dirichlet's principle asserts that for a given choice of w with a finite Dirichlet integral, the function v which vanishes on ∂D and is closest to w in the sense

$$(8.7) \qquad\qquad \|w - v\| = \text{minimum}$$

has the property that the difference $w - v$ is a solution u of (8.1). In a similar way it turns out that for a fixed choice of w the solution u of (8.1) which is singled out by the extremal problem

$$(8.8) \qquad\qquad \|w - u\| = \text{minimum}$$

has the same boundary values as w along ∂D. Therefore it solves the Dirichlet problem corresponding to those data. Indeed, suppose that the difference $w - u$ is a function v vanishing on ∂D. Then the Pythagorean theorem (8.6) gives

$$(8.9) \qquad\qquad \|w - u\| = \|v\| \leq \|w\|.$$

Since altering w through subtraction of an arbitrary solution of (8.1) cannot change the minimum value of the Dirichlet integral specified in (8.8), the form of the right-hand side of (8.9) makes evident the stated extremal property of u.

Our analysis shows that the two components u and v of the orthogonal decomposition (8.5) of a prescribed function w possessing a finite Dirichlet integral over D are determined individually by the two minimum problems (8.7) and (8.8). These results assert that, within the class of solutions of (8.1) and within the class of functions vanishing on ∂D, respectively, u and v are situated closest to the given function w in the sense of the norm

[1] Cf. Courant-Hilbert 2.

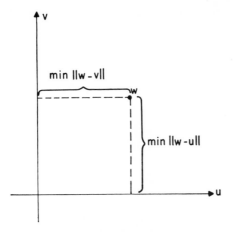

FIG. 28. Orthogonal projection.

defined by the Dirichlet integral. Thus we are led to interpret u and v as the *orthogonal projections* of w onto the above two classes of functions, which span the entire space of functions with a finite Dirichlet integral. Our next objective will be to exploit the extremal properties of u and v within such a function space in order to solve the Dirichlet problem in a rigorous fashion.

The class of functions w possessing a finite Dirichlet integral over D constitutes an infinite-dimensional *vector space* in which the concepts of addition and of multiplication by a numerical factor, or scalar, have an obvious meaning. We cannot expect solutions of extremal problems like (8.7) and (8.8) to exist within such a class, however, unless it is *complete*. In that case it is referred to as a *Hilbert space*.

To be more specific, a Hilbert space H is a set of elements $u, v, w, \ldots$ satisfying the following axioms.[2] It is closed under a commutative and associative addition $u + v$ and under a distributive and associative multiplication au by scalars $a, b, \ldots$. In particular, every linear combination $au + bv$ of elements is also an element. We have

$$u + 0 = u, \qquad 1u = u,$$

where 0 denotes a zero element. A *scalar product* (u, v) and a *norm* $\|u\|$ are defined on H so that

$$(au + bv, w) = a(u, w) + b(v, w), \qquad (u, v) = (v, u),$$

$$\|u\| = \sqrt{(u, u)} \geq 0.$$

[2] Cf. Stone 1.

The equals sign holds in the last relation only for the zero element. The space is supposed to be *separable*, which means that a sequence of elements u_m exists such that to each element u and each $\epsilon > 0$ there correspond scalars $a_1, \ldots, a_\mu$ with

$$\left\| u - \sum_{m=1}^{\mu} a_m u_m \right\| < \epsilon.$$

Finally, H is supposed to be *complete* in the sense that with any Cauchy sequence w_k, i.e., with any sequence such that

$$\lim_{k,l \to \infty} \|w_k - w_l\| = 0,$$

there is associated a limit element w in the space having the property

$$\lim_{k \to \infty} \|w_k - w\| = 0.$$

Actually, the vector space of continuously differentiable functions w possessing a finite Dirichlet integral over D is not complete. However, it has all the other properties attributed to Hilbert space, including separability.[3] We proceed to indicate how it can be extended so that it becomes complete and hence comprises a Hilbert space in the true sense of the term.

If ϕ is any trial function that is continuously differentiable in the closure of D and vanishes on ∂D, then by the divergence theorem

$$\int_D \left[\frac{\partial w}{\partial x_j} \frac{\partial \phi}{\partial x_k} - \frac{\partial w}{\partial x_k} \frac{\partial \phi}{\partial x_j} \right] dx = \int_D \phi \left[\frac{\partial^2 w}{\partial x_j \partial x_k} - \frac{\partial^2 w}{\partial x_k \partial x_j} \right] dx$$
$$- \int_{\partial D} \phi \left[\frac{\partial w}{\partial x_j} \frac{\partial x_k}{\partial \nu} - \frac{\partial w}{\partial x_k} \frac{\partial x_j}{\partial \nu} \right] d\sigma = 0.$$

Reversal of the roles of w and ϕ in the integration by parts shows the above identity remains valid even when w does not have second derivatives, provided that the first derivatives of ϕ vanish along ∂D. Similarly

$$\int_D \left[\frac{\partial w}{\partial x_j} \phi + w \frac{\partial \phi}{\partial x_j} \right] dx = - \int_{\partial D} \phi w \frac{\partial x_j}{\partial \nu} d\sigma = 0.$$

Thus when we set

$$w^{(j)} = \frac{\partial w}{\partial x_j}, \quad j = 1, \ldots, n,$$

[3] Cf. Stone 1.

we find that

$$(8.10) \qquad \int_D \left[w^{(j)} \frac{\partial \phi}{\partial x_k} - w^{(k)} \frac{\partial \phi}{\partial x_j} \right] dx = 0, \qquad j, k = 1, \ldots, n,$$

$$(8.11) \qquad \int_D \left[w^{(j)} \phi + w \frac{\partial \phi}{\partial x_j} \right] dx = 0, \qquad j = 1, \ldots, n,$$

for every trial function ϕ which reduces to zero in some neighborhood of ∂D. These results suggest that we call the quantities $w^{(j)}$ *weak partial derivatives* of w whenever the functions $w, w^{(1)}, \ldots, w^{(n)}$ merely fulfill (8.10) and (8.11) and have squares that are Lebesgue integrable[4] over D. We intend to show that the class of functions with a finite Dirichlet integral becomes a complete Hilbert space provided that elements w are included whose first partial derivatives are only required to exist in the weak sense thus defined.

Let us assume for the sake of simplicity that the coefficient P is bounded from below by a positive number, and suppose that w_k is a Cauchy sequence of functions continuously differentiable in the closure of D. It follows that

$$\lim_{k,l \to \infty} \int_D (w_k - w_l)^2 \, dx = 0,$$

$$\lim_{k,l \to \infty} \int_D (w_k^{(j)} - w_l^{(j)})^2 \, dx = 0, \qquad j = 1, \ldots, n.$$

We now appeal to the *Riesz-Fischer theorem*, for which a concise proof is presented by Stone (cf. Stone 1). It states that there exist limit functions $w, w^{(1)}, \ldots, w^{(n)}$ which are uniquely determined almost everywhere in D, which have Lebesgue integrable squares over that region, and which satisfy

$$\lim_{k \to \infty} \int_D (w_k - w)^2 \, dx = 0,$$

$$\lim_{k \to \infty} \int_D (w_k^{(j)} - w^{(j)})^2 \, dx = 0, \qquad j = 1, \ldots, n.$$

In view of the Schwarz inequality

$$\left[\int_D (w_k - w) \psi \, dx \right]^2 \le \int_D (w_k - w)^2 \, dx \int_D \psi^2 \, dx,$$

we see that

$$(8.12) \qquad \lim_{k \to \infty} \int_D w_k \psi \, dx = \int_D w \psi \, dx$$

[4] Cf. Munroe 1.

for any trial function ψ with an integrable square. A similar result holds, of course, for each of the sequences $w_k^{(j)}$, too. An application of this convergence theorem to the relations (8.10) and (8.11) suffices to establish that they remain valid for the limiting quantities w, $w^{(1)}, \ldots, w^{(n)}$.

In every case of the kind just described we shall add the limit function w defined by the Riesz-Fischer theorem, which has the weak derivatives $w^{(1)}, \ldots, w^{(n)}$, to the class of functions possessing a finite Dirichlet integral over D. Such a process provides us with a *complete class*, or *Hilbert space*, in much the same way as the set of rational numbers is completed to generate the real number system.[5] The Hilbert space in question will be denoted by W_2. We shall let V_2 stand for the subclass of W_2 consisting of limits of functions that vanish on the boundary ∂D of D, which also forms a Hilbert space. Similarly, the notation U_2 will be used to describe the Hilbert space within W_2 generated by solutions of the partial differential equation (8.1) whose Dirichlet integrals are finite. It is a consequence of the convergence theorem (8.12) that each element u in U_2 is orthogonal to every element v of V_2. Our objective in what follows will be to solve the Dirichlet problem for (8.1) by developing in a precise fashion the decomposition (8.5) of an arbitrary element of W_2 into a sum of two orthogonal components lying in the completed subspaces U_2 and V_2. This is an example of the *projection theorem* in Hilbert space.

We wish to verify that for a given choice of w the extremal problems (8.7) and (8.8) have solutions v and u belonging, respectively, to the complete classes of functions V_2 and U_2. Suppose there were no such extremal function v. Then we could find a *minimal sequence* v_k of elements in V_2 such that the norm $\|w - v_k\|$ approaches the greatset lower bound d of the set of numbers $\|w - v\|$ generated by allowing v to range over that class. We maintain that v_k is actually a Cauchy sequence.

For the proof we consider the quite general *parallelogram law*

$$(8.13) \qquad \left\| \frac{w_1 + w_2}{2} \right\|^2 + \left\| \frac{w_1 - w_2}{2} \right\|^2 = \frac{\|w_1\|^2 + \|w_2\|^2}{2},$$

which is established for any two elements w_1 and w_2 in Hilbert space by direct expansion of the squared norms on the left. For any given $\epsilon > 0$, pick k and l so large that

$$\|w - v_k\|^2 \leq d^2 + \epsilon, \qquad \|w - v_l\|^2 \leq d^2 + \epsilon.$$

Then set $\qquad w_1 = w - v_k, \qquad w_2 = w - v_l \qquad$ in (8.13). Since

$$d^2 \leq \left\| w - \frac{v_k + v_l}{2} \right\|^2 = \left\| \frac{w_1 + w_2}{2} \right\|^2$$

[5] Cf. Birkhoff-MacLane 1.

by the very definition of d, it follows that

$$d^2 + \left\| \frac{v_k - v_l}{2} \right\|^2 \leq \frac{d^2 + \epsilon + d^2 + \epsilon}{2} = d^2 + \epsilon.$$

After subtraction of d^2 from both sides, this is enough to prove the desired convergence of the sequence v_k.

According to our earlier analysis there has to exist in the complete space V_2 a limit v of the convergent minimum sequence v_k. We shall establish that the element v so obtained is the extremal function for (8.7) that we have been looking for. We use the *triangle inequality*

$$\| w - v \| \leq \| w - v_k \| + \| v_k - v \|,$$

which is, of course, a consequence of Schwarz's inequality

$$(w - v_k, v_k - v) \leq \| w - v_k \| \, \| v_k - v \|.$$

It serves to show that

·(8.14) $$\| w - v \| \leq d,$$

since

$$\lim_{k \to \infty} \| w - v_k \| = d, \qquad \lim_{k \to \infty} \| v_k - v \| = 0.$$

The estimate (8.14) makes evident the extremal property of v.

A quite analogous argument shows that there is a solution u of the extremal problem (8.8) which is the limit in the sense

(8.15) $$\lim_{k \to \infty} \| u_k - u \| = 0$$

of convergence in the mean of a minimum sequence of solutions u_k of (8.1). The functions u and v thus determined by (8.8) and (8.7) are the *orthogonal projections* of the element w of W_2 onto the perpendicular subspaces U_2 and V_2. What remains to be shown is that Dirichlet's problem for the self-adjoint equation (8.1), with boundary values defined by w, is actually solved by both u and $w - v$, which are therefore identical quantities. Two approaches to an existence proof are placed at our disposal in this way, one based on Dirichlet's principle in the form (8.7) and the other based on the quite different minimum problem (8.8). It should be emphasized that when we come to fill in all the details of these proofs (cf. Sections 2 and 3), the hypothesis must be made that a continuous function w possessing a finite Dirichlet integral over D can be associated with the prescribed boundary values, which will be true, as a matter of fact, if they are smooth enough.

The functions $w - v$ and u obtained from the extremal problems (8.7) and (8.8) are often referred to as *weak solutions* of the Dirichlet problem because they are not known to satisfy either the specified partial differential equation or the appropriate boundary condition. More generally, we might call any Lebesgue integrable function u a weak solution of (8.1) if it fulfills the orthogonality requirement $(u, v) = 0$ for every sufficiently smooth function v that differs from zero only over a closed subset of the open region D. We say that such a function v has *compact support*. Green's theorem and the definition of weak derivatives suggest that we describe u as a weak solution of (8.1) if and only if

$$\iint\limits_{D} [\Delta v - Pv]u \, dx = 0$$

for every trial function v of compact support. In contrast, we say that u is a *strong*, or *strict*, solution of (8.1) if it is twice continuously differentiable and satisfies that equation in the usual sense. Clearly any strong solution is a weak solution. The major difficulty in using the calculus of variations to establish existence theorems for the Dirichlet problem turns out to arise in the proof that the weak solutions found above are also strong solutions (cf. Section 3).

Before proceeding to the main body of the existence theory, we discuss a few practical implications of the ideas involved. First of all, it should be mentioned that in certain of the physical applications in which an equation of the type (8.1) arises, the corresponding Dirichlet integral has an interpretation as the *energy* of the system under consideration. When $P \equiv 0$ and $n = 2$, for example, we can view solutions of (8.1) as representing the small displacements of a thin stationary membrane. The Dirichlet integral is related to the potential energy of that membrane. Thus Dirichlet's principle becomes equivalent to the assertion that stable equilibrium is achieved for a configuration of minimum potential energy. These remarks have a connection with Hamilton's principle (cf. Section 2.5).

From the standpoint of numerical evaluation of solutions of the Dirichlet problem, Dirichlet's principle lends itself to an approximation procedure that is usually referred to in the literature as the *Rayleigh-Ritz method*. One formulation of this technique is based on selecting a set of functions v_j that vanish on the boundary ∂D of our domain and are complete in the Hilbert space V_2. To obtain an approximate solution of the Dirichlet problem for (8.1) corresponding to given boundary values w, we first extend the definition of w throughout D in some reasonable fashion. We then choose parameters $a_1, \ldots, a_\mu$ so that

$$\left\| w - \sum_{j=1}^{\mu} a_j v_j \right\| = \text{minimum},$$

where μ is a conveniently large integer. The approximate solution of the boundary value problem at hand takes the form

$$u = w - \sum_{j=1}^{\mu} a_j v_j,$$

with the above determination of the coefficients a_j. The error presumably approaches zero as $\mu \to \infty$. The answer fulfills the prescribed boundary condition exactly, of course, but it only obeys the partial differential equation (8.1) in some approximate sense when μ is finite. Note that an unusually accurate approximation of the Dirichlet integral, or energy, is obtained.

On the basis of the extremal problem (8.8) a similar approximation to the solution of Dirichlet's problem can be constructed from a complete system of elements u_k of the space U_2. If the u_k happen to be particular solutions of (8.1) that satisfy the orthonormality requirement

$$(u_k, u_l) = \delta_{kl}, \qquad k, l = 1, 2, \ldots,$$

then the analysis of Section 7.3 shows that the best approximation is provided by the formula

$$u = \sum_{j=1}^{\mu} (w, u_j) u_j.$$

Here the Fourier coefficients

$$(w, u_j) = - \int_{\partial D} w \frac{\partial u_j}{\partial \nu} \, d\sigma$$

can be computed directly in terms of the given boundary values w. The answer does not in general fit the assigned boundary condition exactly, but it does fulfill (8.1). We might think of this latter procedure as a variant of the method developed in Section 7.3 for calculating the normal derivative of the Green's function in terms of the infinite series (7.80) of orthonormal functions, which was identified as the kernel function. Finally, the elements of a complete orthonormal system might be interpreted as *coordinate axes* in Hilbert space, whereupon the Fourier coefficients acquire the significance of a set of coordinates.

EXERCISES

1. Develop the formal aspects of orthogonal projection for the general self-adjoint equation

$$\sum_{j,k=1}^{n} \frac{\partial}{\partial x_j}\left(a_{jk}\frac{\partial u}{\partial x_k}\right) - Pu = 0$$

of the elliptic type, using the quadratic integral

$$\|w\|^2 = \int_D \left[\sum_{j,k=1}^{n} a_{jk}\frac{\partial w}{\partial x_j}\frac{\partial w}{\partial x_k} + Pw^2\right] dx$$

as a norm.

2. Use the energy integral

$$\|w\|^2 = \int_D (\Delta w)^2\, dx$$

to derive two different extremal problems characterizing the solution of the first boundary value problem for the biharmonic equation

$$\Delta\Delta u = 0.$$

3. Expand the Poisson integral formula (7.52) into a Fourier series

(8.16) $$u = a_0 + \sum_{k=1}^{\infty}(a_k\rho^k \cos k\phi + b_k\rho^k \sin k\phi)$$

for the solution u of Laplace's equation

$$u_{\rho\rho} + \frac{1}{\rho}u_\rho + \frac{1}{\rho^2}u_{\phi\phi} = 0$$

inside the circle $\rho < R$. Derive the expression

(8.17) $$\int_0^{2\pi}\int_0^R \left[u_\rho^2 + \frac{1}{\rho^2}u_\phi^2\right]\rho\, d\rho\, d\phi = \pi\sum_{k=1}^{\infty} kR^{2k}(a_k^2 + b_k^2)$$

for the Dirichlet integral of u.

4. Use formulas (8.16) and (8.17) to develop a rigorous proof of the Dirichlet principle for Laplace's equation in the unit circle.

5. Show that the gap series

$$f(\theta) = \sum_{k=1}^{\infty}\frac{\sin k^4\theta}{k^2}$$

defines a continuous distribution of boundary values on the circle of radius R such that the corresponding solution (8.16) of Dirichlet's problem has an infinite Dirichlet integral (8.17).

6. Under appropriate hypotheses about the trial functions involved, deduce (8.10) from (8.11). Show that for a continuously differentiable function the notions of weak and strict derivative coincide.

7. Let D be a plane region bounded by two simple closed curves C_1 and C_2, and let w be a smooth function possessing a finite Dirichlet integral over D. Show that among all functions reducing to w on C_1, the one that solves the Euler equation (8.1) in D and satisfies the *natural boundary condition*

$$(8.18) \qquad \frac{\partial u}{\partial v} = 0$$

along C_2 minimizes the Dirichlet integral.

8. In some doubly connected region D of the kind described in Exercise 7, let u denote the harmonic function such that $u = 0$ on C_1 and $u = 1$ on C_2. If (α, β) ranges over the class of vectors defined in D with zero divergence

$$\frac{\partial \alpha}{\partial x} + \frac{\partial \beta}{\partial y} = 0$$

and with

$$\int_{C_2} (\alpha \, dy - \beta \, dx) = 1,$$

show that

$$(8.19) \qquad \int_D (\alpha^2 + \beta^2) \, dx \, dy \geq \frac{1}{\|u\|^2},$$

where equality holds only when

$$\alpha = \frac{1}{\|u\|^2} \frac{\partial u}{\partial x}, \qquad \beta = \frac{1}{\|u\|^2} \frac{\partial u}{\partial y}.$$

Discuss how upper and lower bounds can be found on the Dirichlet integral $\|u\|^2$ by means of Dirichlet's principle and this dual extremal problem, respectively.

9. For Laplace's equation in the unit circle, construct a complete orthonormal system of elements in the Hilbert space W_2 which consists of a single sequence of functions of the class U_2 and a double sequence of functions of the class V_2. The latter should be generated as products of Bessel functions and trigonometric functions that satisfy the partial differential equation

$$\Delta v + \lambda v = 0$$

for appropriate choices of the parameter λ.

10. Derive the equation

$$(1 + u_y^2)u_{xx} - 2u_x u_y u_{xy} + (1 + u_x^2)u_{yy} = 0$$

of minimal surfaces (cf. Section 15.4) as the Euler equation for the *Plateau problem*

$$\int_D \sqrt{1 + u_x^2 + u_y^2}\, dx\, dy = \text{minimum},$$

where the values of u are prescribed along the boundary curve ∂D.

11. Show[6] that the harmonic function

$$u = \text{Im}\, \{\sqrt{\log z}\}$$

has continuous boundary values, but an infinite Dirichlet integral, over any circle $|z - \epsilon| < \epsilon < \tfrac{1}{2}$.

2. EXISTENCE PROOF MOTIVATED BY THE KERNEL FUNCTION

We wish to present a proof of the existence of the solution of Dirichlet's problem for a linear partial differential equation of the elliptic type which is based on the orthogonal projection (8.8) of a given function onto a Hilbert space of solutions of the equation.[7] Such a method is suggested by the similarity between the extremal problem (8.8) and the minimum problem (7.73) characterizing the kernel function (cf. Exercise 3 below), which furnishes in turn the integral formula (7.66) for the desired solution.[8] To simplify matters we shall confine our treatment to the example of Laplace's equation

(8.20) $u_{xx} + u_{yy} = 0$

in the plane. However, since special tools such as conformal mapping are avoided in the exposition, it is not too difficult to work out an extension of the method for a more general equation like (8.1) in a larger number of independent variables.

We consider a domain D whose boundary ∂D is composed of a finite set of simple closed curves with a smoothness property that will be needed for the analysis of boundary values. Namely, corresponding to each point of ∂D there are supposed to be two circles of some preassigned radius which just touch ∂D at that point, but whose closures lie otherwise the first entirely inside and the second entirely outside of D. No generality will be lost if we make the further assumption that the origin is contained in the interior of D, which enables us to restrict the class U_2 of harmonic functions with a finite Dirichlet integral over D to those that vanish at the

---

[6] Cf. Prym 1.
[7] Cf. Garabedian-Schiffer 1, Lax 1.
[8] Cf. Fichera 1, Garabedian-Schiffer 1, Lehto 1.

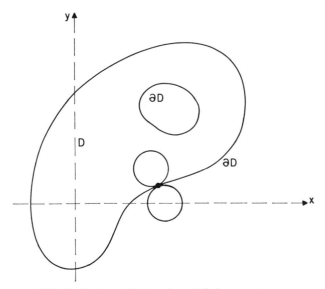

FIG. 29. Region with boundary of finite curvature.

origin. This constraint prevents the occurrence in the solution of the extremal problem (8.8) of an additive constant which would not otherwise be determined uniquely because $P \equiv 0$. Our requirement that

$$(8.21) \qquad\qquad u(0) = 0$$

for every element u in U_2 is equivalent to the normalization imposed on the harmonic kernel function in Section 7.3.

For a given function w that is continuous together with its first and second partial derivatives in D and on ∂D, we are interested in the Dirichlet problem of finding a harmonic function that takes on the same boundary values as w along ∂D. As an approach to the question of the existence of a solution, we shall now consider the minimum problem (8.8) in a formulation that is independent of both the Riesz-Fischer theorem and the notion of a Lebesgue integral. Such an analysis is feasible because it turns out, as we shall see presently, that the complete Hilbert space U_2 consists exclusively of functions that are harmonic throughout the open region D.

Let d^2 stand for the greatest lower bound of the set of numbers

$$(8.22) \qquad \|w - u\|^2 = \int_D [(w_x - u_x)^2 + (w_y - u_y)^2]\, dx\, dy$$

generated by allowing u to range over all harmonic functions of the class U_2. We observe that there is a minimum sequence u_k such that

$$(8.23) \qquad\qquad \lim_{k \to \infty} \| w - u_k \| = d.$$

Our aim is to find an extremal harmonic function minimizing the Dirichlet integral (8.22) and to construct from it the solution of Dirichlet's problem. To do so we proceed to establish that the limit of the sequence u_k exists and is harmonic.

An application of the parallelogram law (8.13) to the pair of functions

$$w_1 = w - u_k, \qquad w_2 = w - u_l$$

suffices to show, in a way we have already described (cf. Section 1), that

$$(8.24) \qquad\qquad \lim_{k,l \to \infty} \| u_k - u_l \| = 0.$$

We now recall the *Heine-Borel theorem* of elementary set theory,[9] which asserts that when the (bounded) region D is covered by a countable system of open circles whose closures lie within it, every closed subdomain of D can be included in a finite union of these circles. The relation (8.24) remains valid with D replaced by such a circle, since the Dirichlet integral over D dominates its value over the circle.

In any one of the above circles that is centered at the origin, where each of the functions u_k vanishes, the estimate (7.103) of Exercise 7.3.4 shows that

$$[u_k(\rho \cos \phi, \rho \sin \phi) - u_l(\rho \cos \phi, \rho \sin \phi)]^2 \leq \left(\frac{1}{\pi} \log \frac{R^2}{R^2 - \rho^2} \right) \| u_k - u_l \|^2.$$

From (8.24) it follows that u_k is a Cauchy sequence in the usual sense throughout this circle and must tend toward a limit function u. The convergence is uniform, too, because some slightly larger circle is still included in D. A similar argument serves to establish that u_k converges uniformly to a limit u in any circle of our system which overlaps the one just considered, in view of the known convergence in the intersection. Finally, the same conclusion is valid for every closed subdomain of D, since, as was observed to start with, such a subdomain can be covered by a finite set of overlapping circles of the above kind. The key to our success here is that any harmonic function whose Dirichlet integral over a circle is small must be nearly constant there; and the result remains true for an arbitrary domain because the latter has a covering by overlapping circles.

---

[9] Cf. Buck 1.

Let us apply the Poisson integral formula (7.52) to the sequence u_k in some closed circle lying inside D where it is uniformly convergent. We find by a passage to the limit that the function

$$u = \lim_{k \to \infty} u_k$$

must also be represented by that formula and is consequently harmonic. Moreover, all the partial derivatives of u_k converge uniformly in any smaller circle. The full statement of the Weierstrass convergence theorem (cf. Section 7.2) asserts that these same results apply throughout the domain D, too. In particular, the first partial derivatives of u_k converge uniformly toward those of u in every closed subdomain $\tilde{D}$ of D.

From the above remarks it becomes clear that over any closed subdomain $\tilde{D}$ of D the Dirichlet integral of $w - u$ is equal to the limit of the Dirichlet integral of $w - u_k$. Hence

$$\int_{\tilde{D}}[(w_x - u_x)^2 + (w_y - u_y)^2]\, dx\, dy \leq \lim_{k \to \infty} \|w - u_k\|^2 = d^2,$$

since the integral of any non-negative function must be larger over D than it is over $\tilde{D}$. Allowing $\tilde{D}$ to expand until it exhausts D, we arrive at the important inequality

$$(8.25) \qquad \|w - u\| \leq d.$$

What underlies this conclusion is the *lower semicontinuity*

$$(8.26) \qquad \|w - \lim u_k\| \leq \underline{\lim} \, \|w - u_k\|$$

of the Dirichlet integral of a convergent sequence of harmonic functions.

Because of (8.25) and the fact that u is harmonic we see that it must be the extremal function from the class U_2 that we have been looking for. Therefore we actually have

$$(8.27) \qquad \|w - u\| = \text{minimum} = d.$$

Now let u' be any trial function of the class U_2. From the definition of d^2 as a minimum value of the Dirichlet integral we obtain

$$(8.28) \qquad d^2 \leq \|w - u - \epsilon u'\|^2 = \|w - u\|^2 - 2\epsilon(w - u, u') + \epsilon^2 \|u'\|^2$$

for every value of the parameter ϵ. Combined with (8.27) this gives

$$0 \leq -2\epsilon(w - u, u') + \epsilon^2 \|u'\|^2.$$

Since the quadratic form on the right is non-negative, the coefficient of ϵ must actually vanish. Hence the variational condition

$$(8.29) \qquad (w - u, u') = 0$$

holds for every choice of u' in U_2.

The statement (8.29) is a variational formulation of the extremal property exhibited by u with respect to the minimum problem (8.27). We might attempt to derive from it the conclusion that u assumes the same boundary values as w along ∂D. It turns out to be awkward to make a direct verification of this conjecture, however. Therefore we resort to developing from (8.29) an alternate expression for the solution of the Dirichlet problem that is easier to study at the boundary. The device we shall exploit to such an end is suggested by the integral identity (7.107).

Let us introduce the quantity

$$(8.30) \qquad v = \frac{1}{2\pi}\left(w - u, \log\frac{1}{r}\right),$$

where

$$r = \sqrt{(x - \xi)^2 + (y - \eta)^2},$$

and where x and y are understood to be the variables of integration in the computation of the scalar product. When the point (ξ, η) lies in the exterior of D, the fundamental solution

$$S = \frac{1}{2\pi}\log\frac{1}{r}$$

of Laplace's equation is regular in the closure of D and belongs to the class U_2. Therefore according to (8.29) we have

$$(8.31) \qquad v(\xi, \eta) \equiv 0.$$

On the other hand, motivated by the notion that v ought to be identical with $w - u$, we intend to show that

$$(8.32) \qquad \Delta v = v_{\xi\xi} + v_{\eta\eta} = \Delta w$$

when (ξ, η) is an interior point of D.

For any smoothly bounded region E situated entirely inside D and not enclosing the singular point (ξ, η) of S we have

$$\int_E [(w_x - u_x)S_x + (w_y - u_y)S_y]\, dx\, dy = -\int_{\partial E}(w - u)\frac{\partial S}{\partial \nu}\, ds.$$

However, if (ξ, η) does lie in E, the term $w(\xi, \eta) - u(\xi, \eta)$ must be added on the right, as can be seen in the usual way by first excluding from E a small circle about (ξ, η) and then allowing its radius to approach zero. Thus we can write

(8.33)

$$v = w - u - \int_{\partial E} (w - u) \frac{\partial S}{\partial \nu} \, ds + \int_{D-E} [(w_x - u_x) S_x + (w_y - u_y) S_y] \, dx \, dy$$

whenever E is a subset of D containing (ξ, η). Differentiation under the integral sign shows that the two integrals on the right are harmonic functions of (ξ, η) in E, as is also the term u. Hence the relation (8.32) follows directly from an application of the Laplace operator to (8.33).

If we could establish that the definite integral (8.30) approaches zero as (ξ, η) tends toward the boundary ∂D of D, it would be evident that the expression $w - v$, which is harmonic by virtue of (8.32), is the solution of Dirichlet's problem. Since the variational condition (8.31) stating that v vanishes identically when (ξ, η) is located anywhere outside D is available to us, it suffices to prove that the integral defining v is continuous across the boundary curves ∂D. For this purpose we shall estimate the difference between the values of that integral at a pair of neighboring points situated on opposite sides of ∂D. For the first time it will be necessary to appeal to our hypotheses concerning the smoothness of ∂D.

Let (ξ_0, η_0) be an arbitrary point on ∂D. Let E denote a circle just touching ∂D at (ξ_0, η_0) but otherwise located entirely inside D, and let $\tilde{E}$ denote a circle of the same radius which lies in the exterior of D, except for the point (ξ_0, η_0), where it, too, just touches ∂D. Suppose that (ξ, η) is a point of E on the normal to ∂D through (ξ_0, η_0), and suppose that its inverse $(\tilde{\xi}, \tilde{\eta})$ in the circle E is a point of $\tilde{E}$. We recall (cf. Section 7.2) that the Green's function of E for Laplace's equation has the form

$$G(x, y; \xi, \eta) = \frac{1}{2\pi} \log \frac{1}{r} - \frac{1}{2\pi} \log \frac{1}{\tilde{r}} + \frac{1}{2\pi} \log \frac{\rho}{R},$$

where r and $\tilde{r}$ are the distances from (x, y) to (ξ, η) and $(\tilde{\xi}, \tilde{\eta})$, respectively, where ρ is the distance between (ξ, η) and the center of E, and where R is the radius of E. It can be seen that we have

(8.34) $$v(\xi, \eta) = v(\xi, \eta) - v(\tilde{\xi}, \tilde{\eta}) = (w - u, G),$$

in view of (8.30), (8.31), and the fact that $(\tilde{\xi}, \tilde{\eta})$ lies outside D.

We shall exploit formula (8.34) to examine the behavior of v as (ξ, η) approaches the boundary point (ξ_0, η_0). Let us treat separately the

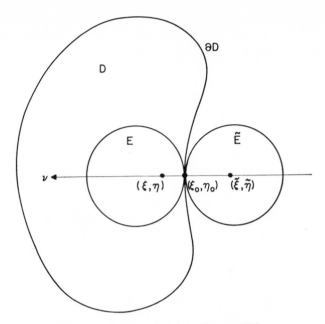

FIG. 30.   Verification of the boundary condition.

contributions to the scalar product $(w - u, G)$ from integrals over E and from integrals over the set $D - E$ that is left when E is removed from D. It is convenient to indicate scalar products or norms evaluated over sub-domains of D by using the letters for those subdomains as subscripts. Thus we may write

$$(8.35) \quad v = (w - u, G)_D = (w, G)_E - (u, G)_E + (w - u, G)_{D-E}.$$

We shall prove that the three terms on the right tend toward zero individually as $(\xi, \eta) \to (\xi_0, \eta_0)$. Moreover, the estimates of convergence leading to that conclusion will be valid uniformly for all choices of the point (ξ_0, η_0) on ∂D.

We begin by establishing the crucial identity

$$(8.36) \quad (u, G)_E = 0$$

for every position of (ξ, η) in E. Consider a pair of level curves $G = \lambda_1$ and $G = \lambda_2$ of the Green's function G that subdivide E into three regions E_0, E_λ and E_∞ defined, respectively, by the sets of inequalities

$$0 < G < \lambda_1, \quad \lambda_1 < G < \lambda_2, \quad \lambda_2 < G < \infty.$$

An application of the divergence theorem gives

$$(u, G)_E = (u, G)_{E_0} + (u, G)_{E_\lambda} + (u, G)_{E_\infty}$$

$$= (u, G)_{E_0} - \int_{G=\lambda_1} G \frac{\partial u}{\partial \nu} \, ds + \int_{G=\lambda_2} G \frac{\partial u}{\partial \nu} \, ds + (u, G)_{E_\infty}.$$

On the other hand, we also have

$$\int_{G=\lambda_j} G \frac{\partial u}{\partial \nu} \, ds = \lambda_j \int_{G=\lambda_j} \frac{\partial u}{\partial \nu} \, ds = 0.$$

It follows that

$$(u, G)_E = \lim_{\lambda_1 \to 0} (u, G)_{E_0} + \lim_{\lambda_2 \to \infty} (u, G)_{E_\infty} = 0,$$

whence we may cancel out the middle term on the right in (8.35).

The first term on the right in (8.35) is even easier to handle. The first partial derivatives of w are bounded in E, or, for that matter, even in D, and the first partial derivatives of G have an integrable majorant there and approach zero as $(\xi, \eta) \to (\xi_0, \eta_0)$. Consequently an application of the Lebesgue convergence theorem[10] gives

$$(8.37) \qquad \lim_{\xi \to \xi_0, \eta \to \eta_0} (w, G)_E = 0.$$

To obtain an analogous result for the last term in (8.35), we resort to the Schwarz inequality

$$|(w - u, G)_{D-E}| \leq \|w - u\|_{D-E} \|G\|_{D-E} \leq d \|G\|_{D-E}.$$

It is important to observe that throughout $D - E$ the integrand $G_x^2 + G_y^2$ that occurs in the norm on the right is dominated by some suitably large constant times r_0^{-2}, where r_0 represents the distance from (x, y) to the boundary point (ξ_0, η_0). Therefore $G_x^2 + G_y^2$ is integrable over the cusped domain $D - E$, for the area involved lies outside of the two tangent circles E and $\tilde{E}$, which are separated at any specific small distance r_0 from (ξ_0, η_0) by an arc whose length is less than some constant times r_0^2. Since $G_x \to 0$ and $G_y \to 0$ as $(\xi, \eta) \to (\xi_0, \eta_0)$, we conclude from the Lebesgue convergence theorem that

$$\lim_{\xi \to \xi_0, \eta \to \eta_0} \|G\|_{D-E} = 0;$$

hence also (cf. Exercises 1 and 11 below)

$$(8.38) \qquad \lim_{\xi \to \xi_0, \eta \to \eta_0} (w - u, G)_{D-E} = 0.$$

[10] Cf. Munroe 1.

The results (8.35) to (8.38) suffice to show that $v \to 0$ uniformly as its argument point (ξ, η) approaches the boundary ∂D of our domain. Thus the function $w - v$ we have constructed takes on the prescribed boundary values w and furnishes a solution of Dirichlet's problem for the Laplace equation in D. We do not go to the trouble of proving the harder theorem that it also coincides with the extremal function u except for an additive constant.

Notice that for the existence proof it was necessary to assume that w could be extended from ∂D to the entire region D in such a way that it at least has a finite norm

$$(8.39) \qquad \|w\| < \infty$$

there (cf. Exercise 1.5). However, this restriction can now be removed by approximating an arbitrary continuous distribution of boundary values by smoother ones and then passing to the limit, for we have already established in Section 7.1 that the solution of the Dirichlet problem varies continuously with the data in such a situation. Finally, we mention that our hypothesis that the boundary curves ∂D have finite curvature can also be weakened by exhausting any more general region by some sequence of subregions of the kind we have treated here. In particular, slits and corners are admissible on the limiting boundary.

EXERCISES

1. Complete details of the proof of (8.38) presented in the text.

2. Work out in rigorous detail the solution of Dirichlet's problem for arbitrary continuous boundary values through an approximation by smooth functions w fulfilling the requirement (8.39). Discuss the difficulties encountered in solving the Dirichlet problem for an arbitrary plane domain that is represented as the limit of an expanding sequence of smoothly bounded subdomains.[11]

3. For any plane domain bounded by smooth curves, prove that a harmonic kernel function $K(x, \xi)$ with the reproducing property (7.71) exists by solving the minimum problem (7.73) within the class U_2. Then use the procedure of the text to show that the expression

$$(8.40) \qquad G(y, \xi) = \frac{1}{2\pi} \log \frac{1}{|\xi - y|} - \frac{1}{2\pi}\left(K(x, \xi), \log \frac{1}{|x - y|}\right)$$

defines the Green's function G of Laplace's equation in its dependence on the point y. Thus establish the existence of the Green's function, too.[12] Why does the proof fail for a domain bounded by slits?

4. Establish the existence of a solution of the first boundary value problem for the biharmonic equation in two independent variables by projecting a

[11] Cf. Kellogg 1.
[12] Cf. Garabedian-Schiffer 1.

prescribed function with uniformly continuous second derivatives orthogonally onto the class of biharmonic functions with a finite energy integral (cf. Exercise 1.2 and Section 7.3).

5. Extend the existence proof given in the text to the case of Laplace's equation in a three-dimensional domain that is bounded by a sufficiently smooth surface.

6. Extend the existence proof given in the text to the case of the equation (8.1) in two independent variables. Assume that the coefficient P is an entire function, so that a fundamental solution will be available in the large (cf. Section 5.1).

7. Use the Riesz-Fischer theorem to simplify the derivation of (8.29) presented in the text.

8. In the case of Laplace's equation, show that every element of the completed Hilbert space U_2 is actually a harmonic function in the interior of the region D.

9. In the case of Laplace's equation, prove that the subset of functions u in U_2 satisfying the inequality

$$\|u\| \leq 1$$

is *equicontinuous* in each closed subdomain $\tilde{D}$ of D. This means (cf. Section 10.3) that for any number $\epsilon > 0$ there should exist a $\delta > 0$ depending only on ϵ and $\tilde{D}$ such that

$$|u(x, y) - u(\xi, \eta)| \leq \epsilon$$

for every u of the subset and for every pair of points (x, y) and (ξ, η) in $\tilde{D}$ with

$$\sqrt{(x - \xi)^2 + (y - \eta)^2} \leq \delta.$$

10. Does the function v defined by the definite integral (8.30) have a finite norm $\|v\| < \infty$? Is the extremal function u for (8.8) related to v by the formula

$$u = w - v,$$

except perhaps for an additive constant?

11. Give an alternate proof of (8.38) based on the formula

$$\|G\|_{D-E}^2 = -\int_{\partial D} G \frac{\partial G}{\partial \nu} \, ds,$$

on the fact that $G \to 0$ uniformly along ∂D as $(\xi, \eta) \to (\xi_0, \eta_0)$, and on the observation that $\partial G/\partial \nu \geq 0$ along ∂D in some fixed neighborhood of (ξ_0, η_0).

12. Give a precise proof of (8.37) based on moving majorants.

3. EXISTENCE PROOF BASED ON DIRICHLET'S PRINCIPLE

We are interested in using the orthogonal projection theorem (8.7) to establish the existence of a solution of Dirichlet's problem for the elliptic partial differential equation

$$(8.41) \qquad \Delta u = Pu$$

in a region D of arbitrary dimension n. We shall not need to assume that the coefficient P is analytic, but for the sake of simplicity we shall suppose that it is bounded from below by a positive number and that it possesses continuous partial derivatives of a high order. In analogy with the hypotheses of Section 2, we shall ask that the boundary ∂D of the region D have the property that to each of its points there corresponds a pair of tangent spheres of a fixed size of which one lies completely inside D and the other lies completely outside D. Furthermore, we shall require that the boundary values w prescribed in the Dirichlet problem be so smooth that they possess an extension over D of finite norm, or, better, an extension which has bounded first derivatives in the closure of D. These restrictions facilitate our analysis of the boundary condition, which, however, will be confined to the special case $n = 3$. They can be relaxed afterwards in that case by means of the maximum principle, as we saw in Section 2.

Among all functions with the assigned boundary values w, we seek one u that minimizes the Dirichlet integral

$$\|u\|^2 = \int_D [(\nabla u)^2 + Pu^2]\, dx.$$

Dirichlet's principle states that the answer to this multiple integral problem in the calculus of variations provides the desired solution of the Dirichlet problem for (8.41). However, in order to establish the existence of a solution we prefer to recast the question in the equivalent form (8.7), which asks for the function v vanishing on ∂D that is closest to a specified extension of w in the sense that the norm $\|w - v\|$ is as small as possible.

We have already seen in Section 1 that there is a solution v of the latter extremal problem in the completed Hilbert space V_2 of functions with zero boundary values. The extremal function v is uniquely determined almost everywhere in D as the limit of a minimum sequence of continuously differentiable elements v_k of the class V_2. It is convenient to introduce the corresponding minimal sequence

$$w_k = w - v_k$$

of continuous competing functions for our original problem in the calculus of variations and to set

$$(8.42) \qquad\qquad u = \lim_{k \to \infty} w_k = w - v.$$

Formula (8.42) defines the quantity u almost everywhere in D as an element of the Hilbert space W_2 possessing first partial derivatives $u^{(1)}, \ldots, u^{(n)}$ in the weak sense. We intend to show that u can actually be identified as the solution of Dirichlet's problem.

The extremal property of u implies that for every element v' of the class V_2 and for every value of the real parameter ϵ we have

$$\|u\|^2 \leq \|u + \epsilon v'\|^2 = \|u\|^2 + 2\epsilon(u, v') + \epsilon^2 \|v'\|^2.$$

Thus the quadratic form on the right becomes a minimum at $\epsilon = 0$, where, in consequence, its derivative with respect to ϵ must vanish. We conclude that the variational condition

$$(8.43) \qquad\qquad (u, v') = 0$$

holds for every choice of v' in V_2. By substituting a convenient special family of functions v' into (8.43) we shall endeavor to deduce that u has continuous second derivatives and is actually a solution of the partial differential equation (8.41), at least after it has been appropriately redefined on a set of Lebesgue measure zero.

A quantity u possessing weak first derivatives in the sense of formula (8.1$\dot{1}$) and fulfilling the integral relation (8.43) for a suitable class of trial functions v' is often referred to in the literature as a *weak solution* of (8.41). What we have to establish is that such a weak solution is, in fact, a *strong solution* (cf. Section 1). In other words, we have to show that its second derivatives exist, are continuous, and satisfy (8.41), which then becomes equivalent to (8.43) in view of the divergence theorem.

Our aim is to make a clever selection of v' that will convert (8.43) into an identity exhibiting more clearly the nature of the extremal function u. One choice which is advantageous from the point of view of elegance and simplicity depends on the construction in Section 5.3 of a local fundamental solution S of (8.41). It should be observed in this connection that the procedure given there becomes quite elementary in the case of an equation like (8.41) in which the derivatives of highest order occur in the form of a Laplacian (cf. Exercise 5.3.4 and Exercise 6 below). To be precise, we recall that around each point ξ we can find a sphere of pre-assigned radius in which a solution $S = S(x, \xi)$ of (8.41) is defined such that $r^{n-2}S$ remains bounded and positive in a neighborhood of ξ, where

$$r = \sqrt{(x_1 - \xi_1)^2 + \cdots + (x_n - \xi_n)^2}$$

stands for the distance between x and ξ. We shall confine our attention to a specific choice of S, such as the one displayed in (5.88), which represents a function of the parameter point ξ possessing continuous partial derivatives of a suitable order when $r > 0$.

For any point ξ of the region D, let $\lambda > 0$ be small enough so that the sphere E_λ of radius λ around ξ is contained in D, and set

$$(8.44) \qquad S_\lambda(x, \xi) = \int_{E_\lambda} S(x, \eta) \, d\eta.$$

For x outside E_λ, but still within the domain of definition of $S(x, \eta)$ for every position of η inside E_λ, we derive the equation

$$(8.45) \qquad \Delta S_\lambda - P S_\lambda = 0$$

by direct differentiation under the integral sign in (8.44). However, when x lies in the interior of E_λ the generalized Poisson equation (5.92) of Chapter 5 shows that

$$(8.46) \qquad \Delta S_\lambda - P S_\lambda = -1.$$

Thus, although its first partial derivatives remain continuous, S_λ has only piecewise continuous second partial derivatives. Now suppose that S_λ has meaning throughout some fixed sphere E_2 which lies in D, and suppose that E_1 is a smaller concentric sphere which, in turn, encloses E_λ.

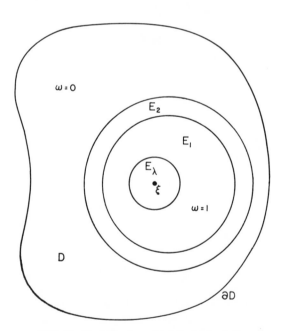

FIG. 31. Variation by a fundamental solution.

We shall set

(8.47) $$v' = \omega S_\lambda$$

in the variational condition (8.43), where $\omega = \omega(x)$ is a sufficiently differentiable factor that vanishes for x in the exterior of E_2 but reduces to unity for x in the interior of E_1.

Since the expression (8.47) for v' not only vanishes in a neighborhood of the boundary ∂D of D but also can be approximated by twice continuously differentiable functions with the same property, the integral relations (8.11) apply to the weak derivatives $u^{(1)}, \ldots, u^{(n)}$ of u as they occur in the scalar product (u, v'). Therefore we may bring the divergence theorem into play to deduce from (8.43) the variational condition

(8.48) $$\int_D u[\Delta v' - Pv']\,dx = 0.$$

Observing that the auxiliary factor ω which appears in (8.47) is constant outside the shell $E_2 - E_1$, and taking (8.46) and (8.45) into account in E_λ and in its complement, respectively, we transform (8.48) into the equivalent result

(8.49) $$\int_{E_\lambda} u\,dx = \int_{E_2-E_1} [\Delta(\omega S_\lambda) - P\omega S_\lambda]u\,dx.$$

If we divide (8.49) through by the volume $\tau_n\lambda^n$ of the sphere E_λ and then allow the radius λ to approach zero, we might expect to obtain

(8.50) $$u(\xi) = \int_{E_2-E_1} [\Delta\{\omega(x)S(x, \xi)\} - P(x)\omega(x)S(x, \xi)]u(x)\,dx.$$

Indeed, the definition (8.44) shows that as $\lambda \to 0$ the quotient $S_\lambda/(\tau_n\lambda^n)$ converges uniformly toward S in $E_2 - E_1$, together with all first and second derivatives. Hence the procedure described does convert the integral on the right in (8.49) into the integral on the right in (8.50). If, for the moment, we let $u_0(\xi)$ stand for the latter quite regular expression, we see, conversely, that its integral over E_λ has the value (8.49). Thus we have

$$\int_{E_\lambda} (u - u_0)\,dx = 0$$

for any sphere E_λ lying inside E_1. This implies, according to an elementary corollary of the Radon-Nikodym theorem (cf. Munroe 1), that u and u_0 are identical almost everywhere in E_1.

The preceding remarks serve to establish (8.50) in E_1, except possibly on a set of measure zero where u is not fixed uniquely by its construction from the Riesz-Fischer theorem. It is natural to redefine u on such a set so that it coincides there with u_0 and is consequently given throughout E_1 by the integral representation (8.50). This exhibits it as a continuous function of ξ possessing partial derivatives of an order commensurate with the differentiability of the fundamental solution $S = S(x, \xi)$ with respect to that parameter. If the coefficient P in the partial differential equation (8.41) is sufficiently smooth, we can at least assert that u has continuous second derivatives. Furthermore, the conclusion is valid throughout the interior of the domain D because the location there of the sphere E_1 is quite arbitrary.

If S were a symmetric fundamental solution of (8.41), we could proceed to prove that u satisfies that equation by differentiating directly under the integral sign in (8.50). It is not necessary to have so much information about the fundamental solution, however, since we are now justified in applying the divergence theorem to the variational condition (8.43) to obtain

$$(8.51) \qquad \int_D [\Delta u - Pu]v' \, dx = 0$$

for every continuously differentiable choice of the trial function v' which reduces to zero in some neighborhood of ∂D. A familiar argument of the calculus of variations suffices here to show that u fulfills the Euler equation (8.41) associated with (8.51). Indeed, if this were not the case $\Delta u - Pu$ would be either positive or negative over some small sphere inside D, and by constructing v' to be positive there, but to vanish elsewhere in D, we could achieve a contradictory situation in which the integral on the left in (8.51) would acquire a value different from zero.

Thus far we have succeeded in establishing that the extremal function u defined in terms of a minimal sequence w_k by (8.42) is a solution of the partial differential equation (8.41) possessing continuous partial derivatives of at least the second order in the interior of the region D. In order to verify that u solves the Dirichlet problem with which we began our discussion, we have still to show that it is continuous in the closure of D and assumes the prescribed boundary values w along ∂D. In view of (8.42), this is equivalent to the contention that the orthogonal projection $v = \lim_{k \to \infty} v_k$ of w onto the Hilbert space V_2 reduces continuously to zero on ∂D. Our procedure will be to prove first that v satisfies the required boundary condition in a weak sense motivated by (7.107), which will be done for an arbitrary number of independent variables n. Then we shall deduce in turn that it vanishes along ∂D in the strict sense, but that will

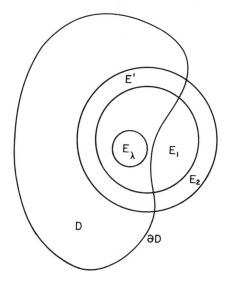

FIG. 32. Derivation of the weak boundary condition.

only be done for $n \leq 3$. For a treatment of the same result in space of higher dimension, see Hellwig 1.

It is a consequence of the Green's identity (8.2) that any element v_k of the class V_2 which vanishes continuously on ∂D must fulfill

$$(8.52) \qquad (\phi, v_k) + \int_D [\Delta \phi - P\phi] v_k \, dx = 0$$

for every continuously differentiable trial function ϕ that has piecewise continuous second derivatives in the closure of D. In particular, (8.52) holds for each function of the minimal sequence v_k used to construct the orthogonal projection v of the extended boundary data w. Because of the convergence theorem (8.12), we can pass to the limit as $k \to \infty$ in (8.52) to establish that it is still valid for the extremal function v.

Now take ϕ to have the special form $\phi = \omega S_\lambda$ encountered previously in (8.47), where, however, the concentric spheres E_1 and E_2 involved in the definition of the damping factor ω are allowed to overlap the boundary surface ∂D. Letting E' stand for the intersection of the shell $E_2 - E_1$ with D and using the customary product notation DE_λ for the intersection of D with E_λ, we find from (8.52) that

$$(8.53) \qquad \int_{DE_\lambda} v \, dx = (\omega S_\lambda, v) + \int_{E'} [\Delta(\omega S_\lambda) - P\omega S_\lambda] v \, dx.$$

Since v is already known to be continuously differentiable inside D, we can divide (8.53) by the volume $\tau_n \lambda^n$ of E_λ and pass to the limit as $\lambda \to 0$ to obtain

$$(8.54) \qquad v(\xi) = (\omega S, v) + \int_{E'} [\Delta(\omega S) - P\omega S] v \, dx,$$

provided that the center ξ of the sphere E_λ lies in both D and E_1. On the other hand, for ξ in the exterior of D the same limit process yields

$$(8.55) \qquad 0 = (\omega S, v) + \int_{E'} [\Delta(\omega S) - P\omega S] v \, dx$$

instead.

The integral over E' which occurs on the right in (8.54) and (8.55) is a continuous function of the parameter ξ across the section of ∂D located in the interior of E_1, because for each ξ inside E_1 the fundamental solution $S = S(x, \xi)$ remains finite as its argument x ranges over E'. Thus if we could show that the scalar product $(\omega S, v)$ is continuous across ∂D, too, then it would follow that the term v on the left in (8.54) approaches the desired value zero of the left-hand side of (8.55) as ξ tends toward ∂D. This observation leads us to interpret the pair of relations (8.54) and (8.55) holding inside and outside of D, respectively, as a satisfactory *weak statement* of the boundary condition that v is expected to fulfill. In the case $n = 3$ we shall go a step further and verify that v actually vanishes on ∂D in the *strict sense*. The proof will be seen to work for $n = 2$ as well, but our success for $n = 3$ indicates that it is independent of special techniques like conformal mapping which might confine its scope to plane problems.

What is still to be shown in the case $n = 3$ is that for an arbitrary point ξ_0 on ∂D, the limits of $(\omega S, v)$ as ξ approaches ξ_0 from the two opposite sides of ∂D are equal. The statement is clearly true when v is replaced by w, because w has bounded first derivatives. Thus, since

$$(\omega S, v) = (\omega S, w) - (\omega S, u)$$

by virtue of (8.42), the desired result will follow if we establish instead the continuity of the scalar product $(\omega S, u)$ across ∂D.

Next, observe that S differs from the fundamental solution $1/(4\pi r)$ of Laplace's equation by a term which has bounded first partial derivatives (cf. Exercise 6 below) and which therefore forms with u a scalar product that depends continuously on ξ. Since, furthermore, contributions to the integral $(\omega S, u)$ from the exterior of any fixed sphere E_1 enclosing ξ_0 are evidently continuous, too, we conclude that the factor ω plays an inessential role here. Indeed, it suffices to prove that the scalar product

$(1/r, u)$ has equal limits from the two sides of ∂D. We shall model the demonstration of this fact on our treatment in Section 2 of the analogous problem for the quantity (8.30). The analysis will hinge on replacing S by what amounts to a parametrix for the boundary condition $v = 0$.

Let E and $\tilde{E}$ be two small spheres of radius R lying inside and outside of D, respectively, but just touching ∂D at the point ξ_0. We denote by ξ a point of E located on the normal to ∂D through ξ_0, and we assume that its inverse $\tilde{\xi}$ in the sphere E is situated in $\tilde{E}$. Imitating the procedure of Section 2, we introduce the Green's function

$$(8.56) \qquad G(x, \xi) = \frac{1}{4\pi r} - \frac{R}{4\pi \rho \tilde{r}}$$

of E for Laplace's equation, where r and $\tilde{r}$ stand for the distances from x to ξ and $\tilde{\xi}$, respectively, and where ρ stands for the distance between ξ and the center of E. We can establish that $(1/r, u)$ has the same limit on ∂D from the interior of D as it has from the exterior of D simply by showing that

$$(8.57) \qquad \lim_{\xi \to \xi_0} (G, u) = 0.$$

This contention is not violated by the appearance in the difference (8.56) defining G of the factor R/ρ, which tends toward unity in the limit as $\xi \to \xi_0$.

Given a number $\epsilon > 0$, we can find a closed subregion $\tilde{D}$ of D such that

$$(8.58) \qquad \|u\|_{D-\tilde{D}} < \epsilon.$$

Now consider a sphere around ξ_0 which is so small that its intersection Ω with $D - E$ is contained in $D - \tilde{D}$. We shall prove (8.57) by showing that in the decomposition

$$(G, u) = (G, u)_{D-E-\Omega} + (G, u)_E + (G, u)_\Omega$$

the first two terms on the right approach zero separately as $\xi \to \xi_0$, while the last term remains less than some fixed factor times ϵ.

The uniform convergence of G and its first partial derivatives toward zero in the truncated region $D - E - \Omega$ is enough to give

$$(8.59) \qquad \lim_{\xi \to \xi_0} (G, u)_{D-E-\Omega} = 0.$$

Furthermore, because u is a solution of (8.41) we can apply the divergence theorem twice over the sphere where $G(x, \xi) > \mu > 0$ to establish

$$(8.60) \qquad (G, u)_E = - \lim_{\mu \to 0} \int_{G=\mu} G \frac{\partial u}{\partial \nu} \, d\sigma$$

$$= \lim_{\mu \to 0} \mu \int_{G>\mu} Pu \, dx = 0,$$

a result almost identical with (8.36). Finally, we obtain from Schwarz's inequality and (8.58) the estimate

$$(8.61) \qquad |(G, u)_\Omega| \leq \|G\|_\Omega \, \|u\|_\Omega$$

$$\leq \epsilon \left(\int_\Omega PG^2 \, dx - \int_{\partial\Omega} G \frac{\partial G}{\partial \nu} \, d\sigma \right).$$

It will be proved that the two integrals enclosed in parentheses on the right remain bounded as $\xi \to \xi_0$. Hence (8.59), (8.60) and (8.61) combine to yield the desired conclusion (8.57).

The square of the Green's function G is integrable for arbitrary choices of the parameter ξ. It follows that the volume integral multiplying ϵ in the inequality (8.61) actually approaches zero as $\xi \to \xi_0$. Moreover, since $G \equiv 0$ along the portion of $\partial\Omega$ situated on the spherical surface ∂E, and

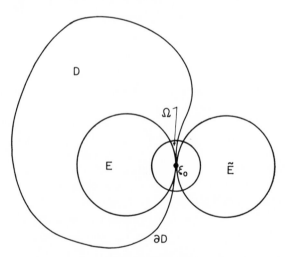

FIG. 33. Strict analysis of boundary values.

since G and $\partial G/\partial \nu$ tend toward zero uniformly on the section of $\partial \Omega$ lying between E and ∂D, we have

(8.62)
$$\overline{\lim_{\xi \to \xi_0}} \int_{\partial \Omega} (-G) \frac{\partial G}{\partial \nu} \, d\sigma = \overline{\lim_{\xi \to \xi_0}} \int_{\partial D} (-G) \frac{\partial G}{\partial \nu} \, d\sigma.$$

In estimating the latter quantity, we need only consider the contribution from a fixed neighborhood of ξ_0 on ∂D. Motivated by our discussion of the Poisson kernel in Section 7.2, we observe that

$$\int_{\partial D} \frac{\partial G}{\partial \nu} \, d\sigma = \int_{\partial E} \frac{\partial G}{\partial \nu} \, d\sigma = 1.$$

Hence if $\partial G/\partial \nu > 0$ in the vicinity of ξ_0, the integral (8.62) will be dominated by any upper bound on $-G$ holding there. Thus our concluding step will be to show that a neighborhood of ξ_0 can be found on ∂D where the kernel $\partial G/\partial \nu$ is positive and where G stays between some negative lower bound and zero (cf. Exercise 2.11).

For each value of τ in the interval

(8.63)
$$\frac{\rho}{R} \le \tau \le \frac{R}{\rho}$$

we consider the expression

$$G_\tau(x, \xi) = \frac{1}{4\pi r} - \frac{\tau}{4\pi \tilde{r}}.$$

For $\tau = R/\rho$ we see that G_τ reduces to the Green's function (8.56) of E, whereas for $\tau = \rho/R$ it represents the Green's function of the exterior of another sphere of equal radius whose center lies beyond the center of $\tilde{E}$ along the outer normal to ∂D through ξ_0. For intermediate choices of τ it is clear that G_τ represents the Green's function of a larger sphere E_τ which moves monotonically as τ increases and whose surface ∂E_τ sweeps out the region located between the interior of the second sphere referred to above and E. In particular, G_τ is the Green's function of a half-space when $\tau = 1$.

Through each point x near ξ_0 on ∂D at which we might wish to evaluate G and $\partial G/\partial \nu$ there passes precisely one spherical surface ∂E_τ. Since $G_\tau = 0$ along ∂E_τ, we can write

$$G = G - G_\tau = \frac{\tau - R/\rho}{4\pi \tilde{r}} = \frac{\tau - R/\rho}{4\pi \tau r}$$

there, whence

$$-\frac{R + \rho}{4\pi \rho^2} \le G \le 0$$

in view of (8.63) and the obvious inequality $r \geq R - \rho$. This furnishes the required upper bound on $-G$.

To estimate $\partial G/\partial \nu$, we note that

$$\frac{\partial G}{\partial \nu} = \frac{\partial G_r}{\partial \nu} + \frac{\tau - R/\rho}{4\pi}\frac{\partial}{\partial \nu}\frac{1}{\tilde{r}} \geq \frac{\partial G_r}{\partial \nu} - \frac{R^2 - \rho^2}{4\pi\rho R}\frac{1}{\tilde{r}^2}.$$

The angle θ at which the surface ∂D cuts ∂E_r is small when x is close to ξ_0; therefore

$$\frac{\partial G_r}{\partial \nu} = \frac{R^2 - \rho^2}{4\pi\tau\rho r^3}\cos\theta$$

is positive, where ν stands for the inner normal to ∂D at its intersection with ∂E_r. It follows that

$$\frac{\partial G}{\partial \nu} \geq \frac{R^2 - \rho^2}{4\pi R r^3}\left[\cos\theta - \frac{R^2 r}{\rho^3}\right] > 0$$

in some fixed neighborhood of ξ_0, which is what was left to be shown.

To summarize, we have used the Dirichlet principle to develop a proof of the existence of a solution of Dirichlet's problem that proceeded in three main stages. First, we established by means of the Riesz-Fischer theorem and the parallelogram law (8.13) that there was an extremal function for the minimum problem (8.7) in the calculus of variations which provided a solution of the problem in the weak sense. Next, we

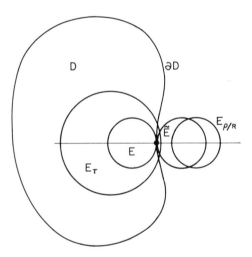

FIG. 34. Estimation of G and $\partial G/\partial \nu$.

showed that the weak solution possessed continuous partial derivatives of an appropriate order in the interior of the region D by deriving for it the representation formula (8.50), which followed from our construction in terms of a local fundamental solution of the special variation (8.47). Finally, we exploited a similar trial function to arrive in the general case at the weak form (8.54), (8.55) of the boundary condition. For $n = 3$, and also for $n = 2$, this led in turn to the more usual strict formulation of the boundary condition because we could verify, by a method already encountered in Section 2, that the scalar product $(\omega S, v)$ had no jump across the boundary surface ∂D. Such an analysis required smoothness assumptions about the boundary data which, however, can be removed afterward through an approximation procedure based on the maximum principle.

It is of interest to observe (cf. Exercise 3 below) that in our choice of a specific variation (8.47) the fundamental solution S may be replaced by a parametrix P. Consequently Dirichlet's principle leads to a procedure for constructing global solutions of a self-adjoint elliptic partial differential equation when none are known in advance even locally. This comment has a significant application (cf. Exercises 9 and 10 below) to the Beltrami equation introduced in Section 3.1. By contrast, the method of Section 2 requires that a solution be available in the large, since it involves projecting a given function onto the class of solutions with a finite Dirichlet integral.

The calculus of variations also suggests an approach to the Dirichlet problem for nonlinear partial differential equations of the elliptic type. However, in the nonlinear case the problem becomes extremely difficult because of its essentially global character. Therefore a solution can be expected to exist only in special circumstances and even then it may not be unique. For examples and indications of a general theory of nonlinear boundary value problems that is beyond the scope of this book[13] we refer to Sections 14.3 and 15.4.

EXERCISES

1. Use Dirichlet's principle to solve the inhomogeneous equation

$$\Delta u - Pu = f.$$

2. Show that the Riesz-Fischer theorem and the concept of a Lebesgue integral actually play an inessential role in the proof of Dirichlet's principle.[14]

---

[13] Cf. Bers-Nirenberg 1, De Giorgi 1, Hopf 2, Lichtenstein 1, Morrey 1, Nash 1.
[14] Cf. Courant 2.

3. Show that it is feasible to replace the fundamental solution S by a parametrix P in the definition (8.47) of a variation leading to a demonstration of the regularity of weak solutions of the Dirichlet problem.

4. Prove the theorem that an integrable function whose average vanishes over any sphere must necessarily be zero almost everywhere.

5. Precisely how smooth must the coefficient in equation (8.41) be if we are to conclude that every weak solution u satisfying (8.50) has continuous second derivatives?

6. In the case $n = 3$, show that a symmetric fundamental solution S of (8.41) can be found of the form

$$S(x, \xi) = \frac{1}{4\pi r(x, \xi)} - \frac{1}{(4\pi)^2} \int \frac{P(\eta) \, d\eta}{r(x, \eta) r(\eta, \xi)}$$

$$+ \frac{1}{(4\pi)^3} \iint \frac{P(\zeta) P(\eta) \, d\zeta \, d\eta}{r(x, \zeta) r(\zeta, \eta) r(\eta, \xi)} - \cdots,$$

where $r(x, \xi)$ stands for the distance between the points x and ξ, and where the triple integrals are extended over a fixed region of space (cf. formulas (5.88) and (5.91) in Chapter 5). Conclude that the first partial derivatives of the difference $4\pi S - 1/r$ are bounded at $x = \xi$.

7. In the plane case $n = 2$ work out the details of the verification that the extremal function (8.42) defined by Dirichlet's principle reduces to the prescribed boundary values w in the strict sense along ∂D. For Laplace's equation in two independent variables, give an alternate proof of this fact based on the mean value theorem for harmonic functions and on applications of Schwarz's inequality to the Dirichlet integral that furnish estimates of length in terms of area (cf. Section 15.4).

8. For Laplace's equation in the plane, show by means of the Schwarz reflection principle (7.105) that the solution of the Dirichlet problem is analytic at any boundary point where both the boundary curve ∂D and the boundary values w are given analytically. Consider at first only the simplest case where ∂D reduces near the boundary point in question to a line segment along which $w = 0$.

9. Use Dirichlet's principle to establish the existence of a solution of the Dirichlet problem for the general self-adjoint partial differential equation

$$\sum_{j,k=1}^{n} \frac{\partial}{\partial x_j} \left(a_{jk} \frac{\partial u}{\partial x_k} \right) = Pu$$

of the elliptic type in two or three independent variables (cf. Exercise 1.1). To facilitate a proof that the weak solution obtained from the Riesz-Fischer theorem is also a strong one, introduce a mollified fundamental solution (8.44) and make a special variation of the form (8.47).

10. Conclude from the result of Exercise 9 that non-trivial solutions of the Beltrami equation (3.43) exist in the large even when the coefficients a, b and c are not analytic. Show in this way how to reduce elliptic partial differential

PROOF BASED ON DIRICHLET'S PRINCIPLE   311

equations of the second order in two independent variables to canonical form globally. Prove that any sufficiently smooth two-dimensional surface can be mapped conformally onto a plane region or at least onto a Riemann surface of several sheets over the plane (cf. Exercise 3.1.8).

11. Show by means of the Dirichlet principle that any plane region bounded by a smooth closed curve can be mapped conformally onto the exterior of a rectilinear slit.

12. Establish the existence of a solution of the Plateau problem (cf. Exercise 1.10 and Section 15.4) when the domain D is convex. Obtain estimates of its first partial derivatives by appealing to the idea that, while volume remains fixed, surface area diminishes under classical Steiner symmetrization (cf. Section 11.3). Use variations of the independent variables to examine the differentiability of the solution.[15]

[15] Cf. Garabedian-Schiffer 3.

9

Existence Theorems
of Potential Theory

1. THE HAHN-BANACH THEOREM

In Chapter 8 we saw how to use extremal problems for the Dirichlet integral to establish the existence of a solution of Dirichlet's problem for the Laplace equation or for more general linear partial differential equations of the elliptic type. Here we shall develop other existence proofs for the same problem that are based either on the maximum principle (cf. Section 7.1) or on integral equations (cf. Chapter 10). It is of interest that the first of these proofs, which will be described immediately, has a relationship to the method of Section 8.2 when the latter is formulated in terms of the kernel function (cf. Exercise 8.2.3).

The existence of the Green's function of a linear elliptic partial differential equation follows when it is known that the Dirichlet problem can be solved. Moreover, the representation of the solution of Dirichlet's problem in terms of the Green's function provides a means for establishing the converse result, although a more difficult analysis is needed to complete the proof (cf. Exercise 7.3.10). Our intention at present is to develop a method for the construction of the Green's function that exploits its dependence on a parameter point. Only an indirect approach to the Dirichlet problem results. The procedure in question will be described in a rather abstract formulation involving the terminology of functional analysis. However, it has its basis in the simple idea that we should be able to approximate the solution of Dirichlet's problem in a region D by linear combinations of fundamental solutions whose singularities lie outside of D, which turn out to be dense in the space of all boundary values defined over ∂D.

To be specific, we shall prove the existence of the Green's function G

for the equation

(9.1) $$L[u] = u_{xx} + u_{yy} - Pu = 0$$

in a multiply connected plane region D whose boundary ∂D has continuous curvature. It will be assumed that the coefficient $P = P(x, y)$ is analytic and even entire, so that the theory of Section 5.1 guarantees the existence of a fundamental solution $S = S(x, y; \xi, \eta)$ in the large. We shall also make the more important hypothesis that

$$P \geq 0,$$

which implies that the maximum principle is valid for (9.1). Here the precise statement, which applies both when P is positive and when it vanishes identically, is that no solution of (9.1) other than a constant can take on a positive maximum or a negative minimum value in the interior of its domain of definition. This permits us to conclude, of course, not only that the solution of Dirichlet's problem for (9.1) is unique but also that the absolute value of any solution inside D is dominated by the maximum modulus of its boundary values. Finally, we add the comment, which will be amplified later on in the exercises, that the restrictions to two independent variables and to an analytic equation are not vital for the proof we have in mind.

Let us consider the class B of continuous functions f defined on the boundary ∂D of the domain D. By introducing the maximum modulus

$$\|f\| = \max_{\partial D} |f|$$

of f over these curves as a *norm*, we can ascribe to B the properties of a *Banach space* (cf. Banach 1). For present purposes all we need to know about the axioms characterizing a Banach space is that they coincide with those for Hilbert space (cf. Section 8.1), with the single exception that the scalar product is omitted. Thus we observe that B is a complete, normed, linear space. In particular, it is obvious how to multiply any element of B by a numerical factor, or scalar; and the sum $f + g$ of every pair of elements f and g in B is a new function belonging to B which satisfies the *triangle inequality*

$$\|f + g\| \leq \|f\| + \|g\|.$$

We shall denote by B' the normed subspace of B consisting of all those functions which represent boundary values of solutions u of (9.1) that are defined throughout D and are continuous in its closure. In particular, any fundamental solution of (9.1) whose singularity is located outside of D generates an element of B'. Furthermore, because of the linearity of the partial differential equation (9.1), every linear combination of elements of B' lies again in that subspace. Consequently B' is closed under the operations of addition and of multiplication by a scalar. Actually, it ought to be identical with the original Banach space B if the Dirichlet problem can be solved, but that is an assertion we do not yet grant.

Corresponding to any fixed interior point (x, y) of D, let us introduce a *linear functional* Λ defined for every element f' of the class B' by the rule

$$(9.2) \qquad \Lambda\{f'\} = u(x, y),$$

where u stands for the solution of (9.1) in D which reduces to f' along ∂D. Since equation (9.1) is linear, Λ is *additive* and *homogeneous* in the sense

$$(9.3) \qquad \Lambda\{\alpha f' + \beta g'\} = \alpha \Lambda\{f'\} + \beta \Lambda\{g'\},$$

where f' and g' are functions in B' and α and β are arbitrary constants. Moreover, the maximum principle shows that $|u|$ cannot exceed in D the norm $\|f'\|$ of its boundary values. Hence we have

$$(9.4) \qquad |\Lambda\{f'\}| \leq \|f'\|.$$

In the language of functional analysis, the estimate (9.4) states that the linear functional Λ is *bounded* and that its *norm*

$$\|\Lambda\| = \sup_{f'} \frac{|\Lambda\{f'\}|}{\|f'\|}$$

is less than or equal to unity.

Now let $S(x, y; \xi, \eta)$ be any fundamental solution of (9.1) which is defined over a domain containing the closure of D, and place its singular point (ξ, η) in the exterior of D. Then the boundary values of S will be in the class B'. According to (9.2) we conclude that

$$(9.5) \qquad 0 = S(x, y; \xi, \eta) - \Lambda\{S(X, Y; \xi, \eta)\},$$

where Λ operates on S with respect to the dummy variables X and Y, which are supposed to indicate the coordinates of a point on ∂D. In contrast with (9.5), we might expect (cf. Exercise 8.2.3) that when the singularity (ξ, η) of S lies inside D the expression

$$(9.6) \qquad G(x, y; \xi, \eta) = S(x, y; \xi, \eta) - \Lambda\{S(X, Y; \xi, \eta)\}$$

would represent the *Green's function* of D for (9.1). However, since we do not know in advance that the boundary values of S belong to B' in this case, formula (9.6) has as yet no meaning. To give it significance we shall extend the definition of the bounded linear functional Λ to the entire Banach space B of continuous functions on ∂D. Thereafter it will be a relatively easy task to verify that (9.6) does, indeed, yield the desired Green's function G.

Our problem is to extend the functional Λ specified on B' by the rule (9.2) to the possibly larger space B in such a way that its basic properties (9.3) and (9.4) are preserved. That such an objective can be achieved is precisely the content of the *Hahn-Banach theorem* in its most elementary form. This states that any bounded linear functional that is defined over some normed subspace of a Banach space can be extended to the whole space without increasing its norm. Rather than refer to the literature for a proof of the extension theorem, we shall develop a direct treatment of it here which brings to light the actual construction of a solution of Dirichlet's problem that is implicit in our more devious and abstract procedure.

Suppose that f is an element of B that does not belong to B'. We wish to define Λ on the vector space composed of all functions of the form $f' + \alpha f$, where f' is any element of B' and α is any constant. First observe that if f' and g' are both in B', then according to (9.3) and (9.4)

$$\Lambda\{f'\} - \Lambda\{g'\} = \Lambda\{f' - g'\}$$

$$\leq \|f' - g'\| = \|(f' - f) - (g' - f)\|$$

$$\leq \|f - f'\| + \|f - g'\|$$

for any f in B. Hence we have

$$\Lambda\{f'\} - \|f - f'\| \leq \Lambda\{g'\} + \|f - g'\|.$$

It follows that there is at least one number λ in the interval

$$(9.7) \qquad \sup_{f'}\,[\Lambda\{f'\} - \|f - f'\|] \leq \lambda \leq \inf_{g'}\,[\Lambda\{g'\} + \|f - g'\|],$$

where f' and g' are allowed to range independently over the class B'. In the applications one usually finds that both equality signs hold in (9.7), so that λ is uniquely determined.

Let λ have a specific value from the interval (9.7) and set

$$\Lambda\{f\} = \lambda.$$

It is then natural to define Λ over the vector space of functions $f' + \alpha f$ by linearity, which leads us to put

$$(9.8) \qquad \Lambda\{f' + \alpha f\} = \Lambda\{f'\} + \alpha\lambda.$$

This extension of Λ clearly still satisfies (9.3), and the property (9.4) is preserved for $\alpha \neq 0$, too, because

$$|\Lambda\{f' + \alpha f\}| = |\alpha| \left|\Lambda\left\{\frac{1}{\alpha}f'\right\} + \lambda\right|$$

$$\leq |\alpha| \left\|f + \frac{1}{\alpha}f'\right\| = \|f' + \alpha f\|$$

in view of (9.7).

To extend Λ over the entire Banach space B we make use of its *separability*, which enables us to avoid appealing to the axiom of choice.[1] To say that B is separable simply means that it contains a sequence of elements $f_1, f_2, f_3, \ldots$ such that to every $\epsilon > 0$ and every f in B there corresponds an integer m for which

$$\|f - f_m\| \leq \epsilon.$$

Such a dense sequence can be constructed, for example, from trigonometric polynomials with rational coefficients, or from polygonal functions whose corners have rational coordinates. We have already developed the formula (9.8) to define Λ over any space generated by adjoining to B' a single additional function. We can now apply the same result in succession to each element of the sequence $f_1, f_2, f_3, \ldots$ that is not included in the space spanned by B' and the previous elements. Thus we are able to specify Λ by mathematical induction over the space generated by all linear combinations of functions from the dense sequence. For any remaining elements in B we can define the bounded linear functional Λ by continuity. Indeed, we have only to set

$$\Lambda\{f\} = \lim_{k \to \infty} \Lambda\{f_{m_k}\}$$

whenever $\|f - f_{m_k}\| \to 0$. This completes our proof of the Hahn-Banach theorem.

We digress for a moment to discuss an interesting heuristic interpretation of the extension of Λ just described. It asserts that the solution u_f of the Dirichlet problem associated with the prescribed boundary values f can be defined at each specific point (x, y) in D by means of the extremal problem

$$(9.9) \qquad u_f(x, y) = \inf_u \left[u(x, y) + \max_{\partial D} |f - u|\right],$$

[1] Cf. Riesz-Nagy 1.

where u may range, for example, over all linear combinations of fundamental solutions of (9.1) whose singularities lie in the exterior of D. Formula (9.9) is seen to result from our proof of the Hahn-Banach theorem when λ is allowed to coincide with the upper estimate given by (9.7). This suggests (cf. Exercise 11 below) that u_f might also be characterized as the smallest solution of (9.1) whose boundary values exceed f.

The maximum principle plays an important role here. In fact, it is the *a priori* inequality (9.4) that furnishes the key to our whole construction. Note, finally, that because the expression (9.9) does not appear at first sight to be a linear functional of the boundary values f it is necessary to detour through a more subtle approach to the Dirichlet problem built around the Hahn-Banach theorem and the representation (9.6) of the Green's function.[2]

Formula (9.6) has acquired a meaning because we have succeeded in defining the bounded linear functional Λ for arbitrary elements of the Banach space B of continuous functions on ∂D. Our aim is to establish that the quantity

$$G(x, y; \xi, \eta) = S(x, y; \xi, \eta) - \Lambda\{S(X, Y; \xi, \eta)\}$$

is actually the Green's function of (9.1) for the region D. Heuristically this would appear to be evident from the fact that the expression on the right consists of a fundamental solution minus a term that ought to be the regular solution with the same boundary values. However, to make such a direct proof rigorous becomes troublesome because it requires a detailed investigation of weak solutions of the Dirichlet problem. Therefore we resort to a less obvious device that simplifies the discussion of G decisively.

Our idea is to examine G in its dependence on the parameters ξ and η rather than in its dependence on the principal variables x and y. Thus we intend to show that as a function of the point (ξ, η), with (x, y) held fixed, G is a fundamental solution of (9.1) in D which vanishes along the boundary curves ∂D. Our demonstration will rely heavily on the boundedness (9.4) of Λ, which is a property that it retained when we extended it to the space B.

Let us take the fundamental solution S occurring in the construction of G to have the special form

$$S = \frac{1}{2\pi}\left[A \log \frac{1}{r} + B\right]$$

given by (5.4) and (5.17), but divided by the normalizing factor 2π. Thus it will be differentiable, and even analytic, in its dependence on (ξ, η),

---

[2] Cf. Caccioppoli 1, Garabedian 1, Lax 2.

provided that $r \neq 0$. It follows that we can apply the differential operator L of equation (9.1) to both sides of (9.6) with respect to the variables ξ and η. Since the functional Λ is bounded and since the difference quotients defining partial derivatives of $S = S(X, Y; \xi, \eta)$ with respect to ξ and η approach their limits uniformly in (X, Y) along ∂D for each particular location of (ξ, η) inside D, we are justified in interchanging the order of the operations L and Λ here (cf. Exercise 2 below). Consequently we obtain

$$(9.10) \quad L[G(x, y; \xi, \eta)] = L[S(x, y; \xi, \eta)] - \Lambda\{L[S(X, Y; \xi, \eta)]\}.$$

Recall that the operator L annihilates the singularity of any fundamental solution (5.4) of (9.1); hence $L[S(x, y; \xi, \eta)]$ is a regular solution of (9.1) throughout D in its dependence on (x, y), and its boundary values belong to the class B'. We conclude that the right-hand side of (9.10) can be evaluated by means of our original rule (9.2) for computing Λ on B', and therefore

$$(9.11) \qquad\qquad L[G] = 0.$$

Since the first term on the right in (9.6) becomes logarithmically infinite at $\xi = x$, $\eta = y$, while the second term remains bounded, (9.11) actually implies that G is a fundamental solution of the partial differential equation (9.1) in D when it is considered as a function of (ξ, η). To establish that it is the Green's function of D we have only to prove that it vanishes on the boundary ∂D. This can be achieved by comparing formula (9.6) with the result (9.5) obtained when (ξ, η) lies in the exterior of D. The analysis required is quite analogous to the procedure we used in Chapter 8 to study boundary conditions. Indeed, it will become clear that

$$(9.12) \qquad\qquad G = 0$$

along the boundary curves ∂D if the expression on the right in (9.5) and (9.6) turns out to vary continuously as the point (ξ, η) crosses ∂D, for then the jump of the terms on the left in (9.5) and (9.6) must reduce to zero, too. Obviously what has to be checked is simply that $\Lambda\{S\}$ exhibits no discontinuity across the boundary ∂D.

Suppose that (ξ, η) is a point of D situated near ∂D. Let $(\bar{\xi}, \bar{\eta})$ be an auxiliary point outside D such that the line segment joining (ξ, η) to $(\bar{\xi}, \bar{\eta})$ coincides with a normal to ∂D that is bisected by one of those boundary curves. In view of what has just been said, the relation

$$(9.13) \qquad \lim_{\xi \to \bar{\xi}, \eta \to \bar{\eta}} \Lambda\{S(X, Y; \xi, \eta) - S(X, Y; \bar{\xi}, \bar{\eta})\} = 0$$

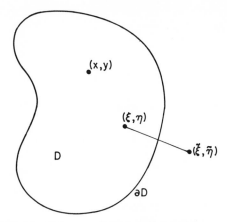

FIG. 35.   Existence of the Green's function.

would be enough to establish (9.12). On the other hand, the boundedness of the linear functional Λ shows that (9.13) would follow from the stronger statement

$$(9.14) \qquad \lim_{\xi \to \bar{\xi}, \eta \to \bar{\eta}} \| S(X, Y; \xi, \eta) - S(X, Y; \bar{\xi}, \bar{\eta}) \| = 0,$$

which we therefore undertake to prove.

First observe that the coefficients A and B in the expansion

$$S = \frac{A}{4\pi} \log \frac{1}{(X - \xi)^2 + (Y - \eta)^2} + \frac{B}{2\pi}$$

of the fundamental solution S are regular at the singular point $\xi = X$, $\eta = Y$, so that (9.14) is equivalent to the quite elementary requirement

$$(9.15) \qquad \lim_{\xi \to \bar{\xi}, \eta \to \bar{\eta}} \left\| \log \frac{(X - \bar{\xi})^2 + (Y - \bar{\eta})^2}{(X - \xi)^2 + (Y - \eta)^2} \right\| = 0.$$

We shall verify (9.15) by direct computation, although it could also be deduced from the estimates at the end of Section 8.3.

Without loss of generality we may place the origin at the midpoint between (ξ, η) and $(\bar{\xi}, \bar{\eta})$ and rotate coordinates so that

$$\xi = \bar{\xi} = 0, \qquad \eta = -\bar{\eta} > 0.$$

It is then possible to express the equation of the arc of ∂D passing through the origin in the form

$$Y = F(X),$$

with $F(0) = F'(0) = 0$. We can prove (9.15) merely by demonstrating that as $\eta \to 0$ the difference

$$\frac{X^2 + [F(X) + \eta]^2}{X^2 + [F(X) - \eta]^2} - 1 = \frac{4\eta F(X)}{X^2 + [F(X) - \eta]^2}$$

tends toward zero uniformly in some interval $|X| < \epsilon$. On the other hand, this is evident because

$$\frac{2|F(X)|}{X^2 + [F(X) - \eta]^2} \leq \frac{2|F(X)|}{X^2},$$

and because the latter quantity approaches $|F''(0)|$ as $X \to 0$. Thus our verification that (9.6) defines the Green's function of (9.1) for the smoothly bounded region D is complete.

In closing we stress that only the existence of the Green's function has been established and that no attempt has been made to solve the Dirichlet problem more generally. In fact, the solution u_f of (9.1) associated with the boundary values f plays a somewhat secondary role in our construction. The situation here may be compared with the contrasting properties of the functions u and $w - v$ introduced in Section 8.2 to solve Dirichlet's problem. In the next section we shall determine u_f quite directly by means of an extremal problem related to (9.9). Although the method to be described there is based on the maximum principle, too, it is built around the Poisson integral formula rather than the Green's function.

EXERCISES

1. Refine the proof of (9.12) so that it becomes feasible to weaken the hypothesis that the boundary curves ∂D have continuous curvature.

2. Justify interchanging the order of application of the bounded linear operation Λ with any uniformly convergent limit process.

3. Use a mean value theorem to establish the validity of the maximum principle for solutions of (9.1) under the less stringent hypothesis $P \geq 0$.

4. Without permitting its gradient to increase, extend any linear homogeneous function defined on the x-axis so that it becomes a linear homogeneous function over the (x,y)-plane.

5. Given that the Green's function of Laplace's equation exists in a smoothly bounded plane region D, prove by means of the integral representation formula (7.34) that Dirichlet's problem can always be solved in D for that equation (cf. Exercise 7.3.10).

6. Show that the Green's function of an elliptic equation

$$u_{xx} + u_{yy} - au_x - bu_y - (a_x + b_y)u = 0$$

with analytic coefficients a and b exists in any region D bounded by a sufficiently smooth set of curves ∂D by applying the procedure of the text to the adjoint equation.

7. Use (9.6) to define the Green's function G of Laplace's equation in a region D of arbitrary dimension whose boundary fulfills the requirement that each of its points can be touched by some sphere E lying inside D and by some sphere $\tilde{E}$ lying outside D. In order to verify the boundary condition (9.12), replace (9.4) by the actually more restrictive condition

$$\Lambda\{f\} \leq \max_{\partial D} f$$

in the discussion of the Hahn-Banach theorem, conclude that $\Lambda\{f\} \geq 0$ whenever $f \geq 0$, and then show that G can be estimated from above and below by the known Green's functions of the exterior of $\tilde{E}$ and of the interior of E, respectively.[3]

8. Consider the class of functions f' of the form

$$(9.16) \qquad f' = \Delta u - Pu,$$

where u is any twice continuously differentiable function in the closure of the region D which vanishes on ∂D. Define a linear functional Λ by the rule

$$(9.17) \qquad \Lambda\{f'\} = u(x, y)$$

for any fixed choice of the point (x, y) in D, and show that it is bounded. Extend Λ to the Banach space of all functions that are continuous in the closure of D, and apply it to a parametrix for (9.16) in order to construct the Green's function of that equation.[4] Assume that $P \geq \epsilon > 0$, but observe that it is not necessary to suppose that P is analytic.

9. Show that (9.16) and (9.17) still define a bounded linear functional when the condition that u vanishes on ∂D is replaced by the hypothesis that its normal derivative is zero there. Use this modified version of the functional (9.17) to establish the existence of the Neumann's function for equation (9.16)

10. Develop a proof of the existence of the Green's function for a self-adjoint equation

$$\sum_{i,j=1}^{n} \frac{\partial}{\partial x_i}\left(a_{ij} \frac{\partial u}{\partial x_j}\right) = 0$$

of the elliptic type in any number of independent variables by applying a linear functional analogous to (9.17) to an appropriate parametrix.

11. In the case of Laplace's equation, discuss how the Hahn-Banach theorem suggests the formula

$$(9.18) \qquad u_f(x, y) = \sup_{u} [u(x, y) + \min_{\partial D} (f - u)]$$

---

[3] Cf. Lax 2.
[4] Cf. Garabedian-Shiffman 1.

analogous to (9.9) for the solution of Dirichlet's problem, where u ranges over the class of harmonic functions in D with continuous boundary values. By subtracting a suitable constant from u, show that (9.18) is equivalent to the extremal problem

$$(9.19) \qquad u_f(x, y) = \sup_u u(x, y)$$

within the class of harmonic functions u over D such that $u \leq f$ on ∂D (cf. Section 2).

12. Let D be a region of the z-plane bounded by a smooth curve ∂D, and let Ω stand for the class of trial functions $\phi = \phi(z)$ that are complex analytic in the closure of D. Denote by H_2 the Hilbert space of complex-valued functions k defined on ∂D such that

$$\int_{\partial D} |k|^2 \, ds < \infty,$$

and such that

$$\int_{\partial D} k\phi \, dz = 0$$

for every ϕ in Ω. Show that for any fixed choice of the point ζ inside D the extremal problem

$$\frac{1}{2\pi i} \int_{\partial D} \frac{k \, dz}{z - \zeta} = 1,$$

$$\int_{\partial D} |k|^2 \, ds = \text{minimum}$$

defines a unique minimal element k_0 in H_2. Prove[5] that the bounded linear functional Λ given for all real-valued continuous functions f on ∂D by the integral formula

$$(9.20) \qquad \Lambda\{f\} = \int_{\partial D} |k_0|^2 f \, ds \bigg/ \int_{\partial D} |k_0|^2 \, ds$$

has the property (9.5) with respect to Laplace's equation; conclude that (9.6) still yields the harmonic Green's function of D when Λ is constructed in this way (cf. Exercise 8.2.3). What is the connection between k_0 and the *Szegö kernel function* described in Exercise 7.3.16?

13. Recalling the notation of Exercise 12, introduce the class H_1 of complex-valued functions F defined on ∂D which have bounded variation

$$\int_{\partial D} |dF| < \infty$$

and satisfy

$$\int_{\partial D} \phi \, dF = 0$$

[5] Cf. Garabedian 1.

for every ϕ in Ω. By an application of *Helly's convergence theorem*,[6] show that there is an extremal function F_0 in H_1 solving the minimum problem

$$\frac{1}{2\pi i} \int_{\partial D} \frac{dF}{z - \zeta} = 1,$$

$$\int_{\partial D} |dF| = \text{minimum}.$$

Use[7] the bounded linear functional

(9.21) $$\Lambda\{f\} = \int_{\partial D} f \, |dF_0| \Big/ \int_{\partial D} |dF_0|$$

to establish the existence of the Green's function of Laplace's equation for the plane region D. Compare the three different constructions (9.2), (9.20) and (9.21) of Λ, which are based, respectively, on the Hahn-Banach theorem, on the Riesz-Fischer theorem, and on Helly's theorem. What is their relationship to the integral representation (7.34) of the solution of Dirichlet's problem?

14. Use the Green's function of Laplace's equation to establish the existence of a conformal mapping of any plane region bounded by a simple closed curve onto the interior of the unit circle. This is one version of the *Riemann mapping theorem* (cf. Section 7.3).

2. SUBHARMONIC FUNCTIONS; BARRIERS

We shall describe next a rather direct method of solution of the Dirichlet problem that is based on the maximum principle. What we have in mind is the *theory of subharmonic functions*, which we intend to develop for the case of two independent variables. The generalization to space of higher dimension can be effected without difficulty, and an extension to linear partial differential equations of the elliptic type with variable coefficients is feasible, too. However, we prefer to relegate such matters to the exercises.

The construction we shall apply is most easily understood in terms of a quite elementary one-dimensional model. Let w range over the class of all convex functions on the interval $a < x < b$ whose values $w(a)$ and $w(b)$ at the endpoints do not exceed given numbers. The supremum u of the functions w is seen to be a straight line joining the prescribed boundary values. Therefore it provides a solution of Dirichlet's problem for the ordinary differential equation

$$u_{xx} = 0$$

[6] Cf. Coddington-Levinson 1.
[7] Cf. Garabedian-Schiffer 1.

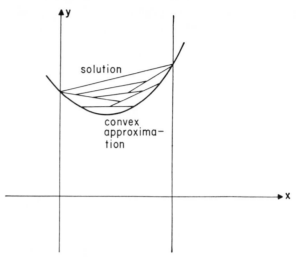

FIG. 36.   Perron process for convex functions.

on the interval $a < x < b$. Moreover, any of the convex functions w can be converted into better approximations to u by replacing them on shorter intervals by the line segments joining their end values. What we have described is a simple example of the *Perron process* and of the *Poincaré sweeping out process*[8] for solving the Dirichlet problem. It is related to the idea of *relaxation* used to treat Laplace's equation by finite differences (cf. Section 13.5). To develop it into a tool for proving existence theorems concerning the Dirichlet problem for a partial differential equation, we shall generalize the notion of a convex function by introducing subharmonic functions.

We shall call a function w defined in an open region D *subharmonic* if it is continuous (or at least upper semicontinuous) there and if for every choice of a trial harmonic function u the difference $w - u$ satisfies the maximum principle in any subdomain $\tilde{D}$ of D where it has meaning. The precise requirement to be imposed is that whenever $w - u$ assumes its absolute maximum over $\tilde{D}$ at an interior point of that subdomain, it should be identically constant there. We emphasize, however, that $w - u$ is permitted to attain minimum values. In fact, we say that w is *superharmonic*, rather than subharmonic, when it satisfies the following minimum principle relative to every harmonic function u. Namely, $w - u$ should be identically constant in any subdomain $\tilde{D}$ of D if it achieves its absolute minimum over $\tilde{D}$ at an interior point of that subdomain. Observe, incidentally, that any harmonic function is both subharmonic and superharmonic.

[8] Cf. Kellogg 1.

Let D be a plane region with boundary ∂D, and suppose that a continuous function f is prescribed on ∂D. We introduce the class W_f of subharmonic functions w in D which have the property

$$(9.22) \qquad \overline{\lim_{x \to X, y \to Y}} \ w(x, y) \le f(X, Y)$$

that their boundary values are dominated by f at all points (X, Y) of ∂D. Any large negative constant that is less than the minimum of f constitutes an element of W_f. On the other hand, no function w in W_f can exceed an upper bound M of f; for the difference $w - M$ obeys the maximum principle and is therefore majorized by its supremum on ∂D, which is less than or equal to zero in view of (9.22). Our aim will be to establish that the expression

$$(9.23) \qquad u_f(x, y) = \sup_{w} w(x, y),$$

in which w is supposed to range over the class W_f, defines a harmonic function u_f in D representing the solution of Dirichlet's problem that assumes the given boundary values f. Motivation for this construction, called the *Perron process*,[9] can be found in the similar extremal problems (9.18) and (9.19) characterizing the desired solution, which, were it to exist, would obviously fulfill (9.23) as a consequence of the maximum principle.

We have to show, first, that the function u_f given by (9.23) is harmonic in the region D and, second, that it reduces to f along the boundary ∂D. As a preliminary we discuss some basic properties of subharmonic functions and, in particular, of functions of the class W_f.

If w_1 and w_2 are in W_f, we maintain that so also is their maximum

$$(9.24) \qquad w = \max(w_1, w_2).$$

Indeed, formula (9.24) defines a continuous function w satisfying the boundary condition (9.22). Suppose there is a harmonic function u and a corresponding subdomain $\tilde{D}$ of D such that $w - u$ achieves its absolute maximum over $\tilde{D}$ at some interior point (ξ, η) of that subdomain, where we can assume with no loss of generality that $w_2 \ge w_1$. Then $w_2 - u$ achieves its maximum over $\tilde{D}$ at (ξ, η), too. Consequently $w_2 - u$ is constant in $\tilde{D}$, so that

$$w(x, y) - u(x, y) \le w(\xi, \eta) - u(\xi, \eta) = w_2(\xi, \eta) - u(\xi, \eta)$$

$$= w_2(x, y) - u(x, y) \le w(x, y) - u(x, y)$$

[9] Cf. Perron 1.

for any choice there of the point (x, y). Thus w must actually coincide with w_2 throughout $\tilde{D}$. Hence $w - u$ remains constant in that subdomain of D, which is exactly what had to be established. Inductive reasoning leads to the more general conclusion that

$$w = \max{(w_1, \ldots, w_m)}$$

is of the class W_f whenever all the individual functions $w_1, \ldots, w_m$ are.

Let E be any circle whose closure is contained in the region D, and let w be any element of W_f. The values of w on the boundary ∂E of E define a Dirichlet problem for Laplace's equation in E that can be solved explicitly by means of the Poisson integral formula (7.52). We introduce a modified function w' which is equal inside E to the Poisson integral representation

$$w'(\rho \cos \phi, \rho \sin \phi) = \frac{1}{2\pi} \int_0^{2\pi} \frac{[R^2 - \rho^2] w(R \cos \theta, R \sin \theta)\, d\theta}{R^2 - 2R\rho \cos (\theta - \phi) + \rho^2}$$

of the solution of the boundary value problem just described, but which coincides with the original element w elsewhere in D. We wish to prove that w' is a subharmonic function of the class W_f.

It is evident from its construction that w' is continuous and fulfills the boundary condition (9.22). Now suppose that the difference $w' - u$ between w' and a harmonic function u assumes its absolute maximum over some subdomain $\tilde{D}$ of D at an interior point of $\tilde{D}$. When E contains the maximum point, $w' - u$ is seen to be constant in the corresponding component of the intersection of $\tilde{D}$ with E because it is harmonic there. Thus the maximum of $w' - u$ is attained on ∂E, too, if $\tilde{D}$ extends that far. Hence it suffices to discuss the case where the maximum occurs in the closed exterior of E.

Since w is subharmonic and w' is harmonic in E, the largest value of $w - w'$ over E is assumed on the circumference ∂E and is therefore zero, which means that $w \leq w'$. Consequently $w - u$ has in $\tilde{D}$ the same maximum as $w' - u$ and must reduce to a constant there, granted that we exclude the trivial case in which $\tilde{D}$ is located entirely inside E. Since, however,

$$w(x, y) - u(x, y) \leq w'(x, y) - u(x, y)$$

$$\leq w'(\xi, \eta) - u(\xi, \eta) = w(\xi, \eta) - u(\xi, \eta),$$

where (ξ, η) stands for a maximum point of $w' - u$ in the closed exterior of E and where (x, y) denotes an arbitrary point of $\tilde{D}$, we are able to conclude that $w' - u$ is constant throughout $\tilde{D}$, too. This completes our verification that the modified function w' belongs to W_f.

Actually, the characteristic property of a subharmonic function is that in any region it cannot exceed the harmonic function with the same boundary values. Moreover, it is enough to consider only circles in making the comparison. Thus the subharmonic functions are those that are dominated in every circle by the Poisson integral of their boundary values. In particular, it becomes clear that the sum of any two subharmonic functions is itself subharmonic.

We proceed to show that the function u_f defined by the extremal problem (9.23) is harmonic at any point (x, y) of D. Consider a sequence of elements w_m from W_f which approaches u_f at that point. We know that max $(w_1, \ldots, w_m)$ is a function belonging to W_f, and accordingly we can use it to replace w_m in the above sequence. Thus there is no loss of generality if we suppose that w_m is an increasing sequence

$$w_1 \leq w_2 \leq \cdots \leq w_m \leq \cdots .$$

Next, select a circle E within D that encloses the point (x, y) and replace each of the functions w_m there by the Poisson integral representation of the harmonic function whose boundary values along ∂E coincide with w_m. This construction yields a new monotonic sequence

$$w'_1 \leq w'_2 \leq \cdots \leq w'_m \leq \cdots$$

of elements of W_f which are harmonic in E and have at (x, y) the limiting value $u_f(x, y)$.

The Harnack convergence theorem (cf. Section 7.2) shows that the limit

$$u = \lim_{m \to \infty} w'_m$$

exists uniformly in every closed subdomain of E, where it defines a harmonic function u that is equal to u_f at (x, y). However, since it is not yet clear that u and u_f are identical throughout E, we are not able to conclude immediately that u_f is harmonic. To examine the relationship between u and u_f, we fix our attention on any point (ξ, η) of E distinct from (x, y). A monotonic sequence of elements w_m^* from W_f can be found that approaches u_f at (ξ, η). First set

(9.25) $w_m^{**} = \max (w_m^*, w_m)$

and then, as before, replace w_m^{**} inside E by the Poisson integral formula associated with its boundary values. There results a monotonic sequence

$$w''_1 \leq w''_2 \leq \cdots \leq w''_m \leq \cdots$$

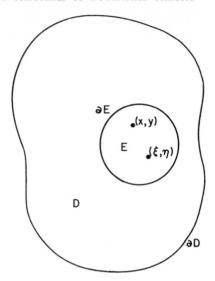

FIG. 37.   The Perron process.

of functions of the class W_f which approach u_f at (ξ, η), satisfy Laplace's equation in E, and dominate the corresponding elements of the sequence w'_m everywhere.

By virtue of Harnack's convergence theorem, the limit function

$$u^* = \lim_{m \to \infty} w''_m$$

is harmonic in E, where it has to satisfy the inequality

$$u \leq u^*$$

to be consistent with (9.25). On the other hand, according to the definition of u_f we must have

$$u^*(x, y) \leq u_f(x, y) = u(x, y).$$

Thus the harmonic function $u^* - u$ assumes its minimum value, namely zero, at the point (x, y) inside E. Hence it has to vanish identically, and, in particular,

$$u(\xi, \eta) = u^*(\xi, \eta) = u_f(\xi, \eta).$$

Since (ξ, η) was picked arbitrarily in E, it follows that u coincides with u_f throughout E. We conclude that the function u_f generated by the Perron process is, indeed, harmonic.

The most natural analysis of whether u_f reduces to the prescribed boundary values f continuously at ∂D is based on the introduction of certain comparison functions that are commonly referred to as *barriers*. To be precise, we call any negative subharmonic function v defined in D a barrier for the boundary point (X, Y) on ∂D if it remains below a negative upper bound outside each circle around (X, Y), but

$$\lim_{x \to X, y \to Y} v(x, y) = 0.$$

We say that the boundary point (X, Y) is *regular* when a barrier of this kind can be associated with it. We shall establish that the harmonic function (9.23) satisfies the boundary condition

$$(9.26) \qquad \lim_{x \to X, y \to Y} u_f(x, y) = f(X, Y)$$

at any regular point. On the other hand, we shall describe (X, Y) as an *exceptional point for the Dirichlet problem* if (9.26) fails to hold for some special choice of the continuous boundary values f.

Given any number $\epsilon > 0$, let E be a circle about (X, Y) so small that

$$|f(X', Y') - f(X, Y)| < \epsilon$$

when (X', Y') lies on the intersection of ∂D with E. Suppose that v is a barrier for (X, Y). Since v has a negative upper bound outside E, we can find a large positive value of the parameter λ such that the subharmonic function

$$(9.27) \qquad w(x, y) = \lambda v(x, y) + f(X, Y) - \epsilon$$

fulfills (9.22) not just at the point (X, Y) but all along ∂D. Hence w belongs to the class W_f. Because of its definition as a supremum the harmonic function u_f must therefore satisfy

$$(9.28) \qquad \lim_{x \to X, y \to Y} u_f(x, y) \geq \lim_{x \to X, y \to Y} [\lambda v(x, y) + f(X, Y) - \epsilon] = f(X, Y) - \epsilon.$$

Our next objective will be to arrive at an analogous estimate on the limit superior of u_f at (X, Y).

The argument used to prove that the comparison function (9.27) belongs to W_f also implies that with a large enough choice of $\lambda > 0$ we have

$$\overline{\lim}_{x \to X', y \to Y'} [w(x, y) + \lambda v(x, y) - f(X, Y) - \epsilon] \leq 0$$

for every element w in W_f, where (X', Y') is allowed to range over all the points of ∂D. Since the sum $w + \lambda v$ is subharmonic (cf. Exercise 1 below) and therefore obeys the maximum principle, it follows that

$$w(x, y) \leq -\lambda v(x, y) + f(X, Y) + \epsilon$$

in D. This uniform upper bound on all the elements w of W_f must apply
to their supremum u_f, too. Hence we obtain

(9.29)
$$\overline{\lim_{x \to X, y \to Y}} \, u_f(x, y) \leq \lim_{x \to X, y \to Y} [-\lambda v(x, y) + f(X, Y) + \epsilon] = f(X, Y) + \epsilon.$$

The results (9.28) and (9.29) combine to yield the desired boundary
condition (9.26) because we may assign arbitrarily small values to the
positive number ϵ.

In order to show that the Perron process actually solves Dirichlet's
problem, it remains to exhibit barriers at all points of ∂D. In the plane
case we can do this readily with the help of analytic functions of a complex
variable. If, for example, near each boundary point $\zeta = X + iY$ another
point $\tilde{\zeta}$ can be found such that the segment joining ζ to $\tilde{\zeta}$ lies outside D,
then the negative branch of the harmonic function

$$v(x, y) = \text{Re} \, \{\sqrt{(z - \zeta)/(z - \tilde{\zeta})}\}$$

furnishes the required barrier, where $z = x + iy$. On the other hand,
under the more stringent hypothesis that a point of ∂D can be touched
by some circle $\tilde{E}$ located in the exterior of D, the truncated Green's
function of that circle provides a convenient barrier at the point of
tangency. The latter construction, which is reminiscent of our analysis
of the boundary condition in Chapter 8, has the advantage that it generalizes
in an obvious fashion to space of arbitrary dimension and assures the
solvability of Dirichlet's problem in regions bounded by smooth surfaces.

It is in space, rather than in the plane, that interesting examples of
exceptional boundary points for the Dirichlet problem are encountered.
To begin with, consider the question whether arbitrary continuous
boundary values can be prescribed along a one-dimensional spike, or
curve, C which penetrates inside a smooth boundary surface in three-
dimensional space. Let u' be the harmonic function that solves Dirichlet's
problem after the spike has been removed. We introduce the comparison
function

$$u = u' + \epsilon \int_C \frac{ds}{r},$$

where ϵ is an arbitrary parameter and where the integral on the right
represents the gravitational potential of a uniform distribution of mass

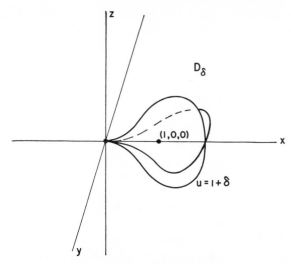

FIG. 38.   Exceptional point for Dirichlet's problem.

along the spike. It is clear that the latter term becomes infinite on C.
Consequently, if the solution of Dirichlet's problem with the spike included
as part of the boundary were to exist, it could not be greater than u for
any $\epsilon > 0$ or less than u for any $\epsilon < 0$, by the maximum principle. Thus
it would have to coincide with u', and its values on the spike could not be
assigned arbitrarily. This leads us to conclude that all the points of such
a curve C are exceptional for Dirichlet's problem.

A more subtle example[10] of an exceptional point is obtained by
examining the potential function

$$(9.30) \qquad u = \int_0^1 \frac{\xi \, d\xi}{\sqrt{(x - \xi)^2 + y^2 + z^2}}.$$

On the segment $0 < x \leq 1, y = z = 0$, u becomes infinite, and everywhere
else it is positive. As the point (x, y, z) approaches the origin along any
ray distinct from the positive x-axis, u tends toward the value 1. On the
other hand, by allowing (x, y, z) to approach the origin along suitable
paths tangent to the positive x-axis, we can achieve any limiting value we
please for u in excess of 1. Thus the level surface

$$(9.31) \qquad u(x, y, z) = 1 + \delta, \qquad \delta > 0,$$

forms a sharp cusp, or needle, at the origin and bounds an infinite region
D_δ from which the segment $0 < x \leq 1, y = z = 0$ is excluded.

[10] Cf. Courant-Hilbert 3.

In the region D_δ just described we wish to interpret the harmonic function (9.30) as the solution of a Dirichlet problem that results from assigning the constant boundary values $1 + \delta$. In this connection it should be observed that in three-dimensional space a harmonic function is only considered to be regular at infinity when it vanishes there, as u does, so that the inversion

$$u^*(x, y, z) = \frac{1}{r} u\left(\frac{x}{r^2}, \frac{y}{r^2}, \frac{z}{r^2}\right), \qquad r = \sqrt{x^2 + y^2 + z^2},$$

leads to a correspondingly regular harmonic function u^* at the origin (cf. Exercise 3.3.1). However, the interesting feature of the question we have posed is that on the cusped surface (9.31) bounding D_δ, the origin turns out to be an exceptional point for Dirichlet's problem. Indeed, along appropriate paths u approaches all limiting values between 1 and $1 + \delta$ there. Furthermore, no harmonic function u' can exist which is continuous in the closure of D_δ and reduces to $1 + \delta$ on the surface (9.31), since the maximum principle implies that any such function lies in the range

$$u - \frac{\epsilon}{r} < u' < u + \frac{\epsilon}{r}$$

for every $\epsilon > 0$. Thus we have succeeded in exhibiting an exceptional point on a boundary surface that is topologically equivalent to a sphere.

In closing it is of interest to mention a physical interpretation of the above examples. This asserts that an exceptional point or spike does not have the capacity to hold electric charge (cf. Exercise 14 below) and may induce sparks to leak off.

EXERCISES

1. Verify that the product of a positive number and a subharmonic function is subharmonic, that the product of a negative number and a subharmonic function is superharmonic, that the sum of two subharmonic functions is subharmonic, and that the difference between a subharmonic function and a superharmonic function is subharmonic.

2. Show that a continuous function of two independent variables is subharmonic if and only if its average over the circumference of every circle in which it is defined is greater than or equal to its value at the center.

3. Show that a function w with continuous second partial derivatives is subharmonic if and only if

$$\Delta w \geq 0.$$

4. Formulate an analogue of the Perron process (9.23) based on superharmonic, rather than subharmonic, functions.

5. To solve the Dirichlet problem in a region D by the *Poincaré sweeping out process*, we cover D with a sequence of circles, extend the boundary values inside D in some arbitrary way, replace the extended function in any circle from the above sequence by the Poisson integral representation of a harmonic function there with the same boundary values, and then iterate this construction successively in all the circles so that it is repeated infinitely often in each of them. Use subharmonic and superharmonic majorants constructed from finite sets of barriers to establish the convergence of the method.

6. Extend the theory of the text to the case of three or more independent variables.

7. Assuming the existence of solutions in the small, generalize the Perron process to solve Dirichlet's problem in the large for a linear elliptic partial differential equation of the second order that obeys the maximum principle.

8. Why is the existence of a barrier for any specific point on the boundary ∂D of a domain D a local property of that point?

9. Show that any isolated boundary point in the plane is exceptional for Dirichlet's problem. What has this to do with removable singularities?

10. Show that all the boundary points of a plane region are regular for the Dirichlet problem if each connected component of the boundary contains more than one point.[11]

11. Describe the asymptotic form of an arc terminating at the origin along which the potential function (9.30) approaches a specific value larger than 1.

12. Prove that when the Dirichlet problem is solvable in a region D for every continuous distribution of boundary values, then a barrier exists at each boundary point.

13. Exhibit a barrier at any boundary point of a region in three-dimensional space that can be touched by a disc located in the exterior of the region.

14. Let D be an infinite region in space bounded by smooth surfaces ∂D. The harmonic function

$$U = \frac{C}{r} + \cdots$$

in D, regular at infinity, which assumes the boundary values 1 on ∂D is called the *conductor potential* of those surfaces. The positive coefficient C is called their *Newtonian capacity*.[12] Show that

$$C = -\frac{1}{4\pi} \int_{\partial D} \frac{\partial U}{\partial \nu}\, d\sigma = \frac{1}{4\pi} \int_D [U_x^2 + U_y^2 + U_z^2]\, dx\, dy\, dz.$$

Establish that if $\tilde{D}$ is contained inside D, the capacity of $\partial\tilde{D}$ exceeds the capacity of ∂D. Introduce the capacity of an arbitrary compact set in space by exhausting the infinite component of its complement with a nested sequence of smoothly bounded subdomains and considering the limit of the capacities of their boundary

[11] Cf. Ahlfors 1.
[12] Cf. Abraham-Becker 1.

surfaces. Prove that the Newtonian capacity of a line segment is zero. What has this to do with exceptional points for the Dirichlet problem?

15. What is the connection between the conductor potential and the Green's function of an infinite domain?

16. Show how to define the harmonic Green's function of a region with or without exceptional boundary points as the smallest positive fundamental solution of Laplace's equation in that region. Establish that it vanishes at all regular boundary points.

17. Use the comparison function arg $[(R + z)/(R - z)]$ to prove the following version of the *Phragmén-Lindelöf theorem*.[13] Let u be a harmonic function of x and y in the upper half-plane $y > 0$, and suppose that $u = 0$ at all finite points of the x-axis. If in addition u fulfills the requirement

$$\overline{\lim_{r \to \infty}} \frac{|u|}{r} = 0$$

at infinity, then it must vanish identically. Compare this result with Liouville's theorem.

3. REDUCTION TO A FREDHOLM INTEGRAL EQUATION

We shall indicate how the Dirichlet and Neumann problems for a linear elliptic partial differential equation can be transformed into *integral equations* through representation of the answer in terms of appropriate integrals of a fundamental solution or of a parametrix. However, we shall postpone until Chapter 10 our treatment of the integral equations that are encountered. The effect of the theory of integral equations is to reduce the question of existence of a solution to the problem of establishing its uniqueness. The method also turns out to be advantageous in cases where the variational principles on which our previous existence proofs were based cease to be valid. Thus in some ways the approach through integral equations is the most powerful available to us, although it is perhaps not the simplest.

We shall begin with a discussion of the Laplace equation in three-dimensional space. In that case the device which enables us to solve the Dirichlet and Neumann problems is a supposition that the answers can be expressed, respectively, as double and single layer potentials of charge distributions over the surface of the region involved. Thus we have to make a preliminary analysis of the behavior of such potentials at the boundary where the charge is distributed.[14] For the sake of simplicity we find it

---

[13] Cf. Coddington-Levinson 1.

[14] Cf. Kellogg 1.

convenient in this connection to confine our attention to a region D bounded by a surface ∂D which is analytic and which is topologically equivalent to a sphere.

We are interested in examining the *single layer potential*

$$V(x, y, z) = \frac{1}{2\pi} \int_{\partial D} \rho(\xi, \eta, \zeta) \, \frac{d\sigma}{r}$$

and the *double layer potential*

$$W(x, y, z) = \frac{1}{2\pi} \int_{\partial D} \mu(\xi, \eta, \zeta) \, \frac{\partial}{\partial \nu} \frac{1}{r} \, d\sigma,$$

where

$$r = \sqrt{(x - \xi)^2 + (y - \eta)^2 + (z - \zeta)^2}$$

and where ν stands, as usual, for the inner normal on ∂D. Obviously V and W are harmonic functions inside and outside of ∂D. Supposing at first that the *density* ρ and the *dipole moment density* μ are real analytic functions on ∂D, we shall evaluate the limits of V and W, and of their normal derivatives, as the point (x, y, z) approaches the surface ∂D. Afterward it will be feasible to extend the results obtained in this way to the case where ρ and μ are, say, twice continuously differentiable, for we can apply Taylor's theorem to approximate them with polynomials and then show directly that the remainder term makes a negligible contribution.

Let E denote the sphere of radius R about a point (X, Y, Z) on the surface ∂D, and let D^* denote the union of D and E. If u is a harmonic

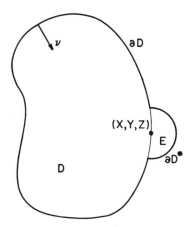

FIG. 39.   Analysis of surface potentials.

function in both E and a neighborhood of ∂D, Green's identity

$$\int_D [u\,\Delta v - v\,\Delta u]\,d\xi\,d\eta\,d\zeta + \int_{\partial D}\left[u\,\frac{\partial v}{\partial \nu} - v\,\frac{\partial u}{\partial \nu}\right]d\sigma = 0$$

implies that for (x, y, z) inside D

$$(9.32) \qquad \int_{\partial D}\left[u\,\frac{\partial}{\partial \nu}\frac{1}{r} - \frac{1}{r}\frac{\partial u}{\partial \nu}\right]d\sigma = \int_{\partial D^*}\left[u\,\frac{\partial}{\partial \nu}\frac{1}{r} - \frac{1}{r}\frac{\partial u}{\partial \nu}\right]d\sigma.$$

We wish to exploit this formula in order to study the behavior of the potentials V and W at (X, Y, Z).

Observe that the Cauchy-Kowalewski theorem (cf. Section 1.2) enables us to construct locally a harmonic function u which fulfills the initial conditions

$$(9.33) \qquad\qquad u = 0, \qquad \frac{\partial u}{\partial \nu} = \rho$$

on ∂D. Hence we can write

$$(9.34) \qquad V(x, y, z) = -\frac{1}{2\pi}\int_{\partial D^*}\left[u\,\frac{\partial}{\partial \nu}\frac{1}{r} - \frac{1}{r}\frac{\partial u}{\partial \nu}\right]d\sigma$$

because of (9.32), which shows that V is regular at (X, Y, Z). Similarly, introducing the harmonic function u such that

$$(9.35) \qquad\qquad u = \mu, \qquad \frac{\partial u}{\partial \nu} = 0$$

on ∂D, we have

$$(9.36) \qquad W(x, y, z) = \frac{1}{2\pi}\int_{\partial D^*}\left[u\,\frac{\partial}{\partial \nu}\frac{1}{r} - \frac{1}{r}\frac{\partial u}{\partial \nu}\right]d\sigma;$$

hence W must be a regular function at (X, Y, Z), too.

We can evaluate

$$W^-(X, Y, Z) = \lim_{(x,y,z)\to(X,Y,Z)} W(x, y, z),$$

with (x, y, z) coming from the *interior* of D, in terms of the moment density μ instead of the auxiliary function u by first setting $x = X, y = Y, z = Z$ in (9.36) and then allowing the radius R of the sphere E to approach zero. The integral in (9.36) over the portion of ∂E located outside D becomes asymptotically equivalent to the average of W over a small hemisphere as $R \to 0$; therefore a residue calculation gives

$$(9.37) \qquad W^-(X, Y, Z) = \mu(X, Y, Z) + \frac{1}{2\pi}\int_{\partial D}\mu\,\frac{\partial}{\partial \nu}\frac{1}{r}\,d\sigma.$$

We emphasize that the improper integral on the right is convergent because differentiation of the function $1/r$ in the normal direction does not increase its rate of growth when both of its arguments lie on the smooth surface ∂D. Indeed, near the singularity of $1/r$ the normals of ∂D are approximately tangent to the spheres on which that function is constant.

To calculate the limit

$$W^+(X, Y, Z) = \lim_{(x,y,z)\to(X,Y,Z)} W(x, y, z),$$

where (x, y, z) is now supposed to approach ∂D from the *exterior* of D, we have only to replace (9.36) by a virtually identical formula in which the deformed surface ∂D^* is replaced by the boundary ∂D^{**} of the difference D^{**} between D and E. Letting $R \to 0$ in the new formula, we obtain

$$(9.38) \qquad W^+(X, Y, Z) = -\mu(X, Y, Z) + \frac{1}{2\pi} \int_{\partial D} \mu \, \frac{\partial}{\partial \nu} \frac{1}{r} \, d\sigma,$$

where the minus sign occurs in front of $\mu(X, Y, Z)$ because the orientation of the normal ν on ∂D^{**} is directed outward from the sphere E. Thus the moment density μ can be computed from the jump

$$(9.39) \qquad\qquad W^+ - W^- = -2\mu$$

of the double layer potential W across the charged surface ∂D.

We shall establish next that the normal derivative of the double layer potential W is continuous across ∂D. We can subtract the normal derivative of (9.36) at the point (X, Y, Z) from the corresponding normal derivative of the representation of W as an integral over ∂D^{**} to derive the expression

$$(9.40) \qquad \frac{\partial W^+}{\partial \nu} - \frac{\partial W^-}{\partial \nu} = -\frac{1}{2\pi} \frac{\partial}{\partial \nu} \int_{\partial E} \left[u \, \frac{\partial}{\partial \nu} \frac{1}{r} - \frac{1}{r} \frac{\partial u}{\partial \nu} \right] d\sigma = -2 \frac{\partial u}{\partial \nu}$$

for the jump in $\partial W/\partial \nu$ across ∂D. As a result of the second initial condition (9.35) that has been imposed on the harmonic function u, (9.40) yields the desired equality

$$(9.41) \qquad\qquad \frac{\partial W^+}{\partial \nu} = \frac{\partial W^-}{\partial \nu}$$

of the normal derivatives of W from the two opposite sides of the surface ∂D, where it is understood, of course, that ν has in both cases the direction of the inner normal with respect to D.

We should like to deduce from (9.34) and from the analogous integral representation involving ∂D^{**} results for the single layer potential V similar to (9.37), (9.38) and (9.41). Defining V^+ and V^- on ∂D to be the limits of V from the exterior and from the interior of D, respectively, we find by a subtraction of the integral formulas just referred to that

$$V^+ - V^- = \frac{1}{2\pi} \int_{\partial E} \left[u \frac{\partial}{\partial \nu} \frac{1}{r} - \frac{1}{r} \frac{\partial u}{\partial \nu} \right] d\sigma = 2u$$

at the boundary point (X, Y, Z). Because of the new initial conditions (9.33) on u required in (9.34), we conclude that

$$(9.42) \qquad\qquad V^+ = V^-.$$

Hence the single layer potential V is seen to be continuous across the charged surface ∂D. A direct passage to the limit under the integral sign would also suffice to prove this less subtle theorem.

To examine the normal derivative of V at (X, Y, Z) calculated from the interior of D, we differentiate (9.34), obtaining

$$(9.43) \qquad \frac{\partial V^-}{\partial \nu} = -\frac{1}{2\pi} \frac{\partial}{\partial \nu} \int_{\partial D^*} \left[u \frac{\partial}{\partial \nu} \frac{1}{r} - \frac{1}{r} \frac{\partial u}{\partial \nu} \right] d\sigma.$$

Similarly, we have

$$(9.44) \qquad \frac{\partial V^+}{\partial \nu} = -\frac{1}{2\pi} \frac{\partial}{\partial \nu} \int_{\partial D^{**}} \left[u \frac{\partial}{\partial \nu} \frac{1}{r} - \frac{1}{r} \frac{\partial u}{\partial \nu} \right] d\sigma$$

for the normal derivative of V based on difference quotients from the exterior of D. Subtracting (9.43) from (9.44), we find

$$\frac{\partial V^+}{\partial \nu} - \frac{\partial V^-}{\partial \nu} = \frac{1}{2\pi} \frac{\partial}{\partial \nu} \int_{\partial E} \left[u \frac{\partial}{\partial \nu} \frac{1}{r} - \frac{1}{r} \frac{\partial u}{\partial \nu} \right] d\sigma = 2 \frac{\partial u}{\partial \nu},$$

or, in view of (9.33),

$$(9.45) \qquad\qquad \frac{\partial V^+}{\partial \nu} - \frac{\partial V^-}{\partial \nu} = 2\rho.$$

Thus the charge density ρ is easily calculated from the jump of the normal derivative of the single layer potential V across the surface ∂D. Furthermore, by symmetry the contributions to the integrals (9.43) and (9.44) from the two sections of the spherical surface ∂E located on opposite sides of ∂D can differ only in sign in the limit as $R \to 0$. Hence we have

$$(9.46) \qquad \frac{\partial V^-}{\partial \nu} = -\rho(X, Y, Z) + \frac{1}{2\pi} \int_{\partial D} \rho \frac{\partial}{\partial \nu} \frac{1}{r} d\sigma$$

and

$$(9.47) \qquad \frac{\partial V^+}{\partial \nu} = \rho(X, Y, Z) + \frac{1}{2\pi} \int_{\partial D} \rho \frac{\partial}{\partial \nu} \frac{1}{r} d\sigma$$

at any point (X, Y, Z) of ∂D, where the normal differentiation of $1/r$ is now made with respect to X, Y and Z, and not with respect to the dummy variables of integration ξ, η and ζ. Note that the improper integral over ∂D is still convergent.

Our hypothesis concerning the analyticity of the surface ∂D, the density ρ and the moment density μ is needed in establishing the basic formulas (9.37) to (9.47) for the Cauchy data of the single and double layer potentials V and W only because of the use we made of the Cauchy-Kowalewski theorem to define the auxiliary function u. A slight modification of our analysis (cf. Exercise 9 below) shows that the results remain valid if ∂D is smooth enough and ρ and μ are merely assumed to be twice continuously differentiable. All we have to do to complete the proof of this assertion is retain in the derivation terms involving the *Newtonian potential*[15]

$$\int \frac{\Delta u}{r} d\xi \, d\eta \, d\zeta,$$

whose gradient is found to be continuous across the boundary of the domain of integration.

We now attempt (cf. Exercise 15 below for motivation of the next step) to solve the *Dirichlet problem* for Laplace's equation in a three-dimensional region D by seeking the answer in the form

$$(9.48) \qquad u = W = \frac{1}{2\pi} \int_{\partial D} \mu \frac{\partial}{\partial \nu} \frac{1}{r} d\sigma$$

of a *double layer potential* of charge distributed over the boundary surface ∂D. Letting f denote the boundary values which the solution u is supposed to assume, we deduce from the formula (9.37) representing on ∂D the limit of W from the interior of D that μ ought to satisfy the integral equation

$$(9.49) \qquad \mu + \frac{1}{2\pi} \int_{\partial D} \mu \frac{\partial}{\partial \nu} \frac{1}{r} d\sigma = f.$$

[15] Cf. Kellogg 1.

Thus if a sufficiently smooth solution μ of this integral equation can be found, the corresponding double layer potential (9.48) will define the harmonic function required in Dirichlet's problem.

An analogous reduction of the *Neumann problem* for Laplace's equation in the region D is obtained by asking that the solution be expressed as a *single layer potential*

$$(9.50) \qquad\qquad u = V = \frac{1}{2\pi} \int_{\partial D} \rho \, \frac{d\sigma}{r} .$$

Because of the relation (9.46) for the inner normal derivative of V on ∂D, the density ρ must fulfill the integral equation

$$(9.51) \qquad\qquad \rho - \frac{1}{2\pi} \int_{\partial D} \rho \, \frac{\partial}{\partial \nu} \frac{1}{r} \, d\sigma = -g,$$

where g stands for the prescribed values of $\partial u / \partial \nu$. We emphasize that, in contrast with (9.49), the normal derivative of $1/r$ is calculated here with respect to the arguments of g and not with respect to the dummy variables of integration. Observe that any smooth solution ρ of the integral equation (9.51) generates a single layer potential (9.50) satisfying all the requirements of the Neumann problem, although no comparable means of finding ρ directly in terms of u seems available.

Both (9.49) and (9.51) have the general form

$$(9.52) \qquad\qquad \phi(s) - \lambda \int K(s, t)\phi(t) \, dt = f(s)$$

of a *Fredholm integral equation*, or *integral equation of the second kind*, with a known *kernel*

$$K = \frac{1}{2\pi} \frac{\partial}{\partial \nu} \frac{1}{r}$$

and with $\lambda = \pm 1$. It is understood that the integral on the left in (9.52) is evaluated over a fixed domain on which dt is the appropriate element of area or volume. In Chapter 10 we shall develop the theory of such integral equations, which will then enable us to treat the Dirichlet and Neumann problems for Laplace's equation by means of (9.49) and (9.51). The theory in question turns out to involve the non-trivial solutions of the *transposed equation*

$$\psi(s) - \lambda \int K(t, s)\psi(t) \, dt = g(s)$$

in the homogeneous case $g = 0$. Therefore we digress briefly to comment on the significance of the transposed integral equations associated with (9.49) and (9.51).

An interchange of arguments in the kernel of equation (9.49) simply has the effect of making the differential operator $\partial/\partial \nu$ apply to the principal variables rather than to the dummy variables of integration. Hence the transpose

$$(9.53) \qquad \rho + \frac{1}{2\pi} \int_{\partial D} \rho \, \frac{\partial}{\partial \nu} \frac{1}{r} \, d\sigma = \tilde{g}$$

of equation (9.49) asserts that the single layer potential V, considered as a harmonic function in the exterior of D, has a normal derivative equal to $\tilde{g}$ on the surface ∂D, in view of (9.47). Thus the transpose of the Fredholm integral equation for Dirichlet's problem *inside* ∂D is the Fredholm integral equation found by using a single layer potential to solve the Neumann problem *outside* ∂D. Similarly, the transpose

$$(9.54) \qquad \mu - \frac{1}{2\pi} \int_{\partial D} \mu \, \frac{\partial}{\partial \nu} \frac{1}{r} \, d\sigma = -\tilde{f}$$

of the equation (9.51) for the Neumann problem in the *interior* of D is, according to (9.38), the Fredholm integral equation to which Dirichlet's problem reduces in the *exterior* of D when the solution is sought as a double layer potential W. Finally, we mention that the exceptionally rapid rate at which W approaches zero at infinity will be shown in Section 10.2 to have a bearing on the compatibility requirement

$$\int_{\partial D} g \, d\sigma = \int_{\partial D} \frac{\partial u}{\partial \nu} \, d\sigma = 0$$

on the data for the Neumann problem (cf. Section 7.1).

The method of single and double layer potentials can be extended to treat the Dirichlet and Neumann problems for Laplace's equation in space of any dimension. Relegating the details of this generalization to the exercises, we turn our attention next to the question of transforming Dirichlet's problem for other linear elliptic partial differential equations into a Fredholm integral equation. We shall discuss the particular example of the equation

$$(9.55) \qquad u_{xx} + u_{yy} + au_x + bu_y + cu = 0$$

in the plane, with variable coefficients a, b and c. Our reduction will be based on substituting the Green's function G of Laplace's equation, which we shall suppose to be known, as a *parametrix* (cf. Section 5.3) for

(9.55). It will be seen that the integral equation obtained using such a parametrix involves integration of the solution u of (9.55) over the domain D where it is to be determined. Only when a fundamental solution is available in the large can we proceed to develop a more direct approach through boundary integrals like V and W.

Let us introduce the differential operator

$$L[u] = u_{xx} + u_{yy} + au_x + bu_y + cu$$

and its adjoint

$$M[v] = v_{xx} + v_{yy} - av_x - bv_y + (c - a_x - b_y)v.$$

We recall from Section 5.1 the important integral identity

(9.56)
$$\int_D (vL[u] - uM[v]) \, dx \, dy = \int_{\partial D} H[u, v],$$

where

$$H[u, v] = (vu_x - uv_x + auv) \, dy - (vu_y - uv_y + buv) \, dx$$

$$= -\left(v \frac{\partial u}{\partial \nu} - u \frac{\partial v}{\partial \nu} + auv \frac{\partial x}{\partial \nu} + buv \frac{\partial y}{\partial \nu} \right) ds.$$

We wish to replace v in (9.56) by the harmonic Green's function

$$G(x, y; \xi, \eta) = \frac{1}{2\pi} \log \frac{1}{\sqrt{(x - \xi)^2 + (y - \eta)^2}} + \cdots$$

of D. In a fashion by now quite familiar to us, we first delete from D a small circle of radius ϵ about the singular point (ξ, η) of G, we set $v = G$ in (9.56), and we then allow ϵ to tend toward zero to derive

(9.57)
$$u(\xi, \eta) - \int_D uM[G] \, dx \, dy = \int_{\partial D} H[u, G],$$

where u is supposed to satisfy (9.55). This is the desired Fredholm integral equation determining the solution u of Dirichlet's problem for equation (9.55) in the region D. Since G is harmonic in D and vanishes on ∂D, we have

$$M[G] = -aG_x - bG_y + (c - a_x - b_y)G,$$

$$H[u, G] = u \frac{\partial G}{\partial \nu} \, ds.$$

It follows that (9.57) can be put in the more explicit form

$$(9.58) \qquad u - \int_D [(c - a_x - b_y)G - aG_x - bG_y]u \, dx \, dy = U,$$

where U stands for the known harmonic function in D which assumes the boundary values prescribed for u.

The kernel in (9.58) becomes infinite no more rapidly than the reciprocal of the distance between the two points (x, y) and (ξ, η); hence the improper integral on the left there is convergent. Since G vanishes identically when (ξ, η) lies on ∂D, a solution u of (9.58) has to have the prescribed boundary values that we have imposed on the term U. Furthermore, an integration by parts transforms (9.58) into

$$(9.59) \qquad u - \int_D (au_x + bu_y + cu)G \, dx \, dy = U.$$

From the Poisson equation

$$\Delta \iint\limits_{D} \rho \log \frac{1}{r} \, dx \, dy = -2\pi\rho$$

for a logarithmic potential (cf. Exercise 5.3.2) it is easy to conclude that every solution of (9.59) satisfies (9.55). Thus the Fredholm integral equation (9.58) is completely equivalent to Dirichlet's problem for the elliptic partial differential equation (9.55).

It is of interest to compare (9.59) with the integrodifferential equation (4.11) used in Chapter 4 to solve Cauchy's problem for the hyperbolic equation (4.1). The main reason why (4.11) was easier to handle than (9.59) will be is that it has variable limits of integration which make the method of successive approximations always convergent. Such *Volterra integral equations* (cf. Section 10.1) arise naturally in initial value problems, for which an answer is sought in the small. In contrast, boundary value problems usually lead to a more difficult analysis based on Fredholm integral equations that must be solved in the large over some preassigned domain. Finally, note the similarity between (9.58) and the integral equation (5.89) applied in Section 5.3 to construct fundamental solutions.

An important special case of (9.55) is the *reduced wave equation*

$$(9.60) \qquad \Delta u + k^2 u = 0,$$

where k is a real constant which may be interpreted as the frequency of vibration of a membrane (cf. Section 11.1). A reasonable problem from the physical point of view is to ask for particular choices of the parameter

k^2, called *eigenvalues*, to which there correspond non-trivial solutions u of (9.60) in D, called *eigenfunctions*, obeying the boundary condition

$$(9.61) \qquad\qquad u = 0$$

on ∂D. According to (9.58), this homogeneous boundary value problem is equivalent to the integral equation

$$(9.62) \qquad\qquad u - k^2 \int_D Gu \; dx \; dy = 0.$$

We shall establish in Section 10.3 that non-trivial solutions u of the homogeneous equation (9.62) do exist for a sequence of values of k. Since the maximum principle does not apply to (9.60), such a contention is not inconsistent with the standard uniqueness theorems for Dirichlet's problem (cf. Exercise 7.1.1).

A correct Dirichlet problem for the elliptic partial differential equation (9.60), quite different from the eigenvalue problem (9.62), can be formulated in an infinite plane region D bounded by a simple closed curve ∂D. Indeed, Exercises 7.1.13 and 7.1.14 serve to establish the uniqueness of a solution u of (9.60) in D which assumes prescribed values on the boundary ∂D and fulfills at infinity the *Sommerfeld radiation condition*[16]

$$(9.63) \qquad \lim_{r \to \infty} \sqrt{r}\left(\frac{\partial u}{\partial r} - iku\right) = 0, \qquad \overline{\lim_{r \to \infty}} \sqrt{r} \, |u| < \infty.$$

We conclude from Exercise 7.2.21 that this radiation problem can be solved by the method of separation of variables when the curve ∂D is a circle of arbitrary radius R and that it possesses a Green's function G_k in the exterior of such a circle. We intend to reduce the Dirichlet problem for (9.60) in a more general infinite domain D to a Fredholm integral equation by mapping D conformally onto the region outside an appropriate circle and then using the Green's function G_k associated with that circle as a parametrix for the transformed problem.[17]

Let

$$(9.64) \qquad\qquad w = z + \sum_{m=0}^{\infty} \frac{a_m}{z^m} = z + q(z)$$

be a normalized conformal transformation of the region $|z| > R$ onto D, where R is a constant known as the *outer mapping radius* of the curve ∂D. Since the coefficient of the leading term in the Laurent series (9.64) is

[16] Cf. Sommerfeld 1.
[17] Cf. Garabedian 5.

unity, the radiation condition (9.63) remains invariant under this substitution of variables. On the other hand, the partial differential equation (9.60) acquires the new form

$$(9.65) \qquad \Delta u + k^2 |1 + q'(z)|^2 u = 0$$

in the z-plane.

Identifying the differential operator L with the expression on the left in (9.65), and putting $v = G_k$ in the corresponding integral formula (9.56) for the circular ring $R < |z| < R^*$, we obtain

$$u - \int_{R < |z| < R^*} (\Delta G_k + k^2 |1 + q'(z)|^2 G_k) u \, dx \, dy$$

$$= \int_{|z|=R} u \frac{\partial G_k}{\partial \nu} \, ds + \int_{|z|=R^*} \left(u \frac{\partial G_k}{\partial \nu} - G_k \frac{\partial u}{\partial \nu} \right) ds,$$

where the singularity of G_k has contributed the residue u in the usual way and where R^* is a large number. The line integral on the right over the circle $|z| = R^*$ approaches zero as $R^* \to \infty$ because both u and G_k satisfy the radiation condition (9.63). Since

$$\Delta G_k + k^2 G_k = 0,$$

it follows that

$$(9.66) \qquad u - k^2 \int_{|z|>R} p G_k u \, dx \, dy = \int_{|z|=R} f \frac{\partial G_k}{\partial \nu} \, ds,$$

where f stands for the boundary values assigned to u and where

$$p = 2 \operatorname{Re} \{q'(z)\} + |q'(z)|^2.$$

Retracing our steps, we ascertain that the Fredholm integral equation (9.66) is equivalent to the exterior Dirichlet problem for solutions of (9.60) satisfying the Sommerfeld radiation condition. That this question is well posed will become clear after the theory of Section 10.1 has been developed.

In closing this chapter it seems appropriate to compare briefly the various methods of solving Dirichlet's problem that we have described. All of them have in common a constructive element involving the fundamental solution, although this may be hidden in the context of a parametrix or of the Poisson integral formula. The methods of Chapter 8 are based on problems in the calculus of variations for Dirichlet's integral, and they are of limited scope in so far as the boundary condition is concerned. Those of Sections 9.1 and 9.2 depend instead on the maximum principle and are more effective at the boundary. Both techniques were used for uniqueness proofs in Section 7.1. We note in passing that the Perron process presented in Section 9.2 is related not only to the classical Poincaré sweeping out process, as was indicated earlier, but also to the *Schwarz alternating procedure,*[18] which was historically the first method to deliver a satisfactory existence proof.

In the present section we have started to discuss the extraordinarily powerful theory of integral equations, which will be completed in Chapter 10. A significant generalization of the latter approach which has important applications to nonlinear boundary value problems, but lies beyond the scope of this book, is the *continuity method.*[19] It has a number of elegant formulations in terms of the fixed point theorems of topology associated with the names of Schauder (cf. Exercise 10.1.13) and of Schauder and Leray. In the solution of Dirichlet's problem by the continuity method a central role is played by certain *a priori estimates* of the answer. Moreover, such estimates lead to another kind of existence proof[20] that is based on finite difference approximations (cf. Section 13.6). Finally, conformal mapping provides a number of interesting existence theorems for Dirichlet's problem in the case of the Laplace equation in two independent variables (cf. also Section 15.4). We have already pointed out, in fact, that this version of the problem can be handled by a variety of fundamental lemmas from the theory of functions of a real variable (cf. Exercise 1.13).

EXERCISES

1. Use double and single layer potentials to develop Fredholm integral equations equivalent to the Dirichlet and Neumann problems for Laplace's equation in space of any dimension. In particular, reduce Dirichlet's problem for the Laplace equation in a plane domain D bounded by a smooth curve ∂D to the Fredholm integral equation

$$(9.67) \qquad \mu + \frac{1}{\pi} \int_{\partial D} \mu \frac{\partial}{\partial \nu} \log \frac{1}{r} \, ds = f.$$

2. Examine in detail the Fredholm integral equations of Exercise 1 for a circle.

3. In the case of Laplace's equation, set up a Fredholm integral equation for the third boundary value problem (cf. Section 7.1). What role do eigenvalues and eigenfunctions play in this question?

4. Formulate a generalization of the integral equation (9.58) in space of arbitrary dimension. Obtain an analogous Fredholm equation for the Neumann problem.

[18] Cf. Courant-Hilbert 3.
[19] Cf. Courant-Hilbert 3.
[20] Cf. Courant-Friedrichs-Lewy 1.

5. Derive a Fredholm integral equation like (9.66) which is equivalent to the exterior Neumann problem for solutions of (9.60) satisfying the Sommerfeld radiation condition.

6. Formulate the Dirichlet and Neumann radiation problems for (9.60) in an infinite domain D as integral equations over the boundary curve ∂D by introducing appropriate double and single layer potentials of a fundamental solution.[21]

7. Show that the integrodifferential equation (9.59) can be solved by successive approximations when the domain D is small enough.

8. Show that the formula

$$ u = - \int_D fG \, dx \, dy $$

solves the inhomogeneous Laplace equation $\Delta u = f$ subject to the homogeneous boundary condition $u = 0$. In the case of a given function F whose first derivative is small enough, solve the same boundary value problem for the nonlinear elliptic equation

$$ \Delta u = F(u) $$

by applying the method of successive approximations to the integral equation

$$ u = - \int_D F(u)G \, dx \, dy. $$

9. Show that the first few terms of the power series expansion in the Cauchy-Kowalewski theorem are all that is needed to prove the basic formulas (9.37) to (9.47). In the case of the single layer potential V, verify these results under a weaker hypothesis which merely asks that the density ρ be Hölder continuous.[22]

10. By considering an iterated integral equation of the form

$$ \mu - \frac{1}{4\pi^2} \int_{\partial D} \mu \int_{\partial D} \frac{\partial}{\partial \nu_1} \frac{1}{r_1} \frac{\partial}{\partial \nu_2} \frac{1}{r_2} \, d\sigma_2 \, d\sigma_1 = f - \frac{1}{2\pi} \int_{\partial D} f \frac{\partial}{\partial \nu} \frac{1}{r} \, d\sigma, $$

establish that a solution μ of (9.49) has continuous derivatives of order m whenever the given function f has continuous derivatives of order $m + 2$. Conclude that the solution (9.48) of Dirichlet's problem is differentiable at the surface ∂D if the prescribed boundary values f are sufficiently smooth.

11. Show that if analytic boundary values are assigned on the analytic surface ∂D, then the corresponding solution of Dirichlet's problem for the Laplace equation is analytic at the boundary ∂D, too.

12. Use complex substitutions like those in Section 5.1, together with the method of contour integration, to show that (4.11) can be derived from (9.59) for an analytic linear partial differential equation in the complex domain (cf. Exercise 16.3.8).

[21] Cf. Müller 1.
[22] Cf. Kellogg 1.

13. Establish that the kernel in (9.67) satisfies

$$\frac{1}{\pi} \frac{\partial}{\partial \nu} \log \frac{1}{r} \geq 0,$$

$$\frac{1}{\pi} \int_{\partial D} \frac{\partial}{\partial \nu} \log \frac{1}{r} \, ds = 1$$

when the curve ∂D is convex. Under the slightly stronger hypothesis that ∂D is strictly convex, obtain positive upper and lower bounds on the kernel. Show that a number $\epsilon > 0$ depending on these bounds can be found such that

(9.68) $$\max_{\partial D} \mu^* - \min_{\partial D} \mu^* \leq (1 - \epsilon) \left\{ \max_{\partial D} \mu - \min_{\partial D} \mu \right\}$$

for any continuous function μ, where μ^* stands for the mean value

$$\mu^* = \frac{1}{\pi} \int_{\partial D} \mu \frac{\partial}{\partial \nu} \log \frac{1}{r} \, ds.$$

14. Solve the Dirichlet problem for Laplace's equation in a strictly convex region of the plane by applying the method of successive approximations to the Fredholm integral equation (9.67). More precisely, use the estimate (9.68) to verify that the *Neumann series*[23]

(9.69) $$\mu = -\gamma + f - f_1 + f_2 - f_3 + \cdots$$

converges to a solution of (9.67), where γ is a suitable constant and

$$f_1 = \frac{1}{\pi} \int_{\partial D} (f - 2\gamma) \frac{\partial}{\partial \nu} \log \frac{1}{r} \, ds,$$

$$f_{m+1} = \frac{1}{\pi} \int_{\partial D} f_m \frac{\partial}{\partial \nu} \log \frac{1}{r} \, ds, \qquad m = 1, 2, \ldots.$$

15. Establish the Green's identity

$$u = \frac{1}{2\pi} \int_{\partial D} \left[u \frac{\partial}{\partial \nu} \frac{1}{r} - \frac{1}{r} \frac{\partial u}{\partial \nu} \right] d\sigma$$

for a harmonic function u at points on the surface ∂D of its domain of definition D. Interpret the result as a Fredholm integral equation for the boundary values of u when the normal derivative $\partial u / \partial \nu$ is prescribed. Compare this equation with (9.54), and rederive it from our previous integral equation (9.51) for the solution of the Neumann problem by multiplying through by $1/r$ and integrating over ∂D.

[23] Cf. Neumann 1.

10

Integral Equations

1. THE FREDHOLM ALTERNATIVE

We indicated in Section 9.3 how a variety of boundary value problems for elliptic partial differential equations can be formulated as *Fredholm integral equations*. In order to take advantage of that analysis, we now proceed to develop briefly the principal aspects of the theory of linear integral equations.[1] This theory plays such a fundamental role in the study of partial differential equations and contributes so heavily to an adequate understanding of their properties that it forms a natural chapter of our main subject.

To be more precise, we shall consider *Fredholm integral equations*

$$(10.1) \qquad \phi(s) - \lambda \int K(s, t)\phi(t)\, dt = f(s)$$

for the determination of an unknown function ϕ. The *kernel K* and the parameter λ are supposed to be given; our aim is to find ϕ for every function f of an appropriate class. The integral appearing in (10.1) is understood to be extended over a *fixed* domain D which also constitutes the range of s, and dt stands for the volume element or any other appropriate measure. The domain D may lie in a space of any dimension, may or may not be bounded, and can, for example, be a curve or surface.

As was pointed out in Section 9.3, boundary value problems for elliptic partial differential equations lead to Fredholm integral equations in which the region of integration has a fixed size. Initial value problems for hyperbolic equations, on the other hand, lead to integral equations in

---

[1] Cf. Mikhlin 1, Petrovsky 2.

which the region of integration coincides with the domain of dependence and therefore does not have a fixed size; an example is the *Volterra integral equation* (4.11) representing the solution of Cauchy's problem for a linear hyperbolic equation (4.1). Volterra integral equations are obviously particular cases of Fredholm integral equations for which the kernel $K(s, t)$ vanishes if t lies outside a region depending on s.

Since we shall have to write out the integral occurring in (10.1) repeatedly and often in an involved context, we prefer to express it in an abbreviated operator notation. Thus we introduce the symbolic product

$$K \circ \phi = \int K(s, t)\phi(t) \, dt$$

to stand for such an integral, so that (10.1) assumes the shorter form

(10.2) $$\phi - \lambda K \circ \phi = f.$$

Using the symbolic notation we can also represent the *iterated kernel* $K^{(n)}$ in the form

$$K^{(n)} = K \circ K \circ \cdots \circ K,$$

where there are supposed to be n factors on the right. Note that the integration should always be performed with respect to the argument adjacent to the symbol $\circ$. For example, the iterated kernel

$$K^{(2)}(s_1, s_2) = K \circ K = \int K(s_1, t)K(t, s_2) \, dt$$

is a function of the two variables s_1 and s_2. Observe that the symbolic product $\circ$ is in general not commutative; in fact, the *transpose*

$$\psi(s) - \lambda \int K(t, s)\psi(t) \, dt = g(s)$$

of the Fredholm integral equation (10.1) has a representation

$$\psi - \lambda \psi \circ K = g$$

in the new notation which involves multiplication of the unknown ψ by the kernel K on the right instead of on the left.

We shall also introduce the *scalar product*

$$\phi \circ \psi = \psi \circ \phi = \int \phi(t)\psi(t) \, dt$$

of two functions ϕ and ψ and write the associated *norm* as

$$\|\phi\| = \sqrt{\phi \circ \phi}.$$

The first method which comes to mind for solving (10.2) is that of successive approximation (cf. Section 4.2). Let us rewrite (10.2) in the equivalent form

(10.3) $$\phi = f + \lambda K \circ \phi,$$

which is solved for ϕ. This expression may be substituted for ϕ on the right to obtain

(10.4) $$\phi = f + \lambda K \circ [f + \lambda K \circ \phi] = f + \lambda K \circ f + \lambda^2 K^{(2)} \circ \phi.$$

A second substitution of (10.3) into (10.4) yields

$$\phi = f + \lambda K \circ f + \lambda^2 K^{(2)} \circ f + \lambda^3 K^{(3)} \circ \phi.$$

Repetition of these steps leads us inductively to suspect that, when it converges, the *Neumann series*[2]

(10.5) $$\phi = f + \lambda K \circ f + \lambda^2 K^{(2)} \circ f + \lambda^3 K^{(3)} \circ f + \cdots$$

should provide a solution of the integral equation (10.2). Indeed, formal insertion of (10.5) into (10.2), combined with a term by term integration, gives the identity

$$f + \lambda K \circ f + \lambda^2 K^{(2)} \circ f + \cdots - \lambda K \circ f - \lambda^2 K^{(2)} \circ f - \cdots = f.$$

A quite general hypothesis under which the above process has meaning is that f be square integrable, that

(10.6) $$B^2 = \iint K(s, t)^2 \, dt \, ds < \infty,$$

and that λ be sufficiently small. In the circumstances described, it follows from (10.6) by Fubini's theorem[3] that

$$\int K(s, t)^2 \, dt$$

---

[2] Cf. Neumann 1.
[3] Cf. Munroe 1.

exists almost everywhere and is a square integrable function of s. Schwarz's inequality shows that

$$(K \circ f)^2 \leq \|f\|^2 \int K(s, t)^2 \, dt,$$

and upon integration with respect to s,

$$\|K \circ f\| \leq \|f\| \, B.$$

Proceeding inductively we find that

$$(K^{(n)} \circ f)^2 = [K \circ (K^{(n-1)} \circ f)]^2 \leq \|f\|^2 B^{2(n-1)} \int K(s, t)^2 \, dt,$$

and hence

$$\|K^{(n)} \circ f\| \leq \|f\| \, B^n.$$

Thus (10.5) is majorized by the geometric series

$$\|f\| \, (1 + |\lambda| \, B + |\lambda|^2 \, B^2 + \cdots),$$

which proves the

THEOREM. *If f is square integrable and*

$$(10.7) \qquad\qquad |\lambda| \, B < 1,$$

the Neumann series (10.5) *converges in the mean to a square integrable function ϕ which satisfies* (10.2) *almost everywhere.*

Under these conditions it is evident that ϕ is unique. Introducing the *reciprocal kernel* or *resolvent*

$$H = K + \lambda K^{(2)} + \lambda^2 K^{(3)} + \cdots,$$

we may thus represent the solution ϕ of (10.2) in terms of the convergent Neumann series as

$$(10.8) \qquad\qquad \phi = f + \lambda H \circ f.$$

Straightforward manipulation leads to a simple necessary condition on f in order that a solution ϕ of the Fredholm integral equation (10.2) should exist. Let ψ stand for any non-trivial solution, or *eigenfunction*, of the homogeneous transposed equation

$$(10.9) \qquad\qquad \psi - \lambda \psi \circ K = 0.$$

Multiplying (10.2) by ψ and integrating, we obtain

$$\psi \circ f = \psi \circ [\phi - \lambda K \circ \phi]$$
$$= [\psi - \lambda \psi \circ K] \circ \phi.$$

Because of (10.9) we conclude that

(10.10) $$\psi \circ f = 0$$

for each eigenfunction ψ. In other words, a necessary condition for the original equation (10.2) to be solvable is that the given function f should be orthogonal to every solution ψ of the homogeneous transposed equation (10.9).

In what follows we shall deal with functions ϕ, ψ, and f that possess integrable squares, and we shall assume that the kernel K is square integrable in the sense (10.6). We shall proceed to prove that the requirement (10.10) on f, which is to hold for every eigenfunction ψ of (10.9), is not only necessary but also sufficient for the existence of a solution of the integral equation (10.2). To be more precise, consider the inhomogeneous integral equation (10.11), the associated transposed equation (10.12), and the two homogeneous equations (10.13) and (10.14) below:

(10.11) $\phi - \lambda K \circ \phi = f,$ (10.12) $\psi - \lambda \psi \circ K = g,$

(10.13) $\phi - \lambda K \circ \phi = 0,$ (10.14) $\psi - \lambda \psi \circ K = 0.$

We shall prove the

FREDHOLM ALTERNATIVE. *Either* (10.13) *and* (10.14) *have only the trivial solutions* $\phi = \psi = 0$ *or they have the same number of linearly independent eigenfunctions (non-trivial solutions)* $\phi_1, \phi_2, \ldots, \phi_m$ *and* $\psi_1, \psi_2, \ldots, \psi_m$.

In the first case (10.11) *and* (10.12) *possess unique solutions* ϕ *and* ψ *for every square integrable f and g.*

In the second case (10.11) *has a solution if and only if f is orthogonal to the m eigenfunctions* ψ_i *of the transposed equation,*

(10.15) $\psi_i \circ f = 0,$ $i = 1, 2, \ldots, m;$

and similarly (10.12) *has a solution if and only if*

(10.16) $\phi_i \circ g = 0,$ $i = 1, 2, \ldots, m.$

In particular, (10.11) can be solved for every assigned f whenever the homogeneous equation (10.13) has the unique solution $\phi = 0$. However, in the alternate case when uniqueness fails, (10.11) is solvable if and only if the orthogonality conditions (10.15) are satisfied.

Our demonstration of the theorem just stated will be developed first for a *degenerate kernel*

$$(10.17) \qquad K_n(s, t) = \sum_{j=1}^{n} \alpha_j(s)\beta_j(t)$$

which consists of a finite sum of products of functions α_j and β_j of a single argument. The more general problem will be treated afterward by approximating the relevant kernel by a degenerate kernel. We shall then introduce a Neumann series based on the difference between the given kernel and its approximation to convert the original equation into a new one displaying the degenerate form. Thus we shall be able to deduce the Fredholm alternative from the analogous alternative for a system of simultaneous linear algebraic equations.

Without loss of generality put $\lambda = 1$, and consider the Fredholm equation

$$(10.18) \qquad \phi - K_n \circ \phi = f$$

with a degenerate kernel K_n. A rearrangement of the representation (10.17) of K_n shows we may assume that $\beta_1, \ldots, \beta_n$ comprise a linearly independent, and even orthonormal, set of functions. The homogeneous transposed equation can be written in the form

$$(10.19) \qquad \psi - \sum_{i=1}^{n} (\psi \circ \alpha_i)\beta_i = 0,$$

which implies that all its solutions are linear combinations of $\beta_1, \ldots, \beta_n$. Therefore we decompose f into the sum $f = f_1 + f_2$ of a component f_1 that is a linear combination of the functions $\beta_1, \ldots, \beta_n$ and a component f_2 that is orthogonal to the β_i in the sense

$$\beta_i \circ f_2 = 0, \qquad i = 1, 2, \ldots, n.$$

Since

$$K_n \circ f_2 = 0,$$

the substitution $\phi = \phi_1 + f_2$ reduces the integral equation (10.18) to the relation

$$(10.20) \qquad \phi_1 - \sum_{j=1}^{n} \alpha_j(\beta_j \circ \phi_1) = f_1.$$

Thus it suffices to discuss the special case $f_2 = 0$, which means that $f = f_1$, $\phi = \phi_1$. This enjoys the property that the n numbers

$$(10.21) \qquad y_i = \beta_i \circ f$$

completely characterize the given function f.

Multiplying (10.20) by β_i and integrating, we find that

(10.22)
$$x_i - \sum_{j=1}^{n} a_{ij} x_j = y_i, \qquad i = 1, \ldots, n,$$

where

(10.23)
$$x_i = \beta_i \circ \phi, \qquad a_{ij} = \beta_i \circ \alpha_j.$$

It follows that assigning f in (10.18) is equivalent to prescribing $y_1, \ldots, y_n$ and that whenever a corresponding solution $x_1, \ldots, x_n$ of the system of simultaneous linear equations (10.22) exists we can determine a function

$$\phi = f + \sum_{j=1}^{n} x_j \alpha_j$$

satisfying not only (10.23) but also (10.18). Thus the integral equation (10.18) can be solved if and only if the associated linear algebraic equations (10.22) have a solution.

Observe that multiplication of equation (10.19) by α_j, followed by appropriate integration, yields the homogeneous set of linear equations

(10.24)
$$z_j - \sum_{i=1}^{n} z_i a_{ij} = 0, \qquad j = 1, \ldots, n,$$

where

(10.25)
$$z_j = \psi \circ \alpha_j.$$

Clearly any solution ψ of (10.19) which does not vanish identically cannot be orthogonal to all the functions α_j, and it must therefore generate a non-trivial solution $z_1, \ldots, z_n$ of (10.24). Conversely, any non-trivial solution of (10.24) defines a function

(10.26)
$$\psi = \sum_{i=1}^{n} z_i \beta_i$$

which is different from zero somewhere and satisfies both (10.25) and (10.19). Furthermore, given any eigenfunction (10.26) of (10.19) we can multiply (10.22) by the coefficient z_i and sum over i to obtain

(10.27)
$$\sum_{i=1}^{n} z_i y_i = \sum_{i=1}^{n} z_i x_i - \sum_{i,j=1}^{n} z_i a_{ij} x_j$$
$$= \sum_{j=1}^{n} \left[z_j - \sum_{i=1}^{n} z_i a_{ij} \right] x_j = 0,$$

because of (10.24).   Finally, according to (10.21) we have

$$(10.28) \qquad \sum_{i=1}^{n} z_i y_i = \left[\sum_{i=1}^{n} z_i \beta_i \right] \circ f = \psi \circ f,$$

so that (10.27) is merely a restatement of the orthogonality requirement
(10.10) occurring in the Fredholm alternative.

Consider now the question of solving the integral equation (10.18),
which we have seen to be equivalent to the system of simultaneous linear
equations (10.22).   The usual criterion that a solution can be found for
any f, or, in other words, for arbitrary $y_1, \ldots, y_n$, is that the determinant
of the matrix of coefficients appearing on the left in (10.22) be different
from zero.   Such a situation is also characterized by the uniqueness of the
solution.   In particular, we can conclude that if $\phi = 0$ is the only function
satisfying (10.18) for $f = 0$, then the above determinant differs from zero,
and (10.18) is solvable for every choice of f.   Furthermore, the determinant
of the transposed system of homogeneous equations (10.24) has the same
value as the determinant of (10.22), so that in the case under discussion
non-trivial solutions of (10.24) and of the associated integral equation
(10.19) do not exist, and (10.10) entails no restriction on f.   Similarly the
inhomogeneous transposed equation is uniquely solvable for every choice
of g.   Thus the first case in the Fredholm alternative occurs when the
determinant in question is non-zero.

When the determinant of (10.22) vanishes, on the other hand, so that
the corresponding matrix of coefficients has a rank $r < n$, the vectors
$(y_1, \ldots, y_n)$ for which (10.22) possesses solutions only span a subspace of
n-dimensional Euclidean space of the smaller dimension r.   Since the
matrix of coefficients of the transposed system (10.24) of homogeneous
linear equations has the same rank r, the eigenvectors $(z_1, \ldots, z_n)$ of
(10.24) generate another subspace of dimension exactly $m = n - r$.
According to (10.27), the r-dimensional and m-dimensional subspaces
just described are orthogonal, whence they must together span the entire
Euclidean space of dimension n.   It follows that the vectors $(y_1, \ldots, y_n)$
for which (10.22) is solvable are precisely those which fulfill the orthogonality condition (10.27) for every solution $(z_1, \ldots, z_n)$ of (10.24).

The established equivalence of (10.18) to (10.22) and of (10.19) to (10.24)
implies, in conjunction with the relation (10.28), that a solution of the
Fredholm integral equation (10.18) exists if and only if the right-hand side
f is orthogonal to every solution ψ of the homogeneous transposed
equation (10.19).   Since the eigenvectors $(z_1, \ldots, z_n)$ span a subspace of
dimension m, this really amounts to the m conditions $\psi_i \circ f = 0$ for
$i = 1, 2, \ldots, m$, where the ψ_i are m linearly independent eigenfunctions
of the homogeneous transposed equation (10.19).

We have now completed a proof of the validity of the Fredholm alternative in the case of a degenerate kernel (10.17). It remains to show how a Neumann series representation (10.8) can be used to reduce the case of an integral equation with a square integrable kernel, in which we may again put $\lambda = 1$ with no loss of generality, to a degenerate problem like (10.18).[4]

Given an arbitrary kernel K with the integrability property (10.6), it is always possible to determine a degenerate kernel K_n such that the difference

$$\tilde{K} = K - K_n$$

satisfies the inequality

(10.29)
$$\iint \tilde{K}(s, t)^2 \, dt \, ds < 1.$$

For example, we could choose the approximating kernel K_n to be a suitable polynomial or to be the partial sum of a Fourier expansion for K. Then according to (10.8) the solution $\tilde{\phi}$ of the Fredholm integral equation

(10.30)
$$\tilde{\phi} - \tilde{K} \circ \tilde{\phi} = \tilde{f}$$

has the representation

(10.31)
$$\tilde{\phi} = \tilde{f} + \tilde{H} \circ \tilde{f}$$

in terms of the resolvent kernel

$$\tilde{H} = \tilde{K} + \tilde{K}^{(2)} + \tilde{K}^{(3)} + \cdots$$

for any square integrable function $\tilde{f}$. We shall use the representation (10.31) to convert the unrestricted Fredholm equation

(10.32)
$$\phi - K \circ \phi = f$$

into a degenerate integral equation.

In terms of $\tilde{K}$ we can rewrite (10.32) in the form

(10.33)
$$\phi - \tilde{K} \circ \phi = f + K_n \circ \phi.$$

On the other hand, since the two relations (10.30) and (10.31) are equivalent, the substitution $\phi = \tilde{\phi}, f + K_n \circ \phi = \tilde{f}$ shows that (10.33) can be reduced to

$$\phi = f + K_n \circ \phi + \tilde{H} \circ (f + K_n \circ \phi),$$

or

(10.34)
$$\phi - (K_n + \tilde{H} \circ K_n) \circ \phi = f + \tilde{H} \circ f.$$

[4] Cf. Petrovsky 2.

The kernel K_n is degenerate in the sense (10.17), and therefore so also is the symbolic product

$$\tilde{H} \circ K_n = \sum_{j=1}^{n} (\tilde{H} \circ \alpha_j)\beta_j.$$

We conclude that (10.34) is a degenerate Fredholm integral equation whose solutions ϕ are identical with those of (10.32). Since we have already established the Fredholm alternative for degenerate kernels, it becomes clear that (10.34), together with (10.32), is solvable if and only if

$$(10.35) \qquad \tilde{\psi} \circ (f + \tilde{H} \circ f) = 0$$

for every eigenfunction $\tilde{\psi}$ of the homogeneous transposed equation

$$(10.36) \qquad \tilde{\psi} - \tilde{\psi} \circ (K_n + \tilde{H} \circ K_n) = 0.$$

Furthermore, no such eigenfunctions exist if (10.32) and (10.34) have only the trivial solution $\phi = 0$ when $f = 0$.

We have yet to prove that (10.35) can be translated into the orthogonality requirement (10.10), which ought to hold for all non-trivial solutions ψ of

$$(10.37) \qquad \psi - \psi \circ K = 0.$$

This is achieved by setting

$$(10.38) \qquad \psi = \tilde{\psi} + \tilde{\psi} \circ \tilde{H}$$

and using the definition of the resolvent $\tilde{H}$ to derive the inverse expression

$$(10.39) \qquad \psi - \psi \circ \tilde{K} = \tilde{\psi}$$

for $\tilde{\psi}$ in terms of ψ. The pair of identities (10.38) and (10.39) imply that ψ vanishes if and only if $\tilde{\psi}$ does, and substitution of (10.39) into (10.36) gives

$$(10.40) \quad \psi - \psi \circ (\tilde{K} + K_n) + \psi \circ [\tilde{K} - (\tilde{H} - \tilde{K} \circ \tilde{H})] \circ K_n = 0.$$

From the definition of $\tilde{H}$ we see that

$$\tilde{K} = \tilde{H} - \tilde{K} \circ \tilde{H},$$

so that the last term on the left in (10.40) vanishes and (10.40) becomes synonymous with (10.37). Thus the eigenfunctions $\tilde{\psi}$ of (10.36) transform into the eigenfunctions ψ of (10.37). It follows that (10.35) has the same significance as (10.10), which suffices to finish our demonstration of the Fredholm alternative for the most general integral equation (10.32) with a kernel satisfying (10.6).

THE FREDHOLM ALTERNATIVE 359

Many of the kernels that arise in applications of the Fredholm theory of integral equations to boundary value problems are not square integrable. We shall now indicate how the Fredholm alternative may be extended to such kernels.

The crucial step in the proof of the Fredholm alternative for the general integral equation (10.32) was the approximation of the kernel K by a degenerate kernel K_n in such a way that the difference $\tilde{K}$ satisfied (10.29), which ensured the convergence of the Neumann series representation (10.31). However, a glance at our analysis of the Neumann series shows that the square integrability of the kernel was really unessential. Instead of (10.7) we could have required

$$(10.41) \qquad |\lambda|\,\|K\| < 1,$$

where $\|K\|$ is the *norm* of the kernel, defined by

$$(10.42) \qquad \|K\| = \sup_{\phi} \frac{\|K \circ \phi\|}{\|\phi\|} = \sup_{\|\phi\|=1} \|K \circ \phi\|.$$

It follows from the proof we have already given that the reduction of an integral equation with an arbitrary kernel K to a degenerate one like (10.18) may be carried out provided K can be approximated arbitrarily well in norm by degenerate kernels. Thus the essential requirement is that, for any preassigned $\epsilon > 0$, K can be written in the form

$$(10.43) \qquad K = K_n + \tilde{K}, \qquad \|\tilde{K}\| < \epsilon,$$

where K_n is degenerate. Such kernels K are called *completely continuous* (see, however, Exercise 7 below).

We conclude that the Fredholm alternative is valid for any equation

$$\phi - \lambda K \circ \phi = f$$

whose kernel K is completely continuous in the above sense. It remains to verify that the kernels which arise in applications to boundary value problems are completely continuous. This will be accomplished in an ingenious fashion that enables us to stay in the class of square integrable functions.[5] A more straightforward approach based on the maximum norm is developed in Exercise 8 below.

The kernels in question (cf. Section 9.3) are those with a *weak singularity* of the form

$$(10.44) \qquad K(s, t) = \frac{\kappa(s, t)}{r^l}, \qquad 0 \leq l < n,$$

[5] Cf. Mikhlin 1.

where $\kappa(s, t)$ is a bounded function, r is the distance between s and t, and n is the dimension of D. In contrast to the case of square integrable kernels, it now becomes necessary to assume that the domain D is *bounded* (cf. Exercise 3.16). We shall proceed to prove that the kernels (10.44) have the complete continuity property (10.43).

By setting

$$K'(s, t) = \begin{cases} K(s, t), & r \geq \delta \\ 0, & r < \delta \end{cases}$$

and

$$K''(s, t) = \begin{cases} 0, & r \geq \delta \\ K(s, t), & r < \delta \end{cases}$$

we truncate the singularity r^{-l} and obtain the decomposition

$$K(s, t) = K'(s, t) + K''(s, t).$$

Since D is bounded, the truncated kernel $K'(s, t)$ is square integrable and may be approximated by a degenerate kernel $K_n(s, t)$ as before. Schwarz's inequality shows, moreover, that

$$\|K' - K_n\|^2 \leq \iint [K' - K_n]^2 \, dt \, ds.$$

Hence the first term $K'(s, t)$ in the decomposition of $K(s, t)$ can be approximated arbitrarily well in norm by a degenerate kernel. It remains to show that the norm of the kernel $K''(s, t)$ can be made arbitrarily small.

First observe that

$$|K'' \circ \phi|^2 \leq M^2 \left[\int_{r < \delta} \frac{|\phi|}{r^l} \, dt \right]^2,$$

where M is a bound on $\kappa(s, t)$. Schwarz's inequality then gives

$$\left[\int_{r < \delta} \frac{|\phi|}{r^l} \, dt \right]^2 = \left[\int_{r < \delta} \frac{|\phi|}{r^{l/2}} \cdot \frac{dt}{r^{l/2}} \right]^2$$

$$\leq \left[\int_{r < \delta} \frac{|\phi|^2}{r^l} \, dt \right] \left[\int_{r < \delta} \frac{dt}{r^l} \right],$$

from which it follows that

$$\|K'' \circ \phi\|^2 = \int_D |K'' \circ \phi|^2 \, ds \leq M^2 \left[\int_D \int_D \frac{|\phi|^2}{r^l} \, dt \, ds \right] \left[\int_{r < \delta} \frac{dt}{r^l} \right].$$

But by Fubini's theorem we have

$$\left[\int_D \int_D \frac{|\phi|^2}{r^l}\, dt\, ds\right] = \int_D |\phi|^2 \int_D \frac{ds}{r^l}\, dt$$

$$\leq \left[\int_D |\phi|^2\, dt\right]\left[\int_{r<d} \frac{ds}{r^l}\right],$$

where d is the diameter of D. Hence we obtain

$$(10.45) \qquad \|K'' \circ \phi\|^2 \leq M^2 \|\phi\|^2 \left[\int_{r<d} \frac{ds}{r^l}\right]\left[\int_{r<\delta} \frac{dt}{r^l}\right].$$

According to the definition (10.42) of the norm of a kernel, (10.45) is equivalent to

$$(10.46) \qquad \|K''\|^2 \leq M^2 \left[\int_{r<d} \frac{ds}{r^l}\right]\left[\int_{r<\delta} \frac{dt}{r^l}\right].$$

Because $l < n$, the first integral on the right in (10.46) is bounded and the second may be made arbitrarily small for sufficiently small δ. Hence $\|K''\|$ can be made as small as desired and $K(s, t)$ may thus be approximated arbitrarily closely in norm by degenerate kernels. This is enough to complete our proof of the validity of the Fredholm alternative in the case of kernels that have a weak singularity.

EXERCISES

1. Solve the Fredholm integral equation

$$\phi(s) - \lambda \int_{-1}^{1} st\phi(t)\, dt = f(s)$$

explicitly.

2. Show that the Volterra integral equation

$$(10.47) \qquad \phi(s) - \lambda \int_0^s K(s, t)\, \phi(t)\, dt = f(s)$$

can be solved by the method of successive approximations, regardless of the value of the parameter λ, provided the kernel $K(s, t)$ is bounded.

3. Verify that all the kernels appearing in Section 9.3 are kernels with a weak singularity and that the kernel in equation (9.62) is square integrable.

4. Establish that for a kernel with a weak singularity some iterated kernel from the sequence

$$K, K^{(2)}, K^{(3)}, \dots$$

remains bounded. Show that the iterated kernel in question is continuous provided $\kappa(s, t)$ is continuous.

5. Prove that the solution ϕ of a Fredholm integral equation whose kernel has a weak singularity is bounded if the prescribed function f is bounded. Show that if $f(s)$ and $\kappa(s, t)$ are continuous in the (closed) domain D, then $\phi(s)$ is also continuous.

6. Establish that the resolvent H is an analytic function of λ for

$$|\lambda| < \frac{1}{\|K\|} \qquad .$$

in the sense that $\psi \circ H \circ \phi$ is an analytic function of λ for arbitrary square integrable ψ and ϕ. Use this result to prove that a Fredholm integral equation with a completely continuous kernel has no more than a countable set of eigenvalues, which can accumulate only at infinity.

7. An alternate definition of complete continuity is that K transforms any bounded set into a compact set; in other words, for any infinite sequence $\phi^{(j)}$ such that $\|\phi^{(j)}\| \leq c$, the sequence $\psi^{(j)} = K \circ \phi^{(j)}$ should contain a convergent subsequence. Show that for square integrable functions $\phi^{(j)}$ and $\psi^{(j)}$ the above definition is equivalent to the one given in the text.

8. Work out a more direct proof of the Fredholm alternative for a kernel with a weak singularity by applying the maximum norm to the case where f and ϕ are bounded functions and $\kappa(s, t)$ is uniformly continuous. Show that $K(s, t)$ can be approximated by a degenerate kernel such that the difference $\tilde{K}$ satisfies

$$\int |\tilde{K}(s, t)| \, dt \leq \epsilon < 1, \qquad \int |\tilde{K}(s, t)| \, ds \leq \epsilon < 1.$$

Then use estimates like

$$|K \circ f| \leq \max |f| \int |K(s, t)| \, dt \leq M\epsilon$$

and

$$|K^{(2)} \circ f| \leq M\epsilon \int |K(s, t)| \, dt \leq M\epsilon^2$$

to prove the Neumann series converges to a uniformly bounded limit.

9. Prove *Hadamard's inequality*

(10.48) $$D^2 \leq \prod_{i=1}^{n} \left[\sum_{j=1}^{n} a_{ij}^2 \right]$$

for an n-by-n matrix (a_{ij}) with determinant D. Conclude that

(10.49) $$D \leq n^{n/2} N^n$$

if $|a_{ij}| \leq N$.

10. Let the range of s and t in (10.1) be the unit interval, and set

$$s_j = t_j = \frac{j}{n}, \qquad j = 0, 1, \ldots, n.$$

Consider the system of n simultaneous linear equations

$$(10.50) \qquad \phi_n(s_i) - \frac{\lambda}{n} \sum_{j=1}^{n} K(s_i, t_j)\, \phi_n(t_j) = f(s_i), \qquad i = 1, \ldots, n,$$

approximating (10.1), and form determinants $D_n(\lambda)$ and $D_n(s_i, t_j; \lambda)$ from the coefficients such that

$$(10.51) \qquad \phi_n(s_i) = \frac{\sum_{j=1}^{n} D_n(s_i, t_j; \lambda) f(t_j)}{D_n(\lambda)}$$

whenever $D_n(\lambda) \neq 0$. Show that *Fredholm determinants*[6] $D(\lambda)$ and $D(s, t; \lambda)$ can be introduced by means of the formulas

$$(10.52) \qquad D(\lambda) = \lim_{n \to \infty} D_n(\lambda) = 1 - \lambda \int_0^1 K(s, s)\, ds$$

$$+ \frac{\lambda^2}{2!} \int_0^1 \int_0^1 \begin{vmatrix} K(s,s) K(s,t) \\ K(t,s) K(t,t) \end{vmatrix} ds\, dt - \cdots$$

and

$$(10.53) \qquad D(s, t; \lambda) = \lim_{\substack{n \to \infty \\ s_i \to s, t_j \to t}} n D_n(s_i, t_j; \lambda)$$

$$= \lambda K(s, t) - \frac{\lambda^2}{1!} \int_0^1 \begin{vmatrix} K(s,t) K(s,\sigma) \\ K(\sigma,t) K(\sigma,\sigma) \end{vmatrix} d\sigma + \cdots$$

if the kernel K is continuous, and if $s \neq t$.

11. In the notation of Exercise 10, show that (10.1) has the solution

$$(10.54) \qquad \phi(s) = f(s) + \int_0^1 \frac{D(s, t; \lambda)}{D(\lambda)} f(t)\, dt$$

when $D(\lambda) \neq 0$. Use this result to establish the Fredholm alternative.

12. Show that for small enough λ

$$(10.55) \qquad \frac{D(s, t; \lambda)}{D(\lambda)} = \lambda K + \lambda^2 K^{(2)} + \lambda^3 K^{(3)} + \ldots,$$

so that (10.54) is synonymous with (10.8).

13. The Schauder fixed point theorem asserts that every continuous transformation of any closed convex set of elements from a Banach space into a compact subset possesses at least one fixed point. Derive this result from the

[6] Cf. Whittaker-Watson 1.

Brouwer fixed point theorem, which states that every continuous mapping of a closed solid sphere of arbitrary finite dimension into itself has a fixed point.[7] What connection has Schauder's theorem with nonlinear integral equations?

2. APPLICATIONS

In Section 9.3 we formulated a variety of boundary value problems as Fredholm integral equations. The Fredholm alternative can now be exploited to investigate the existence of solutions of those problems, which as a rule follows from the corresponding uniqueness theorem (cf. Exercise 1.3). In the more complicated cases, such as that of the Neumann problem for Laplace's equation, the Fredholm alternative serves to establish the validity of compatibility requirements on the data. We shall begin our discussion of these matters with the example of Dirichlet's problem for Laplace's equation in a three-dimensional region D bounded by an analytic surface ∂D.

In Section 9.3 we asked for a solution of Dirichlet's problem in the form of a *double layer potential*. It was established that the boundary condition reduces in such an approach to the Fredholm integral equation

$$\mu + \frac{1}{2\pi} \int_{\partial D} \mu \frac{\partial}{\partial \nu} \frac{1}{r} \, d\sigma = f$$

for the dipole moment density μ of the potential. In order to deduce from the Fredholm alternative that this integral equation can be solved for arbitrary smooth choices of the prescribed boundary values f, we have to show that the homogeneous equation

$$(10.56) \qquad \mu + \frac{1}{2\pi} \int_{\partial D} \mu \frac{\partial}{\partial \nu} \frac{1}{r} \, d\sigma = 0$$

has only the obvious solution

$$(10.57) \qquad \mu = 0.$$

The results of Exercise 9.3.10, which were based on the consideration of an iterated integral equation, imply that any solution μ of (10.56) must have continuous second derivatives, so that the formulas (9.37), (9.39) and (9.41) apply to the corresponding double layer potential

$$(10.58) \qquad W = \frac{1}{2\pi} \int_{\partial D} \mu \frac{\partial}{\partial \nu} \frac{1}{r} \, d\sigma$$

[7] Cf. Courant-Hilbert 3.

along the boundary surface ∂D. We shall obtain the desired conclusion (10.57) from uniqueness theorems for the interior and exterior Dirichlet and Neumann problems satisfied by W.

When we consider W as a harmonic function in the interior of D, the integral equation (10.56) asserts that it has the boundary values zero on ∂D. From the uniqueness of the solution of Dirichlet's problem, it follows that W vanishes identically inside D. The continuity

$$\frac{\partial W^+}{\partial \nu} = \frac{\partial W^-}{\partial \nu}$$

of the normal derivative of the double layer potential W across the charged surface ∂D therefore serves to establish that

$$\frac{\partial W^+}{\partial \nu} = 0$$

on ∂D. Thus W solves a homogeneous Neumann problem in the exterior of D, and consequently it must reduce to a constant there. Since the integral (10.58) obviously vanishes at infinity, we conclude that

$$W = 0$$

everywhere. Finally, the expression

$$W^+ - W^- = -2\mu$$

for μ in terms of the jump of W across ∂D yields (10.57), which is what we set out to prove. This completes our verification, based on the method of integral equations, of the existence of a solution to Dirichlet's problem for Laplace's equation inside the surface ∂D. Observe that the argument involved a uniqueness theorem for the Neumann problem in the exterior region, which is equivalent to the transposed integral equation (9.53).

The Neumann problem for harmonic functions inside D is slightly more difficult to treat than the Dirichlet problem because its solution is not unique and because the data must obey a compatibility condition which brings the full force of the Fredholm alternative into play. In Section 9.3 the Neumann problem in question was transformed into an integral equation

$$\rho - \frac{1}{2\pi}\int_{\partial D} \rho\, \frac{\partial}{\partial \nu}\frac{1}{r}\, d\sigma = -g$$

for the density ρ of a single layer potential V which was supposed to represent the harmonic function u being sought. This integral equation can be solved if and only if the prescribed values

$$g = \frac{\partial u}{\partial \nu}$$

of the normal derivative of u are orthogonal to every eigenfunction u of the homogeneous transposed equation

(10.59) $$\mu - \frac{1}{2\pi} \int_{\partial D} \mu \frac{\partial}{\partial \nu} \frac{1}{r} \, d\sigma = 0.$$

We shall establish that each such eigenfunction is actually a constant. Thus the compatibility requirement

(10.60) $$\int_{\partial D} g \, d\sigma = 0$$

on the data g, which has already been encountered in Chapter 7, is seen to be not only necessary but also sufficient for the existence of a solution of the Neumann problem.

The significance of the relation (10.59) is that upon approach to the surface ∂D from its exterior the double layer potential W defined by (10.58) assumes the boundary values zero. Consequently W vanishes outside ∂D, because of the uniqueness of the solution of Dirichlet's problem. The continuity of the normal derivative of W across ∂D thus leads to the boundary condition

$$\frac{\partial W^-}{\partial \nu} = 0,$$

which shows that W solves a homogeneous Neumann problem for Laplace's equation inside ∂D. It follows in turn that W remains constant in the interior of D. Therefore the jump formula (9.39) yields the desired result

$$\mu = \text{const.}$$

for all non-trivial solutions μ of the homogeneous transposed integral equation (10.59). In this somewhat indirect fashion the Fredholm alternative provides us with an answer to the question of existence for the Neumann problem in the case of Laplace's equation, where difficulties would be encountered with other methods due to the presence of the awkward side condition (10.60).

We pass next to an analysis of the Dirichlet problem for an elliptic equation

$$u_{xx} + u_{yy} + au_x + bu_y + cu = 0$$

with variable coefficients, under the hypothesis that $c \leq 0$. The maximum principle shows that the solution is unique for every continuous assignment of boundary values. On the other hand, we found in Section 9.3 that the problem is equivalent to the Fredhom integral equation

$$u - \int_D [(c - a_x - b_y)G - aG_x - bG_y]u \, dx \, dy = U.$$

Since the foregoing remark about uniqueness implies that the homogeneous integral equation

$$(10.61) \qquad u - \int_D [(c - a_x - b_y)G - aG_x - bG_y]u \, dx \, dy = 0$$

has only the obvious answer

$$u \equiv 0,$$

we are able to conclude that a solution always exists for continuous choices of the prescribed boundary values. Thus the solvability of Dirichlet's problem for our equation with variable coefficients is an immediate consequence of the Fredholm theory of integral equations, granted that we have previously established the necessary uniqueness theorem and have available the Green's function G of Laplace's equation to use as a parametrix. The existence of G is readily deduced, of course, from our earlier results based on single and double layer potentials.

Many boundary value problems of the kind just cited can be treated by the method of integral equations after they have been formulated in terms of some related problem for which an answer is known. Another example of such a phenomenon is the Dirichlet problem for radiating solutions of the reduced wave equation

$$\Delta u + k^2 u = 0$$

in an infinite plane region. We have seen in Section 9.3 how the Green's function for the latter equation in the exterior of a circle can be transformed into a parametrix which enables us to convert this radiation problem into the Fredholm equation

$$u - k^2 \int_{|z|>R} pG_k u \, dx \, dy = \int_{|z|=R} f \frac{\partial G_k}{\partial \nu} \, ds.$$

It is convenient to multiply on both sides here by the square root of the weight factor

$$p = 2 \operatorname{Re}\{q'(z)\} + |q'(z)|^2$$

and to perform the substitution

$$v = \sqrt{p}\, u,$$

so that a new integral equation

$$(10.62) \qquad v - k^2 \int_{|z|>R} \sqrt{p}\, G_k \sqrt{p}\, v \, dx \, dy = \sqrt{p} \int_{|z|=R} f \, \frac{\partial G_k}{\partial \nu} ds$$

is obtained which has a symmetric and square integrable kernel. The Fredholm theory then applies, and the existence of a solution of the exterior Dirichlet radiation problem becomes a consequence of its uniqueness, which was established in Exercise 7.1.14.

We have confined our attention here to a few important special cases which should nevertheless give an indication of the variety of boundary value problems that can be handled through reduction to an appropriate integral equation. Such a technique must be counted among the most powerful and flexible tools in the theory of partial differential equations. We shall see in what follows (cf. Sections 3 and 4) how it can be used to study eigenvalue problems, and it should also be mentioned that certain nonlinear boundary value problems yield to a similar approach. In the nonlinear context, however, the Fredholm alternative has to be abandoned in favor of a more general analysis based on fixed point theorems (cf. Exercise 1.13).

EXERCISES

1. Use the Fredholm alternative to establish the existence of a solution of Dirichlet's problem for the Laplace equation in any number of independent variables.

2. Use the method of integral equations to prove that the Dirichlet problem can be solved for a partial differential equation analogous to (9.55) in space of arbitrary dimension.

3. Verify by direct computation that any constant does, indeed, solve the homogeneous integral equation (10.59). What connection has the eigenfunction of the transposed equation with the conductor potential of the surface ∂D?

4. By means of a Fredholm equation similar to (9.66), treat the exterior Neumann problem for solutions of (9.60) that satisfy the radiation condition at infinity.

5. Formulate the exterior Dirichlet problem for (9.60) as a one-dimensional integral equation based on double layer potentials of a fundamental solution

(cf. Exercise 9.3.6). What difficulties are encountered when eigenfunctions of the associated interior problems appear?

6. Show that the exterior Neumann problem is solvable for Laplace's equation in three-dimensional space without the restriction of a compatibility requirement. Where can a source occur to make this possible? Show that the solution of the corresponding exterior Dirichlet problem cannot be represented as a double layer potential unless a compatibility condition is fulfilled. What role does the conductor potential play here?

7. Motivated by the Fredholm determinants (10.52) and (10.53), develop a procedure to solve the Dirichlet problem numerically for harmonic functions in the plane.

8. Assume that W solves Dirichlet's problem for Laplace's equation in a three-dimensional region and that the associated exterior Neumann problem is known to have a solution. Find from these results the dipole moment density μ occurring in equation (10.58). How does the solvability of the exterior Neumann problem enter here?

3. EIGENVALUES AND EIGENFUNCTIONS OF A SYMMETRIC KERNEL

In the case of a *symmetric kernel*

$$K(s, t) = K(t, s)$$

the integral equation

(10.63) $$\phi(s) - \lambda \int K(s, t)\phi(t)\, dt = 0$$

is identical with its transpose. We shall see how this will enable us to develop the theory of the equation in much greater depth than is feasible in the non-symmetric case. It turns out that investigating the eigenvalues and eigenfunctions of (10.63), as well as the expansion of a given function in terms of them, is of primary importance (the case of a vibrating membrane is a typical example). The next section is concerned with expansion and completeness theorems; the present section is devoted to the question of existence of eigenvalues and eigenfunctions and a study of their properties.

We shall see that the integral equation (10.63) with a symmetric kernel K always has at least one eigenvalue and that the totality of its eigenvalues and eigenfunctions can be determined by a sequence of maximum problems. We shall develop the proof under the hypothesis that the kernel K has a weak singularity of the form

$$K(s, t) = \frac{\kappa(s, t)}{r^l},$$

where the numerator $\kappa(s, t)$ is symmetric and continuous in D. We shall further assume that D is bounded and that $0 \leq l < n/2$ (see, however, Exercises 6 and 16 below). Under these assumptions K will also be square integrable, and moreover

$$(10.64) \qquad \int K(s, t)^2 \, dt = \int K(s, t)^2 \, ds \leq A < \infty.$$

Before proceeding with the proof of the existence of eigenvalues and eigenfunctions of (10.63) we shall discuss several of their elementary properties. These are analogous to the corresponding properties of the eigenvalues and eigenvectors of symmetric matrices[8] and follow from the square integrability of K alone.

Let λ_1 and λ_2 be distinct eigenvalues

$$\lambda_1 \neq \lambda_2$$

of the integral equation (10.63), and let ϕ_1 and ϕ_2 be the corresponding eigenfunctions. We shall prove that ϕ_1 and ϕ_2 are orthogonal in the sense

$$(10.65) \qquad \phi_1 \circ \phi_2 = \int \phi_1(s)\phi_2(s) \, ds = 0.$$

Multiplying the equation

$$\phi_2 = \lambda_2 K \circ \phi_2$$

by ϕ_1 and integrating, we find that

$$\phi_1 \circ \phi_2 = \lambda_2 \phi_1 \circ K \circ \phi_2,$$

whereas a multiplication of the analogous equation

$$\phi_1 = \lambda_1 K \circ \phi_1$$

by ϕ_2, together with an integration, yields

$$\phi_2 \circ \phi_1 = \lambda_1 \phi_2 \circ K \circ \phi_1.$$

Since the kernel K is symmetric, we have

$$\frac{1}{\lambda_2} \phi_1 \circ \phi_2 = \phi_1 \circ K \circ \phi_2 = \phi_2 \circ K \circ \phi_1 = \frac{1}{\lambda_1} \phi_2 \circ \phi_1 ,$$

whence

$$(10.66) \qquad \left(\frac{1}{\lambda_2} - \frac{1}{\lambda_1}\right) \phi_1 \circ \phi_2 = 0.$$

[8] Cf. Birkhoff-MacLane 1.

The desired result (10.65) now follows because, by hypothesis, the first factor on the left in (10.66) differs from zero.

A slight modification of the foregoing argument serves to show that all the eigenvalues of a real symmetric kernel K are themselves real. This is easily seen by setting

$$\lambda_1 = \lambda, \quad \lambda_2 = \bar{\lambda}, \quad \phi_1 = \phi, \quad \phi_2 = \bar{\phi}$$

in (10.66) to obtain

$$\left(\frac{1}{\bar{\lambda}} - \frac{1}{\lambda}\right) \int |\phi|^2 \, dt = 0.$$

Since the integral is positive, its coefficient must vanish, giving $\lambda = \bar{\lambda}$. Because the eigenvalues are thus real, the real and imaginary parts of each complex eigenfunction are again separate eigenfunctions. Hence it is permissible to restrict our attention to eigenfunctions that are real, too.

If $\phi_1, \ldots, \phi_m$ are m linearly independent eigenfunctions of (10.63) corresponding to one and the same eigenvalue

$$\lambda = \lambda_1 = \lambda_2 = \cdots = \lambda_m,$$

we can orthonormalize them by the Gram-Schmidt process introduced in Section 7.3. Since we have already shown that eigenfunctions corresponding to distinct eigenvalues are orthogonal, no generality will be lost if we consider exclusively systems of orthonormal eigenfunctions ϕ_i, which means that

(10.67) $\phi_i \circ \phi_j = \delta_{ij}.$

In this context it is possible to conclude, first, that to no single eigenvalue can there correspond more than a finite number of linearly independent eigenfunctions and, second, that the set of all eigenvalues must be countable. The proof consists in an application of Bessel's inequality to the formal expansion

(10.68) $K(s, t) \sim \sum (K \circ \phi_i)\phi_i = \sum \dfrac{\phi_i(s)\phi_i(t)}{\lambda_i}$

of the kernel K in terms of the orthonormal system of eigenfunctions ϕ_i. To be more precise, observe that the orthonormality relations (10.67) and the integral equations

(10.69) $\phi_i = \lambda_i K \circ \phi_i$

combine to transform the evident inequality

$$0 \leq \int \left[K(s, t) - \sum_{i=1}^{m} \frac{\phi_i(s)\phi_i(t)}{\lambda_i} \right]^2 dt$$

into the more useful result

(10.70) $$\sum_{i=1}^{m} \frac{\phi_i(s)^2}{\lambda_i^2} \leq \int K(s, t)^2 dt \leq A.$$

A final integration of (10.70) with respect to s leads to

(10.71) $$\sum_{i=1}^{m} \frac{1}{\lambda_i^2} \leq B^2 = \iint K(s, t)^2 dt \, ds,$$

which holds for any m eigenvalues $\lambda_1, \ldots, \lambda_m$.
   Setting first

$$\lambda = \lambda_1 = \lambda_2 = \cdots = \lambda_m$$

in equation (10.71), we obtain

$$m \leq \lambda^2 B^2,$$

from which it follows that no eigenvalue of (10.63) can have more than a finite multiplicity. Also, if $|\lambda_i| \leq M$, we have

$$m \leq M^2 B^2,$$

so that there are only a finite number of eigenvalues in any interval $[-M, M]$. Thus the integral equation (10.63) possesses at most countably many eigenvalues. If the set of eigenvalues λ_i is indeed infinite, we clearly have

(10.72) $$\sum_{i=1}^{\infty} \frac{1}{\lambda_i^2} \leq B^2,$$

from which we deduce the simple growth estimate

$$\lim_{m \to \infty} |\lambda_m| = \infty.$$

We summarize the above results in the following

THEOREM. *The eigenvalues of an integral equation* (10.63) *with a real symmetric kernel are real and form a discrete spectrum, that is, they are isolated and can accumulate only at infinity. Eigenfunctions corresponding to distinct eigenvalues are orthogonal, and any system of linearly independent eigenfunctions may be orthonormalized in a natural way.*

We now proceed to prove the existence of at least one eigenfunction of the integral equation (10.63) under the hypothesis mentioned above. We shall work with the *symmetric bilinear form*

$$J(\phi, \psi) = \phi \circ K \circ \psi = J(\psi, \phi)$$

and the associated *quadratic form*

$$J(\phi) = J(\phi, \phi),$$

which has the property

$$(10.73) \qquad J(\phi + \psi) = J(\phi) + 2J(\phi, \psi) + J(\psi).$$

The existence of the eigenfunction in question will be established by considering the variational problem

$$(10.74) \qquad \frac{1}{|\lambda|} = \sup \frac{|J(\phi, \phi)|}{\|\phi\|^2} = \sup \frac{|\phi \circ K \circ \phi|}{\phi \circ \phi}$$

within the class of square integrable functions ϕ. This can be given the equivalent formulation

$$(10.75) \qquad \frac{1}{|\lambda|} = \sup_{\|\phi\|=1} |J(\phi)|.$$

The extremal value of the quadratic form $J(\phi)$ turns out to be equal to the reciprocal of the eigenvalue λ, and the extremal function ϕ is the associated eigenfunction. The variational problem

$$(10.76) \qquad \|K\| = \sup \frac{\|K \circ \phi\|}{\|\phi\|} = \sup \frac{\sqrt{\phi \circ K \circ K \circ \phi}}{\sqrt{\phi \circ \phi}}$$

defining the norm (cf. Section 1) of the kernel K, on the other hand, will be found to be equivalent to (10.74) because of the following lemma:

$$(10.77) \qquad \frac{1}{|\lambda|} = \|K\|.$$

This lemma will play a central role in our analysis.

In order to prove (10.77), we first observe that

$$|J(\phi)| = |\phi \circ K \circ \phi| \le \|K \circ \phi\| \, \|\phi\| \le \|K\| \, \|\phi\|^2,$$

whence

$$(10.78) \qquad \frac{1}{|\lambda|} = \sup_{\|\phi\|=1} |J(\phi)| \le \|K\|.$$

It remains to establish the reversed inequality

(10.79) $$\frac{1}{|\lambda|} \geq \|K\| = \sup_{\|\phi\|=1} \|K \circ \phi\|.$$

Let

$$\psi = K \circ \phi, \qquad \mu^2 = \frac{\|K \circ \phi\|}{\|\phi\|},$$

so that we have

$$\|K \circ \phi\|^2 = \phi \circ K \circ \psi = J\left(\mu\phi, \frac{1}{\mu}\psi\right).$$

Consider the identity

$$J\left(\mu\phi, \frac{1}{\mu}\psi\right) = \frac{1}{4}\left[J\left(\mu\phi + \frac{1}{\mu}\psi\right) - J\left(\mu\phi - \frac{1}{\mu}\psi\right)\right]$$

$$\leq \frac{1}{4\,|\lambda|}\left[\left\|\mu\phi + \frac{1}{\mu}\psi\right\|^2 + \left\|\mu\phi - \frac{1}{\mu}\psi\right\|^2\right]$$

$$= \frac{1}{2\,|\lambda|}\left[\mu^2\,\|\phi\|^2 + \frac{1}{\mu^2}\,\|\psi\|^2\right],$$

which follows from (10.73). Because of our special choice of ψ and μ, we conclude that

$$\|K \circ \phi\|^2 \leq \frac{1}{|\lambda|}\,\|K \circ \phi\|\,\|\phi\|,$$

which yields the desired result (10.79).

An immediate corollary of the above lemma is that

$$\frac{1}{|\lambda|} = \|K\| > 0$$

unless the kernel K vanishes identically. Suppose in fact that $\|K\| = 0$, which means that

$$K \circ \beta = 0$$

for every square integrable function β. Setting $\beta = K$, we obtain

$$K \circ K = \int K(s, t)^2\,dt = 0.$$

In view of the continuity of $\kappa(s, t)$ the kernel K would have to vanish identically. Thus in what follows we may assume $1/|\lambda| > 0$.

Returning to our extremal problem (10.75), we may without loss of generality let $\phi^{(j)}$ stand for a maximal sequence of functions such that

$$(10.80) \qquad \frac{1}{\lambda} = \lim_{j \to \infty} J(\phi^{(j)}) > 0, \qquad \|\phi^{(j)}\| = 1.$$

Our purpose will be to establish that a subsequence of the smoothed functions

$$(10.81) \qquad \psi^{(j)} = K \circ \phi^{(j)}$$

can be found which converges uniformly to an eigenfunction ψ of (10.63) corresponding to the eigenvalue λ. The normalized eigenfunction ϕ corresponding to λ will then be given by

$$\phi = \psi/\|\psi\|.$$

The proof hinges on the fact that the functions $\psi^{(j)}$ are *equicontinuous*. By this we mean that for any $\epsilon > 0$ there exists a number $r_\epsilon > 0$ independent of both the index j and the arguments s_1 and s_2 such that whenever the distance between s_1 and s_2 is less than r_ϵ we have

$$(10.82) \qquad |\psi^{(j)}(s_1) - \psi^{(j)}(s_2)| < \epsilon.$$

To verify the equicontinuity of the sequence $\psi^{(j)}$, we first deduce from Schwarz's inequality that

$$[\psi^{(j)}(s_1) - \psi^{(j)}(s_2)]^2 \leq \int [K(s_1, t) - K(s_2, t)]^2 \, dt$$

and then estimate the integral on the right by separating it into a contribution from the vicinity of the singular points s_1 and s_2 and a contribution from the remainder of the range of t. The details of such a computation, based on the uniform continuity of $\kappa(s, t)$, are so elementary that we relegate them to an exercise. Incidentally, the Schwarz inequality also yields

$$\psi^{(j)}(s)^2 \leq \int K(s, t)^2 \, dt \leq A,$$

which shows that the functions $\psi^{(j)}$ are uniformly bounded.

The importance of the concept of equicontinuity stems from

ARZELA'S THEOREM. *From any sequence of functions which are uniformly bounded and equicontinuous on a closed, bounded domain it is possible to select a uniformly convergent subsequence.*

A proof of the above theorem will be given at the end of this section; we prefer not to interrupt our present train of thought for it, but to proceed directly with our proof of the existence of an eigenfunction of (10.63).

Applying Arzela's theorem to the specific case of the functions (10.81), we see that there is a subsequence $\psi^{(j_k)}$, which we may again denote by $\psi^{(j)}$, such that

$$(10.83) \qquad \lim_{j \to \infty} \psi^{(j)}(s) = \psi(s)$$

exists uniformly and defines a continuous function $\psi(s)$ on the entire range of s. We wish to show that $\psi(s)$ is an eigenfunction of the integral equation (10.63).

By Schwarz's inequality and the lemma (10.77) we have

$$|J(\phi^{(j)})| \leq \|K \circ \phi^{(j)}\| \leq \frac{1}{|\lambda|}.$$

It therefore follows from (10.80) that

$$(10.84) \qquad \|\psi\| = \lim_{j \to \infty} \|K \circ \phi^{(j)}\| = \frac{1}{|\lambda|} > 0,$$

whence the function ψ does not vanish identically. Thus our task is merely to prove that

$$(10.85) \qquad \psi - \lambda K \circ \psi = 0,$$

where λ is given by (10.80).

It turns out to be convenient to introduce an operator δ defined on the class of all square integrable functions ϕ by the rule

$$(10.86) \qquad \delta\phi = \frac{1}{\lambda} \phi - K \circ \phi.$$

Because the convergence in (10.83) is uniform, (10.85) will be a consequence of the more general assertion

$$(10.87) \qquad \lim_{j \to \infty} \delta\psi^{(j)} = 0.$$

According to (10.81) we can write

$$\delta\psi^{(j)} = K \circ \delta\phi^{(j)},$$

and hence an appeal to Schwarz's inequality gives

$$[\delta\psi^{(j)}(s)]^2 \leq \|\delta\phi^{(j)}\|^2 \int K(s, t)^2 \, dt \leq \|\delta\phi^{(j)}\|^2 A.$$

The desired result (10.87) now follows from the observation that

$$\|\delta\phi^{(j)}\|^2 = \left\| \frac{1}{\lambda} \phi^{(j)} - \psi^{(j)} \right\|^2$$

$$= \frac{1}{\lambda^2} - 2\frac{1}{\lambda} J(\phi^{(j)}) + \|\psi^{(j)}\|^2,$$

which tends to zero as $j \to \infty$ because of (10.80) and (10.84). This completes our proof of the

THEOREM. *Every symmetric kernel* K *which does not vanish identically and which satisfies the hypotheses mentioned at the beginning of this section has at least one eigenvalue* λ *and one eigenfunction* ϕ *that solve the maximum problem*

$$(10.88) \qquad \frac{1}{|\lambda|} = \max \frac{|\phi \circ K \circ \phi|}{\phi \circ \phi} = \max \frac{|J(\phi, \phi)|}{\|\phi\|^2}.$$

Moreover, the less evident relation

$$\frac{1^{\cdot}}{|\lambda|} = \|K\| = \max \frac{\|K \circ \phi\|}{\|\phi\|}$$

is also satisfied.

We wish next to find higher eigenvalues and eigenfunctions. Suppose we have already obtained n eigenvalues $\lambda_1, \ldots, \lambda_n$ and n corresponding orthonormal eigenfunctions $\phi_1, \ldots, \phi_n$ for K. Motivated by the formal expansion (10.68), we introduce the new symmetric kernel

$$(10.89) \qquad K_{(n)}(s, t) = K(s, t) - \sum_{i=1}^{n} \frac{\phi_i(s)\phi_i(t)}{\lambda_i}.$$

If $K_{(n)}$ is not identically zero, the maximum problem

$$(10.90) \qquad \frac{1}{|\lambda_{n+1}|} = \max \frac{|\phi \circ K_{(n)} \circ \phi|}{\phi \circ \phi}$$

furnishes, according to the above theorem, an eigenvalue λ_{n+1} and a normalized eigenfunction ϕ_{n+1} of the integral equation

$$(10.91) \qquad \phi_{n+1} = \lambda_{n+1} K_{(n)} \circ \phi_{n+1}.$$

We maintain that ϕ_{n+1} is orthogonal to $\phi_1, \ldots, \phi_n$,

$$(10.92) \qquad \phi_i \circ \phi_{n+1} = 0, \qquad i = 1, \ldots, n,$$

and is therefore an eigenfunction

(10.93) $$\phi_{n+1} = \lambda_{n+1} K \circ \phi_{n+1}$$

of the original integral equation (10.63).
  It is apparent from (10.89) that

$$\phi_i \circ K_{(n)} = K_{(n)} \circ \phi_i = K \circ \phi_i - \frac{\phi_i}{\lambda_i} = 0$$

for $i = 1, \ldots, n$. Hence multiplication of (10.91) by ϕ_i and integration give

$$\phi_i \circ \phi_{n+1} = \lambda_{n+1} \phi_i \circ K_{(n)} \circ \phi_{n+1} = 0,$$

which proves (10.92). Therefore

$$K_{(n)} \circ \phi_{n+1} = K \circ \phi_{n+1},$$

so that (10.91) actually combines in one statement both the integral equation (10.93) and the n side conditions (10.92). Thus λ_{n+1} and ϕ_{n+1} may also be determined by the variational problem

$$\frac{1}{|\lambda_{n+1}|} = \max \frac{|\phi \circ K \circ \phi|}{\phi \circ \phi}$$

subject to the n constraints

$$\phi_i \circ \phi = 0, \qquad i = 1, 2, \ldots, n.$$

  The above construction can now be repeated either indefinitely or until a stage is reached where the auxiliary kernel $K_{(n)}$ vanishes identically. In the latter case the kernel K has the expansion

$$K(s, t) = \sum_{i=1}^{n} \frac{\phi_i(s)\phi_i(t)}{\lambda_i}$$

and is degenerate in the sense of (10.17). Induction shows, however, that when K is not degenerate there exists an infinite sequence

(10.94) $$0 < |\lambda_1| \leq |\lambda_2| \leq \cdots \leq |\lambda_n| \leq \cdots$$

of eigenvalues of the integral equation (10.63) which are associated with a corresponding orthonormal sequence of eigenfunctions

(10.95) $$\phi_1, \phi_2, \ldots, \phi_n, \ldots.$$

As we saw previously, the sequence (10.94) must tend to infinity. Hence the lemma (10.77) applied to the extremal problems (10.90) yields

(10.96) $$\|K_{(n)}\| = \frac{1}{|\lambda_{n+1}|} \to 0$$

as $n \to \infty$. With the aid of (10.96) we shall deduce that the formal expansion (10.68) of the kernel K in terms of its orthonormal eigenfunctions converges in the mean, that is,

(10.97) $$\lim_{n \to \infty} \iint K_{(n)}(s, t)^2 \, dt \, ds = 0.$$

Consequently we shall have

(10.98) $$K(s, t) = \sum_{i=1}^{\infty} \frac{\phi_i(s)\phi_i(t)}{\lambda_i}$$

almost everywhere in the product space of s and t.

The system of functions $\phi_i(s)\phi_i(t)$ is orthonormal in the sense

$$\iint \phi_i(s)\phi_i(t)\phi_j(s)\phi_j(t) \, dt \, ds = \delta_{ij}.$$

It therefore follows from Bessel's inequality (10.72) that the formal expansion

(10.99) $$\sum_{i=1}^{\infty} \frac{\phi_i(s)\phi_i(t)}{\lambda_i} = \sum_{i=1}^{\infty} \left(\iint K \phi_i \phi_i \, dt \, ds \right) \phi_i(s)\phi_i(t)$$

of the kernel K in terms of the orthonormal functions $\phi_i(s)\phi_i(t)$ converges in the mean to a square integrable function. It remains to show that the limit function (10.99) coincides almost everywhere with the kernel K. Observe that

$$\left\| K - \sum_1^{\infty} \frac{\phi_i \phi_i}{\lambda_i} \right\| = \left\| K_{(n)} - \sum_{n+1}^{\infty} \frac{\phi_i \phi_i}{\lambda_i} \right\|$$

$$\leq \|K_{(n)}\| + \left\| \sum_{n+1}^{\infty} \frac{\phi_i \phi_i}{\lambda_i} \right\|.$$

The latter expression tends to zero as $n \to \infty$ because of (10.96) and the convergence of (10.99). Consequently

$$\left\| K - \sum_1^{\infty} \frac{\phi_i \phi_i}{\lambda_i} \right\| = 0,$$

which is enough to establish (10.97) and (10.98) because it implies in particular that the integral of the difference between K and its expansion vanishes over any product of sets in the ranges of s and t.

It should be noted here that we have made no assumption concerning the completeness of the system of eigenfunctions (10.95) and that (10.98) holds even if the eigenfunctions ϕ_i of K are not complete. The question of completeness will be taken up in the next section; the remainder of this section is devoted to proving Arzela's theorem.

Let $\psi^{(j)}$ be a sequence of functions that are uniformly bounded and equicontinuous on the closed, bounded domain D. We shall construct a uniformly convergent subsequence whose limit is similarly continuous.

Let us first consider a dense sequence of points $s_1, s_2, s_3, \ldots$ in D and appeal to a standard diagonalization process to achieve convergence on that dense set. More precisely, we denote by $\psi^{(j_{1n})}$ a subsequence of $\psi^{(j)}$ such that the numbers $\psi^{(j_{1n})}(s_1)$ have a limit, and then we denote by $\psi^{(j_{2n})}$ a further subsequence from $\psi^{(j_{1n})}$ such that $\psi^{(j_{2n})}(s_2)$ converges, too. Sequences of the kind referred to exist because of the uniform boundedness of the functions concerned. Repeating this procedure of selecting subsequences from one another in succession, we arrive inductively at a double array of functions $\psi^{(j_{mn})}$ such that each row is a subsequence of the previous row, and such that the mth row $\psi^{(j_{mn})}$ converges at the first m points $s_1, \ldots, s_m$ as $n \to \infty$. The diagonal sequence $\psi^{(j_{nn})}$ therefore approaches a limit at every individual point of the dense set $s_1, s_2, s_3, \ldots$. We maintain, furthermore, that $\psi^{(j_{nn})}$, which we shall once more indicate by $\psi^{(j)}$ to simplify the notation, converges uniformly throughout D.

To establish the uniform convergence of the diagonal sequence $\psi^{(j)}$, let $\epsilon > 0$ be given, and let $r_\epsilon > 0$ be the corresponding radius associated with the equicontinuity statement (10.82). Pick an integer M so large that any point s in D has a distance less than r_ϵ from at least one element s_i of the set $s_1, \ldots, s_M$. Next choose N so that whenever $j > N$ and $k > N$ we have

$$|\psi^{(j)}(s_i) - \psi^{(k)}(s_i)| < \epsilon$$

for $i = 1, \ldots, M$. From (10.82) and the triangle inequality we obtain

$$|\psi^{(j)}(s) - \psi^{(k)}(s)| \leq |\psi^{(j)}(s) - \psi^{(j)}(s_i)|$$
$$+ |\psi^{(j)}(s_i) - \psi^{(k)}(s_i)| + |\psi^{(k)}(s_i) - \psi^{(k)}(s)| < 3\epsilon,$$

which is enough to establish the uniform convergence of the diagonal sequence $\psi^{(j)}$. This completes the proof of Arzela's theorem.

EXERCISES

1. Give all the details of the proof that the auxiliary functions (10.81) are equicontinuous.

2. State and prove an analogue of Arzela's convergence theorem under the weaker hypothesis that the functions concerned are equicontinuous in each closed, bounded subdomain of some open region where they are supposed to be defined.

3. Show that the eigenfunction (10.83) may also be obtained by considering one or the other of the two extremal problems

$$\frac{\phi \circ K \circ \phi}{\phi \circ \phi} = \text{maximum},$$

$$\frac{\phi \circ K \circ \phi}{\phi \circ \phi} = \text{minimum},$$

depending on whether λ is positive or negative.

4. For a symmetric kernel satisfying the hypotheses of the beginning of this section, prove that the Neumann series (10.5) converges when λ lies inside a circle of radius equal to the absolute value $|\lambda_1|$ of the smallest eigenvalue in the sequence (10.94).

5. Show that the Green's function of Laplace's equation in any bounded region of the plane or of three-dimensional space is a symmetric kernel satisfying the hypotheses mentioned at the beginning of this section.

6. Prove that the results of this section about the existence of eigenvalues and eigenfunctions hold for an arbitrary completely continuous symmetric kernel. In particular, consider the sequence of functions

$$\psi_j = K_n \circ \phi_j + \tilde{K} \circ \phi_j$$

and apply a diagonal process to the double array $K_n \circ \phi_j$ instead of appealing to Arzela's theorem (cf. Exercise 1.7). Verify these results for the particular case of a symmetric kernel with a weak singularity of the form (10.44), where l lies in the range $l < n$ rather than $l < n/2$.

7. Use the integral equation (9.62) to establish the existence of eigenvalues and eigenfunctions for the partial differential equation (9.60). Generalize this result to space of arbitrary dimension with the aid of Exercise 6.

8. Use the Fredholm alternative to verify that the interior Dirichlet problem is correctly set for the elliptic equation (9.60), except for a certain discrete set of values of the parameter k.

9. Show how the theory of the text applies to the eigenfunctions of the one-dimensional problem

$$y'' + k^2 y = 0, \qquad 0 < x < \pi, \qquad y(0) = y(\pi) = 0$$

of the vibrating string. What is the appropriate kernel in this example, and what are its eigenvalues?

10. Discuss the eigenvalues and eigenfunctions of the Poisson kernel (7.51).

11. Exhibit a sequence of eigenvalues and eigenfunctions connected with the third boundary value problem for Laplace's equation in a circle.

12. Develop the theory of the characteristic roots and characteristic vectors of a symmetric matrix. Discuss geometrical aspects of the two-dimensional case.

13. Show that the eigenvalues of a symmetric kernel are the zeros of the corresponding Fredholm determinant (10.52).

14. Show that the Neumann series (9.69) generates a solution of Dirichlet's problem even when the region D involved is no longer convex. Prove that the rate of convergence of this method is described by the ratio of the first two eigenvalues of the relevant kernel. Establish that the first eigenvalue in question is unity, and estimate the second eigenvalue. What conceivable applications does such an analysis have to the numerical treatment of the Dirichlet problem?

15. Prove that the eigenvalues $\lambda_i = k_i^2$ of the integral equation (9.62) are positive and that λ_1 has multiplicity one.

16. In the case of an unbounded domain, show that if the kernel K is not square integrable its spectrum may not be discrete by considering the example

$$(10.100) \qquad \phi(s) = \lambda \int_{-\infty}^{\infty} K(s - t)\phi(t)\, dt$$

of a *convolution* (cf. Exercise 1.6). *Hint:* Try to solve (10.100) by applying the *Fourier transform*

$$(10.101) \qquad \hat{g}(t) = \frac{1}{\sqrt{2\pi}} \int_{-\infty}^{\infty} e^{its}g(s)\, ds.$$

4. COMPLETENESS OF EIGENFUNCTION EXPANSIONS

In the applications of the theory of integral equations one often wishes to expand a given function f in a series

$$(10.102) \qquad f(s) \sim \sum (\phi_i \circ f)\, \phi_i(s)$$

of orthonormalized eigenfunctions ϕ_i. It is therefore necessary to ask in what sense such a formal expansion might represent the function f, and under what circumstances we can expect the system of eigenfunctions to be complete. A partial answer to the latter question is given by the

COMPLETENESS THEOREM. *Any square integrable function f is orthogonal to all the eigenfunctions ϕ_i of the symmetric kernel K if and only if $K \circ f \equiv 0$. For the system of eigenfunctions (10.95) to be complete it is necessary and sufficient that $K \circ f \not\equiv 0$ for every $f \not\equiv 0$.*

If a square integrable function f is orthogonal to the kernel,

$$K \circ f = 0,$$

it must be orthogonal to any eigenfunction ϕ_i, since

$$\phi_i \circ f = \lambda_i \phi_i \circ K \circ f = 0.$$

If, on the other hand, f is orthogonal to every eigenfunction ϕ_i, we have

$$K \circ f = \left[K - \sum_{i=1}^{n} \frac{\phi_i \phi_i}{\lambda_i} \right] \circ f = K_{(n)} \circ f.$$

Therefore

$$\| K \circ f \| = \| K_{(n)} \circ f \| \leq \| K_{(n)} \| \, \| f \| \to 0,$$

so that f is orthogonal to the kernel. Thus the symmetric kernel K and its system of eigenfunctions (10.95) span the same space, which is the essence of the completeness theorem.

In the special case where the kernel is degenerate, its system of eigenfunctions certainly cannot be complete, since they span only a finite-dimensional space (cf. Exercise 1 below). When the kernel is not degenerate, however, its system of eigenfunctions spans an infinite-dimensional space which may or may not be the whole Hilbert space of all square integrable functions. Hence it is still necessary to ask in what sense the formal expansion (10.102) represents f, which we now know ought to be in the space spanned by the kernel. The principal result in this direction is the

HILBERT-SCHMIDT THEOREM. *Any function f which can be expressed in the form*

$$(10.103) \qquad\qquad f = K \circ g,$$

where g is some square integrable function and K is a symmetric kernel satisfying the hypotheses of Section 3, has an absolutely and uniformly convergent representation

$$(10.104) \qquad f(s) = \sum_{i=1}^{\infty} (\phi_i \circ f) \phi_i(s) = \sum_{i=1}^{\infty} \frac{(\phi_i \circ g)}{\lambda_i} \phi_i(s)$$

in terms of the eigenfunctions ϕ_i of K.

To prove the uniform convergence of the series (10.104), observe that by Schwarz's inequality

$$\left[\sum_{i=m}^{n} (\phi_i \circ f)\phi_i \right]^2 = \left[\sum_{i=m}^{n} (\phi_i \circ g) \frac{\phi_i}{\lambda_i} \right]^2$$

$$\leq \left[\sum_{i=m}^{n} (\phi_i \circ g)^2 \right]\left[\sum_{i=m}^{n} \frac{\phi_i^2}{\lambda_i^2} \right].$$

Bessel's inequality

$$\sum_{i=1}^{\infty} (\phi_i \circ g)^2 \leq \|g\|^2 < \infty$$

shows that the first factor on the right can be made as small as we please by taking m and n large enough. Similarly, the second factor remains below the bound (10.70). It follows that the infinite series on the right in (10.104) converges absolutely and uniformly, but we still must verify that its limit coincides with the function f.

Note that because they are given by formulas (10.69) and (10.103), which are analogous to (10.81), the functions ϕ_i and f are uniformly continuous (cf. Exercise 3.1). Thus it will be sufficient to establish the convergence of the series (10.104) toward f in the mean. But that follows from (10.96) because

$$\left\| f - \sum_{1}^{n} (\phi_i \circ f)\phi_i \right\| = \left\| K \circ g - \sum_{1}^{n} \frac{\phi_i}{\lambda_i} (\phi_i \circ g) \right\|$$

$$= \| K_{(n)} \circ g \| \leq \| K_{(n)} \| \, \| g \| \to 0.$$

This completes our proof of the Hilbert-Schmidt theorem.

As an illustration of the usefulness of the Hilbert-Schmidt theorem, we shall establish that the orthonormal system of eigenfunctions $u_1, u_2, u_3, \ldots$ of the reduced wave equation (9.60), subject to the homogeneous boundary condition (9.61), is complete for any bounded region D of the plane. We have already seen that the u_i are identical with the eigenfunctions of the integral equation (9.62), which can be expressed in the alternate form

$$(10.105) \qquad u = \lambda \iint_D Gu \, dx \, dy,$$

with λ replacing the square of the frequency k. In order to demonstrate the completeness of the corresponding eigenfunctions u_i, we have only to examine the class of functions v in D which have a representation

(10.106)
$$v = \iint\limits_{D} Gg \, dx \, dy,$$

where g is any square integrable, or even continuous, function. In fact, according to the Hilbert-Schmidt theorem the eigenfunctions of (10.105) must be complete if every function on D can be approximated arbitrarily well by integrals of this kind.

Green's theorem

$$v = -\iint\limits_{D} (G \, \Delta v - v \, \Delta G) \, dx \, dy - \int_{\partial D} \left(G \frac{\partial v}{\partial v} - v \frac{\partial G}{\partial v} \right) ds$$

shows that if v is any twice continuously differentiable function in D which reduces to zero on the boundary ∂D, then (10.106) is valid with g given by the Poisson equation

(10.107)
$$g = -\Delta v.$$

Consequently the class of functions representable as integrals like (10.106) includes all those that have continuous second derivatives in D and vanish on ∂D. Since there are enough of the latter functions to approximate any square integrable function over D in the mean, we conclude that the eigenfunctions u_i do, indeed, comprise a complete system.

An alternate proof of the completeness of the eigenfunctions of the Green's function will be developed in Exercises 6 and 7 below. Also note that a few of the most basic properties of the eigenvalues λ_j of (10.105) have been described in Exercises 3.5, 3.7 and 3.15. The next chapter will be devoted to a more thorough study of this system of eigenvalues and eigenfunctions.

EXERCISES

1. Show that the eigenfunctions of a degenerate kernel cannot comprise a complete system. What form does the Hilbert-Schmidt theorem take for degenerate kernels?

2. Prove that the Green's function for Laplace's equation is not a degenerate kernel.

3. Apply the Hilbert-Schmidt theorem to the iterated kernels $K^{(n)}$ to show they have the absolutely and uniformly convergent expansions

(10.108)
$$K^{(n)}(s, t) = \sum_{i=1}^{\infty} \frac{\phi_i(s)\phi_i(t)}{\lambda_i^n}, \qquad n = 2, 3, \dots .$$

4. If $0 < |\lambda_1| < |\lambda_2| \leq \cdots$ and if f is any square integrable function that is not orthogonal to the first eigenfunction ϕ_1 of the symmetric kernel K, show that the sequence

$$\lambda_1 K \circ f, \qquad \lambda_1^2 K^{(2)} \circ f, \qquad \lambda_1^3 K^{(3)} \circ f, \quad \cdots$$

converges to a constant times ϕ_1. Describe the rate of convergence in terms of the ratio $|\lambda_1/\lambda_2|$. What modifications of this scheme are necessary if the first eigenvalue λ_1 does not have multiplicity one?

5. Verify that the *Schmidt formula*

$$(10.109) \qquad \qquad \phi = f + \lambda \sum \frac{(\phi_i \circ f)}{\lambda_i - \lambda} \phi_i$$

for the solution of the Fredholm integral equation (10.2) in terms of the eigenvalues λ_i and eigenfunctions ϕ_i of the symmetric kernel K is valid if λ is not an eigenvalue. What changes must be made if λ is an eigenvalue?

6. Show that the eigenfunctions ϕ_i of the symmetric kernel K are complete if K is *strictly positive*, that is, if

$$J(\phi, \phi) = \phi \circ K \circ \phi > 0$$

for any non-trivial square integrable function ϕ.

7. Verify that the Green's function for the vibrating membrane is strictly positive,

$$\phi \circ G \circ \phi > 0,$$

and use the result of Exercise 6 to show completeness of the corresponding eigenfunctions.

8. When are the eigenfunctions of the *Sturm-Liouville problem*

$$(p(x)u')' - q(x)u + \lambda \rho(x)u = 0, \qquad u(0) = u(1) = 0,$$

complete?

9. Determine explicitly the eigenvalues and eigenfunctions for (9.60) and (9.61) in a rectangle. Use (10.108) with $n = 2$ to find an expression for the harmonic Green's function of a rectangle in terms of elliptic functions.

10. Prove the standard completeness theorem for the double Fourier series expansion

$$f(x, y) \sim \sum_{m,n=1}^{\infty} a_{mn} \sin mx \sin ny.$$

11. Establish that if the boundary condition (9.61) is replaced by

$$\frac{\partial u}{\partial \nu} = 0,$$

a complete system of eigenfunctions still exists for the partial differential equation (9.60). Prove that the first eigenfunction for the new problem is a constant associated with the degenerate eigenvalue zero.

12. Use the Green's function of the partial differential equation

$$\Delta u - Pu = 0, \qquad P(x, y) > 0,$$

in two independent variables to obtain a complete orthonormal system of eigenfunctions for the boundary value problem defined by

$$\Delta u - Pu + k^2 u = 0$$

in D and

$$u = 0$$

on ∂D.

13. Discuss the existence of eigenvalues and eigenfunctions for the problem of a vibrating clamped plate, which consists of solving the fourth order partial differential equation

$$\Delta\Delta u - k^4 u = 0$$

in a plane region D, subject to the boundary conditions

$$u = \frac{\partial u}{\partial \nu} = 0$$

along ∂D. Do these eigenfunctions generate a complete orthonormal system in D?.

14. Show that the eigenfunctions u_i of (9.60) and (9.61) are regular at the boundary ∂D of the plane region D, provided that ∂D is an analytic curve.

15. Compare the eigenfunction expansion (10.98) of the Green's function G with the series representation (7.80) of the Bergman kernel function (7.65).

16. Show that the nth eigenvalue λ_n and the nth eigenfunction u_n of the integral equation (10.105) can be characterized by the following *minimax principle*, which does not make reference to the previous eigenvalues and eigenfunctions. First determine

$$\Lambda(v_1, \dots, v_{n-1}) = \max_u \frac{u \circ G \circ u}{u \circ u}$$

within the class of all square integrable functions u which are orthogonal to $n - 1$ other square integrable functions $v_1, \dots, v_{n-1}$,

$$v_i \circ u = 0, \qquad i = 1, 2, \dots, n - 1,$$

and then pick the v_i so that they solve the extremal problem

$$\frac{1}{\lambda} = \min_{v_i} \Lambda(v_1, \dots, v_{n-1}).$$

17. Prove that the nth eigenvalue λ_n of (9.60) and (9.61) decreases as the region D where the boundary value problem is posed becomes larger. What is the physical significance of this assertion?

11

Eigenvalue Problems

1. THE VIBRATING MEMBRANE; RAYLEIGH'S QUOTIENT

In Chapters 9 and 10 we have already had occasion to deal with the *eigenvalues* λ_n and the *eigenfunctions* u_n of the equation

$$(11.1) \qquad \Delta u + \lambda u = 0$$

over a plane region D bounded by smooth curves ∂D along which

$$u = 0.$$

We intend here to make a detailed study of this question and to discuss it as a typical example of the theory of eigenvalue problems for a linear elliptic partial differential equation. To motivate the analysis and render it more understandable, we begin by reviewing the physical background of equation (11.1).

The infinitesimally small vibrations of a thin membrane spread out over the plane region D and pinned down along its edge ∂D are governed by the wave equation

$$(11.2) \qquad \Delta U = U_{tt},$$

subject to the boundary condition

$$U = 0$$

at points (x, y) located on ∂D. Here $U = U(x, y, t)$ represents the deflection of the membrane at the time t. In Section 6.1 we proved that U is uniquely determined by its values and by the values of its time derivative at any initial time, say at $t = 0$.

Our present interest is in the *periodic motions* of the membrane. They are found by asking for solutions of (11.2) that have the special form

(11.3) $$U(x, y, t) = u(x, y) \, e^{ikt},$$

where k is a real parameter related to the *frequency* of the vibration, and where it is now only the real or imaginary part of U that should be interpreted as an actual displacement. Substitution of (11.3) into (11.2) following the standard procedures of the method of separation of variables shows that the factor u, which depends on x and y only, must fulfill the elliptic partial differential equation (11.1) with

$$\lambda = k^2.$$

We therefore refer to (11.1) as the *reduced wave equation* or *equation of the vibrating membrane*.

We conclude from the foregoing remarks that the eigenvalues λ_n and eigenfunctions u_n of the equation

(11.4) $$\Delta u_n + \lambda_n u_n = 0$$

for the region D, subject to the requirement

(11.5) $$u_n = 0$$

on ∂D, serve to describe the natural modes of vibration of a membrane shaped to fit that plane region. According to the maximum principle, the coefficient λ_n in equation (11.4) must be positive if a non-trivial solution u_n satisfying the homogeneous boundary condition (11.5) is to exist. Hence it is legitimate to set

$$\lambda_n = k_n^2 > 0.$$

The integral equation

$$u = \lambda \iint_D Gu \, dx \, dy,$$

in which the kernel G is the Green's function of Laplace's equation, has been used in Section 10.3 to establish that there are infinitely many eigenvalues λ_n of (11.4) and (11.5) for every bounded region of the plane. The theorems of Section 10.4 enabled us afterward to prove that the corresponding orthonormal system of eigenfunctions u_n is *complete*. In what follows we shall be primarily concerned with other constructions, and with further properties, of these eigenvalues and eigenfunctions.

An important application of the complete orthonormal system of eigenfunctions u_n occurs in connection with the *mixed initial and boundary*

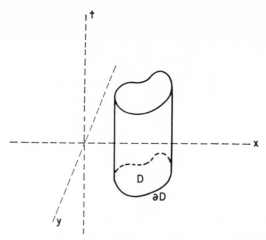

FIG. 40. Mixed initial and boundary value problem.

value problem for the wave equation. Let us determine throughout a cylinder of (x,y,t)-space projecting onto the plane region D a solution U of the wave equation (11.2) which obeys the homogeneous boundary condition

$$U(x, y, t) = 0$$

for (x, y) on ∂D and which satisfies initial conditions of the form

$$U(x, y, 0) = f(x, y), \qquad U_t(x, y, 0) = g(x, y)$$

for (x, y) in D. We assume that the given functions f and g have, in the operator notation of Chapter 10, representations

(11.6)
$$f = G \circ G \circ \phi, \qquad g = G \circ G \circ \psi$$

by means of the Green's function G, where ϕ and ψ are square integrable functions. This amounts to imposing on f and g obvious smoothness hypotheses (cf. Section 10.4) that will ensure the convergence of an eigenfunction expansion of U.

Setting

$$a_n = u_n \circ \phi, \qquad b_n = u_n \circ \psi$$

we conclude from the Hilbert-Schmidt theorem (cf. Section 10.4) that f and g have uniformly convergent series expansions

(11.7)
$$f = \sum_{n=1}^{\infty} \frac{a_n u_n}{\lambda_n^2}$$

and

(11.8)
$$g = \sum_{n=1}^{\infty} \frac{b_n u_n}{\lambda_n^2}$$

in terms of the eigenfunctions u_n. What we intend now to prove is that the solution of our mixed initial and boundary value problem is given by the formula

(11.9)
$$U = \sum_{n=1}^{\infty} \left\{ \frac{a_n}{\lambda_n^2} \cos k_n t + \frac{b_n}{\lambda_n^2 k_n} \sin k_n t \right\} u_n.$$

The expression (11.9) fulfills the boundary condition $U = 0$ because each of the factors u_n vanishes on ∂D. Furthermore, since the Hilbert-Schmidt theorem implies that both of the infinite series

(11.10)
$$G \circ \phi = \sum_{n=1}^{\infty} \frac{a_n u_n}{\lambda_n}$$

and

(11.11)
$$G \circ \psi = \sum_{n=1}^{\infty} \frac{b_n u_n}{\lambda_n}$$

converge absolutely and uniformly, and since $k_n \to \infty$ as $n \to \infty$ according to (10.96), we can differentiate (11.9) term by term to verify that the wave equation (11.2) is satisfied. The first of the initial conditions at $t = 0$ is a consequence of (11.7), and the second initial condition follows from (11.8) after a term-by-term differentiation with respect to t that is again justified because of the convergence of (11.10) and (11.11). Thus the formal expansion (11.9) based on the idea of separating variables does actually solve our mixed initial and boundary value problem for the wave equation in the strict sense, provided that the rather stringent hypothesis (11.6) about the data holds.

We turn our attention next to a characterization of the eigenvalues and eigenfunctions of the vibrating membrane by means of extremal problems in the calculus of variations that are based on the Dirichlet integral (cf. Section 8.1). More specifically, among differentiable functions u defined in D which vanish continuously along the smooth boundary curves ∂D, we ask for one such that

(11.12)
$$\frac{\displaystyle\iint_D (u_x^2 + u_y^2) \, dx \, dy}{\displaystyle\iint_D u^2 \, dx \, dy} = \text{minimum}.$$

We must exclude from competition functions that are identically zero, for otherwise the expression on the left, which is called the *Rayleigh quotient*, would have no meaning. Furthermore, we shall find it convenient in our analysis of the minimum problem (11.12) to consider at first only functions u that are four times continuously differentiable in D with $u = \Delta u = 0$ on ∂D, which means that they have a representation

$$(11.13) \qquad u = G \circ G \circ \phi$$

in terms of some square integrable factor ϕ. Under these circumstances it will be established that the minimum value of the Rayleigh quotient is equal to the lowest eigenvalue λ_1 of (11.4) and (11.5) and that the associated extremal functions u are constant multiples of the corresponding first eigenfunction u_1.

The series expansion

$$(11.14) \qquad u = G \circ G \circ \phi = \sum_{n=1}^{\infty} \frac{a_n u_n}{\lambda_n^2},$$

with coefficients $a_n = u_n \circ \phi$, and its Laplacian

$$(11.15) \qquad \Delta u = -G \circ \phi = -\sum_{n=1}^{\infty} \frac{a_n u_n}{\lambda_n},$$

which has been evaluated by applying the Poisson equation (5.93) to (11.14), both converge uniformly whenever the function u can be written in the form (11.13). Also, since u vanishes on ∂D, Green's theorem gives

$$\iint\limits_{D} (u_x^2 + u_y^2)\, dx\, dy = -\iint\limits_{D} u \Delta u\, dx\, dy.$$

Thus substitution of the representations (11.14) and (11.15), together with an appeal to the orthonormality of the complete system of functions u_n, leads to the pair of identities

$$\iint\limits_{D} (u_x^2 + u_y^2)\, dx\, dy = \sum_{n=1}^{\infty} \frac{a_n^2}{\lambda_n^3}$$

and

$$\iint\limits_{D} u^2\, dx\, dy = \sum_{n=1}^{\infty} \frac{a_n^2}{\lambda_n^4}.$$

Putting

$$c_n = \frac{a_n}{\lambda_n^2} = u_n \circ u,$$

we now find for Rayleigh's quotient the illuminating development

$$
(11.16) \qquad \frac{\displaystyle\iint_D (u_x^2 + u_y^2)\, dx\, dy}{\displaystyle\iint_D u^2\, dx\, dy} = \frac{\displaystyle\sum_{n=1}^{\infty} \lambda_n c_n^2}{\displaystyle\sum_{n=1}^{\infty} c_n^2}.
$$

This makes it clear that the lowest eigenvalue λ_1 is the smallest value that the Rayleigh quotient can assume. Moreover, the first eigenfunction u_1 furnishes at least one choice for the desired minimizing function, since it has, of course, a representation (11.13) with

$$
\phi = \lambda_1^2 u_1.
$$

It is important to eliminate the smoothness requirement (11.13), which involves the Green's function G, from our formulation of the extremal problem (11.12). We wish to prove, indeed, that u_1 minimizes the Rayleigh quotient within the more extensive class of continuous functions in D which have piecewise continuous first derivatives there and which have the boundary values zero. Such a conclusion is drawn by observing (cf. Section 10.4) that any function which has continuous fourth derivatives in D and vanishes together with its Laplacian along ∂D can be expressed in the form (11.13). A sequence of functions of the latter type can be found that approximate an arbitrary piecewise continuously differentiable function defined over D and vanishing on ∂D, and the fit can be arranged so that the first partial derivatives converge in the mean (cf. Exercise 7.2.9). Consequently all results based on formula (11.16) apply to the extremal problem (11.12) even when it is posed within the broader class of competing functions specified above.

We mention in passing that when a smooth function u minimizing the Rayleigh quotient

$$
\lambda = \frac{\displaystyle\iint_D (\nabla u)^2\, dx\, dy}{\displaystyle\iint_D u^2\, dx\, dy}
$$

is known to exist, it is easy to conclude that the reduced wave equation (11.1) is the Euler equation corresponding to this extremal problem in the calculus of variations. The assertion follows if we replace u by $u + \epsilon h$

in the Rayleigh quotient and make use of the formula

$$\frac{1}{2}\frac{\partial \lambda}{\partial \epsilon}\bigg|_{\epsilon=0} = \frac{\displaystyle\iint_D \nabla u \, \nabla h \, dx \, dy}{\displaystyle\iint_D u^2 \, dx \, dy} - \frac{\displaystyle\iint_D (\nabla u)^2 \, dx \, dy \iint_D uh \, dx \, dy}{\left[\displaystyle\iint_D u^2 \, dx \, dy\right]^2}$$

$$= -\frac{\displaystyle\iint_D (\Delta u + \lambda u)h \, dx \, dy}{\displaystyle\iint_D u^2 \, dx \, dy}$$

for the first variation of λ. Also, the orthogonality of higher eigenfunctions of the vibrating membrane can be established by applying Green's theorem appropriately to the Dirichlet integral (cf. Exercise 1 below). An alternate proof of the existence of the complete system of eigenfunctions u_n can be developed from these remarks (cf. Exercise 19 below).

There are several ways of using the Rayleigh quotient to characterize higher eigenfunctions. Perhaps the simplest method of determining the nth eigenfunction u_n is to solve the minimum problem (11.12) within the class of functions u that are orthogonal to the $n - 1$ previous eigenfunctions,

(11.17) $u_1 \circ u = \cdots = u_{n-1} \circ u = 0.$

Under the assumption that the eigenvalues have been arranged in the increasing order

$$\lambda_1 \le \lambda_2 \le \cdots \le \lambda_n \le \cdots,$$

the relation (11.16) shows that u_n, or, more generally, any constant multiple of u_n, is an extremal function for this problem. The corresponding minimum value of the Rayleigh quotient is λ_n. To be sure, if several linearly independent eigenfunctions are associated with the same eigenvalue λ_n, such a procedure does not define u_n uniquely but merely singles it out as some linear combination of the eigenfunctions in question. Note, finally, that our earlier remarks serve to justify omitting the regularity condition (11.13) from the formulation of the extremal problem for λ_n and u_n, too, even though use has been made of (11.16) in the analysis.

In the applications it is convenient to have a characterization of the nth eigenvalue λ_n that does not depend on knowledge of the previous eigenfunctions $u_1, \ldots, u_{n-1}$. With this in mind, we formulate a new definition of λ_n that is based on a *minimax problem*.[1]

_____
[1] Cf. Courant-Hilbert 2.

Given $n - 1$ linearly independent, piecewise continuous functions $v_1, \ldots, v_{n-1}$, we set

$$(11.18) \qquad \Lambda_n(v_1, \ldots, v_{n-1}) = \inf \iint_D (u_x^2 + u_y^2) \, dx \, dy,$$

where u is restricted to the class of functions which are piecewise continuously differentiable in D, which reduce to zero along ∂D, and which fulfill the normalization requirements

$$(11.19) \qquad v_1 \circ u = \cdots = v_{n-1} \circ u = 0$$

and

$$(11.20) \qquad u \circ u = 1.$$

Our aim is to show that

$$(11.21) \qquad \lambda_n = \sup \Lambda_n(v_1, \ldots, v_{n-1})$$

over all possible choices of the auxiliary functions $v_1, \ldots, v_{n-1}$. It will turn out that the eigenfunctions

$$(11.22) \qquad u = u_n, v_1 = u_1, \ldots, v_{n-1} = u_{n-1}$$

constitute a set of extremals for the minimax problem (11.18) to (11.21) despite the fact that no boundary conditions have been imposed on $v_1, \ldots, v_{n-1}$.

Our investigation of the minimum problem (11.12) under the side condition (11.17) suffices to establish that

$$\Lambda_n(u_1, \ldots, u_{n-1}) = \lambda_n.$$

In order to verify (11.21), we must also examine situations in which the various linear combinations of $v_1, \ldots, v_{n-1}$ do not generate the space spanned by the eigenfunctions $u_1, \ldots, u_{n-1}$. In any such case a linear combination

$$(11.23) \qquad u = c_1 u_1 + \cdots + c_{n-1} u_{n-1}$$

of those eigenfunctions can be found which satisfies both (11.19) and the normalization

$$u \circ u = c_1^2 + \cdots + c_{n-1}^2 = 1.$$

Inserting (11.23) into the integral on the right in (11.18), we conclude that

$$\Lambda_n(v_1, \ldots, v_{n-1}) \leq \lambda_1 c_1^2 + \cdots + \lambda_{n-1} c_{n-1}^2 \leq \lambda_{n-1} \leq \lambda_n,$$

whence (11.21) follows. The analysis makes it evident that (11.22) defines one possible system of extremal functions for (11.21), but observe that rearrangements of the v_j are permissible. Moreover, the minimum function u is not uniquely determined when λ_n is a multiple eigenvalue associated with several different eigenfunctions. Finally, note that our earlier discussion enables us to omit regularity requirements of the type (11.13) from the statement of the minimax characterization of λ_n.

The extremal properties of the eigenvalues λ_n that we have just described will provide a most useful tool for their estimation and for the study of their dependence on the domain D and on the index n (cf. Sections 2 and 3). In particular, we wish to examine here the feasibility of calculating the lowest eigenvalue λ_1 numerically by means of the minimum problem (11.12). The procedure we shall develop along these lines is commonly referred to as the *Rayleigh-Ritz method* (cf. Sections 7.3 and 8.1).

What we have in mind is to insert for u in the Rayleigh quotient a linear combination
$$u = a_1 w_1 + \cdots + a_m w_m$$
of known continuously differentiable functions $w_1, \ldots, w_m$ that vanish on ∂D, and then to minimize with respect to the parameters $a_1, \ldots, a_m$. The extremal combination ought to furnish an approximation to the first eigenfunction u_1, according to the theory we have outlined above. The most striking feature of such a technique is, however, that the Rayleigh quotient

$$\frac{\displaystyle\iint_D (\nabla u)^2 \, dx \, dy}{\displaystyle\iint_D u^2 \, dx \, dy} = \frac{\displaystyle\iint_D (a_1 \nabla w_1 + \cdots + a_m \nabla w_m)^2 \, dx \, dy}{\displaystyle\iint_D (a_1 w_1 + \cdots + a_m w_m)^2 \, dx \, dy}$$

thus formed must yield twice as accurate an approximation to the desired eigenvalue λ_1. Indeed, because of its stationary behavior, the Rayleigh quotient of a trial function u that differs from u_1 by a term of the order ϵ ought to deviate from λ_1 by an error of the order of magnitude ϵ^2. Furthermore, the error is non-negative. These properties are shared by many of the evaluations of energy that are obtained from the variational principles of mathematical physics.

EXERCISES

1. Use Green's theorem to give a direct proof of the orthogonality

$$\iint_D \nabla u_m \, \nabla u_n \, dx \, dy = 0 = \iint_D u_m u_n \, dx \, dy$$

of any pair of eigenfunctions u_m and u_n of the vibrating membrane that correspond to different eigenvalues $\lambda_m \neq \lambda_n$.

2. Show that for the region D the eigenvalues μ_n of the reduced wave equation

$$(11.24) \qquad \Delta u + \mu u = 0$$

subject to the *natural boundary condition*

$$(11.25) \qquad \frac{\partial u}{\partial \nu} = 0$$

along ∂D are obtained by solving the minimax problem (11.18) to (11.21) without imposing the additional constraint that the competing functions u should vanish on D.

3. Verify that the first eigenfunction of (11.24) and (11.25) is a constant associated with the degenerate eigenvalue

$$\mu_1 = 0.$$

Prove that

$$\mu_n < \lambda_n$$

for every choice of the index n.

4. Show that for the rectangular membrane $0 < x < a$, $0 < y < b$, the eigenvalues λ_n obtained when the edges are pinned down are given by the relations

$$(11.26) \qquad \lambda_n = \frac{l^2 \pi^2}{a^2} + \frac{m^2 \pi^2}{b^2}$$

with $l, m = 1, 2, 3, \ldots$, whereas the eigenvalues μ_n obtained when the edges are free are given by the same relations

$$(11.27) \qquad \mu_n = \frac{l^2 \pi^2}{a^2} + \frac{m^2 \pi^2}{b^2},$$

but with $l, m = 0, 1, 2, \ldots$ (cf. Exercise 10.4.9).

5. Use an integral equation based on a Neumann's function whose Laplacian is constant to establish the existence of a complete orthonormal system of eigenfunctions for the problem (11.24), (11.25) of a freely vibrating membrane D.

6. Develop a theory of the eigenvalues and eigenfunctions of the linear partial differential equation

$$L[u] + \lambda P u = 0,$$

where L stands for a second order self-adjoint differential operator of the elliptic type and P is a positive variable coefficient.[2]

7. Show formally how to find the eigenvalues and eigenfunctions of the equation

$$\Delta \Delta u - \lambda u = 0$$

[2] Cf. Courant-Hilbert 2.

of a vibrating plate D, subject to the boundary condition

(11.28)
$$u = \frac{\partial u}{\partial \nu} = 0$$

asserting that the edge ∂D is clamped, by solving the problem

$$\frac{\displaystyle\iint_D (\Delta u)^2 \, dx \, dy}{\displaystyle\iint_D u^2 \, dx \, dy} = \text{minimum}$$

in the calculus of variations, subject, of course, to (11.28).

8. Under appropriate hypotheses about the given function f, use an eigenfunction expansion to solve the mixed initial and boundary value problem (cf. Section 13.6)

$$U(x, y, 0) = f(x, y), \qquad (x, y) \text{ in } D,$$

$$U(x, y, t) = 0, \qquad t > 0, (x, y) \text{ on } \partial D,$$

for the heat equation

$$\Delta U = U_t.$$

In what way does the eigenvalue λ_1 describe the rate at which U approaches zero as $t \to \infty$?

9. Show that the variational principle (11.12), considered within the class of functions (11.13), is equivalent to the extremal problem

$$\frac{\phi \circ G \circ G \circ G \circ G \circ \phi}{\phi \circ G \circ G \circ G \circ \phi} = \text{maximum}.$$

Compare this result with the maximum problem used in Section 10.3 to construct eigenfunctions.

10. Give all details of the justification for eliminating the regularity requirement (11.13) from the formulation of the extremal problem (11.12).

11. Use the Rayleigh-Ritz method to find an elementary upper bound on the lowest eigenvalue λ_1 for an ellipse with major axis a and minor axis b.

12. Show that if $v_1, \ldots, v_n$ are n linearly independent continuous functions vanishing on ∂D, then

$$\lambda_n \leq \max_{c_i} \frac{\displaystyle\iint_D (c_1 \nabla v_1 + \cdots + c_n \nabla v_n)^2 \, dx \, dy}{\displaystyle\iint_D (c_1 v_1 + \cdots + c_n v_n)^2 \, dx \, dy},$$

provided that the Rayleigh quotient on the right makes sense.

13. Formulate a relationship between Hamilton's principle (cf. Section 2.5) and the minimum problem (11.12) for the Rayleigh quotient.

14. Use the Fredholm alternative (cf. Section 10.1) to show that Dirichlet's problem can be solved for the partial differential equation (11.1) in a plane region D even when the coefficient λ coincides with an eigenvalue, provided only that the prescribed boundary values f fulfill the compatibility requirement

$$\int_{\partial D} f \frac{\partial u}{\partial \nu} \, ds = 0$$

for every eigenfunction u associated with λ.

15. If the possibility of multiplication by a numerical factor is excluded, show that the first eigenfunction u_1 is the unique solution of the extremal problem (11.12) within the class of smooth functions in D that vanish on ∂D. Start by proving that it is of one sign.

16. By applying Schwarz's inequality to an appropriate line integral, establish the *Poincaré inequality*

$$(11.29) \qquad \iint_D u^2 \, dx \, dy \leq \frac{1}{A} \left[\iint_D u \, dx \, dy \right]^2 + A \iint_D (u_x^2 + u_y^2) \, dx \, dy,$$

where D stands for a square of area A and u is any piecewise continuously differentiable function defined over D.

17. Let u be any piecewise continuously differentiable function which is defined in a bounded plane region D and which vanishes along ∂D, and let a number $\epsilon > 0$ be given. Show that a finite set of piecewise continuous functions $w_1, \ldots, w_m$ not depending on u can be found such that

$$u \circ u = \iint_D u^2 \, dx \, dy \leq \sum_{j=1}^m (w_j \circ u)^2 + \epsilon \iint_D (u_x^2 + u_y^2) \, dx \, dy.$$

18. Let $u^{(n)}$ denote a sequence of piecewise continuously differentiable functions which vanish on the boundary ∂D of their domain of definition D, and suppose that there is a number I such that

$$\iint_D u^{(n)2} \, dx \, dy \leq I, \qquad \iint_D (u_x^{(n)2} + u_y^{(n)2}) \, dx \, dy \leq I$$

for every choice of the index n. Show that a subsequence $u^{(n_j)}$ of the functions $u^{(n)}$ exists with the property

$$\lim_{i,j \to \infty} \iint_D [u^{(n_i)} - u^{(n_j)}]^2 \, dx \, dy = 0$$

that it converges in the mean. This is the *Rellich selection principle*.

19. Use the direct method of the calculus of variations described in Section 8.3, together with the results of Exercises 16, 17 and 18, to develop a proof[3] of the existence of a complete system of eigenfunctions for the vibrating membrane equations (11.4) and (11.5). The proof should be based on the minimum problem for the Rayleigh quotient that is defined by (11.12) and (11.17), subject to the boundary condition $u = 0$.

20. Let k_{mn} stand for the sequence of positive roots of the Bessel function $J_m(k)$. Show that the expressions $J_m(k_{mn}r)\cos m\theta$ and $J_m(k_{mn}r)\sin m\theta$ generate the eigenfunctions of a circular membrane of unit radius which is pinned down at its boundary, and show that the corresponding eigenvalues are identical with the double array of numbers k_{mn}^2. In particular, prove that the lowest eigenvalue λ_1 of the circle is given by the square k_{01}^2 of the first root k_{01} of $J_0(k)$. What modifications are necessary to find the eigenvalues of a circular membrane whose circumference is free?

2. ASYMPTOTIC DISTRIBUTION OF EIGENVALUES

We are interested both in examining the dependence of the eigenvalues λ_n on their domain of definition D and in studying their asymptotic behavior as the index n approaches infinity. For this purpose we shall make essential use of the minimax characterization (11.18) to (11.21) of λ_n. Our analysis is actually motivated by a physical principle which asserts that the fundamental frequency and overtones of any vibrating mechanism should increase when new constraints are imposed on the system, whereas they should diminish if constraints are removed. For the example of the vibrating membrane the constraints with which we shall be concerned are the boundary conditions and continuity requirements on the deflection u. A rigorous mathematical treatment of the theorems just described about monotonicity entails, in the case of the eigenvalues λ_n, an application of the minimax problem referred to above.

We begin by showing that if the region D is contained within a second region D', a relationship that is usually expressed in the form

$$D \subset D',$$

then the nth eigenvalue λ_n of D is not less than the nth eigenvalue λ'_n of the larger region D',

$$(11.30) \qquad \lambda_n \geq \lambda'_n.$$

For the proof it may be assumed that the functions $v_1, \ldots, v_{n-1}$ which occur in the construction (11.18) of the intermediate quantity Λ_n for the

_____
[3] Cf. Courant-Hilbert 1.

region D are defined not only in D but also in D'. The significance of the corresponding quantity

$$\Lambda_n' = \Lambda_n'(v_1, \ldots, v_{n-1})$$

associated with D' is evident. If each competing function u for the extremal problem (11.18) in D is extended continuously to the portion of D' located outside D by setting it equal to zero there, then the normalizations (11.19) and (11.20) remain valid with respect to D' as well as to D. Functions of this kind have identical Dirichlet integrals

$$\iint\limits_{D} (u_x^2 + u_y^2) \, dx \, dy = \iint\limits_{D'} (u_x^2 + u_y^2) \, dx \, dy$$

over D and D'. Since, however, additional trial functions might appear when the integral over the bigger domain D' is minimized to calculate Λ_n', our conclusion is that

$$\Lambda_n'(v_1, \ldots, v_{n-1}) \leq \Lambda_n(v_1, \ldots, v_{n-1}).$$

Thus the maximum of Λ_n' over all choices of $v_1, \ldots, v_{n-1}$ cannot exceed the corresponding maximum of Λ_n; hence the desired result (11.30) follows by virtue of the extremal characterization (11.21) of the nth eigenvalue.

Next let us suppose that the region D has been divided up into a class D^* of subdomains by means of a system of smooth curves or lines C^*. We may consider the combined set of eigenvalues of all the various regions D^* simultaneously as one single ordered sequence

(11.31) $$\lambda_1^* \leq \lambda_2^* \leq \cdots \leq \lambda_n^* \leq \cdots.$$

Each of the associated eigenfunctions coincides with an eigenfunction of some particular domain of the class D^* and reduces to zero outside that domain. Also, any specific eigenvalue λ_n^* of the chain (11.31) can be found by first introducing the greatest lower bound

(11.32) $$\Lambda_n^*(v_1, \ldots, v_{n-1}) = \inf \iint\limits_{D} (u_x^2 + u_y^2) \, dx \, dy$$

of the Dirichlet integral of functions u which are subject to (11.19) and (11.20), and which are in addition required to vanish on ∂D and along the set of curves C^* separating the individual regions D^*, and by afterward evaluating the least upper bound

$$\lambda_n^* = \sup \Lambda_n^*(v_1, \ldots, v_{n-1})$$

of Λ_n^* with respect to all choices of the auxiliary functions $v_1, \ldots, v_{n-1}$. Since the class of competing functions u allowed in (11.32) is more exclusive than that used in the extremal problem (11.18), we have

$$\Lambda_n^*(v_1, \ldots, v_{n-1}) \geq \Lambda_n(v_1, \ldots, v_{n-1}).$$

Maximizing with respect to $v_1, \ldots, v_{n-1}$, we conclude, as in the case of (11.30), that

(11.33) $$\lambda_n^* \geq \lambda_n.$$

The estimate (11.33) of λ_n from above has been obtained by pinning the membrane D down along the arcs C^*. An analogous lower bound on λ_n can be derived by cutting the membrane along the curves C^* instead and releasing it at its edge. To be precise, we introduce the ordered sequence

$$\mu_1^* \leq \mu_2^* \leq \cdots \leq \mu_n^* \leq \cdots$$

of all the eigenvalues of the *free membrane problem* (11.24), (11.25) for the whole class of subdomains D^* simultaneously. Each of the corresponding eigenfunctions satisfies (11.24) and (11.25) in some individual region of the set D^* but reduces to zero elsewhere. Moreover, as in previous examples we have at our disposal the minimax characterization

$$\mu_n^* = \sup_{v_1, \ldots, v_{n-1}} \inf_u \iint_D (u_x^2 + u_y^2)\, dx\, dy$$

of μ_n^*, where the competing function u is still subject to the constraints (11.19) and (11.20) but is now permitted to be discontinuous across the curves C^* and does not have to vanish on ∂D. Since the latter specifications are fulfilled by more functions u than we considered in the definition (11.21) of the eigenvalue λ_n, we have

(11.34) $$\mu_n^* \leq \lambda_n.$$

The inequality of Exercise 1.3 is seen to be a special case of this result.

It is convenient to gather the two estimates (11.33) and (11.34) together in the form of a single statement

(11.35) $$\mu_n^* \leq \lambda_n \leq \lambda_n^*.$$

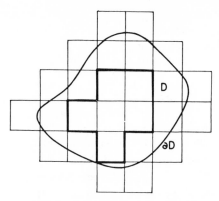

FIG. 41. Rectangular comparison domains.

about the range to which the eigenvalue λ_n is confined. For the applications it is important to observe that because of the monotonicity theorem (11.30) we are permitted to compute the lower bound μ_n^* in (11.35) for a set of regions D^* which do not just overlap D but actually extend beyond that domain. Similarly, for the calculation of the upper bound λ_n^* any disjoint system of domains D^* that are contained entirely inside D can be used, even if they do not include every point of D. In particular, we can pick D^* in the two different cases to be, respectively, a set of rectangles covering D and a set of rectangles enclosed by D. Choices of this kind for D^* turn out to be helpful because they enable us to evaluate λ_n^* and μ_n^* explicitly by means of formulas (11.26) and (11.27).

The comparison theorem (11.35) will enable us to derive an asymptotic expression for the nth eigenvalue λ_n of the membrane D in the limit as $n \to \infty$. The result we have in mind, which was first discovered by Rayleigh and Jeans and was later proved with mathematical rigor by Weyl and Courant,[4] asserts that

$$(11.36) \qquad \lim_{n \to \infty} \frac{\lambda_n}{n} = \frac{4\pi}{A},$$

where A stands for the area of the region D. We shall begin by establishing (11.36) for a rectangle, and then we shall proceed to discuss the general case by bringing the upper and lower bounds (11.35) into play.

The eigenvalues λ_n of the rectangle $0 < x < a$, $0 < y < b$ are specified explicitly in terms of pairs of positive integers l and m by the relation

$$(11.26) \qquad \lambda_n = \frac{l^2 \pi^2}{a^2} + \frac{m^2 \pi^2}{b^2}.$$

[4] Cf. Courant-Hilbert 2, Weyl 1.

We shall find it convenient to consider instead of λ_n an inverse quantity $n(\lambda)$ defined to be the number of eigenvalues λ_n which do not exceed λ. Formula (11.26) makes it clear that $n(\lambda)$ is equal to the number of lattice points with positive integral coordinates that are contained in the ellipse

$$(11.37) \qquad \frac{l^2}{a^2} + \frac{m^2}{b^2} \leq \frac{\lambda}{\pi^2}$$

in the (l,m)-plane. It follows that $n(\lambda)$ is dominated by the area $\lambda ab/4\pi$ of the sector E_λ of the ellipse (11.37) located in the first quadrant. Furthermore, the difference between this area and $n(\lambda)$ has the order of magnitude $\sqrt{\lambda}$ of the perimeter of the ellipse, since it coincides with an area between the boundary of E_λ and a system of unit squares exhausting that region from within. Therefore we can write

$$(11.38) \qquad n(\lambda) = \frac{\lambda ab}{4\pi} + O(\sqrt{\lambda}),$$

where $O(\sqrt{\lambda})$ stands for an expression with the property

$$\varlimsup_{\lambda \to \infty} \frac{|O(\sqrt{\lambda})|}{\sqrt{\lambda}} < \infty.$$

The relation (11.27) implies that for the rectangle $0 < x < a, 0 < y < b$, the number $\tilde{n}(\mu)$ of eigenvalues μ_n of the free membrane problem (11.24), (11.25) that do not exceed μ is identical with the number of lattice points with *non-negative* integral coordinates located in the ellipse

$$(11.39) \qquad \frac{l^2}{a^2} + \frac{m^2}{b^2} \leq \frac{\mu}{\pi^2}.$$

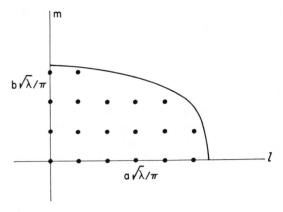

FIG. 42. Asymptotic estimation of $n(\lambda)$.

An analysis similar to our derivation of (11.38) now serves to show that

$$(11.40) \qquad \tilde{n}(\mu) = \frac{\mu ab}{4\pi} + O(\sqrt{\mu}).$$

In the present case, however, the error term $O(\sqrt{\mu})$ includes contributions due to the area of strips of unit width along the major and minor axes of (11.39) corresponding to extra lattice points situated there.

From (11.38) and (11.40) we are able to deduce the laws

$$(11.41) \qquad \lim_{\lambda \to \infty} \frac{n(\lambda)}{\lambda} = \lim_{\mu \to \infty} \frac{\tilde{n}(\mu)}{\mu} = \frac{A}{4\pi}$$

describing asymptotically how the eigenvalues λ_n and μ_n for a rectangle of area $A = ab$ are distributed over a given interval. It is to be noted that for such a rectangle the theorem (11.36) we are trying to establish is an immediate consequence of (11.41).

The result (11.41) remains valid when λ_n and μ_n stand for the combined sequences of eigenvalues of some finite system of disjoint rectangles, so that $n(\lambda)$ and $\tilde{n}(\mu)$ become, respectively, the number of eigenvalues λ_n not exceeding λ and the number of eigenvalues μ_n not exceeding μ for all the rectangles simultaneously. Indeed, the more general statement is obtained merely by adding together the corresponding formulas for each of the individual rectangles of the system. We intend to apply it to a configuration of rectangles D^* contained inside an arbitrary plane region D and to another set of rectangles D^{**} that cover D.

Since the three sets D^*, D and D^{**} stand in the relationship

$$D^* \subset D \subset D^{**}$$

to each other, their areas A^*, A and A^{**} fulfill the corresponding chain of inequalities

$$(11.42) \qquad A^* \leq A \leq A^{**}.$$

If $n^*(\lambda)$ represents the number of eigenvalues λ_n not exceeding λ for all the rectangles D^*, while $n^{**}(\lambda)$ indicates the number of eigenvalues μ_n not exceeding λ for all the rectangles D^{**}, then the estimates (11.35) imply that

$$(11.43) \qquad n^*(\lambda) \leq n(\lambda) \leq n^{**}(\lambda),$$

where, of course, $n(\lambda)$ is the number of eigenvalues λ_n of D that are not larger than λ. Dividing (11.43) through by λ and passing to the limit as

$\lambda \to \infty$, we conclude from (11.41) that

$$\frac{A^*}{4\pi} \leq \varliminf_{\lambda \to \infty} \frac{n(\lambda)}{\lambda} \leq \varlimsup_{\lambda \to \infty} \frac{n(\lambda)}{\lambda} \leq \frac{A^{**}}{4\pi} ,$$

which is consistent with (11.42). Since the rectangles comprising both D^* and D^{**} can be taken arbitrarily small, it is feasible to approximate D so well by D^* and D^{**} that the corresponding areas A^* and A^{**} come as close as we please to A. Hence

$$(11.44) \qquad\qquad \lim_{\lambda \to \infty} \frac{n(\lambda)}{\lambda} = \frac{A}{4\pi}$$

quite generally, and the validity of our original asymptotic formula (11.36) for λ_n follows.

In our derivation of (11.36) we have been able to make a surprisingly successful use of estimates of the eigenvalue λ_n based on subdivision of the membrane D into rectangles at whose edges we either pinned or cut it. The failure of such processes to alter the asymptotic behavior of λ_n in an essential way may seem less remarkable after an examination of the *nodal lines* of the nth eigenfunction u_n, which are defined to be the loci where

$$u_n(x, y) = 0.$$

These level curves separate the region D into subdomains which, while they are not necessarily rectangular, must nevertheless diminish in size as $n \to \infty$, for λ_n grows without limit and therefore cannot remain the principal eigenvalue of any region of fixed proportions. Although the subdomains into which the nodal lines $u_n = 0$ of the nth eigenfunction subdivide D thus become ever smaller or narrower as $n \to \infty$, we shall establish here that they are not more than n in number.[5]

Suppose that there were $m > n$ distinct subregions $D_1, \ldots, D_m$ of D bounded by the nodes $u_n = 0$. In such a situation we could define m new functions $u^{(1)}, \ldots, u^{(m)}$ by specifying that $u^{(j)}$ should coincide with the nth eigenfunction u_n of D in the subdomain D_j but should vanish elsewhere. It would also be possible to find a non-trivial set of parameters $a_1, \ldots, a_n$ with the property that the linear combination

$$(11.45) \qquad\qquad u = a_1 u^{(1)} + \cdots + a_n u^{(n)}$$

[5] Cf. Courant-Hilbert 2.

of just the first n functions $u^{(1)}, \ldots, u^{(n)}$ fulfills all the normalization conditions (11.17). The corresponding value

(11.46)
$$\frac{\iint\limits_D (u_x^2 + u_y^2)\, dx\, dy}{\iint\limits_D u^2\, dx\, dy} = -\frac{\sum\limits_{j=1}^n a_j^2 \iint\limits_{D_j} u^{(j)} \Delta u^{(j)}\, dx\, dy}{\sum\limits_{j=1}^n a_j^2 \iint\limits_{D_j} u^{(j)^2}\, dx\, dy} = \lambda_n$$

of the Rayleigh quotient would then coincide with its minimum subject to (11.17). Consequently we might expect the trial function (11.45) to reduce to a regular eigenfunction of the reduced wave equation (11.1), in contradiction to the fact that it vanishes identically in the subregions $D_{n+1}, \ldots, D_m$ of D and hence has merely piecewise continuous partial derivatives.

The detailed verification of our contention that (11.45) should not define an extremal function u for (11.46) is achieved by substituting a suitably varied function of the form $u + \epsilon h$ into the Rayleigh quotient and observing that the derivative with respect to the parameter ϵ of the quantity so obtained ought to vanish at $\epsilon = 0$. This procedure yields the relation

(11.47)
$$\iint\limits_D (u_x h_x + u_y h_y)\, dx\, dy = \lambda_n \iint\limits_D uh\, dx\, dy$$

for every choice of h obeying the normalization requirements

(11.48)
$$u_1 \circ h = \cdots = u_{n-1} \circ h = 0.$$

On the other hand, an application of Green's theorem based on the Euler equation

(11.49)
$$\Delta u + \lambda_n u = 0$$

satisfied by (11.45) in each of the individual domains $D_1, \ldots, D_m$ transforms (11.47) into the statement

(11.50)
$$\sum\limits_{j=1}^n a_j \int_{\partial D_j} h \frac{\partial u^{(j)}}{\partial \nu}\, ds = 0$$

about integrals over the nodal lines $u_n = 0$. Variations h can be found that take on arbitrarily prescribed values along the nodes $u_n = 0$ and yet fulfill (11.48) because of the way they are specified inside the regions D_j. Due to (11.50) it follows that the normal derivative of the function u given by (11.45) must be continuous across all the nodal lines $u_n = 0$.

Therefore u itself is a real analytic eigenfunction of (11.49) throughout D, because of the results described in Section 5.1 about the regularity of solutions of a linear elliptic partial differential equation. Since, however, u vanishes identically in some, but not all, of the regions D_j, we arrive at a contradiction and must conclude, finally, that there could actually have been no more than n such regions in the first place, which is what we set out to prove.

EXERCISES

1. Show that the asymptotic formula analogous to (11.44) for the eigenvalue problem (11.4), (11.5) in a three-dimensional region of volume V is

$$\lim_{\lambda \to \infty} \frac{n(\lambda)}{\lambda^{3/2}} = \frac{V}{6\pi^2}.$$

2. Derive the asymptotic expression

$$\lim_{n \to \infty} \frac{\lambda_n}{n^2} = \left(\frac{4\pi}{A}\right)^2$$

for the eigenvalues λ_n of the problem described in Exercise 1.7 of a vibrating clamped plate of area A.

3. Show that the analogue of (11.44) for the eigenvalues of the self-adjoint partial differential equation

$$(\rho u_x)_x + (\rho u_y)_y + \lambda P u = 0$$

in a plane region D, subject to the boundary condition

$$u = 0$$

on ∂D, is

$$\lim_{\lambda \to \infty} \frac{n(\lambda)}{\lambda} = \frac{1}{4\pi} \iint_D \frac{P}{\rho} \, dx \, dy,$$

where the coefficients P and ρ are supposed to be positive.

4. Establish the sharper form

$$n(\lambda) = \frac{A\lambda}{4\pi} + O(\sqrt{\lambda} \log \lambda)$$

of the result (11.44) in the case of a smoothly bounded domain.

5. Show that the eigenvalues of the vibrating membrane problem (11.4), (11.5) depend continuously on their domain of definition.

6. Establish that the monotonicity theorem (11.30) fails to hold for the eigenvalues μ_n of a free membrane. For a counterexample choose the larger comparison region D' to be the rectangle $-1 < x < 1$, $-\pi < y < \pi$, and take the smaller region D to be the same rectangle slit along the segment $0 \leq x < 1$ of the x-axis.

7. Examine the nodal lines of the nth eigenfunction of a rectangular membrane. Show that they are not uniquely determined in the case of a square.

8. Show that every eigenfunction u_n of (11.4) and (11.5) with $n > 1$ has at least one nodal line.

9. Work out in more detail our argument that the function (11.45) must be regular because it satisfies the variational condition (11.50).

10. Given a membrane D of area A, show that when ϵ is small enough, each square inside D of area ϵ must contain nodes of every eigenfunction u_n such that

$$n > \frac{\pi A}{\epsilon}.$$

3. UPPER AND LOWER BOUNDS; SYMMETRIZATION

Thus far we have discussed the existence of the eigenfunctions of the vibrating membrane problem by more or less constructive procedures and have studied the asymptotic behavior of the associated eigenvalues by means of their minimax characterization. Our interest now turns to the more specific question of finding explicit estimates of the principal eigenvalue λ_1 from both above and below. Such upper and lower bounds are of obvious physical significance, and another reason they are important is that they serve to describe the rate at which solutions of the heat equation approach a steady state and become harmonic functions (cf. Exercise 1.8).

The monotonicity theorems (11.30) and (11.35) provide upper and lower bounds of the kind we are seeking if rectangles or circles either containing or else entirely included inside the given membrane D are used as domains of comparison. In most cases these estimates are not sufficiently accurate to be satisfactory, however. Better upper bounds can be determined directly on the basis of the extremal problem (11.12) and the Rayleigh-Ritz method. On the other hand, we shall see that it is a more subtle matter to establish lower bounds that furnish an adequate approximation to λ_1, since λ_1 is obtained by minimizing the Rayleigh quotient. Thus our deepest result, whose proof entails the notion of *symmetrization*, will be an isoperimetric inequality asserting that the dimensionless product $\lambda_1 A$ of the first eigenvalue λ_1 times the area A of the region D is least when D is a circle. Incidentally, the technique of symmetrization will be developed in considerable detail here because it has so many other useful applications (cf. Section 15.2).

According to (11.12) we can derive upper bounds on the principal eigenvalue λ_1 of D by inserting into the inequality

$$(11.51) \qquad \lambda_1 \leq \frac{\displaystyle\iint_D (u_x^2 + u_y^2) \, dx \, dy}{\displaystyle\iint_D u^2 \, dx \, dy}$$

any trial function u which vanishes on ∂D. The accuracy of the estimate found in such a way depends on how appropriate the choice of the test function u is. The Rayleigh-Ritz method consists in a special set of rules that determine u as a linear combination of known functions reducing to zero on ∂D. We shall illustrate here how (11.51) can be exploited somewhat differently through a procedure in which we first prescribe the level curves of u and then minimize the Rayleigh quotient with respect to the function of a single variable that remains free to be specified. Particular care will be taken in our selection of the level curves to achieve a sharp estimate of λ_1 with the property that the equality sign holds when D is a circle. Thus an unusually good approximation to λ_1 will be obtained which is stationary with respect to perturbations of any circular domain.[6]

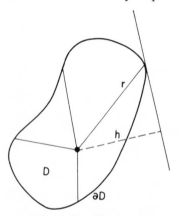

FIG. 43. Star-like domain.

We confine our attention to a domain D bounded by a single *star-like* curve ∂D. By this we mean that in terms of polar coordinates r and θ with a suitably located origin ∂D has a non-parametric representation of the form

$$(11.52) \qquad r = f(\theta).$$

We impose the requirement that all the level lines of the trial function u in (11.51) should be similar to ∂D; thus they are simply obtained by

[6] Cf. Pólya-Szegö 1.

contracting that curve toward the origin. It follows that a function v of one variable can be found such that

(11.53)
$$u(x, y) = v\left(\frac{r}{f(\theta)}\right).$$

Because of equation (11.52), the condition that u vanishes along ∂D transforms into the assertion

(11.54)
$$v(1) = 0$$

about v.

In order to compute the Rayleigh quotient of the function (11.53), we observe that

(11.55)
$$\iint_D u^2 \, dx \, dy = \int_0^{2\pi} \int_0^f v\left(\frac{r}{f}\right)^2 r \, dr \, d\theta$$

$$= \int_0^{2\pi} f(\theta)^2 \, d\theta \int_0^1 v(\rho)^2 \rho \, d\rho$$

$$= 2A \int_0^1 v(\rho)^2 \rho \, d\rho,$$

where we have put $r = f\rho$, and where A stands, of course, for the area of D. Similarly, we have

$$\iint_D (u_x^2 + u_y^2) \, dx \, dy = \int_0^{2\pi} \int_0^f \left(u_r^2 + \frac{1}{r^2} u_\theta^2\right) r \, dr \, d\theta$$

$$= \int_0^{2\pi} \int_0^f \left[\frac{1}{f(\theta)^2} + \frac{f'(\theta)^2}{f(\theta)^4}\right] v'\left(\frac{r}{f}\right)^2 r \, dr \, d\theta$$

$$= \int_0^{2\pi} \left[1 + \frac{f'(\theta)^2}{f(\theta)^2}\right] d\theta \int_0^1 v'(\rho)^2 \rho \, d\rho.$$

If we let s denote the arc length along the boundary curve ∂D, we can write

$$ds^2 = dr^2 + r^2 \, d\theta^2.$$

Hence it becomes clear that

$$\left[1 + \frac{f'(\theta)^2}{f(\theta)^2}\right] d\theta = \frac{ds}{r \, d\theta} \frac{ds}{r} = \frac{r}{h} \frac{ds}{r} = \frac{ds}{h},$$

where $h = h(\theta)$ represents the distance from the origin to the tangent of ∂D at the point with polar coordinates $r = f(\theta)$ and θ. Therefore

(11.56)
$$\iint_D (u_x^2 + u_y^2) \, dx \, dy = \int_{\partial D} \frac{ds}{h} \int_0^1 v'(\rho)^2 \rho \, d\rho.$$

Formulas (11.55) and (11.56) combine to show that under the assumption (11.53) about u, the inequality (11.51) reduces to

$$(11.57) \qquad \lambda_1 \leq \frac{1}{2A} \int_{\partial D} \frac{ds}{h} \frac{\int_0^1 v'(\rho)^2 \rho \, d\rho}{\int_0^1 v(\rho)^2 \rho \, d\rho},$$

provided that the region D is star-like.

An application of standard procedures of the calculus of variations[7] to the problem of minimizing the right-hand side of (11.57) within the class of functions v satisfying the boundary condition (11.54) suggests that we take

$$(11.58) \qquad v(\rho) = J_0(k_{01}\rho),$$

where k_{01} stands for the smallest positive root of the Bessel function $J_0 = J_0(k)$. Since (11.58) defines a solution of Bessel's ordinary differential equation

$$\rho v'' + v' + k_{01}^2 \rho v = 0,$$

an integration by parts exploiting (11.54) yields the relation

$$\int_0^1 [k_{01}^2 v(\rho)^2 - v'(\rho)^2] \rho \, d\rho = \int_0^1 [\rho v'' + v' + k_{01}^2 \rho v] v \, d\rho = 0.$$

Inserting this result into (11.57), we obtain the elementary upper bound

$$(11.59) \qquad \lambda_1 \leq \frac{k_{01}^2}{2A} \int_{\partial D} \frac{ds}{h}$$

on the first eigenvalue λ_1 of any star-shaped domain D. Observe that the equality sign holds for circular membranes. Indeed, when D is the unit circle, with the origin located at its center,

$$\frac{1}{2A} \int_{\partial D} \frac{ds}{h} = \frac{2\pi}{2\pi} = 1,$$

and, according to the results of Exercise 1.20,

$$\lambda_1 = k_{01}^2,$$

too.

[7] Cf. Courant-Hilbert 2.

We shall establish next the more subtle *Faber-Krahn inequality*[8]

$$(11.60) \qquad \lambda_1 \geq \frac{\pi k_{01}^2}{A},$$

which gives a lower bound on the principal eigenvalue λ_1 of a plane region D of arbitrary shape. As in the previous example, the equality sign holds in (11.60) for every circular membrane, and for neighboring domains the estimate is exceptionally good. Our method of proving (11.60) consists in seeking those domains which minimize the dimensionless product $\lambda_1 A$ and in verifying that they must be circles. More precisely, we shall associate with the first eigenfunction u_1 of the most general domain D a comparison function which is defined symmetrically over a circle of the same area as D but which yields a smaller value of the Rayleigh quotient. This construction, which is a special case of *Schwarz symmetrization*,[9] will enable us to deduce the Faber-Krahn inequality (11.60) from the basic estimate (11.51) of the first eigenvalue of a circle.

Denote by D_t the subregions of D where

$$u_1(x, y) > t,$$

and let it be understood that u_1 is positive inside D, which is reasonable because u_1 cannot change sign there (cf. Exercise 10.3.15). For t in the interval between zero and the maximum of u_1, the boundary ∂D_t of D_t is composed of the set of level curves

$$(11.61) \qquad u_1(x, y) = t$$

of the eigenfunction u_1 and is therefore piecewise analytic. Now introduce the circle E_t about the origin which has an area identical with that of D_t. We denote by u the function that assumes the constant value

$$(11.62) \qquad u = t$$

on the circumference ∂E_t of E_t. Thus u is a *symmetrization* of u_1 in the sense that each of its level curves (11.62) is a circle enclosing the same area as the corresponding set of level lines (11.61) of u_1. In particular, the domain of definition of u is a circle $E = E_0$ whose area A is equal to that of D. Also, u vanishes on the boundary ∂E of E. It will be our purpose to show that the Rayleigh quotient of the symmetrized function u over the circular disc E does not exceed the Rayleigh quotient of u_1 with respect to the original domain D.

---

[8] Cf. Faber 1, Krahn 1.
[9] Cf. Pólya-Szegö 1.

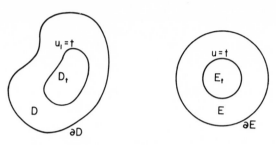

FIG. 44. Symmetrization with respect to a point.

From the construction of u we immediately deduce the identity

$$(11.63) \qquad \iint\limits_{D} u_1^2 \, dx \, dy = - \int_0^T t^2 \, dA(t) = \iint\limits_{E} u^2 r \, dr \, d\theta$$

for the denominators of the Rayleigh quotients of u_1 and u, where T stands for the maximum of u_1 in D, where $A(t)$ indicates the area of D_t and of E_t, and where r and θ represent polar coordinates throughout E. It remains to compare the corresponding numerators. Denoting by s the arc length and by v the inner normal along the level curves (11.61), we find for the element of area in D the expression

$$(11.64) \qquad dx \, dy = ds \, dv = \frac{ds \, dt}{\partial u_1 / \partial v}.$$

Hence

$$(11.65) \qquad \iint\limits_{D} \left[\left(\frac{\partial u_1}{\partial x} \right)^2 + \left(\frac{\partial u_1}{\partial y} \right)^2 \right] dx \, dy = \int_0^T \int_0^{L(t)} \left(\frac{\partial u_1}{\partial v} \right)^2 \frac{ds \, dt}{\partial u_1 / \partial v}$$

$$= \int_0^T \int_0^{L(t)} \frac{\partial u_1}{\partial v} \, ds \, dt,$$

where $L(t)$ stands for the length of the boundary ∂D_t of the point set D_t. We shall proceed to estimate the integral on the right from below by means of Schwarz's inequality and by means of the classical *isoperimetric inequality*[10]

$$(11.66) \qquad 4\pi A(t) \le L(t)^2,$$

in which the equality sign holds only when the locus ∂D_t is a circle (cf. Exercise 7 below).

[10] Cf. Pólya-Szegö 1.

415 UPPER AND LOWER BOUNDS; SYMMETRIZATION

According to the Schwarz inequality we can write

$$(11.67) \qquad L(t)^2 = \left[\int_0^{L(t)} ds \right]^2 \leq \int_0^{L(t)} \frac{\partial u_1}{\partial v} \, ds \int_0^{L(t)} \frac{ds}{\partial u_1 / \partial v} \, .$$

Furthermore, (11.64) implies that

$$A(t) = \int_t^T \int_{0_-}^{L(\tau)} \frac{ds \, d\tau}{\partial u_1 / \partial v},$$

and consequently

$$\int_0^{L(t)} \frac{ds}{\partial u_1 / \partial v} = -A'(t).$$

Inserting this result into (11.67) and taking (11.66) into account, we conclude that

$$-\frac{4\pi A(t)}{A'(t)} \leq \int_0^{L(t)} \frac{\partial u_1}{\partial v} \, ds.$$

Hence an integration with respect to t gives

$$(11.68) \qquad -4\pi \int_0^T \frac{A(t)}{A'(t)} \, dt \leq \iint_D \left[\left(\frac{\partial u_1}{\partial x} \right)^2 + \left(\frac{\partial u_1}{\partial y} \right)^2 \right] dx \, dy,$$

because of (11.65).

It is now essential to observe that the equality sign holds in both (11.66) and (11.67), and therefore also in (11.68), if and only if for each choice of t the region D_t is a circle on whose perimeter ∂D_t the normal derivative $\partial u_1 / \partial v$ reduces to a constant. In particular, note that when u_1 is replaced by the symmetrized function u and D_t is replaced by E_t in the derivation of (11.68), the identity

$$(11.69) \qquad \iint_E \left(u_r^2 + \frac{1}{r^2} u_\theta^2 \right) r \, dr \, d\theta = \iint_E u_r^2 r \, dr \, d\theta = -4\pi \int_0^T \frac{A(t)}{A'(t)} \, dt$$

is obtained instead.

The two relations (11.68) and (11.69) combine with (11.63) to show that

$$\frac{\iint_D \left[\left(\frac{\partial u_1}{\partial x} \right)^2 + \left(\frac{\partial u_1}{\partial y} \right)^2 \right] dx \, dy}{\iint_D u_1^2 \, dx \, dy} \geq \frac{\iint_E \left(u_r^2 + \frac{1}{r^2} u_\theta^2 \right) r \, dr \, d\theta}{\iint_E u^2 r \, dr \, d\theta},$$

or, in other words, that

$$(11.70) \qquad \lambda_1 \geq \frac{\displaystyle\int_0^R u_r^2 r \, dr}{\displaystyle\int_0^R u^2 r \, dr},$$

where $R = \sqrt{A/\pi}$ is the radius of the circle E. If D is not a circle to begin with, we cannot expect that the right-hand side of (11.70) will coincide with the first eigenvalue of E, which we denote by $\tilde\lambda_1$. However, the basic estimate (11.51) implies that the Rayleigh quotient of u dominates $\tilde\lambda_1$, and therefore

$$(11.71) \qquad \lambda_1 \geq \tilde\lambda_1.$$

The dimensionless product $\tilde\lambda_1 A = \pi R^2 \tilde\lambda_1$ is independent of the radius R of E and can easily be evaluated in the specific case of the unit circle. Thus according to Exercise 1.20 we have

$$\tilde\lambda_1 A = \pi k_{01}^2.$$

The latter step is comparable to an explicit minimization of the expression on the right in (11.70) such as we encountered in our derivation of (11.59) from (11.57). Finally, (11.71) yields the desired lower bound

$$\lambda_1 A \geq \pi k_{10}^2,$$

equivalent to (11.60), on the principal eigenvalue λ_1 of a vibrating membrane. The proof just presented makes it clear that the equality sign holds only for circular regions D.

A somewhat different demonstration of the Faber-Krahn inequality (11.60) can be developed from the concept of *Steiner symmetrization*.[11] We shall outline here the central ideas involved in such an analysis, for they have important applications in the theory of partial differential equations and will be needed in Section 15.2.

The process we have in mind can be described as symmetrization of the level curves of the eigenfunction u_1 in a straight line. No generality is lost if we pick the line in question to be the x-axis. For almost all values of x the set of piecewise analytic curves $u_1 = t$ associated with a fixed choice of t decomposes into an even number $2m$ of branches which have non-parametric representations of the form

$$y = y_j(x), \qquad j = 1, 2, \ldots, 2m,$$

with

$$y_1 > y_2 > \cdots > y_{2m}.$$

[11] Cf. Pólya-Szegö 1.

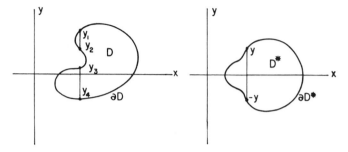

FIG. 45. Steiner symmetrization.

The various ordinates y_j specify intersections of the locus $u_1 = t$ with a line parallel to the y-axis. We introduce a symmetrized function u whose level curves $u = t$ consist of precisely two branches defined by the equations

$$y = y(x), \qquad y = -y(x),$$

where

(11.72)
$$2y(x) = \sum_{j=1}^{2m}(-1)^{j-1}y_j(x) = \sum_{j=1}^{m}(y_{2j-1} - y_{2j}).$$

The new locus $u = t$ cuts from each parallel to the y-axis a single symmetrically placed segment whose length is given by the sum of the m intervals from that same parallel enclosed by the original level curve $u_1 = t$.

In particular, the above transformation takes the perimeter ∂D of D, along which $u_1 = 0$, into a simple closed curve ∂D^* bounding a region D^* with the property that any parallel to the y-axis intersects it in at most one segment. Each such segment is bisected by the x-axis and has a length equal to the linear measure of the intersection of the same parallel with the original domain D.

The symmetrized function u is specified throughout D^* and vanishes on ∂D^*. We shall proceed to establish the fundamental inequality

(11.73)
$$\frac{\iint_D \left[\left(\frac{\partial u_1}{\partial x}\right)^2 + \left(\frac{\partial u_1}{\partial y}\right)^2 \right] dx\, dy}{\iint_D u_1^2\, dx\, dy} \geq \frac{\iint_{D^*} \left[\left(\frac{\partial u}{\partial x}\right)^2 + \left(\frac{\partial u}{\partial y}\right)^2 \right] dx\, dy}{\iint_{D^*} u^2\, dx\, dy}.$$

Hence it will follow from the quite general estimate (11.51) that

$$\lambda_1 \geq \lambda_1^*,$$

where λ_1^* stands for the first eigenvalue of D^*. Thus we shall be led to the interesting conclusion that Steiner symmetrization of the membrane D diminishes its principal eigenvalue, except for those cases where no change at all occurs.

Observe that according to its very definition (11.72) the symmetrized level curve $u = t$ encloses an area

$$2 \int y \, dx = \sum_{j=1}^{2m} (-1)^{j-1} \int y_j \, dx$$

equal to that of the domain D_t where $u_1 > t$. In particular, D has the same area A as D^*. Therefore we can assert that the dimensionless product $\lambda_1 A$ either decreases or else remains invariant under Steiner symmetrization. One might expect the statement that $\lambda_1 A$ achieves its minimum value for a circle to be a consequence of repeated applications of this lemma in a variety of rotated coordinate systems.

Just as in our derivation of (11.63), the fact that the construction (11.72) preserves the area of D_t for any meaningful choice of t suffices to show that

$$\iint_D u_1^2 \, dx \, dy = \iint_{D*} u^2 \, dx \, dy.$$

Thus for a proof of (11.73) we have only to verify that the Dirichlet integral of u_1 does not increase under Steiner symmetrization.

Introducing x and t as new independent variables to replace x and y, we can write

$$\left(\frac{\partial u_1}{\partial y}\right)^2 dx \, dy = \frac{dx \, dt}{|\partial y_j/\partial t|}, \quad \left(\frac{\partial u_1}{\partial x}\right)^2 dx \, dy = \left(\frac{\partial y_j}{\partial x}\right)^2 \frac{dx \, dt}{|\partial y_j/\partial t|}$$

at each point on the jth branch $y = y_j(x)$ of the level curve $u_1 = t$. Let $X(t)$ stand for the projection of that curve onto the x-axis. In analogy with (11.65) we now obtain the expression

$$\iint_D \left[\left(\frac{\partial u_1}{\partial x}\right)^2 + \left(\frac{\partial u_1}{\partial y}\right)^2 \right] dx \, dy = \int_0^T \int_{X(t)} \sum_{j=1}^{2m} \frac{1 + (\partial y_j/\partial x)^2}{|\partial y_j/\partial t|} \, dx \, dt$$

for the Dirichlet integral of u_1. Because of the similar formula

$$\iint_{D*} \left[\left(\frac{\partial u}{\partial x}\right)^2 + \left(\frac{\partial u}{\partial y}\right)^2 \right] dx \, dy = 2 \int_0^T \int_{X(t)} \frac{1 + (\partial y/\partial x)^2}{|\partial y/\partial t|} \, dx \, dt$$

for the Dirichlet integral of u, we see that (11.73) will follow if we can establish the key inequality

(11.74)
$$\sum_{j=1}^{2m} \frac{1 + (\partial y_j/\partial x)^2}{|\partial y_j/\partial t|} \geq 2 \frac{1 + (\partial y/\partial x)^2}{|\partial y/\partial t|}.$$

Geometrical considerations show that

$$\left|\frac{\partial y_j}{\partial t}\right| = (-1)^j \frac{\partial y_j}{\partial t},$$

and therefore

(11.75)
$$\left|\frac{\partial y}{\partial t}\right| = \left|\frac{1}{2} \sum_{j=1}^{2m} (-1)^{j-1} \frac{\partial y_j}{\partial t}\right| = \frac{1}{2} \sum_{j=1}^{2m} \left|\frac{\partial y_j}{\partial t}\right|$$

on account of (11.72). On the other hand, Schwarz's inequality yields

$$\left[\sum_{j=1}^{2m} \alpha_j \frac{\sqrt{|\partial y_j/\partial t|}}{\sqrt{|\partial y_j/\partial t|}}\right]^2 \leq \left[\sum_{j=1}^{2m} \left|\frac{\partial y_j}{\partial t}\right|\right]\left[\sum_{j=1}^{2m} \frac{\alpha_j^2}{|\partial y_j/\partial t|}\right]$$

for every choice of the parameters $\alpha_1, \ldots, \alpha_{2m}$. We add the result found with $\alpha_j = 1$ to that found with $\alpha_j = (-1)^{j-1} \partial y_j/\partial x$ to obtain

(11.76)
$$\sum_{j=1}^{2m} \frac{1 + (\partial y_j/\partial x)^2}{|\partial y_j/\partial t|} \geq \frac{\left[\sum_{j=1}^{2m} 1\right]^2 + \left[\sum_{j=1}^{2m} (-1)^{j-1} \partial y_j/\partial x\right]^2}{\sum_{j=1}^{2m} |\partial y_j/\partial t|}.$$

Comparing (11.72), (11.75) and (11.76), we conclude that because $m \geq 1$ and $\left[\sum_{j=1}^{2m} 1\right]^2 \geq 4$ the statement (11.74) is, indeed, correct. This completes our proof that both the Dirichlet integral of u_1 and the principal eigenvalue λ_1 decrease, or at least remain unaltered, under Steiner symmetrization.

EXERCISES

1. What does it mean to say that both $\lambda_1 A$ and $\lambda_1 L^2$ are dimensionless, where L stands for the length of ∂D?

2. Show that
$$\lambda_1 L^2 \geq 4\pi^2 k_{01}^2.$$

3. Generalize the Faber-Krahn inequality (11.60) to obtain the estimate

$$\lambda_1 V^{2/3} \geq \pi^2 \left(\frac{4\pi}{3}\right)^{2/3}$$

for the principal eigenvalue λ_1 of a three-dimensional domain of volume V.

4. Examine numerically the accuracy of the combined estimates

$$\pi k_{01}^2 \leq \lambda_1 A \leq \frac{k_{01}^2}{2} \int_{\partial D} \frac{ds}{h}$$

for the first eigenvalue λ_1 of a square membrane D.

5. Find upper and lower bounds on the first eigenvalue of the problem of Exercise 2.3 in terms of upper and lower bounds on the coefficients P and ρ.

6. Verify that the Bessel function (11.58) actually minimizes the ratio of integrals on the right in (11.57), subject to the constraint (11.54).

7. Prove the isoperimetric inequality (11.66).

8. Use the triangle inequality to establish that the surface area

$$S = \iint \sqrt{1 + u_x^2 + u_y^2}\, dx\, dy$$

of the boundary $z = u(x, y)$ of any region in space decreases, or at least does not increase, under Steiner symmetrization of that region with respect to an arbitrary plane. Next let u stand for a function defined in a doubly connected plane domain D, with values in the range $0 < u < 1$, and suppose that u vanishes on one boundary component of D and reduces to unity on the other. By applying the previous result about surface area S to the representation

$$(11.77) \quad \iint_D (u_x^2 + u_y^2)\, dx\, dy = \lim_{\epsilon \to 0} \frac{2}{\epsilon^2} \iint_D [\sqrt{1 + \epsilon^2 u_x^2 + \epsilon^2 u_y^2} - 1]\, dx\, dy$$

of the Dirichlet integral of u, show that Steiner symmetrization of u and of D in any line either diminishes this integral or leaves it unchanged.

9. If u is harmonic, the Dirichlet integral (11.77) coincides with the *electrostatic capacity* C of the *condenser* D. Show that for a fixed choice of the areas enclosed by the two boundary components of D, the capacity C assumes its minimum value when D is a circular ring.

10. Let D be a simply connected region of the (x, y)-plane that contains the point at infinity and possesses symmetry in both the coordinate axes. Let u stand for the regular harmonic function in D such that

$$\psi = y + u$$

vanishes on ∂D. The Dirichlet integral

$$M = \iint_D (u_x^2 + u_y^2)\, dx\, dy$$

represents the *virtual mass* of a potential flow described by the *stream function* ψ. Show[12] that M decreases, or at least does not increase, when ψ and D are symmetrized with respect to either the x-axis or the y-axis (cf. Section 15.2).

[12] Cf. Garabedian-Spencer 1.

11. Give a complete proof of the Faber-Krahn inequality (11.60) by means of repeated applications of Steiner symmetrization with respect to lines inclined at arbitrary angles.

12. If μ_2 stands for the second eigenvalue of the free membrane problem (11.24), (11.25), show[13] that the dimensionless product $\mu_2 A$ becomes a maximum in the case of a circle by inserting into the Rayleigh quotient an appropriate trial function suggested by the extremal configuration.

13. Discuss symmetrization with respect to the (x, y)-plane as a tool to prove uniqueness of the solution of the Plateau problem in non-parametric form for the nonlinear partial differential equation

$$(1 + u_y^2)u_{xx} - 2u_x u_y u_{xy} + (1 + u_x^2)u_{yy} = 0$$

over a convex plane region (cf. Exercise 7.1.15 and Section 15.4).

14. Show in detail that when analytic level curves are symmetrized according to either the rule (11.62) or the rule (11.72), the discrepancies introduced by tangent planes that slice through critical points or by tangent lines that possess odd numbers of intercepts cause difficulty only at sets of measure zero which do not contribute to the integrals considered.

[13] Cf. Weinberger 1.

12

Tricomi's Problem; Formulation of Well Posed Problems

1. EQUATIONS OF MIXED TYPE; UNIQUENESS

The theory we have developed thus far illustrates the influence that the type of a partial differential equation has on the auxiliary conditions needed to determine a solution. From a physical point of view we have found it natural to examine those boundary or initial value problems which are well posed in the sense that existence, uniqueness and continuous dependence on data are assured for the solutions. In particular, we have learned that the Cauchy problem is appropriate for hyperbolic equations, of which the wave equation is the most familiar example, whereas the Dirichlet problem is correctly set for equations of the elliptic type like Laplace's equation. Furthermore, a reasonable question concerning parabolic equations is the mixed initial and boundary value problem for the heat equation.

For more general partial differential equations it is not always clear how to formulate auxiliary conditions leading to a well posed problem. In fact, it is easy to think of examples totally lacking in physical interest, such as the ultrahyperbolic equation

$$\frac{\partial^2 u}{\partial x_1 \, \partial x_2} = \frac{\partial^2 u}{\partial x_3 \, \partial x_4},$$

for which no appropriate boundary conditions have as yet been found.

More indicative of the difficulties that can arise is the case of the *Tricomi equation*

(12.1) $$yu_{xx} + u_{yy} = 0.$$

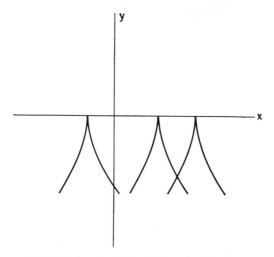

FIG. 46. Characteristics of Tricomi's equation.

It plays a central role in the mathematical analysis of transonic flow because it is of mixed elliptic and hyperbolic type in any region where the coefficient y changes sign. As the simplest equation with that property, it provides a useful mathematical model of the transition from subsonic to supersonic speeds in aerodynamics. The mysteries of transonic flow lead to an unusual situation in which the formulation of a correct problem for (12.1) furnishes the most satisfactory guide to an understanding of what should be expected of the associated physical phenomena. We shall elaborate on this matter by introducing and investigating the *Tricomi problem*[1] for (12.1), whose significance for gas dynamics will become apparent later in connection with our treatment of the hodograph method (cf. Section 14.4).

We prefer to formulate Tricomi's problem for the somewhat more general partial differential equation

$$(12.2) \qquad K(y)u_{xx} + u_{yy} = 0$$

with a coefficient $K = K(y)$ that is positive when $y > 0$ and negative when $y < 0$. Like (12.1), the equation (12.2) is of the elliptic type in the upper half-plane $y > 0$ and is of the hyperbolic type in the lower half-plane $y < 0$. We shall refer to the x-axis as the *parabolic line*.

In order to formulate the Tricomi problem for equation (12.2), we first have to determine its characteristics. According to the rules given in

[1] Cf. Tricomi 1.

Section 3.1, they consist of the two-parameter family of curves in the lower half-plane satisfying the ordinary differential equation

$$K(y)\, dy^2 + dx^2 = 0.$$

We can represent them in the explicit form

(12.3)
$$x = \pm \int_0^y \sqrt{-K(\eta)}\, d\eta + \text{const.}$$

Each individual locus (12.3) is composed of a pair of curves issuing from a cusp situated on the x-axis, provided that K is a sufficiently regular function.

The Tricomi problem for (12.2) applies to a domain D that overlaps the parabolic line $y = 0$ and is bounded by a simple arc C_3 in the upper half-plane joining two points z_1 and z_2 of the x-axis, together with a pair of characteristics C_1 and C_2 descending from z_1 and z_2 until they terminate at a common point of intersection z_3 in the lower half-plane (cf. Figure 47). Equation (12.2) is of the elliptic type in the portion of D bounded by C_3 and the segment of the x-axis located between z_1 and z_2, whereas it is of the hyperbolic type in the remaining curvilinear triangle bounded by that same parabolic segment and the two characteristics arcs C_1 and C_2. Tricomi's problem asks for a solution u of the partial differential equation (12.2) that assumes prescribed values both on the characteristic C_2 in the hyperbolic half-plane and on the boundary curve C_3 in the elliptic half-plane. It should be emphasized that no data are assigned along the characteristic C_1.

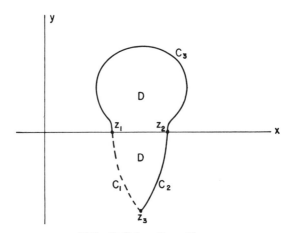

FIG. 47. Tricomi's problem.

Observe that when the points z_1 and z_2 on the parabolic line coincide, C_3 becomes a closed loop in the upper half-plane and Tricomi's problem degenerates into a Dirichlet problem. On the other hand, if C_3 collapses onto the segment of the x-axis joining z_1 and z_2, then the requirements of the Tricomi problem reduce to a limiting case of the Goursat problem for a hyperbolic equation (cf. Section 4.2). These two simplified situations provide preliminary evidence in favor of our contention that the Tricomi problem is correctly set. Moreover, they suggest that if we assign tentative values of u along the parabolic line in the general case, then the continuity of the normal derivative u_y there gives just enough information to determine those values.

We shall develop further justification that Tricomi's problem is the right question to pose for the mixed equation (12.2) by establishing rigorously the uniqueness of the answer when the coefficient K and the boundary curve C_3 are subject to certain hypotheses concerning their regularity. Our uniqueness theorem for the Tricomi problem will be based on a manipulation of suitable quadratic *energy integrals*.[2] The technique we have in mind may be viewed as a generalization of the uniqueness proof for Cauchy's problem that followed from a consideration of the energy inequality (6.16). At the same time it is an extension of the analysis of uniqueness for Dirichlet's problem that hinged on the positive-definite nature of the Dirichlet integral (7.4).

In the case at hand we are led to examine the identity

$$(12.4) \qquad \iint_D (au + bu_x + cu_y)[K(y)u_{xx} + u_{yy}]\, dx\, dy = 0,$$

which is obviously valid for every solution u of (12.2) and for virtually arbitrary choices of the factors a, b and c. Our aim will be to select the functions a, b and c in such a way that the expression on the left in (12.4) can be transformed, through applications of the divergence theorem, into a sum of generalized energy integrals whose values are found by inspection to be non-negative whenever u reduces to zero on the boundary arcs C_2 and C_3 of D. Since the energy integrals in question would have to vanish under these circumstances, we might expect to be able to conclude that u itself is identically zero, which would establish the desired uniqueness theorem.

The chief difficulty in the procedure we have just described, which is often referred to as the (a, b, c)-*method*, arises in making an appropriate selection of the undetermined coefficients a, b and c. Some guidance in

[2] Cf. Morawetz 1, Protter 1.

such matters can be extracted from the earlier examples of the Cauchy problem for the wave equation and of the Dirichlet problem for Laplace's equation (cf. Sections 6.1 and 7.1). In the former case our point of departure was the formula

$$\iint u_t [u_{xx} - u_{tt}] \, dx \, dt = 0,$$

which corresponds essentially to setting $a = 0$, while taking either b or c different from zero. In connection with Dirichlet's problem, on the other hand, we used the relation

$$\iint u [u_{xx} + u_{yy}] \, dx \, dy = 0$$

for harmonic functions, which is analogous to putting $a = 1$ and $b = c = 0$. These remarks suggest that for the Tricomi problem we might have to make a non-trivial choice of the factor a in the upper half-plane, where (12.2) is of the elliptic type, whereas in the hyperbolic region $y < 0$ the coefficients b and c ought to play a dominating role.

There is no loss of generality if we place the origin of our coordinate system at z_1, with z_2 located on the positive x-axis. In order to develop a proof of the uniqueness of the solution of Tricomi's problem involving a quite simple choice of the three functions a, b and c, we make important additional hypotheses.[3] Let us assume that K has a continuous derivative satisfying

(12.5) $K'(y) > 0$

and that the boundary arc C_3 is continuously differentiable and *star-like* in the sense that as it is traversed with the region D lying on the left we have

(12.6) $x \, dy - y \, dx = r^2 \, d\theta \geq 0,$

where r and θ indicate polar coordinates. Furthermore, let us consider only solutions u of (12.2) that are continuous together with their first partial derivatives u_x and u_y throughout the closure of D.

Because of the linearity of the Tricomi problem, the uniqueness of the solution will follow if we can establish that in the homogeneous case where zero boundary values are prescribed along both C_2 and C_3 the

---

[3] Cf. Morawetz 1.

answer u vanishes identically. To arrive at such a conclusion, we begin by putting $a \equiv 0$ in (12.4) and rearranging the integrand to derive

$$\iint_D (bu_x + cu_y)[Ku_{xx} + u_{yy}]\, dx\, dy = \frac{1}{2}\iint_D \left\{ \frac{\partial}{\partial x}\, [Kbu_x^2 + 2Kcu_xu_y - bu_y^2] \right.$$

$$- \frac{\partial}{\partial y}\, [Kcu_x^2 - 2bu_xu_y - cu_y^2] + \left(\frac{\partial}{\partial y}\, [Kc] - Kb_x\right)u_x^2$$

$$\left. - 2(Kc_x + b_y)u_xu_y + (b_x - c_y)u_y^2 \right\}\, dx\, dy = 0.$$

An application of the divergence theorem now yields the relation

(12.7)

$$\int_{C_1+C_2+C_3} \left\{(Kbu_x^2 + 2Kcu_xu_y - bu_y^2)\, dy + (Kcu_x^2 - 2bu_xu_y - cu_y^2)\, dx\right\}$$

$$+ \iint_D \left\{ \left(\frac{\partial}{\partial y}\, [Kc] - Kb_x\right)u_x^2 - 2(Kc_x + b_y)u_xu_y + (b_x - c_y)u_y^2 \right\}\, dx\, dy = 0$$

among quadratic expressions in the partial derivatives u_x and u_y. It is our intention to specify the coefficients b and c so that these quadratic forms are, if not positive-definite, then at least non-negative. (Cf. Exercise 3 below for a less artificial approach taking $a \neq 0$.)

Observe that because C_1 is a characteristic of (12.2) with negative slope it satisfies the ordinary differential equation

$$dx = -\sqrt{-K}\, dy,$$

and observe that along C_2 and C_3 we have

(12.8) $$u_x\, dx + u_y\, dy = 0$$

because u vanishes there. Thus the line integral on the left in (12.7) simplifies and becomes

(12.9) $$\int_{C_1+C_2+C_3} \{b[(Ku_x^2 - u_y^2)\, dy - 2u_xu_y\, dx]$$

$$+ c[(Ku_x^2 - u_y^2)\, dx + 2Ku_xu_y\, dy]\}$$

$$= \int_{C_1} \left(\frac{du}{dy}\right)^2(-b\, dy - c\, dx) + \int_{C_2+C_3}\left[K + \left(\frac{dx}{dy}\right)^2\right]u_x^2(b\, dy - c\, dx),$$

where it is understood that

$$\frac{du}{dy} = u_x \frac{dx}{dy} + u_y = u_y - \sqrt{-K}\, u_x$$

along C_1. The terms on the right in (12.9) will be non-negative provided that

$$(12.10) \qquad\qquad -b\, dy - c\, dx \geq 0$$

on C_1 and provided that

$$(12.11) \qquad\qquad b\, dy - c\, dx \geq 0$$

on C_2 and, more significantly, on C_3. Furthermore, the integrand of the double integral in (12.7) will be positive-definite, or at least non-negative, if

$$(12.12) \qquad (Kc_x + b_y)^2 \leq (b_x - c_y)\left(\frac{\partial}{\partial y}[Kc] - Kb_x\right)$$

and either

$$(12.13) \qquad\qquad \frac{\partial}{\partial y}[Kc] - Kb_x \geq 0$$

or

$$(12.14) \qquad\qquad b_x - c_y \geq 0$$

in D.

A choice of b and c designed to fulfill the requirements (12.10) to (12.14) that all the integrals in (12.7) assume values which are either positive or zero is defined by setting

$$(12.15) \qquad\qquad b = x, \qquad c = y$$

in the intersection D_+ of D with the upper half-plane, where (12.2) is elliptic, and by setting

$$(12.16) \qquad\qquad b = x, \qquad c = 0$$

in the intersection D_- of D with the lower half-plane, where (12.2) is hyperbolic. Notice that in the upper half-plane (12.11) now becomes a consequence of our hypothesis (12.6) that C_3 is star-like, and that it also remains true on C_2 because $x > 0$ and $dy > 0$ there. On the other hand, (12.10) is satisfied because $x \geq 0$ and $dy < 0$ along C_1. The inequality (12.13) follows from (12.5) when $y > 0$, whereas for $y < 0$ it is valid because $K < 0$. Moreover, the rules (12.15) and (12.16) yield (12.12) and (12.14) directly. Finally, we emphasize that the discontinuity of the partial derivative c_y across the parabolic line $y = 0$ does not affect our application of the divergence theorem to derive (12.7), since the line integral involved still has a continuous integrand.

In summary, the specifications (12.15) and (12.16) transform (12.7) into the more cogent identity

$$(12.17) \quad \int_{C_1} \left(\frac{du}{dy}\right)^2 (-x \, dy) + \int_{C_2} \left[K + \left(\frac{dx}{dy}\right)^2\right] u_x^2 x \, dy$$

$$+ \int_{C_3} \left[K + \left(\frac{dx}{dy}\right)^2\right] u_x^2 (x \, dy - y \, dx) + \iint_{D_+} yK'u_x^2 \, dx \, dy$$

$$+ \iint_{D_-} [u_y^2 - Ku_x^2] \, dx \, dy = 0$$

in which all terms are seen by inspection to be either positive or zero. It follows that each of these generalized energy integrals must vanish separately. Incidentally, the second integral on the left in (12.17) could have been omitted because the ordinary differential equation for the characteristics shows that it reduces to zero. More important, however, is the conclusion that

$$u_x = 0$$

in D_+ and

$$u_x = u_y = 0$$

in D_- due to the vanishing of the double integrals in (12.17) over those two portions of D. Since integration of u_x along parallels to the x-axis issuing from C_2 and C_3 now implies that u itself must be identically zero, our proof of the uniqueness of the solution of Tricomi's problem is complete.

We have presented the (a,b,c)-method in a formulation that is easily modified to establish uniqueness of the solution of an extension of the Tricomi problem usually associated with the name of Frankl (cf. Section 14.4). For the statement of Frankl's problem it is convenient to suppose that the point z_1 lies on the negative x-axis and that the point z_2 lies on the positive x-axis. Let Γ_1 and Γ_2 denote two curves descending from z_1 and z_2, respectively, to terminate at the lower ends z_1^* and z_2^* of a pair of characteristic arcs C_1^* and C_2^* for (12.2) that together form a cusp at the origin

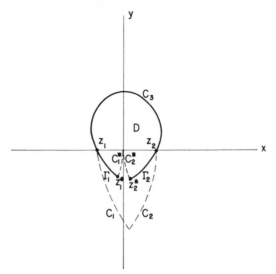

FIG. 48. Frankl's problem.

(cf. Figure 48). We require that Γ_1 and Γ_2 be monotonic, with slopes nowhere exceeding the characteristic slopes in the sense that the differential inequality

(12.18)
$$K + \left(\frac{dx}{dy}\right)^2 \geq 0$$

holds on both of them. With these preliminaries in mind we let D stand for the domain bounded in the lower half-plane by Γ_1, C_1^*, C_2^* and Γ_2, and bounded in the upper half-plane by an arc C_3 joining z_2 to z_1. It will still be assumed, for the purposes of our later analysis, that C_3 is star-like.

The *Frankl problem* asks for a solution u of the mixed equation (12.2) in D which takes on prescribed boundary values along Γ_1, Γ_2 and C_3, but which is not restricted along the pair of characteristics C_1^* and C_2^*. We shall prove the uniqueness of the answer under the same hypotheses (12.5) and (12.6) about the coefficient K and the arc C_3 that arose in our discussion of the Tricomi problem.

In view of the linearity of (12.2), what has to be shown is that u vanishes identically in the homogeneous case where its boundary values along Γ_1, Γ_2 and C_3 are zero. To establish this, we again consider the integral formula (12.4) with $a \equiv 0$ and with b and c defined by (12.15) and (12.16). Taking into account the ordinary differential equation

$$\left(\frac{dx}{dy}\right)^2 = -K$$

satisfied by C_1^* and C_2^* and the relation (12.8) valid on Γ_1, Γ_2 and C_3, we are led to transform (12.4) by means of the divergence theorem into the identity

$$(12.19) \quad \int_{C_1^*+C_2^*} \left(\frac{du}{dy}\right)^2 (-x\,dy) + \int_{\Gamma_1+\Gamma_2} \left[K + \left(\frac{dx}{dy}\right)^2\right] u_x^2 x\,dy$$

$$+ \int_{C_3} \left[K + \left(\frac{dx}{dy}\right)^2\right] u_x^2 (x\,dy - y\,dx) + \iint_{D_+} yK' u_x^2\,dx\,dy$$

$$+ \iint_{D_-} [u_y^2 - Ku_x^2]\,dx\,dy = 0$$

analogous to (12.17).

The geometry of C_1^* and C_2^* shows that they both satisfy

$$-x\,dy \geq 0,$$

whereas on Γ_1 and Γ_2 the two differential inequalities (12.18) and

$$x\,dy \geq 0$$

hold. Thus it is easily concluded that each of the generalized energy integrals on the left in (12.19) is non-negative. Therefore these integrals all have to vanish, since a sum of positive quantities could not reduce to zero. In particular, we have

$$\iint_{D_+} yK' u_x^2\,dx\,dy = \iint_{D_-} |K|\,u_x^2\,dx\,dy = 0.$$

It follows that $u_x = 0$ in both D_+ and D_-, whence an integration with respect to x along line segments starting from Γ_1, Γ_2 or C_3 implies that u vanishes throughout the interior of D. This completes our proof of the uniqueness of the solution of Frankl's problem.

The uniqueness theorems we have developed for the Tricomi and Frankl problems provide partial justification of our contention that they are the right questions to pose for the mixed equation (12.2) in a region overlapping the parabolic line $y = 0$. We relegate to the exercises a discussion of the existence of the solutions, for a thorough analysis here would lead us too far. The principal idea that we wish to bring out is how the method of energy integrals, or, in other words, the (a,b,c)-method, serves as a general guide in determining what problems are correctly set. We shall elaborate further on such matters in the next section, where the role of energy inequalities will be described in the development of existence proofs. In this context we stress that it is the energy inequality implicit in our analysis of Tricomi's problem rather than the actual uniqueness theorem we have established that makes the problem plausible.

EXERCISES

1. Prove the uniqueness of the solution of the Dirichlet problem for Laplace's equation in a star-like domain D by means of the (a,b,c)-method with $a = 0$, $b = x$ and $c = y$.

2. We call (12.2) the *Bitsadze-Lavrentiev equation*[4] when $K \equiv 1$ for $y > 0$ and $K \equiv -1$ for $y < 0$. Show how to solve Tricomi's problem explicitly for this equation in the case where the region D is a square $|x| + |y| < $ const. Continuity of the first derivatives of the solution should be required across the parabolic line, of course.

3. In the statement of our uniqueness theorem for Tricomi's problem, replace the assumption (12.6) about the boundary curve C_3 by the new hypothesis[5] that the coefficient K satisfies the requirement

$$2 \frac{d}{dy} \frac{K}{K'} + 1 > 0$$

in the lower half-plane. To prove the revised theorem by the (a,b,c)-method, put $a = 1$ everywhere, put $b = c = 0$ for $y \geq 0$, but put

$$b = c\sqrt{-K}, \qquad c = 4K/K'$$

for $y < 0$.

4. Show that in any region D where u is a solution of the mixed equation (12.2) the line integral

$$\psi = \int_{z_0}^{z} \{-2u_x u_y \, dx + (Ku_x^2 - u_y^2) \, dy\}$$

from a fixed position z_0 to a variable point z defines a function of the coordinates of z which has to assume its maximum on the boundary of D. Use this maximum principle to establish the uniqueness theorem for Frankl's problem that we described in the text.[6]

5. Let u be a solution of (12.2) in a curvilinear triangle bounded by a segment of the x-axis and two characteristic arcs C_1 and C_2. If u does not vanish identically and, furthermore, represents along C_2 a non-decreasing function of y, while

(12.20) $$5(K')^2 \geq 4KK'', \qquad K' \geq 0$$

[4] Cf. Lavrentiev-Bitsadze 1.
[5] Cf. Protter 1.
[6] Cf. Morawetz 2.

for all $y \leq 0$, show that u must achieve its maximum over the above triangle at some point on the parabolic line where the partial derivative u_y is positive.[7] Use this maximum principle to establish the uniqueness of the solution of Tricomi's problem when the hypothesis (12.20) replaces (12.6).

6. Use the (a,b,c)-method to establish the uniqueness of the solution of the mixed initial and boundary value problem for the parabolic equation

$$(\rho u_x)_x + (\rho u_y)_y = u_t, \qquad \rho = \rho(x, y) > 0,$$

in which $u = u(x, y, t)$ is prescribed over a finite region D of the (x,y)-plane when $t = 0$ and is prescribed on ∂D for all $t > 0$.

7. Deduce by the (a,b,c)-method the uniqueness of a solution of the wave equation

$$u_{xx} + u_{yy} = u_{tt}$$

whose values are prescribed on the circular disc

$$t = 0, \qquad x^2 + y^2 \leq 1,$$

and on the mantle of the characteristic cone

$$t = 1 - \sqrt{x^2 + y^2}$$

located above the plane $t = 0$, but below a second cone

$$t = \sqrt{(x - x_0)^2 + (y - y_0)^2}, \qquad x_0^2 + y_0^2 < 1.$$

8. Find an explicit representation in terms of hypergeometric functions for the Green's function $G = G(z;\zeta)$ of Tricomi's equation (12.1) for the elliptic half-plane $y > 0$. To do this, first bring (12.1) into canonical form and then appeal to results from the exercises at the end of Section 5.1. Solve Cauchy problems with data assigned on the parabolic line to extend the definition of G into the lower half-plane. Thus obtain in a natural way closed formulas for a fundamental solution of (12.1) specified everywhere.

9. Seek the solution of the Tricomi problem for Tricomi's equation (12.1) in the form[8]

$$(12.21) \quad u(z) = \left(\frac{3}{2}\right)^{1/3} \int_{C_3} \mu(s) \, \eta^{1/3} \frac{\partial}{\partial \nu} \, G(z;\zeta) \, ds$$

$$+ \left(\frac{3}{2}\right)^{1/3} \int_{C_2} [2(1 - \xi)^{1/3} \rho'(\xi) - \frac{1}{3}(1 - \xi)^{-2/3} \rho(\xi)] G(z;\zeta) \, d\xi$$

of potentials of the fundamental solution G introduced in Exercise 8, where ξ and η stand for the coordinates of the parameter point ζ and s and ν denote the arc length and unit normal along C_3, and where we have placed the endpoints

[7] Cf. Agmon-Nirenberg-Protter 1.
[8] Cf. Agmon 1, Germain-Bader 1.

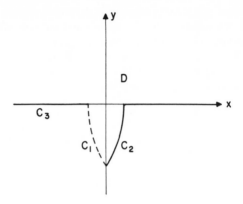

FIG. 49. Tricomi's problem for a half-plane.

z_1 and z_2 of C_3 at $(-1, 0)$ and $(1, 0)$ for the sake of simplicity. Use the procedures of Section 9.3 to derive integral equations for the unknown densities μ and ρ in (12.21). Apply the Fredholm alternative to conclude that at least for the special equation (12.1) a solution of the Tricomi problem always exists.

10. Solve the Tricomi problem for Tricomi's equation (12.1) in closed form in the degenerate case where D_+ coincides with the upper half-plane, while D_- is still the curvilinear triangle bounded from below by characteristics emanating from $(-1, 0)$ and $(1, 0)$. More specifically, suppose that u is prescribed to be zero along the two semi-infinite segments $x \leq -1$ and $x \geq 1$ of the x-axis comprising C_3, and let the boundary values of u along C_2 be described by a function $f(x)$ defined on the interval $0 \leq x \leq 1$. Show that formula (12.21) provides a solution of this particular Tricomi problem if we put $\mu = 0$ and

$$\rho(\xi) = \frac{f(\xi)}{2} - \frac{\sqrt{3}}{2\pi} \int_0^1 f(x) \left(\frac{\xi}{x}\right)^{1/3} \left(\frac{1-\xi}{1-x}\right)^{1/2} \frac{dx}{x-\xi},$$

with the improper integral interpreted in the sense of a Cauchy principal value.

2. ENERGY INTEGRAL METHOD FOR SYMMETRIC HYPERBOLIC SYSTEMS

Our intention here is to show how the (a,b,c)-method, or method of energy inequalities, can be utilized to establish existence theorems about solutions of partial differential equations. We shall illustrate the basic procedure by applying it to what is perhaps the simplest example that still indicates its broader scope and generality. The case we have in mind is that of the Cauchy problem for a symmetric hyperbolic system of linear partial differential equations in space of arbitrary dimension (cf. Section 3.5). The existence of a solution in the large will be deduced by appealing to *a priori* estimates of its behavior that are derived from a consideration of suitable energy integrals.

To be precise, let $A_1, \ldots, A_n$ and B stand for $n + 1$ matrices, each consisting of m rows and m columns, whose elements are given functions of a single time variable t and of n space variables $x_1, \ldots, x_n$ forming an n-dimensional vector x. Moreover, suppose that $A_1, \ldots, A_n$ are symmetric. We denote by u an m-dimensional column vector depending on x and t which satisfies the matrix equation

$$(12.22) \qquad \frac{\partial u}{\partial t} = \sum_{j=1}^{n} A_j \frac{\partial u}{\partial x_j} + Bu.$$

As already indicated in Section 3.5, (12.22) defines a linear hyperbolic system of m partial differential equations of the first order for the components $u_1, \ldots, u_m$ of the unknown vector u. More specifically, because the matrices A_j are symmetric we call the system (12.22) *symmetric hyperbolic*. In this connection it is not necessary to impose any requirement of symmetry on the coefficient B of the undifferentiated term u.

We shall be interested in the Cauchy problem which asks for a solution $u = u(x, t)$ of (12.22) that takes on prescribed initial values

$$(12.23) \qquad u(x, 0) = F(x)$$

at $t = 0$. To facilitate our analysis we shall make the hypothesis that the coefficient matrices A_j and B, together with the given vector F, are real analytic functions of their arguments in the sense that they possess convergent Taylor series expansions everywhere. In these circumstances the Cauchy-Kowalewski theorem (cf. Section 1.2) implies that a unique analytic solution u of the Cauchy problem defined by (12.22) and (12.23) exists locally. Our aim will be to establish that the solution referred to is valid in the large and that its value at each point involves only data specified over a domain of dependence bounded by characteristic surfaces through that point.

A *characteristic* of equation (12.22) is an n-dimensional surface

$$\phi(x, t) = 0$$

on which the values of u are not enough to determine the normal derivative $\partial u / \partial \nu$ of that vector. Thus the condition for $\phi = 0$ to be a characteristic is the relation

$$\left| \sum_{j=1}^{n} A_j \frac{\partial \phi}{\partial x_j} - I \frac{\partial \phi}{\partial t} \right| = 0$$

asserting that the determinant of the coefficient of $\partial u/\partial \nu$ in (12.22) vanishes, where, of course, I stands for the identity matrix. The system (12.22) is said to be *hyperbolic* when it has real characteristics and when, furthermore, for every real choice of the parameters $\lambda_1, \ldots, \lambda_n$ the polynomial equation

$$\left| \sum_{j=1}^{n} \lambda_j A_j - \lambda I \right| = 0$$

in λ has m distinct real characteristic roots, or, if not, the matrix $\sum_{j=1}^{n} \lambda_j A_j$ at least possesses a full set of m linearly independent real characteristic vectors. The symmetry requirement imposed above on the matrices A_j is seen to be the simplest available sufficient condition that (12.22) be of the hyperbolic type.

In Exercise 6.1.8 we developed a uniqueness theorem for symmetric hyperbolic systems which shows that for each point in space of dimension $n + 1$ the value of the solution u of the Cauchy problem (12.22), (12.23) is determined by a portion of the initial data prescribed over a certain domain of dependence within the hyperplane $t = 0$. In particular when A_j, B and F are all analytic the power series solution u of (12.22) and (12.23) is the only one that exists. More precisely, at any specific point (x, t) the data on which $u(x, t)$ depends are confined to a closed domain cut out of the hyperplane $t = 0$ by a conoid whose vertex is situated at (x, t) and whose normal at each point forms with the t-axis an angle not exceeding those made by the normals of the characteristic surfaces there. We wish to establish that whenever analytic initial values F are given over a domain of dependence of the kind just described, the associated analytic

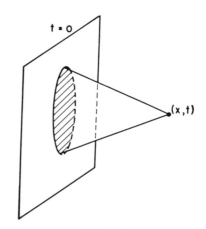

FIG. 50. Domain of dependence.

solution of the Cauchy problem (12.22), (12.23) is defined throughout the $(n + 1)$-dimensional pyramid bounded by that domain and the above conoid through (x, t). The Cauchy-Kowalewski theorem does not yield such a conclusion directly because the infinite series on which it is based can only be expected to converge in the small.

We might attempt to construct u in the pyramid where we have asserted that it should exist by either performing an analytic continuation on it or, in a similar way, by iterating the Cauchy-Kowalewski procedure at a succession of levels of the time variable t. In order to justify such techniques we would have to derive strong *a priori* estimates on the growth of the higher partial derivatives of u. However, we shall find that it is easier to replace estimates of that kind in our analysis by bounds on integrals of $|u|^2$ in the complex domain which also provide the desired restriction on the derivatives of u. Thus our next step will be to extend the definition of u to complex values of its arguments and to formulate a new symmetric hyperbolic system of partial differential equations that its extension has to fulfill (cf. Section 16.1). Energy inequalities obtained from this complex system will yield the estimates of u needed for our existence proof.

It is of interest to compare the existence proof just referred to with the one we developed in Section 6.4 for hyperbolic equations of the second order. The technique of iteration at successive levels of t appears in both approaches. Moreover, they are both built around *a priori* estimates of the solution. The principal difference lies in the manner in which these estimates are obtained. Here we shall derive them from a basic energy inequality. There they were deduced from a representation of the answer in terms of a fundamental solution, which was advantageous because it furnished an explicit solution of Cauchy's problem whenever the fundamental solution could be calculated in closed form. Finally, note that the energy integral method is related to Dirichlet's principle, whereas the approach through a fundamental solution is more analogous to the method of integral equations.

We can continue u analytically into the complex domain by merely substituting for the real arguments $x_1, \ldots, x_n$ in its Taylor series expansion

$$u(x, t) = \sum_{j_0, \ldots, j_n=0}^{\infty} a_{j_0 \ldots j_n}(t - t^{(0)})^{j_0}(x_1 - x_1^{(0)})^{j_1} \cdots (x_n - x_n^{(0)})^{j_n}$$

complex variables

$$z_1 = x_1 + iy_1, \ldots, z_n = x_n + iy_n$$

with sufficiently small imaginary parts $y_1, \ldots, y_n$. Incidentally, for the purposes we have in mind it will never be necessary to make t complex.

Considered as a complex analytic function

$$u(z, t) = \xi + i\eta$$

of the complex vector $z = (z_1, \ldots, z_n)$ and of the real time t, u still has to satisfy the partial differential equation (12.22) as an identity

$$(12.24) \qquad \frac{\partial u}{\partial t} = \sum_{j=1}^{n} A_j \frac{\partial u}{\partial z_j} + Bu$$

in which the coefficients A_j and B have also been extended analytically into the complex domain. Similarly, the initial condition

$$(12.25) \qquad u(z, 0) = F(z)$$

remains valid for complex values of z if the function on the right is understood to represent the analytic continuation of the right-hand side of (12.23).

We wish to reformulate (12.24) as a symmetric hyperbolic system of $2m$ partial differential equations for the real components $\xi_1, \ldots, \xi_m$ and $\eta_1, \ldots, \eta_m$ of the unknown vectors ξ and η in the space of dimension $2n + 1$ spanned by the real coordinates $x_1, \ldots, x_n, y_1, \ldots, y_n$ and t. To this end we observe that according to the Cauchy-Riemann equations we can write

$$(12.26) \qquad \frac{\partial u}{\partial z_j} = \frac{1}{2} \left(\frac{\partial u}{\partial x_j} - i \frac{\partial u}{\partial y_j} \right), \qquad j = 1, \ldots, n,$$

and

$$(12.27) \qquad \frac{\partial u}{\partial \bar{z}_j} = \frac{1}{2} \left(\frac{\partial u}{\partial x_j} + i \frac{\partial u}{\partial y_j} \right) = 0, \qquad j = 1, \ldots, n,$$

where, of course,

$$\bar{z}_1 = x_1 - iy_1, \ldots, \bar{z}_n = x_n - iy_n.$$

We multiply (12.27) by the transposed complex conjugate $\bar{A}_j'$ of the matrix A_j, sum on the index j, and then add the resulting expression

$$\sum_{j=1}^{n} \bar{A}_j' \frac{\partial u}{\partial \bar{z}_j} = 0$$

to (12.24) in order to derive the quite equivalent relation

$$(12.28) \qquad \frac{\partial u}{\partial t} = \sum_{j=1}^{n} A_j \frac{\partial u}{\partial z_j} + \sum_{j=1}^{n} \bar{A}_j' \frac{\partial u}{\partial \bar{z}_j} + Bu$$

for the analytic vector u. Introducing (12.26) and (12.27) into (12.28), we obtain, finally, the new system of partial differential equations

$$(12.29) \qquad \frac{\partial u}{\partial t} = \sum_{j=1}^{n} \frac{A_j + \bar{A}_j'}{2} \frac{\partial u}{\partial x_j} + \sum_{j=1}^{n} \frac{A_j - \bar{A}_j'}{2i} \frac{\partial u}{\partial y_j} + Bu$$

involving exclusively real independent variables.

We assert that the system (12.29) is symmetric hyperbolic. Indeed, for every real choice of the parameters $\lambda_1, \ldots, \lambda_n, \mu_1, \ldots, \mu_n$ the matrix

$$(12.30) \qquad A = \sum_{j=1}^{n} \lambda_j \frac{A_j + \bar{A}_j'}{2} + \sum_{j=1}^{n} \mu_j \frac{A_j - \bar{A}_j'}{2i}$$

is Hermitian,

$$\bar{A}' = A.$$

Hence it possesses a full set of m linearly independent characteristic vectors and has strictly real characteristic roots. This property of (12.29) will enable us to develop a generalized energy inequality (cf. Exercise 6.1.8) for the estimation of u over the complex domain in terms of the initial data F.

Motivated by the (a,b,c)-method, we multiply (12.29) on the left by the row vector $\bar{u}'$ and find that

$$\bar{u}' \frac{\partial u}{\partial t} = \sum_{j=1}^{n} \bar{u}' \frac{A_j + \bar{A}_j'}{2} \frac{\partial u}{\partial x_j} + \sum_{j=1}^{n} \bar{u}' \frac{A_j - \bar{A}_j'}{2i} \frac{\partial u}{\partial y_j} + \bar{u}'Bu.$$

Addition of the latter scalar formula to its complex conjugate yields

$$\frac{\partial}{\partial t} |u|^2 = \frac{\partial}{\partial t} \bar{u}'u = \sum_{j=1}^{n} \frac{\partial}{\partial x_j} \left[\bar{u}' \frac{A_j + \bar{A}_j'}{2} u \right] + \sum_{j=1}^{n} \frac{\partial}{\partial y_j} \left[\bar{u}' \frac{A_j - \bar{A}_j'}{2i} u \right]$$

$$+ \bar{u}' \left[B + \bar{B}' - \sum_{j=1}^{n} \frac{\partial}{\partial x_j} \frac{A_j + \bar{A}_j'}{2} - \sum_{j=1}^{n} \frac{\partial}{\partial y_j} \frac{A_j - \bar{A}_j'}{2i} \right] u.$$

Integration over a $(2n + 1)$-dimensional region D, followed by an application of the divergence theorem, now shows that

$$(12.31) \qquad \int_{\partial D} \bar{u}' \left[\frac{\partial t}{\partial \nu} I - \sum_{j=1}^{n} \frac{\partial x_j}{\partial \nu} \frac{A_j + \bar{A}_j'}{2} - \sum_{j=1}^{n} \frac{\partial y_j}{\partial \nu} \frac{A_j - \bar{A}_j'}{2i} \right] u \, d\sigma$$

$$+ \int_D \bar{u}'Cu \, dx \, dy \, dt = 0,$$

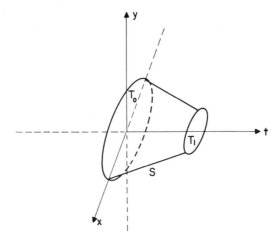

FIG. 51. Geometry of the energy inequality.

where

$$C = B + \bar{B}' - \sum_{j=1}^{n} \frac{\partial}{\partial x_j} \frac{A_j + \bar{A}_j'}{2} - \sum_{j=1}^{n} \frac{\partial}{\partial y_j} \frac{A_j - \bar{A}_j'}{2i},$$

and where

$$dx = dx_1 \cdots dx_n, \qquad dy = dy_1 \cdots dy_n.$$

We shall derive the desired energy inequality from (12.31) by choosing D in a manner suggested by our knowledge about domains of dependence for (12.29).

Let T_0 and T_1 stand, respectively, for sections of the hyperplanes $t = 0$ and $t = \tau > 0$ in $(2n + 1)$-dimensional space, and let S denote a sheath joining the edge of T_0 to the edge of T_1. We shall suppose that the region D bounded by T_0, S and T_1 has the topological structure of a cell, or solid sphere. Moreover, we shall assume that the inner normal ν on S makes such a small angle with the negative t-axis that the matrix

$$(12.32) \quad \frac{\partial t}{\partial \nu} [A - I] = \sum_{j=1}^{n} \frac{\partial x_j}{\partial \nu} \frac{A_j + \bar{A}_j'}{2} + \sum_{j=1}^{n} \frac{\partial y_j}{\partial \nu} \frac{A_j - \bar{A}_j'}{2i} - \frac{\partial t}{\partial \nu} I$$

is not only Hermitian but also positive-definite along S because of the preponderance of the term involving I. In other words, S is supposed to be *space-like* (cf. Section 6.1). In this connection it is understood, incidentally, that we have put

$$\lambda_j = \frac{\partial x_j}{\partial \nu} \bigg/ \frac{\partial t}{\partial \nu}, \qquad \mu_j = \frac{\partial y_j}{\partial \nu} \bigg/ \frac{\partial t}{\partial \nu}, \qquad j = 1, \ldots, n,$$

in the definition (12.30) of A.

Inserting the domain D just described as the region of integration in (12.31), we are led to the inequality

$$(12.33) \quad \int_{T_0} \bar{u}'u \, dx \, dy - \int_{T_1} \bar{u}'u \, dx \, dy + \int_D \bar{u}'Cu \, dx \, dy \, dt$$

$$= \int_S \bar{u}' \frac{\partial t}{\partial \nu} [A - I]u \, d\sigma \geq 0.$$

Observe that in an appropriate physical interpretation formula (12.33) relates the energy contained in T_1 to the initial energy distributed over T_0.

Since the matrix C is analytic, and therefore certainly continuous, a number γ can be found so large that

$$\bar{u}'Cu \leq \gamma \bar{u}'u = \gamma |u|^2$$

for every choice of the m-dimensional vector u. Hence it follows from (12.33) that

$$(12.34) \quad \int_{T_1} |u|^2 \, d\sigma \leq \int_{T_0} |u|^2 \, d\sigma + \gamma \int_D |u|^2 \, d\sigma \, dt,$$

where we have utilized the abbreviation

$$d\sigma = dx \, dy.$$

Since the result (12.34) is actually valid when T_0 is located in any auxiliary hyperplane $t = $ const. to the left of T_1, rather than just in the hyperplane $t = 0$, we are able to conclude by a passage to the limit as $T_0 \to T_1$ that the differential inequality

$$(12.35) \quad \frac{\partial}{\partial \tau} \int_{T_1} |u|^2 \, d\sigma \leq \gamma \int_{T_1} |u|^2 \, d\sigma$$

holds, too. Multiplying (12.35) by $e^{-\gamma \tau}$ and integrating with respect to the parameter τ, and then applying the initial condition (12.25) at $\tau = 0$, we obtain the final *energy inequality*

$$(12.36) \quad \int_{T_1} |u|^2 \, d\sigma \leq e^{\gamma \tau} \int_{T_0} |F|^2 \, d\sigma.$$

This is the fundamental *a priori* estimate of u on which our existence theorem for the Cauchy problem is to be based.

The Cauchy-Riemann equations (12.27) imply that u is a harmonic function of the variables $x_1, \ldots, x_n, y_1, \ldots, y_n$. Hence its value at the center of any sphere E of dimension $2n$ and volume V contained within T_1 coincides with its average

$$u = \frac{1}{V} \int_E u \, d\sigma$$

over E, according to the mean value theorem of Section 7.2. Thus from Schwarz's inequality and (12.36) we are able to derive the bound

$$(12.37) \qquad |u|^2 \leq \frac{1}{V} \int_E |u|^2 \, d\sigma \leq \frac{e^{\gamma \tau}}{V} \int_{T_0} |F|^2 \, d\sigma$$

on u at the center of E. A representation of u by means of the Cauchy integral formula now yields in each closed subdomain of T_1 or of D uniformly valid estimates in terms of F on u and all its partial derivatives with respect to $x_1, \ldots, x_n, y_1, \ldots, y_n$. Similar estimates hold for all the other partial derivatives of u, also, because the system of equations (12.29) can be applied to eliminate differentiations with respect to t. In particular, we conclude from the linearity of (12.29) that if the analytic initial data F are made to approach a limit throughout T_0, then u and its derivatives converge uniformly to analytic limits over any closed set covered by the interior of T_1.

Our next aim will be to deduce from the energy inequality (12.36) a proof that a solution of the Cauchy problem (12.22), (12.23) exists in the large at least when the data are analytic. For this purpose we shall find it convenient[9] to introduce polynomial approximations to the initial values F. Since our hypothesis is merely that F is analytic over some region of the real domain, its analytic extension can only be assumed to have meaning throughout some thin neighborhood of that region in the complex space spanned by $x_1, \ldots, x_n, y_1, \ldots, y_n$. However, in any closed subdomain of such a complex neighborhood, polynomials in $z_1, \ldots, z_n$ can be constructed which furnish a uniformly valid approximation to F. To verify the latter assertion we have only to make a preliminary real substitution of variables mapping the real region of definition of F onto a product of intervals and then apply Runge's theorem[10] to each of the transformed complex variables z_j separately in a $(2n)$-dimensional product domain surrounding those intervals.

Because the equation (12.29) is linear, superposition of its solutions is possible. Consequently the Cauchy-Kowalewski power series solution

[9] Cf. Schauder 1.
[10] Cf. Nehari 1.

associated with any polynomial initial values converges in a region whose size depends only on properties of the coefficients A_j and B (cf. Exercise 1.2.3). In particular, within the sheath S such solutions are valid throughout slabs perpendicular to the t-axis of a fixed width δt. Now let us approximate the arbitrary complex analytic initial values F on T_0 by polynomials in $z_1, \ldots, z_n$. According to the results about convergence derived from the energy inequality (12.36) above, the corresponding solutions of Cauchy's problem will converge to a complex analytic limit solution throughout a slab of width δt enclosed by S. Thus an analytic solution of the Cauchy problem can be found in any such slab of D when the data are prescribed analytically.

In order to obtain a global solution of Cauchy's problem in the pre-assigned region D bounded by T_0, S and T_1, we subdivide D into a finite number of slabs perpendicular to the t-axis whose widths do not exceed δt. We have left merely to solve analytic initial value problems for (12.29) at a succession of levels of t thus specified, which can be accomplished in the fashion just described. Fitting together the answers so determined, we arrive at the desired global solution. In this connection observe that for real arguments the extent of the solution depends only on the real characteristics of (12.22) and the size of the region over which F is defined as a real analytic function of x. Indeed, since we consider only matrices A_j with the symmetry property

$$A_j = A_j' = \bar{A}_j', \qquad j = 1, \ldots, n,$$

in the real domain, the coefficients $(A_j - \bar{A}_j')/2i$ of the partial derivatives $\partial u / \partial y_j$ in (12.29) remain small when the imaginary parts $y_1, \ldots, y_n$ of $z_1, \ldots, z_n$ are small. Therefore the inner normal ν of the sheath S that joins the edges of the plane sections T_0 and T_1 occurring in (12.36) can have large components $\partial y_j / \partial \nu$ along the axes of $y_1, \ldots, y_n$ without violating the important hypothesis that (12.32) is a positive-definite Hermitian matrix. This means that in the derivation of the energy inequality (12.36) we are justified in using narrow sheaths S that surround and nearly touch a characteristic pyramid of dependence in the real domain associated with the original symmetric hyperbolic system (12.22).

Our construction of a solution of the Cauchy problem (12.22), (12.23) in the large, even though it only succeeds in the case of analytic data, still serves to confirm that the problem is reasonable for a symmetric hyperbolic system. However, a modification of our argument is required if the assumption about analyticity of $A_1, \ldots, A_n$, B and F is dropped. Since analytic continuation is then impossible, it becomes necessary to estimate the higher derivatives of u by applying the energy inequality (12.36) to new

Cauchy problems for them obtained by differentiation of (12.22) and (12.23) with respect to the space variables $x_1, \ldots, x_n$ (cf. Exercise 10 below). Instead of going into that analysis in detail, we shall outline another, more abstract, method of constructing solutions of quite general linear symmetric hyperbolic systems. The procedure in question, while it avoids explicit mention of power series, does exploit lemmas from the theory of Hilbert space that may seem a trifle deep.[11]

In the notation of (12.22) we introduce the differential operator

$$L[u] = \frac{\partial u}{\partial t} - \sum_{j=1}^{n} A_j \frac{\partial u}{\partial x_j} - Bu,$$

where the coefficients A_j and B may no longer be analytic but the A_j are still supposed to be symmetric. We shall be interested in finding solutions of the inhomogeneous equation

(12.38) $L[u] = f$

for arbitrary choices of the given vector f. For the purposes of the present investigation it will be convenient to assume that B is a predominantly negative matrix in the sense that it has the form

(12.39) $B = B_0 - \alpha I$

with B_0 small, but with the parameter α extremely large and positive. We note that (12.39) does not impose any genuine restriction on the problem of solving (12.38), since the substitution $u = U e^{\alpha t}$ could always be used to convert that partial differential equation into an equivalent one

$$\frac{\partial U}{\partial t} - \sum_{j=1}^{n} A_j \frac{\partial U}{\partial x_j} - (B - \alpha I)U = f e^{-\alpha t}$$

for U which has the desired property. For the sake of simplicity we shall make the further hypothesis that the data A_j, B and f in (12.38) are periodic functions of each of the independent variables $x_1, \ldots, x_n$ and t. This does not limit the generality of the Cauchy problem for (12.38) in an essential way provided that the period parallelepiped D involved is taken large enough.

We intend to establish here the existence of a weak solution u of the symmetric hyperbolic system (12.38) which has the same periods as the coefficients A_j and B and the term f. The connection between such a periodic solution and the notion of an eigenfunction is clear from the

[11] Cf. Lax 3.

remarks at the beginning of Section 11.1. Although it actually follows that the Cauchy problem can be solved for (12.38), too, we shall relegate the proof of that more important result to the exercises.[12] Our construction of u will be based on an energy inequality satisfied by periodic solutions v of the inhomogeneous *adjoint equation*

$$(12.40) \qquad\qquad M[v] = g,$$

where

$$M[v] = -\frac{\partial v}{\partial t} + \sum_{j=1}^{n} A_j \frac{\partial v}{\partial x_j} - \left(B' - \sum_{j=1}^{n} \frac{\partial A_j}{\partial x_j}\right)v.$$

The divergence theorem shows that for any pair of periodic vectors u and v defined over the period parallelepiped D we have

$$(12.41) \quad \int_D \{v'L[u] - u'M[v]\} \, dx \, dt$$

$$= \int_D \left\{\frac{\partial}{\partial t} v'u - \sum_{j=1}^{n} \frac{\partial}{\partial x_j} v'A_ju\right\} dx \, dt$$

$$= -\int_{\partial D} \left\{\frac{\partial t}{\partial \nu} v'u - \sum_{j=1}^{n} \frac{\partial x_j}{\partial \nu} v'A_ju\right\} d\sigma = 0,$$

since the integrals over opposite faces of ∂D cancel each other out. In particular, observe that a periodic solution u of the partial differential equation (12.38) has to fulfill

$$(12.42) \qquad\qquad \int_D \{v'f - u'M[v]\} \, dx \, dt = 0$$

for every sufficiently smooth choice of the periodic trial function v. Thus we are led to call any square integrable vector u satisfying the integral identity (12.42) for all such choices of v a *weak solution* of the symmetric hyperbolic system (12.38). Similar concepts have been encountered before in Section 8.1.

We introduce the Hilbert space H of all square integrable, periodic, m-dimensional vectors defined on the period parallelepiped D. The scalar product of two elements u and v in H is given by the integral

$$(u, v) = \int_D u'v \, dx \, dt,$$

---

[12] Cf. Lax 3.

and the corresponding norm is

$$\|u\| = \sqrt{(u, u)}.$$

Putting $u = v$ in (12.41), we are able to conclude that

$$(v, M[v]) = \frac{1}{2}(v, L[v]) + \frac{1}{2}(v, M[v])$$

$$= \int_D v' \left(\frac{1}{2} \sum_{j=1}^{n} \frac{\partial A_j}{\partial x_j} - B \right) v \, dx \, dt$$

for any continuously differentiable element v of H. On the other hand, if an adequate bound is assumed on the derivatives $\partial A_j / \partial x_j$, the requirement (12.39) that the matrix B should be predominantly negative gives

$$\int_D v' \left(\frac{1}{2} \sum_{j=1}^{n} \frac{\partial A_j}{\partial x_j} - B \right) v \, dx \, dt \geq (v, v).$$

It follows that

(12.43) $(v, v) \leq (v, M[v]).$

Taking (12.40) into account and applying the Schwarz inequality to (12.43), we obtain

$$\|v\|^2 \leq (v, g) \leq \|v\| \, \|g\|,$$

or

(12.44) $\|v\| \leq \|g\|.$

It is from the energy inequality (12.44) that we shall develop our existence theorem about periodic solutions of the symmetric hyperbolic system (12.38). To do so we consider the scalar product (f, v) as a linear functional of elements v in the Hilbert space H. For those vectors v in H that are continuously differentiable, (12.44) shows that the rule

(12.45) $\Lambda\{g\} = (f, v)$

defines a bounded linear functional Λ of the term g on the right in the partial differential equation (12.40), as well as of v itself. Indeed, by Schwarz's inequality we have

(12.46) $|\Lambda\{g\}| \leq \|f\| \, \|v\| \leq \|f\| \, \|g\|,$

which means that Λ has a norm $\|\Lambda\|$ not exceeding $\|f\|$. Now the Hahn-Banach theorem (cf. Section 9.1) asserts that Λ can be extended from the linear subspace of continuous expressions $g = M[v]$ to the full Hilbert space H of arbitrary square integrable elements g in such a fashion that it retains the property (12.46). We shall prove that the extended functional can be represented as the scalar product

$$(12.47) \qquad \Lambda\{g\} = (u, g)$$

of g with some element u in H which will turn out to be the desired weak solution of (12.38).

Let ϕ_k denote a complete orthonormal system of elements in H. According to (12.46) we can write

$$\sum_{k=1}^{N} \Lambda\{\phi_k\}^2 = \Lambda\left\{\sum_{k=1}^{N} \Lambda\{\phi_k\}\phi_k\right\} \le \|f\| \left\|\sum_{k=1}^{N} \Lambda\{\phi_k\}\phi_k\right\| = \|f\| \sqrt{\sum_{k=1}^{N} \Lambda\{\phi_k\}^2},$$

and therefore

$$\sum_{k=1}^{\infty} \Lambda\{\phi_k\}^2 < \infty.$$

By the Riesz-Fischer theorem (cf. Section 8.1) it follows that the infinite series

$$(12.48) \qquad u = \sum_{k=1}^{\infty} \Lambda\{\phi_k\}\phi_k$$

defines a square integrable vector u in H. Moreover, since

$$(u, g) = \sum_{k=1}^{\infty} \Lambda\{\phi_k\}(\phi_k, g) = \Lambda\left\{\sum_{k=1}^{\infty} (\phi_k, g)\phi_k\right\},$$

(12.48) is the required element u of H for which the representation (12.47) of Λ is valid.

It remains to show that u is a weak solution of (12.38). But a comparison of (12.45) with (12.47) yields

$$(f, v) = \Lambda\{g\} = (u, g),$$

or in other words

$$(f, v) - (u, M[v]) = 0,$$

for every continuously differentiable vector v in H. This is precisely the statement (12.42) that u solves (12.38) in the weak sense.

The Hahn-Banach theorem is actually not so important in the argument just presented, for what has really been used is simply the *Riesz representation theorem* for bounded linear functionals on a Hilbert space.[13] It has provided us in a more or less abstract way with a generalized

[13] Cf. Riesz-Nagy 1.

solution of the integrated formulation (12.42) of the symmetric hyperbolic system (12.38). The more difficult step of proving that the item u so obtained actually possesses continuous partial derivatives and fulfills (12.38) in the strict sense will be left to the exercises because it would carry us too far considering the attention we have already devoted to initial value problems for equations of the hyperbolic type in Chapters 4 and 6.

EXERCISES

1. Solve the Cauchy problem for Maxwell's equations

$$\dot{B} + \nabla \times E = 0, \qquad \dot{E} - \nabla \times B = 0.$$

2. If $A, A_1, \ldots, A_n$ stand for symmetric matrices, and A is also positive-definite, bring the general linear symmetric hyperbolic system

$$A \frac{\partial u}{\partial t} = \sum_{j=1}^{n} A_j \frac{\partial u}{\partial x_j} + Bu$$

into the normal form (12.22). Extend the theory of the text to this case.

3. Verify that the Hermitian symmetric system (12.29) for the complex-valued unknown vector $u = \xi + i\eta$ is equivalent to a real symmetric hyperbolic system of partial differential equations for the real and imaginary parts ξ and η of u.

4. Show that the complex system (12.29) is symmetric hyperbolic even when the original analytic system (12.22) from which it was derived is not (cf. Section 16.1).

5. Verify that when the A_j are constants and B and F are polynomials, the Cauchy-Kowalewski power series solution of the initial value problem (12.22), (12.23) converges in the large.

6. Use a uniqueness theorem to show that every solution of the complex system (12.29) associated with complex analytic initial data (12.25) must also solve the real system (12.22).

7. Use an energy inequality in the complex domain to establish the existence in the large of solutions of Cauchy's problem for a single linear analytic partial differential equation of the second order and of the hyperbolic type, provided that the initial data are analytic, too.

8. Let u be a real analytic function defined over some n-dimensional domain D, and let D^* stand for a closed subdomain of D. According to a lemma of Sobolev, the partial derivatives of u of all orders not exceeding m can be estimated uniformly in D^* by means of integrals over D of the squares of its partial derivatives of orders not exceeding l, where l is a suitable integer depending only on m and n. Prove Sobolev's lemma by using the Poisson equation (cf. Section 5.3) to represent u as an (iterated) potential of its Laplacian.[14]

[14] Cf. Courant-Hilbert 3.

9. Find estimates of the square integrals of all partial derivatives of the solution u of the Cauchy problem defined by (12.22) and (12.23) by differentiating those relations repeatedly and then deriving energy inequalities applicable to the solutions of the new symmetric hyperbolic systems so obtained.

10. Using the combined results of Exercises 8 and 9 to estimate an approximation by analytic problems, prove that the Cauchy problem (12.22), (12.23) can be solved in the large even when A_j, B and F are no longer analytic.[15]

11. Establish the uniqueness of a periodic solution u of (12.38) by applying the (a,b,c)-method.

12. Show that any weak solution u of (12.42) must actually be a strong solution of (12.38), too, if $A_1, \ldots, A_n$, B and f are analytic.

13. Use the existence of a periodic solution of (12.38) to show that the Cauchy problem can be solved for that equation.[16]

14. Give an alternate proof of the Cauchy-Kowalewski theorem that avoids power series completely by applying the more abstract construction (12.48) to the complex symmetric hyperbolic system (12.29).

15. If the matrix B is sufficiently negative in the sense described by (12.39), establish the existence of a weak solution u of a periodic symmetric system

$$\sum_{j=1}^{n} A_j \frac{\partial u}{\partial x_j} + Bu = f$$

that is not necessarily symmetric hyperbolic.[17]

16. Use the approach of Exercise 15 to obtain a weak solution of the Frankl problem.[18] More precisely, assume the notation and hypotheses of formula (12.19) in Section 1, and solve the inhomogeneous mixed equation

$$(12.49) \qquad K(y)\psi_{xx} + \psi_{yy} = f_1$$

with $\psi = 0$ on Γ_1, Γ_2 and C_3. First set $u_1 = \psi_x$ and $u_2 = \psi_y$ to reduce (12.49) to the symmetric system

$$(12.50) \quad L[u] = \begin{pmatrix} K & 0 \\ 0 & -1 \end{pmatrix} \frac{\partial}{\partial x} \begin{pmatrix} u_1 \\ u_2 \end{pmatrix} + \begin{pmatrix} 0 & 1 \\ 1 & 0 \end{pmatrix} \frac{\partial}{\partial y} \begin{pmatrix} u_1 \\ u_2 \end{pmatrix} = \begin{pmatrix} f_1 \\ 0 \end{pmatrix} = f.$$

Next introduce the Hilbert space H^* of two-dimensional column vectors $v, w, \ldots$ whose scalar product is given by

$$(v, w)^* = \iint_D \left\{ \frac{v_1 w_1}{\sqrt{x^2 + y^2}} + v_2 w_2 \right\} dx\, dy$$

[15] Cf. Courant-Hilbert 3.
[16] Cf. Lax 3.
[17] Cf. Friedrichs 3.
[18] Cf. Morawetz 3.

and whose norm is $\|v\|^* = \sqrt{(v, v)^*}$. Then, with an appropriate constant β, establish the *energy inequality*

$$(v, w) = \iint_D \left\{ v_1 w_1 + v_2 w_2 \right\} dx\, dy \leq \beta \, \|L[v]\|^* \, \|w\|^*$$

for every w in H^* and every continuously differentiable v which vanishes at the origin and which fulfills the boundary conditions that $v_1 = 0$ on Γ_1, Γ_2 and C_3, that $\sqrt{-K}\, v_1 - v_2 = 0$ on C_1^*, and that $\sqrt{-K}\, v_1 + v_2 = 0$ on C_2^*. Now define a bounded linear functional Λ of $g = L[v]$ by the rule

$$\Lambda\{g\} = (f, v),$$

and extend it to H^* by the Hahn-Banach theorem. Show that because of the Riesz-Fischer theorem there exists in H^* an element w for which

$$\Lambda\{g\} = (w, g)^*,$$

and prove that

$$u = -\begin{pmatrix} \dfrac{w_1}{\sqrt{x^2 + y^2}} \\ w_2 \end{pmatrix}$$

is the desired weak solution of (12.50).

17. Imitate the construction of Exercise 16 in the case of the Dirichlet problem for Laplace's equation.

3. INCORRECT PROBLEMS; EQUATIONS WITH NO SOLUTION

Thus far we have concentrated most of our efforts on determining and solving well posed initial or boundary value problems for specific classes of partial differential equations. We have found that the type of a given equation plays a central role in the analysis of such questions. Let us now turn our attention to other problems which might at first sight seem reasonable but which turn out on closer examination to have no solutions at all. To start with, we shall take up again the example of Cauchy's problem for the Laplace equation in two independent variables, and we shall reconfirm that it is incorrectly set by exhibiting cases in which it is not solvable. With this background material at our disposal, we shall then proceed to construct a linear partial differential equation in three independent variables which does not have any solutions whatever, even when no boundary conditions are imposed. Thus we shall see that, while, to be sure, every analytic equation can be solved by the Cauchy-Kowalewski power series method, the question of type must be expected to feature to some extent in deciding whether there exists even one solution of a more general partial differential equation.

We consider the Cauchy problem of determining for small $x > 0$ a solution $u = u(x, y)$ of the Laplace equation

(12.51) $$u_{xx} + u_{yy} = 0$$

which fulfills special initial conditions of the form

(12.52) $$u(0, y) = 0, \qquad u_x(0, y) = f(y)$$

along a segment $-\epsilon < y < \epsilon$ of the y-axis. It will be shown that when $f(y)$ is not analytic at $y = 0$ no solution of (12.51) and (12.52) can exist in the intersection of the right half-plane with any neighborhood of the origin, no matter how small. For example, if

$$f(y) = |y|$$

there is no function u satisfying (12.51) in a semicircle about the origin, and located in the right half-plane $x > 0$, along whose diameter (12.52) holds. Notice, however, that the Cauchy-Kowalewski theorem yields a solution of (12.51) and (12.52) whenever $f(y)$ is analytic.

To prove the above assertions, we apply the Schwarz principle of reflection (cf. Section 7.2) to extend the solution of our Cauchy problem backward across the initial line. Indeed, any function u which is harmonic

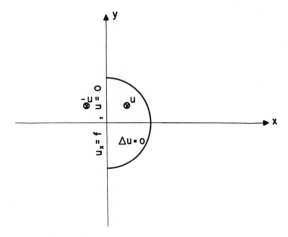

FIG. 52. Irregular Cauchy problem.

in a section D^+ of the half-plane $x > 0$ bordering on a segment of the y-axis, and which vanishes along that segment, can be continued analytically into the reflected image D^- of D^+ with respect to the y-axis by the rule

$$u(-x, y) = -u(x, y).$$

In particular, any solution u of the Cauchy problem (12.51), (12.52) has to be regular in a complete neighborhood of each point which, like the origin, lies on the segment of the y-axis where the first initial condition $u = 0$ is imposed. Consequently the partial derivative u_x is regular on the segment in question, which means that the given function $f(y)$ appearing in the second initial condition (12.52) has to be analytic in at least some finite interval about the origin. Thus analytic continuation across the y-axis serves to establish our contention that the Cauchy problem (12.51), (12.52) can never be solved near a point where the prescribed normal derivative u_x is not analytic, such as the origin in the case $f = |y|$.

We recall here the counterexample due to Hadamard which was used in Section 4.1 to show that solutions of the Cauchy problem for Laplace's equation do not necessarily depend in a continuous way on the initial data even when they do exist. That example combined trigonometric and exponential functions to achieve a situation where the Cauchy data approached zero, whereas the solution itself became infinite away from the initial line. The conclusion that the problem is incorrectly set is brought out in a quite different fashion by the theorem we have proved above concerning analyticity of the prescribed function $f(y)$ in (12.52), which excludes the possibility that a solution could exist at all for less regular data.

The latter technique will now be extended to produce a linear partial differential equation that has no solutions whatever in the neighborhood of a certain point.[19] Moreover, both methods used to investigate the peculiarities of the Cauchy problem for Laplace's equation can be exploited to generate unreasonable questions for various other classes of partial differential equations (cf. the exercises below).

We ask whether there exists in the neighborhood of the origin a complex-valued function $u = v + iw$ of the three real independent variables x, y and ξ which satisfies the first order partial differential equation

(12.53) $$\frac{\partial u}{\partial x} + i \frac{\partial u}{\partial y} - 2i(x + iy) \frac{\partial u}{\partial \xi} = g'(\xi),$$

where g' is a given function of ξ alone. Actually, the real and imaginary parts of (12.53) constitute a system of two separate equations of the first order for the real and imaginary parts v and w of the unknown u. It will

[19] Cf. Lewy 4.

be shown here that if the inhomogeneous term $g'(\xi)$ is real everywhere and fails to be analytic at $\xi = 0$, then (12.53) has no continuously differentiable solution in any neighborhood of the origin. This rather startling conclusion will be drawn from the reversed observation that the very existence of such a solution for real $g'(\xi)$ entails analyticity of that function at $\xi = 0$. Our proof of the latter assertion has features in common with the previous analysis of the problem (12.51), (12.52).

The method of separation of variables is applicable to equation (12.53). Putting $z = x + iy = re^{i\theta}$, we shall be able to reduce it to an equation in only two independent variables by integrating both sides with respect to θ from 0 to 2π.

Consider the Fourier coefficient

$$(12.54) \qquad V = ir \int_0^{2\pi} u e^{i\theta} \, d\theta = \oint_{|z|=r} u(x, y, \xi) \, dz$$

of any solution u of (12.53) as a function of the two independent variables ξ and $\eta = r^2$. Because of the divergence theorem we can write

$$V = i \iint_{|z|<r} \left(\frac{\partial u}{\partial x} + i \frac{\partial u}{\partial y} \right) dx \, dy.$$

Therefore

$$\frac{\partial V}{\partial r} = i \oint_{|z|=r} \left(\frac{\partial u}{\partial x} + i \frac{\partial u}{\partial y} \right) ds$$

$$= \oint_{|z|=r} \left(\frac{\partial u}{\partial x} + i \frac{\partial u}{\partial y} \right) \frac{r \, dz}{z},$$

whence it follows from (12.53) that

$$\frac{\partial V}{\partial \eta} = \frac{1}{2r} \frac{\partial V}{\partial r} = \oint_{|z|=r} \frac{1}{2} \left(\frac{\partial u}{\partial x} + i \frac{\partial u}{\partial y} \right) \frac{dz}{z}$$

$$= i \oint_{|z|=r} \frac{\partial u}{\partial \xi} \, dz + g'(\xi) \oint_{|z|=r} \frac{dz}{2z}.$$

Taking (12.54) into account we obtain

$$\frac{\partial V}{\partial \xi} + i \frac{\partial V}{\partial \eta} + \pi g'(\xi) = 0,$$

which shows that the expression

$$(12.55) \qquad U = V + \pi g(\xi)$$

satisfies the Cauchy-Riemann equations

$$\frac{\partial U}{\partial \xi} + i \frac{\partial U}{\partial \eta} = 0.$$

Consequently it must be an analytic function of the complex variable $\zeta = \xi + i\eta$.

If u is defined in some neighborhood of the origin, formula (12.55) specifies $U = U(\zeta)$ for all sufficiently small values of the real part ξ of ζ and for all sufficiently small positive values of the imaginary part $\eta = r^2$ of ζ. On the other hand, (12.54) shows that $V = 0$ when $\eta = 0$, so that on the real axis we have

(12.56) $$U(\xi) = \pi g(\xi).$$

Because we assumed g to be real-valued, (12.56) implies that the imaginary part of the analytic function $U(\zeta)$ vanishes along the real axis. Hence the Schwarz reflection principle applies and furnishes the rule

$$U(\bar{\zeta}) = \overline{U(\zeta)}$$

for continuing U analytically across the line $\eta = 0$ into the lower half-plane. Since, in particular, U must thus be regular on the real axis itself, we conclude from (12.56) that g is an analytic function of ξ in some interval around $\xi = 0$. This completes our proof that when there exists near the origin a continuously differentiable solution u of (12.53) corresponding to a real choice of the inhomogeneous term $g'(\xi)$, then $g'(\xi)$ has to be analytic at $\xi = 0$.

The analysis we have just presented shows that the linear partial differential equation (12.53) possesses no continuously differentiable solutions near the origin when the prescribed function $g'(\xi)$ appearing on the right is real, but not analytic, in an interval around $\xi = 0$. The conclusion holds no matter how smooth g' might otherwise be. We propose next to establish that with an arbitrary point (x_0, y_0, ξ_0) in space one can associate choices of the inhomogeneous term in (12.53) for which no continuously differentiable solutions of that equation exist in any neighborhood of (x_0, y_0, ξ_0). More precisely, we shall prove that

(12.57) $$\frac{\partial u}{\partial x} + i \frac{\partial u}{\partial y} - 2i(x + iy)\frac{\partial u}{\partial \xi} = g'(\xi - \xi_0 - 2y_0 x + 2x_0 y)$$

has no such solution u near (x_0, y_0, ξ_0) when the real function $g'(\xi^*)$ fails to be analytic at $\xi^* = 0$.

To verify this contention, we observe that the change of variables

$$x^* = x - x_0, \quad y^* = y - y_0, \quad \xi^* = \xi - \xi_0 - 2y_0 x + 2x_0 y$$

transforms (12.57) into an equation

(12.58) $$\frac{\partial u}{\partial x^*} + i \frac{\partial u}{\partial y^*} - 2i(x^* + iy^*) \frac{\partial u}{\partial \xi^*} = g'(\xi^*)$$

more directly analogous to (12.53), since

$$\frac{\partial u}{\partial x} = \frac{\partial u}{\partial x^*} - 2y_0 \frac{\partial u}{\partial \xi^*}, \quad \frac{\partial u}{\partial y} = \frac{\partial u}{\partial y^*} + 2x_0 \frac{\partial u}{\partial \xi^*}, \quad \frac{\partial u}{\partial \xi} = \frac{\partial u}{\partial \xi^*}.$$

Our earlier results for (12.53) imply that (12.58) cannot have a continuously differentiable solution near the origin. It follows that the same is true of (12.57) in every neighborhood of the point (x_0, y_0, ξ_0), as was stated above. Incidentally, the parameter ξ_0 plays a quite inessential role here, since it merely represents a translation of the variable ξ. We should mention, too, in closing, that there are sequences of numbers a_j, x_j and y_j such that for some real periodic function $g'(\xi)$ which is infinitely differentiable everywhere, but analytic nowhere, the equation

(12.59) $$\frac{\partial u}{\partial x} + i \frac{\partial u}{\partial y} - 2i(x + iy) \frac{\partial u}{\partial \xi} = \sum_{j=1}^{\infty} a_j g'(\xi - 2y_j x + 2x_j y)$$

has no twice continuously differentiable solution in any open region whatever, although the inhomogeneous term on the right is infinitely differentiable. It is a routine matter to prove the latter theorem by condensing singularities in a standard fashion and exploiting category arguments from set theory (cf. Exercise 4 below).

EXERCISES

1. Show that Laplace's equation (12.51) cannot be solved subject to initial conditions of the form

$$u(0, y) = g(y), \quad u_x(0, y) = 0$$

unless the given function g is analytic.

2. Show that the formula

$$u(x, t) = \lambda e^{-t/\lambda^2} \sin \frac{x}{\lambda}$$

defines a solution of the heat equation

(12.60) $$u_{xx} = u_t$$

that approaches zero for $t \geq 0$, but not for $t < 0$, in the limit as $\lambda \to 0$.

3. Prove that the initial value problem

$$u(x, 0) = f(x)$$

for the heat equation (12.60) does not have a solution for negative time $t < 0$ unless the given function f is analytic.

4. Verify that the parameters a_j, x_j and y_j can be chosen so that (12.59) has in no open set a solution with continuous second derivatives, or even with first derivatives satisfying a Hölder condition.[20]

5. Let L stand for a linear differential operator

$$L[u] = \sum_{0 \le j_1 + \cdots + j_n \le m} a_{j_1 \cdots j_n} \frac{\partial^{j_1 + \cdots + j_n} u}{\partial x_1^{j_1} \cdots \partial x_n^{j_n}}$$

of order m with complex-valued coefficients $a_{j_1 \cdots j_n}$ whose real and imaginary parts are real analytic functions of the independent variables $x_1, \ldots, x_n$, and consider the *commutator*[21]

$$C = \bar{L}L - L\bar{L},$$

which is a differential operator of order $2m - 1$ or less. Denote by L_m and C_{2m-1} the principal parts of L and C, respectively. In other words, let L_m indicate the sum of the terms of order precisely m in L, and let C_{2m-1} indicate the sum of the terms of order precisely $2m - 1$ in C. Finally, let $L_m(\xi)$ be the homogeneous polynomial of degree m in the variables $\xi_1, \ldots, \xi_n$ obtained by substituting $\xi_1^{j_1} \cdots \xi_n^{j_n}$ for the partial derivative $\partial^{j_1 + \cdots + j_n} u / \partial x_1^{j_1} \cdots \partial x_n^{j_n}$ in the representation of $L_m[u]$, and let $C_{2m-1}(\xi)$ be the analogous homogeneous polynomial of degree $2m - 1$ derived from $C_{2m-1}[u]$. Show that if the linear partial differential equation

$$L[u] = f$$

can be solved for every infinitely differentiable choice of the inhomogeneous term f, then

$$C_{2m-1}(\xi) = 0$$

for every real vector $(\xi_1, \ldots, \xi_n)$ with the characteristic property

$$L_m(\xi) = 0.$$

Use this theorem to verify once again that we can insert an inhomogeneous term on the right in (12.53) such that no smooth solutions of the resulting linear partial differential equation exist.

6. Imitating Hadamard's counterexample concerning the Cauchy problem for Laplace's equation, piece together exponential and trigonometric functions to construct[22] an infinitely differentiable coefficient $p = p(x, y)$ such that the

[20] Cf. Lewy 4.
[21] Cf. Hörmander 1.
[22] Cf. Pliś 1.

elliptic partial differential equation

$$\left(\frac{\partial}{\partial x} + i\frac{\partial}{\partial y}\right)^6 u - ix^6 \frac{\partial^5 u}{\partial y^5} - \frac{\partial^4 u}{\partial y^4} + pu = 0$$

has an infinitely differentiable solution u of the general form

$$u(x, y) = \phi(x)\, e^{ik(x)y} + \psi(x)\, e^{il(x)y}$$

which vanishes identically in the left half-plane $x \leq 0$ but differs from zero somewhere in the right half-plane $x > 0$.

7. Use the *closed graph theorem*[23] to show that if a unique solution of Cauchy's problem for a linear analytic (not necessarily hyperbolic) system (12.22) exists for all infinitely differentiable initial data (12.23), then the solution depends continuously on the data in an appropriate topology. Thus conclude that in cases where the solution is unique, but does not depend continuously on the data, it cannot exist for every choice of the initial values. In particular, deduce in this way from Hadamard's counterexample that the Cauchy problem cannot always be solved for Laplace's equation.

8. Show that the infinite series $\sum_0^\infty f^{(k)}(t)x^{2k}/(2k)!$, where $f(t) = e^{-1/t^2}$, defines a solution of the heat equation $u_{xx} = u_t$ in the upper half-plane $t > 0$, but approaches zero as $t \to 0$.

---

[23] Cf. Lax 4.

13

Finite Differences

1. FORMULATION OF DIFFERENCE EQUATIONS

For the most part the methods we have presented until now are not by themselves adequate for really computing the solution of a particular initial or boundary value problem encountered in mathematical physics. It therefore becomes desirable to discuss more specific ways of finding numerical solutions. Almost any method for solving a partial differential equation numerically reduces ultimately to the computation of discrete sets of data. Moreover, tabulation of the solution over a finite lattice of mesh points appears as perhaps the most natural way in which to describe it in terms of numbers. Thus a convenient and quite general procedure for calculating numerical solutions of initial or boundary value problems is to approximate them by corresponding problems for *finite difference equations*. Actually, the simplest example of such a technique is the computation of a definite integral by means of the trapezoid rule or Simpson's rule. These considerations lead us to examine here analogues for finite difference equations of some of the more significant aspects of the theory of partial differential equations that we have developed.

The study of difference equations will provide us with an opportunity to review various salient features of our knowledge about initial and boundary value problems in the context of a more elementary discrete model. Furthermore, we shall find that a thorough understanding of what questions are appropriate for a given partial differential equation and of the techniques that lead to a solution furnishes the best guide to an effective and sophisticated treatment of the related finite difference problems. Thus it turns out that in order to formulate convergent difference schemes for initial value problems it is essential to take into account the geometry of

458

the relevant characteristics and of the corresponding domains of dependence. On the other hand, a principal difficulty for examples of the elliptic type is to solve the large numbers of simultaneous linear equations that occur in any adequate difference approximation. It will be observed that the relaxation procedure frequently used to treat such systems bears some similarity to the Poincaré sweeping out process that motivated our analysis of the Dirichlet problem by means of subharmonic functions (cf. Section 9.2). Also, for the Laplace difference equation a direct analogue of the maximum principle enables us to estimate errors conveniently and to exploit the relationship between existence and uniqueness of solutions.

Difference approximations to differential equations are based on elementary representations of the derivatives of a smooth function in terms of appropriate difference quotients. We shall deal here primarily with examples involving functions of two independent variables that possess continuous partial derivatives of a suitably high order. Consider the Taylor series expansion

$$u(x + h, y + k) = u(x, y) + u_x(x, y)h + u_y(x, y)k$$
$$+ \tfrac{1}{2}u_{xx}(x, y)h^2 + u_{xy}(x, y)hk + \tfrac{1}{2}u_{yy}(x, y)k^2 + O(|h|^3 + |k|^3)$$

of such a function. This formula yields the expressions

$$(13.1) \qquad u_x(x, y) = \frac{u(x + h, y) - u(x, y)}{h} + O(|h|)$$

and

$$(13.2) \qquad u_y(x, y) = \frac{u(x, y + k) - u(x, y)}{k} + O(|k|)$$

for the first partial derivatives u_x and u_y of u in terms of so-called *forward differences*. Similarly, the second derivatives of u have representations

$$(13.3) \quad u_{xx}(x, y) = \frac{u(x + h, y) - 2u(x, y) + u(x - h, y)}{h^2} + O(h^2),$$

$$(13.4) \quad u_{yy}(x, y) = \frac{u(x, y + k) - 2u(x, y) + u(x, y - k)}{k^2} + O(k^2),$$

and

$$u_{xy}(x, y) = \frac{1}{4hk} [u(x + h, y + k) - u(x + h, y - k)$$
$$+ u(x - h, y - k) - u(x - h, y + k)] + O(h^2 + k^2)$$

by means of *central differences* which, it should be observed, involve remainders of a higher order of magnitude because of their symmetric behavior in dependence on the increments h and k. Finally, we make a note, too, of the central difference approximations

$$u_x(x, y) = \frac{u(x + h, y) - u(x - h, y)}{2h} + O(h^2),$$

$$u_y(x, y) = \frac{u(x, y + k) - u(x, y - k)}{2k} + O(k^2)$$

to the first derivatives of u.

We introduce a lattice of mesh points whose coordinates x and y have the form

$$x = mh, \qquad y = nk,$$

where m and n are integers and where h and k are positive numbers known as the *mesh sizes* of the grid. Let us consider a discrete function $u_{m,n}$ defined only at the lattice points. Our aim is to formulate difference equations for $u_{m,n}$ which are valid approximations to the wave equation

(13.5) $$u_{xx} - u_{yy} = 0,$$

the heat equation

(13.6) $$u_{xx} - u_y = 0,$$

and Laplace's equation

(13.7) $$u_{xx} + u_{yy} = 0.$$

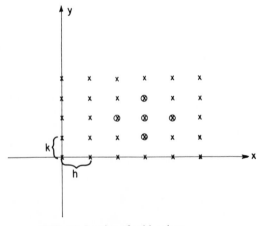

FIG. 53. Lattice of grid points.

Formulas (13.2), (13.3) and (13.4) suggest that the desired difference analogues of (13.5), (13.6) and (13.7) are, respectively,

(13.8) $$\frac{u_{m+1,n} - 2u_{m,n} + u_{m-1,n}}{h^2} - \frac{u_{m,n+1} - 2u_{m,n} + u_{m,n-1}}{k^2} = 0,$$

(13.9) $$\frac{u_{m+1,n} - 2u_{m,n} + u_{m-1,n}}{h^2} - \frac{u_{m,n+1} - u_{m,n}}{k} = 0,$$

(13.10) $$\frac{u_{m+1,n} - 2u_{m,n} + u_{m-1,n}}{h^2} + \frac{u_{m,n+1} - 2u_{m,n} + u_{m,n-1}}{k^2} = 0.$$

For example, the form of the last equation is justified because (13.3) and (13.4) yield the relation

$$u_{xx} + u_{yy} = \frac{u(mh + h, nk) - 2u(mh, nk) + u(mh - h, nk)}{h^2}$$

(13.11)

$$+ \frac{u(mh, nk + k) - 2u(mh, nk) + u(mh, nk - k)}{k^2} + O(h^2 + k^2)$$

for the Laplacian of a continuously specified smooth function $u = u(x, y)$.

Substitution of (13.11) into (13.7) shows that any harmonic function $u(mh, nk)$ of two independent variables satisfies the Laplace difference equation (13.10) except for an error of the order of magnitude $h^2 + k^2$. Consequently we can expect that when (13.10) is applied to solve Laplace's differential equation (13.7) numerically, the errors in the results obtained will be of a similar order of magnitude. The same conclusion is valid for the wave equation and the heat equation. For more general partial differential equations it turns out that the kind of interpolation formula used to represent the derivatives that appear plays a decisive role in determining the size of the error involved in a finite difference approximation. Instead of examining such complicated situations in detail, we prefer to study here the simplest examples (13.5), (13.6) and (13.7) of hyperbolic, parabolic and elliptic equations and to deduce from them the most important facts about numerical solution of partial differential equations by the method of finite differences.

The main issues with which we shall be concerned are those, first, of convergence $u_{m,n} \to u(mh, nk)$ of a difference approximation as the associated mesh sizes approach zero and, second, of stability of the errors that accumulate due to rounding in numerical computations of $u_{m,n}$. The analysis of convergence has much in common with the analysis of stability, and both are closely connected with the question of whether the initial or boundary value problem to be solved has been correctly formulated. Thus a firm grasp of the theory of partial differential equations is required if we are to achieve a successful numerical treatment of even the most elementary initial and boundary value problems of mathematical physics.

EXERCISES

1. Use Taylor's theorem to verify that the order of magnitude of the errors in the interpolation formulas (13.3) and (13.4) has been assessed correctly.

2. Show that because the coefficients involved are constant it is possible to find exponential solutions

$$u_{m,n} = \alpha^m \beta^n$$

of the difference equations (13.8), (13.9) and (13.10).

3. Derive a first order difference equation that yields a convergent, stable approximation to the solution of the initial value problem

$$\frac{du}{dx} + u = 0, \qquad u(0) = 1.$$

If the second order difference equation

$$u_{m+1} + 2hu_m - u_{m-1} = 0$$

is applied to the same problem, in what respect is the answer obtained unstable? How might it be made to converge better?

4. Find a finite difference approximation to the equation

$$u_{xx} + u_{yy} + \frac{1}{y} u_y = 0$$

of axially symmetric potentials that entails errors of the order of magnitude of the squares of the mesh sizes.

5. Derive a difference equation of the fourth order for the numerical analysis of the biharmonic equation

$$u_{xxxx} + 2u_{xxyy} + u_{yyyy} = 0.$$

6. Work out an analogue of (13.10) that can be used to approximate Laplace's equation near a curved boundary by introducing unequal mesh sizes that bring the outside grid points onto the boundary.

7. In Figure 54 spell out the word "mathematics" by starting on the main diagonal and proceeding one step at a time either downward or to the right until the opposite corner is reached. In precisely how many different ways can this be done?

2. HYPERBOLIC EQUATIONS

The first question we shall take up is the Cauchy problem for the wave equation

$$(13.5) \qquad u_{xx} - u_{yy} = 0$$

with initial data

$$(13.12) \qquad u(x, 0) = f(x), \qquad u_y(x, 0) = g(x),$$

where f and g are supposed to be functions possessing several continuous derivatives. To formulate a finite difference approximation to this initial value problem we replace (13.5) by

$$(13.8) \qquad \frac{u_{m+1,n} - 2u_{m,n} + u_{m-1,n}}{h^2} - \frac{u_{m,n+1} - 2u_{m,n} + u_{m,n-1}}{k^2} = 0,$$

and we replace (13.12) by the discrete conditions

$$(13.13) \qquad u_{m,0} = f_m$$

```
           M
          MA
         MAT
        MATH
       MATHE
      MATHEM
     MATHEMA
    MATHEMAT
   MATHEMATI
  MATHEMATIC
 MATHEMATICS
```

FIG. 54. Spelling puzzle.

and

$$(13.14) \qquad u_{m,1} = f_m + g_m k + \frac{f_m''}{2} k^2,$$

where $f_m = f(mh)$, $g_m = g(mh)$ and $f_m'' = f''(mh)$.

The expression on the right in (13.14) is based on the interpolation formula

$$u_y(mh, 0) = \frac{u(mh, k) - u(mh, 0)}{k} - \frac{u_{yy}(mh, 0)}{2} k + O(k^2)$$

$$= \frac{u(mh, k) - u(mh, 0)}{k} - \frac{u_{xx}(mh, 0)}{2} k + O(k^2),$$

which follows in turn from Taylor's theorem and an application of (13.5). Thus our approximation to the second initial condition (13.12) entails a truncation error of the order of magnitude k^2. Observe that omission in (13.14) of the somewhat subtle term involving f_m'' would lead to an error of the lowest order k instead.

Assume that the functions f and g have been prescribed over an interval $0 \le x \le 2Nh$ of the x-axis, so that f_m and g_m are known in the corresponding range $0 \le m \le 2N$. Then the initial conditions (13.13) and (13.14) define $u_{m,0}$ and $u_{m,1}$ for $0 \le m \le 2N$, and it is natural to try to determine $u_{m,n}$ for larger choices of the index n by applying the difference equation (13.8). For this purpose we have only to rearrange (13.8) as a representation

$$(13.15) \quad u_{m,n+1} = 2u_{m,n} - u_{m,n-1} + \frac{k^2}{h^2}(u_{m+1,n} - 2u_{m,n} + u_{m-1,n})$$

of the desired solution at the level $y = (n + 1)k$ of grid points in terms of its values at the two previous levels $y = nk$ and $y = (n - 1)k$. Observe that a star of precisely five mesh points occurs in (13.15), and that data from three different columns $x = (m - 1)h$, $x = mh$ and $x = (m + 1)h$ of points must be inserted on the right for the evaluation of $u_{m,n+1}$. In particular, substitution into (13.15) of the portion of initial data (13.13), (13.14) with $0 \le m \le 2N$ fixes $u_{m,2}$ over the slightly narrower interval $1 \le m \le 2N - 1$. Another appeal to (13.15) then yields $u_{m,3}$ in the range $2 \le m \le 2N - 2$, and so on. Finally, mathematical induction shows that from the given initial values $u_{m,n}$ can be found step by step at all the lattice points of a triangle in the upper half-plane described by the inequalities

$$(13.16) \qquad n - 1 \le m \le 2N - n + 1, \qquad 1 \le n \le N + 1.$$

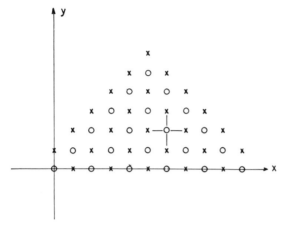

FIG. 55. The initial value problem.

When the ratio of mesh sizes k/h reduces to unity, the difference equation (13.15) acquires the much simpler form

$$(13.17) \qquad u_{m,n+1} = u_{m+1,n} - u_{m,n-1} + u_{m-1,n},$$

from which the term involving $u_{m,n}$ has disappeared. In the case of (13.17) we shall be able to obtain a closed expression for the answer to the above initial value problem. It is our intention to verify by a direct passage to the limit that this result converges toward the corresponding solution of Cauchy's problem for the wave equation (13.5) as $h = k \to 0$.

Motivated by our knowledge of characteristic coordinates, we put $m = p + q$ and $n = p - q$ in (13.17) and rearrange the terms to establish

$$(13.18) \qquad u_{p+q,p-q+1} - u_{p+q+1,p-q} = u_{p+q-1,p-q} - u_{p+q,p-q-1}.$$

Then we observe that replacing p by $p - 1$ in the quantity on the left transforms it into the difference on the right. It follows that the difference in question is actually independent of p. Hence, in particular, we are justified in substituting $q + 1$ for p on the right in (13.18) to derive

$$u_{p+q,p-q+1} - u_{p+q+1,p-q} = u_{2q,1} - u_{2q+1,0}.$$

Finally, we can sum on the index q between the limits $m - p$ and $p - 1$ to obtain

$$u_{m,2p-m+1} - u_{2p,1} = \sum_{q=m-p}^{p-1} (u_{2q,1} - u_{2q+1,0}).$$

Setting $2p - m + 1 = n$ and taking the conditions (13.13) and (13.14) into account, we find that the solution $u_{m,n}$ of our initial value problem for the difference equation (13.17) is given explicitly by the formula

$$(13.19) \quad u_{m,n} = \sum_{j=m-n+1}^{m+n-1} (-1)^j f_j + \frac{k^2}{2} \sum_{j=\frac{m-n+1}{2}}^{\frac{m+n-1}{2}} f''_{2j} + k \sum_{j=\frac{m-n+1}{2}}^{\frac{m+n-1}{2}} g_{2j}.$$

Since

$$h^2 \sum_{j=\frac{m-n+1}{2}}^{\frac{m+n-1}{2}} f''_{2j} = \sum_{j=\frac{m-n+1}{2}}^{\frac{m+n-1}{2}} [f_{2j+1} - 2f_{2j} + f_{2j-1} + O(h^4)]$$

$$= f_{m-n} + f_{m+n} - 2 \sum_{j=m-n+1}^{m+n-1} (-1)^j f_j + O(h^3)$$

and

$$h \sum_{j=\frac{m-n+1}{2}}^{\frac{m+n-1}{2}} g_{2j} = \sum_{j=\frac{m-n+1}{2}}^{\frac{m+n-1}{2}} \left[\frac{1}{2} \int_{(2j-1)h}^{(2j+1)h} g(x) \, dx + O(h^3) \right]$$

$$= \frac{1}{2} \int_{(m-n)h}^{(m+n)h} g(x) \, dx + O(h^2),$$

and since $h = k$, we conclude that

$$u_{m,n} = \frac{1}{2} \left[f(mh - nk) + f(mh + nk) + \int_{mh-nk}^{mh+nk} g(x) \, dx \right] + O(h^2).$$

On the other hand, d'Alembert's formula (4.17) yields the expression

$$u(x, y) = \frac{1}{2} \left[f(x - y) + f(x + y) + \int_{x-y}^{x+y} g(\xi) \, d\xi \right]$$

for the solution of the Cauchy problem (13.5), (13.12). Thus the truncation error in our finite difference approximation $u_{m,n}$ to the wave function $u(x, y)$ has at any fixed mesh point (mh, nk) the order of magnitude

$$u_{m,n} - u(mh, nk) = O(h^2)$$

of the square of the mesh size h, whence it tends toward zero rapidly as $h \to 0$.

The kind of convergence that might be expected of the finite difference method is exhibited in a quite specific way by the example we have just presented. However, the facility with which we obtained the elegant solution (13.19) of the difference equation (13.17) hinged on intermediate steps that actually amount to integrations, or, rather, summations, along a pair of characteristics of the wave equation. A much more difficult situatión presents itself when the ratio of mesh sizes k/h differs from unity. In fact, when

$$\frac{k}{h} > 1$$

the triangle (13.16) in which a given interval $0 \leq m \leq 2N$ of initial data (13.13), (13.14) determines the solution $u_{m,n}$ of the difference equation (13.15) extends beyond the domain of dependence

$$y \leq x \leq 2Nh - y, \qquad 0 \leq y \leq Nh$$

where $u_{m,n}$ could have any relationship to the solution $u(x, y)$ of the wave equation (13.5) defined by the corresponding segment $0 \leq x \leq 2Nh$ of Cauchy data (13.12). Hence the convergence of our difference scheme as $h \to 0$ becomes questionable if the ratio of mesh sizes k/h is larger than 1. We shall proceed to establish that it does indeed fail whenever, for the same sets of initial data, the domains of dependence for the differential equation do not contain the analogous domains of dependence for the approximating difference equation, but that it succeeds in the opposite case. This is the substance of the *Courant-Friedrichs-Lewy criterion* for convergence and stability of finite difference schemes.[1]

The predicted divergence of the finite difference procedure when $k/h > 1$ will be brought out by establishing that the solution $u_{m,n}$ of (13.8) then has an unstable dependence on the initial data (13.13), (13.14) in the limit as

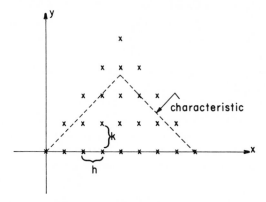

FIG. 56. Unstable grid.

[1] Cf. Courant-Friedrichs-Lewy 1.

$h \rightarrow 0$. The verification of such a contention is most readily achieved by examining the behavior of the exponential solutions of (13.8) mentioned in Exercise 1.2. More specifically, we investigate solutions of the form

$$(13.20) \qquad u_{m,n} = e^{im\theta} e^{in\lambda},$$

where θ is any real number and $\lambda = \mu + i\nu$ is a possibly complex parameter yet to be determined. Substitution of (13.20) into (13.8) and division by $e^{im\theta} e^{in\lambda}$ yields the trigonometric relation

$$(13.21) \qquad \sin^2 \frac{\lambda}{2} = \frac{k^2}{h^2} \sin^2 \frac{\theta}{2}$$

for λ. Since $u_{m,0} = e^{im\theta}$ represents the typical term of a Fourier series into which the initial values (13.13) might have been developed, we are led to assert that the difference scheme (13.15) is stable when (13.20) remains bounded as $n \rightarrow \infty$ for every real choice of θ but is unstable when under the same circumstances (13.20) can increase indefinitely. It is thus apparent from the elementary properties of the exponential function $e^{in\lambda}$ that (13.15) will be stable if the roots λ of (13.21) have non-negative imaginary parts for all real θ, whereas it will be unstable if one of these roots has an imaginary part less than zero.

With any ratio of mesh sizes in the range $k/h > 1$ we can associate a real number θ such that the term on the right in (13.21) exceeds unity. The corresponding roots λ of (13.21) must occur in conjugate complex pairs, and hence at least one of them has a negative imaginary part. The resulting solution (13.20) of the difference equation (13.15) grows exponentially as the index n becomes infinite. If we now fix the coordinates $x = mh$ and $y = nk > 0$, but allow the mesh sizes h and k to approach zero in a specific ratio $k/h > 1$, we observe that because $n \rightarrow \infty$ the quantity $u_{m,n}$ can assume indefinitely large values even for arbitrarily small choices of the initial data (13.13), (13.14). In this sense $u_{m,n}$ depends in an unstable fashion on its initial values. In particular, truncation errors or other perturbations in the assignment of the data (13.13), (13.14) may generate contributions that ultimately dominate the rest of the solution $u_{m,n}$ of (13.15). Thus convergence of $u_{m,n}$ to the solution $u(x, y) = u(mh, nk)$ of Cauchy's problem for our original hyperbolic differential equation (13.5) cannot be expected when the time increment k is taken larger than the space increment h in our lattice of mesh points (mh, nk).

Note that when

$$(13.22) \qquad \frac{k}{h} \leq 1$$

the difference scheme (13.15) is stable in the sense that all the roots λ of (13.21) are real, so that (13.20) remains bounded as $n \to \infty$. Thus it is not unreasonable to presume that in the case (13.22) the solution $u_{m,n}$ of (13.13), (13.14) and (13.15) will converge to the solution $u(mh, nk)$ of the Cauchy problem (13.5), (13.12) as $h \to 0$. Postponing for the moment further substantiation of this conjecture, we generalize it to formulate the quite useful and important *von Neumann criterion* for the stability of finite difference approximations to a much wider class of initial value problems.

For difference equations with constant coefficients, von Neumann's test consists in examining all exponential solutions to determine whether they grow exponentially in the time variable even when their initial values are bounded functions of the space variables (cf. Section 6.1). If any of them do increase without limit, the difference scheme in question is said to be *unstable*, whereas it is called *stable* if every admissible solution of the kind described remains bounded. We apply von Neumann's criterion to examples with variable coefficients, too, by simply introducing new, constant coefficients equal to the frozen values of the original ones at some specific point that is of interest and then testing the modified problem instead.

According to the von Neumann criterion, (13.15) is, as we have seen, unstable for a ratio of mesh sizes $k/h > 1$, but stable when $k/h \leq 1$. We have already remarked that such a distinction of cases is equivalent to a decision whether the triangle where a solution of the difference equation (13.15) can be determined from appropriate initial data extends beyond, or is contained within, the triangle in which the solution of the corresponding Cauchy problem (13.5), (13.12) exists. Thus the issue of stability of finite difference approximations to a hyperbolic partial differential equation is closely linked with the geometry of the associated characteristics. More specifically, it turns out in practice that any ratio of mesh sizes satisfying the von Neumann stability criterion is always small enough so that the results it yields are only defined inside the relevant domain of dependence which appears in the underlying Cauchy problem. To provide at least partial justification for the restrictions that these considerations lead us to impose on grids for initial value problems, we establish next a convergence theorem for difference formulations of certain canonical hyperbolic systems which follows directly from the hypothesis (13.22).

We shall develop a quite general convergence theorem for the *canonical hyperbolic system*

(13.23) $$U_t = U_x + aU + bV,$$

(13.24) $$V_t = -V_x + cU + dV$$

with constant coefficients a, b, c and d. Its characteristics (cf. Section 3.5)

(13.25) $$t = \pm x + \text{const.}$$

are the same as those of the wave equation (13.5), provided that we agree to identify the variables y and t. Moreover, the substitutions

$$v = \int (u_y \, dx + u_x \, dy) = \int (u_t \, dx + u_x \, dt)$$

and

$$U = u + v, \qquad V = u - v$$

actually transform the wave equation (13.5) into the special case

(13.26) $$U_t = U_x, \qquad V_t = -V_x$$

of (13.23) and (13.24). A unique solution of the system (13.23), (13.24) is determined (cf. Section 4.3) by prescribing sufficiently smooth initial values

(13.27) $$U(x, 0) = F(x), \qquad V(x, 0) = G(x).$$

In particular, for the example (13.26) it is seen that the functions F and G are connected with the data (13.12) by the relations

$$F'(x) = f'(x) + g(x), \qquad G'(x) = f'(x) - g(x).$$

A natural finite difference approximation to the Cauchy problem (13.23), (13.24), (13.27), based on the rectangular grid of mesh points

$$x = mh, \qquad t = nk,$$

is defined by the equations

(13.28) $$U_{m,n+1} = \left(1 - \frac{k}{h}\right) U_{m,n} + \frac{k}{h} U_{m+1,n} + kaU_{m,n} + kbV_{m,n},$$

(13.29) $$V_{m,n+1} = \left(1 - \frac{k}{h}\right) V_{m,n} + \frac{k}{h} V_{m-1,n} + kcU_{m,n} + kdV_{m,n},$$

and

(13.30) $$U_{m,0} = F_m = F(mh), \qquad V_{m,0} = G_m = G(mh).$$

These can readily be solved step by step marching forward with the index n. First order differences have been selected here so that in the important case $h = k$ of a square grid the increments $U_{m,n+1} - U_{m+1,n}$ and $V_{m,n+1} - V_{m-1,n}$ that occur are measured in the directions of the characteristics

(13.25) of the hyperbolic system (13.23), (13.24). Such an arrangement, which is suggested by the *method of characteristics* (cf. Sections 3.5 and 4.3), has the advantage that when $k < h$ the various weight factors in front of the more essential terms on the right in (13.28) and (13.29) all turn out to be positive.[2] Coefficients which are thus non-negative for a ratio of mesh sizes k/h in the range (13.22) permit us to interpret (13.28) and (13.29) as an averaging process. Consequently we shall be able to obtain estimates on the solution $U_{m,n}$, $V_{m,n}$ of these difference equations that assure its convergence to the solution $U(mh, nk)$, $V(mh, nk)$ of the Cauchy problem (13.23), (13.24), (13.27) in the limit as $h \to 0$.

Taylor's theorem with remainder shows that the solution of the latter Cauchy problem fulfills the requirements

$$(13.31) \quad U(mh, nk + k) = \left(1 - \frac{k}{h}\right)U(mh, nk) + \frac{k}{h} U(mh + h, nk)$$

and

$$+ kaU(mh, nk) + kbV(mh, nk) + O(hk)$$

$$(13.32) \quad V(mh, nk + k) = \left(1 - \frac{k}{h}\right)V(mh, nk) + \frac{k}{h} V(mh - h, nk)$$

$$+ kcU(mh, nk) + kdV(mh, nk) + O(hk).$$

We introduce the error terms

$$\delta_{m,n} = U(mh, nk) - U_{m,n}, \qquad \epsilon_{m,n} = V(mh, nk) - V_{m,n}$$

and subtract (13.28), (13.29) from (13.31), (13.32) to derive for them the new difference equations

$$(13.33) \quad \delta_{m,n+1} = \left(1 - \frac{k}{h}\right)\delta_{m,n} + \frac{k}{h} \delta_{m+1,n} + ka\delta_{m,n} + kb\epsilon_{m,n} + O(hk)$$

and

$$(13.34) \quad \epsilon_{m,n+1} = \left(1 - \frac{k}{h}\right)\epsilon_{m,n} + \frac{k}{h} \epsilon_{m-1,n} + kc\delta_{m,n} + kd\epsilon_{m,n} + O(hk).$$

Finally, a comparison of (13.30) with (13.27) yields the homogeneous initial conditions

$$(13.35) \qquad \qquad \delta_{m,0} = 0, \qquad \epsilon_{m,0} = 0$$

for $\delta_{m,n}$ and $\epsilon_{m,n}$. What we aim to prove is that $\delta_{m,n}$ and $\epsilon_{m,n}$ approach zero as $h \to 0$.

[2] Cf. Courant-Isaacson-Rees 1.

Let

$$\epsilon_n = \sup_m \max \left(|\delta_{m,n}|, |\epsilon_{m,n}| \right)$$

over some range of the index m depending appropriately on n. Under our hypothesis (13.22) about the ratio of mesh sizes k/h, formulas (13.33) and (13.34) imply that

$$|\delta_{m,n+1}| \leq \left(1 - \frac{k}{h} \right) \epsilon_n + \frac{k}{h} \epsilon_n + kM\epsilon_n + Rhk$$

and

$$|\epsilon_{m,n+1}| \leq \left(1 - \frac{k}{h} \right) \epsilon_n + \frac{k}{h} \epsilon_n + kM\epsilon_n + Rhk,$$

where

$$M = 2 \max \left(|a|, |b|, |c|, |d| \right),$$

and where R is a sufficiently large positive number. It follows that

$$(13.36) \qquad \epsilon_{n+1} \leq (1 + Mk)\epsilon_n + Rhk.$$

Because of (13.35) we have $\epsilon_0 = 0$. Hence (13.36) yields the further estimates

$$\epsilon_1 \leq Rhk, \qquad \epsilon_2 \leq Rhk + (1 + Mk)Rhk,$$

and by mathematical induction we deduce that more generally

$$(13.37) \quad \epsilon_n \leq [1 + (1 + Mk) + (1 + Mk)^2 + \cdots + (1 + Mk)^{n-1}]Rhk$$

$$\leq \frac{(1 + Mk)^n}{Mk} Rhk = \frac{R}{M} (1 + Mk)^n h.$$

Writing $nk = t$ and observing that

$$(1 + Mk)^n = e^{n \log (1 + Mk)} \leq e^{nMk} = e^{Mt},$$

we derive from (13.37) the more tangible inequality

$$\epsilon_n \leq \frac{Re^{Mt}}{M} h$$

which our upper bound ϵ_n on the truncation errors $\delta_{m,n}$ and $\epsilon_{m,n}$ has to satisfy. Thus we conclude that as $h \to 0$, the solution $U_{m,n}$, $V_{m,n}$ of our finite difference approximation (13.28), (13.29), (13.30) to the Cauchy problem (13.23), (13.24), (13.27) does, indeed, converge toward the exact answer $U(mh, nk)$, $V(mh, nk)$ to the latter problem, whose existence and regularity have, of course, been assumed. We have even succeeded in establishing that the error involved in the approximation has the same order of magnitude as the mesh size h, although it may grow exponentially with the time variable t.

The stability for $k \leq h$ of the difference scheme defined by (13.28) and (13.29) also enables us to assess the rounding errors generated in a step by step numerical solution of that system. For example, if only l decimal places are taken into account in computing $U_{m,n}$ and $V_{m,n}$, then round-off errors of the order of magnitude 10^{-l} occur in (13.28) and (13.29) and play there a role quite analogous to that of the terms $O(hk)$. Thus in the basic estimate (13.36) a contribution of the form $10^{-l}R$ should be added on the right if the effect of rounding is to be included. This leads in our final inequality for the total error to an extra expression involving the factor $10^{-l}/k$ instead of the factor h. The true bound on the error is therefore

$$(13.38) \qquad \epsilon = \frac{Re^{Mt}}{M}\left(h + \frac{10^{-l}}{k}\right).$$

We emphasize once more that the first term on the right here is a *truncation error* caused by replacing the differential equations (13.23) and (13.24) by finite difference equations, whereas the second term represents *rounding errors* of the kind that necessarily feature in any explicit numerical calculation. Note that because k appears in the denominator in (13.38), for each fixed value of l the choices of the mesh sizes k and h which minimize the sum of both errors, subject to the constraint $k \leq h$, are of the intermediate order of magnitude $10^{-l/2}$. Finally, for an unstable grid with $k > h$ the errors due to truncation and rounding must be expected to increase without limit as $h \to 0$, whence such a grid is of little practical interest.

EXERCISES

1. Verify that every function u satisfying the wave equation (13.5) is an exact solution of the special difference equation (13.17), which is based on a grid of characteristic squares.

2. Show how to formulate and how to solve finite difference equations approximating the characteristic initial value problem or the simplest mixed initial and boundary value problem for the wave equation (13.5).

3. Apply the von Neumann stability criterion to the system of difference equations (13.28), (13.29).

4. Extend our convergence theorem concerning finite difference approximations for the Cauchy problem (13.23), (13.24), (13.27) to the case of variable coefficients a, b, c and d.

5. Introduce a finite difference equation of the form

(13.39)
$$\frac{u_{l,m,n+1} - 2u_{l,m,n} + u_{l,m,n-1}}{(\Delta t)^2}$$

$$= \frac{u_{l+1,m,n} - 2u_{l,m,n} + u_{l-1,m,n}}{(\Delta x)^2} + \frac{u_{l,m+1,n} - 2u_{l,m,n} + u_{l,m-1,n}}{(\Delta y)^2}$$

to solve Cauchy's problem for the wave equation

(13.40)
$$u_{tt} = u_{xx} + u_{yy}$$

numerically; show that it is stable for mesh sizes Δx, Δy and Δt satisfying

$$\Delta x = \Delta y = \sqrt{2}\,\Delta t;$$

then make a comparison of the domains of dependence associated with (13.39) and (13.40).

6. To solve initial value problems for a symmetric hyperbolic system

$$u_t = Au_x + Bu_y$$

of partial differential equations in three independent variables, develop a finite difference scheme suggested by the formula

(13.41)
$$u_{l,m,n+1} = \frac{u_{l+1,m,n} + u_{l,m+1,n} + u_{l-1,m,n} + u_{l,m-1,n}}{4}$$

$$+ \frac{\Delta t}{\Delta x} A_{l,m,n} \frac{u_{l+1,m,n} - u_{l-1,m,n}}{2} + \frac{\Delta t}{\Delta y} B_{l,m,n} \frac{u_{l,m+1,n} - u_{l,m-1,n}}{2}.$$

Use von Neumann's criterion to verify that this numerical procedure is stable when the ratios $\Delta t/\Delta x$ and $\Delta t/\Delta y$ do not exceed one half the reciprocal of the largest among the absolute values of the characteristic roots of the two (symmetric) matrices A and B.

7. Give a justification of von Neumann's criterion for the stability of the finite difference scheme (13.41) by developing[3] an energy inequality for a suitable sum of squares (cf. Section 12.2).

8. For a nonlinear hyperbolic system in the canonical form

(13.42)
$$u_t^{(j)} = \lambda^{(j)}(x, t, u)u_x^{(j)} + \beta^{(j)}(x, t, u), \qquad j = 1, \dots, N,$$

of Section 3.5, consider a numerical method of solution of the initial value problem which is based on using the difference equation

(13.43)
$$u_{m,n+1}^{(j)} = \left[1 - \frac{k}{h}\lambda_{m,n}^{(j)}\right]u_{m,n}^{(j)} + \frac{k}{h}\lambda_{m,n}^{(j)}u_{m+1,n}^{(j)} + k\beta_{m,n}^{(j)}$$

---

[3] Cf. Friedrichs 2, Lees 1.

to approximate (13.42) whenever $\lambda^{(j)} > 0$, but on using

(13.44) $\qquad u_{m,n+1}^{(j)} = \left[1 + \dfrac{k}{h}\,\lambda_{m,n}^{(j)} \right] u_{m,n}^{(j)} - \dfrac{k}{h}\,\lambda_{m,n}^{(j)} u_{m-1,n}^{(j)} + k\beta_{m,n}^{(j)}$

instead whenever $\lambda^{(j)} < 0$. Show that such a scheme is both convergent and stable provided that the ratio of mesh sizes k/h is kept so small that

$$\frac{k}{h}\max |\lambda^{(j)}| \le 1.$$

9. Assuming that all the coefficients $\lambda^{(j)}$ in (13.42) reduce to either $+1$ or -1, develop[4] a finite difference approximation to that system with a truncation error of the order of magnitude h^2 by putting $k = h$ and by substituting the averages $(\beta_{m,n+1} + \beta_{m+1,n})/2$ and $(\beta_{m,n+1} + \beta_{m-1,n})/2$ for $\beta_{m,n}$ in (13.43) and (13.44), respectively. How can a set of nonlinear difference equations of this kind be solved?

10. Use the method of finite differences, together with a limit process in which the mesh sizes h and k approach zero in a fixed ratio satisfying (13.22), to establish the existence of a solution of the Cauchy problem (13.26), (13.27).

11. Show that according to von Neumann's criterion the second order central difference scheme

$$\frac{u_{l,m,n+1} - u_{l,m,n-1}}{\Delta t} = A_{l,m,n}\frac{u_{l+1,m,n} - u_{l-1,m,n}}{\Delta x} + B_{l,m,n}\frac{u_{l,m+1,n} - u_{l,m-1,n}}{\Delta y}$$

should furnish a stable solution of the first order symmetric hyperbolic system of Exercise 6 provided that the ratios $\Delta t/\Delta x$ and $\Delta t/\Delta y$ are taken small enough and provided that the initial data are handled with sufficient accuracy. What advantage does this scheme have over that of Exercise 6?

3. PARABOLIC EQUATIONS

The numerical analysis of initial value problems for equations of the parabolic type has much in common with that for equations of the hyperbolic type. We shall illustrate the basic phenomena that occur in the parabolic case by investigating the example of the finite difference analogue

(13.9) $\qquad \dfrac{u_{m+1,n} - 2u_{m,n} + u_{m-1,n}}{h^2} - \dfrac{u_{m,n+1} - u_{m,n}}{k} = 0$

of the heat equation

(13.6) $\qquad\qquad\qquad u_{xx} - u_y = 0.$

[4] Cf. Forsythe-Wasow 1.

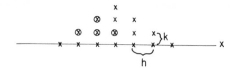

FIG. 57. Difference scheme for the heat equation.

Here we confine our attention to the technical difficulties that arise in computing solutions of (13.6) by means of the approximation (13.9). However, in Section 6 we shall discuss the heat equation again and shall describe an existence theorem for mixed initial and boundary value problems that stems from a finite difference scheme.

We are interested in finding a numerical method to solve the initial value problem

$$(13.45) \qquad u(x, 0) = f(x)$$

for (13.6), where $f(x)$ is any sufficiently differentiable function defined for all x. In order to use the difference equation (13.9) we replace the requirement (13.45) by the initial condition

$$(13.46) \qquad u_{m,0} = f(mh) = f_m.$$

The solution $u_{m,n}$ of the discrete problem (13.9), (13.46) can then be determined successively for choices of the index n larger than zero by applying (13.9) in the more convenient form

$$(13.47) \qquad u_{m,n+1} = u_{m,n} + \frac{k}{h^2}(u_{m+1,n} - 2u_{m,n} + u_{m-1,n})$$

similar to (13.15).

To investigate the stability of the difference scheme (13.47) we subject it to the von Neumann test. Substituting the exponential expression $u_{m,n} = e^{im\theta}e^{in\lambda}$ into (13.47), and simultaneously dividing through by $e^{im\theta}e^{in\lambda}$, we find that λ must be a root of the equation

$$e^{i\lambda} = 1 - 4\frac{k}{h^2}\sin^2\frac{\theta}{2}.$$

The stability criterion asks that the imaginary part of λ be non-negative, or, in other words, that $|e^{i\lambda}| \leq 1$, for every real choice of the parameter θ. This is the case if and only if

$$(13.48) \qquad \frac{k}{h^2} \leq \frac{1}{2}.$$

We observe in particular that in order to refine a grid (mh, nk) on which the stability requirement (13.48) has been imposed, it is necessary to allow the time increment k to approach zero like the square of the space mesh size h. Consequently the data (13.46) needed to determine a stable solution $u_{m,n}$ of the finite difference equation (13.47) at some fixed time $y = nk$ are located in a domain of dependence which becomes infinitely wide as $h \to 0$. Such a phenomenon is consistent with the fact that a solution $u = u(x, y)$ of the heat equation (13.6) depends at any time $y > 0$ in an essential way on its initial values (13.45) along the entire x-axis (cf. Exercise 5.1.14).

Under the hypothesis (13.48) concerning the ratio of mesh sizes k/h^2, we shall present here a rigorous proof that not only do the round-off errors generated by the finite difference scheme (13.47) have a stable behavior but also the solution $u_{m,n}$ of (13.46) and (13.47) converges toward the corresponding solution $u(mh, nk)$ of the parabolic initial value problem (13.6), (13.45) as $h \to 0$.

For the present purpose we transform (13.47) into the expression

$$(13.49) \qquad u_{m,n+1} = \frac{k}{h^2} u_{m+1,n} + \left(1 - \frac{2k}{h^2}\right) u_{m,n} + \frac{k}{h^2} u_{m-1,n}$$

for $u_{m,n+1}$ as an average of $u_{m+1,n}$, $u_{m,n}$ and $u_{m-1,n}$ with non-negative weight factors. According to Taylor's theorem with remainder, the differential equation (13.6) provides us with the analogous relation

$$(13.50) \quad u(mh, nk + k) = \frac{k}{h^2} u(mh + h, nk) + \left(1 - \frac{2k}{h^2}\right) u(mh, nk)$$

$$+ \frac{k}{h^2} u(mh - h, nk) + O(kh^2)$$

for u. Subtracting (13.49) from (13.50) and introducing the truncation errors

$$\epsilon_{m,n} = u(mh, nk) - u_{m,n},$$

we conclude that

$$(13.51) \quad \epsilon_{m,n+1} = \frac{k}{h^2} \epsilon_{m+1,n} + \left(1 - \frac{2k}{h^2}\right) \epsilon_{m,n} + \frac{k}{h^2} \epsilon_{m-1,n} + O(kh^2).$$

On the other hand, we have

$$(13.52) \qquad\qquad\qquad \epsilon_{m,0} = 0$$

because of the initial conditions (13.45) and (13.46).

Motivated by the analysis of Section 2, we define

$$\epsilon_n = \sup_m |\epsilon_{m,n}| .$$

The difference equation (13.51), together with our hypothesis (13.48), shows that

(13.53) $$\epsilon_{n+1} \leq \epsilon_n + Rkh^2,$$

where R is some sufficiently large positive constant. Moreover, (13.52) implies that $\epsilon_0 = 0$, whence we derive from (13.53) the basic inequality

(13.54) $$\epsilon_n \leq Rnkh^2 = Ryh^2$$

for the truncation error ϵ_n. Also, if rounding errors occur because only l decimal places are used in computing an actual numerical solution of the difference equation (13.49), then an additional term of the form $10^{-l}R$ ought to be included on the right in (13.53). Therefore our final estimate of the total error ϵ becomes

(13.55) $$\epsilon \leq Ry\left(h^2 + \frac{10^{-l}}{k}\right) .$$

The result (13.54) implies that for any ratio of mesh sizes in the range (13.48) the solution $u_{m,n}$ of the finite difference approximation (13.46), (13.47) to our initial value problem (13.45) for the heat equation (13.6) converges toward the exact answer $u(mh, nk)$ in the limit as $h \to 0$. However, (13.55) indicates that in order to retain control over the round-off errors generated by a numerical calculation of $u_{m,n}$, we should avoid picking the time increment k of our grid to be unnecessarily small. In this connection we wish to emphasize once more the close relationship between estimations of the truncation and of the rounding errors despite the diversity of the sources from which they arise.

EXERCISES

1. Discuss the convergence and stability of the most obvious finite difference scheme for solving the partial differential equation

$$u_t = u_{xx} + u_{yy}.$$

2. Show that the *implicit* finite difference approximation

$$\frac{u_{m+1,n+1} - 2u_{m,n+1} + u_{m-1,n+1}}{h^2} = \frac{u_{m,n+1} - u_{m,n}}{k}$$

to the heat equation (13.6) is stable regardless of the choice of the ratio k/h^2. In the case of the mixed initial and boundary value problem, develop a specific procedure for calculating the solution of such a system of difference equations.

3. Use Taylor's theorem with remainder to establish that when $k/h^2 = \frac{1}{6}$, the estimate (13.54) of the truncation error associated with the difference scheme (13.47) can be replaced by a better upper bound of the form Ryh^4.

4. Explain in what sense the solution of a mixed initial and boundary value problem for the heat equation can be expected to approach the corresponding solution of Dirichlet's problem for the Laplace equation as the time variable concerned becomes infinite.

5. Develop a classification of systems of partial differential equations with constant coefficients according to type by examining exponential solutions in a fashion suggested by von Neumann's stability criterion for difference equations (cf. Section 6.1). Extend the classification to more general cases by applying Fourier analysis to appropriately modified systems whose coefficients have been frozen so that they retain everywhere the values they had at a specific point.

4. ELLIPTIC EQUATIONS

We shall investigate the finite difference approximation

$$(13.10) \quad \frac{u_{m+1,n} - 2u_{m,n} + u_{m-1,n}}{h^2} + \frac{u_{m,n+1} - 2u_{m,n} + u_{m,n-1}}{k^2} = 0$$

to Laplace's equation in considerable detail here in order to give some indication of the difficulties that are encountered in the elliptic case. In general, boundary value problems lead to large sets of simultaneous difference equations that cannot be solved in a direct way numerically. Therefore we shall devote the next section especially to practical methods of treating such equations and, in particular, to an analysis of the discrete analogue of the Dirichlet problem. First, however, we wish to study the convergence properties of solutions of (13.10).

Since the ratio of mesh sizes turns out to play a quite insignificant role in the theory of boundary value problems, it will suffice for the present purposes to confine our attention to the case $k = h$ of a square grid, so that (13.10) can be expressed in the simplified form

$$(13.56) \quad u_{m,n} = \frac{u_{m+1,n} + u_{m,n+1} + u_{m-1,n} + u_{m,n-1}}{4}.$$

The latter relation, which is often referred to as *Laplace's difference equation*, asserts that at each grid point in the interior of our lattice $u_{m,n}$ coincides with the average of its values at the four neighboring grid points. It follows that not all the neighboring values can exceed $u_{m,n}$, nor can every one of them be less than $u_{m,n}$ either. Consequently $u_{m,n}$ cannot assume a strong maximum or a strong minimum in the interior of the lattice. Indeed, only when it is identically constant can it achieve even a weak maximum or a weak minimum there. This *maximum principle* for solutions of the Laplace difference equation will constitute the basis for our estimation of the errors introduced by the finite difference approximation of harmonic functions.

In order to put our ideas about finite difference equations in perspective it is helpful to compare them here with the two kinds of uniqueness theorems for Dirichlet's problem that we developed in Section 7.1 by means of the maximum principle and of energy integrals. Throughout our treatment of convergence and stability questions in this chapter we prefer to exploit analogues of the maximum principle rather than the more subtle technique of energy inequalities. Thus we usually attempt to formulate our difference equations as averaging processes in which the loss of significance due to cancellations will be minimized. However, certain problems, such as those connected with symmetric hyperbolic systems (cf. Exercises 2.6 and 2.7) are beyond the capacity of such an elementary approach and demand the broader scope of the method of energy inequalities. It is to be noted that the latter is also fruitful for proving existence theorems by passage to the limit from a finite difference approximation.[5]

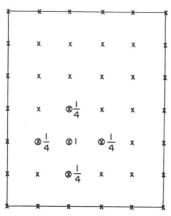

FIG. 58. The Dirichlet problem.

In the formulation of Dirichlet's problem for (13.56) we shall consider lattices each of which is composed of a finite block of grid points just filling out a polygonal domain D, such as a rectangle, bounded exclusively by vertical and horizontal line segments. We restrict our analysis to simple domains of this kind so that each lattice point lies on the boundary ∂D of D or else is an interior point in the sense that its four immediate neighbors also belong to the grid. At each mesh point of the boundary ∂D let us prescribe for $u_{m,n}$ the same value that is assigned there in some Dirichlet problem for Laplace's differential equation (13.7). In order to obtain an approximate solution of the latter problem numerically, we need to determine $u_{m,n}$ at the interior points of the grid, too. With each such point we can associate

[5] Cf. Epstein 1, Friedrichs 2.

a specific difference equation (13.56). Therefore what we actually have to find is the solution of a system of N simultaneous linear equations (13.56) in N unknowns, where N stands for the total number of interior mesh points of the lattice exhausting D.

We are already familiar with the Fredholm alternative (cf. Section 10.1). In its most elementary form this states that any system of simultaneous linear equations can either be solved in a unique way for every choice of the inhomogeneous terms or it must possess non-trivial solutions in the homogeneous case. In the foregoing example the role of the inhomogeneous terms is played by the boundary values we assign to $u_{m,n}$. Consequently we can prove that a solution of the discrete Dirichlet problem always exists merely by verifying that when the data are made to vanish, the only answer available is the obvious one $u_{m,n} = 0$. On the other hand, an appeal to the maximum principle established earlier for solutions of the Laplace difference equation assures us of the validity of the latter uniqueness assertion, as is clear from arguments developed in Section 7.1. Thus our formulation of the Dirichlet problem for difference equations is meaningful, and it makes sense for us to investigate the convergence of the answer toward the solution of the corresponding problem for the elliptic partial differential equation (13.7).

Taylor's theorem shows that any harmonic function $u = u(x, y)$ with bounded derivatives of the fourth order must satisfy the relation

$$(13.57) \quad u(mh, nh) = \tfrac{1}{4}\{u(mh + h, nh) + u(mh, nh + h) + u(mh - h, nh) + u(mh, nh - h)\} + O(h^4).$$

When the solution u of Dirichlet's problem that we are trying to calculate possesses the necessary derivatives, a subtraction of (13.56) from (13.57) yields the inhomogeneous difference equation

$$(13.58) \quad \epsilon_{m,n} - \frac{\epsilon_{m+1,n} + \epsilon_{m,n+1} + \epsilon_{m-1,n} + \epsilon_{m,n-1}}{4} = O(h^4)$$

for the truncation error

$$\epsilon_{m,n} = u(mh, nh) - u_{m,n}.$$

Furthermore, our specification of the boundary condition to be imposed on $u_{m,n}$ shows that we have

$$(13.59) \quad \epsilon_{m,n} = u(mh, nh) - u_{m,n} = 0$$

at each mesh point located on ∂D. We shall now proceed to apply the maximum principle in order to establish that as we refine our lattice of grid points the error $\epsilon_{m,n}$ approaches zero like the square h^2 of the mesh size.

Let Mh^4 stand for an upper bound on the absolute value of the expression $O(h^4)$ appearing on the right in (13.58). Since $\epsilon_{m,n}$ could be estimated more easily if this expression turned out to be of one sign, we choose to introduce a comparison function into (13.58) that leads to such a situation. Putting

$$\omega(x, y) = x^2 + y^2, \qquad \omega_{m,n} = \omega(mh, nh),$$

we note that

$$\omega_{m,n} - \frac{\omega_{m+1,n} + \omega_{m,n+1} + \omega_{m-1,n} + \omega_{m,n-1}}{4} = -h^2.$$

Hence we have

$$v_{m,n} - \frac{v_{m+1,n} + v_{m,n+1} + v_{m-1,n} + v_{m,n-1}}{4} \le Mh^4 - Mh^4 = 0,$$

where

$$v_{m,n} = \epsilon_{m,n} + Mh^2\omega_{m,n}.$$

Observe that at each interior point of our lattice $v_{m,n}$ is dominated by the average of its values at the four adjacent grid points. Consequently it must achieve its maximum on the boundary ∂D of D. Since (13.59) holds on ∂D, we may conclude that

$$v_{m,n} \le Mh^2\Omega,$$

where Ω is the square of the radius of a circle about the origin enclosing the region D. A similar analysis of the quantity

$$w_{m,n} = \epsilon_{m,n} - Mh^2\omega_{m,n}$$

shows that

$$w_{m,n} \ge -Mh^2\Omega.$$

These results combine to yield the final inequality

$$|\epsilon_{m,n}| \le M\Omega h^2$$

for the truncation error $\epsilon_{m,n}$ itself.

We stress that to take into account rounding errors due to retaining only l decimal places in the computation of $u_{m,n}$ it is necessary to add a term on the right in (13.58) which possesses a bound of the form $10^{-l}M$. Consequently the estimate we obtain for the total error ϵ occurring in our approximate solution of the Dirichlet problem is

(13.60) $$\epsilon \le M\Omega \left\{ h^2 + \frac{10^{-l}}{h^2} \right\}.$$

In particular, the numerical answer has to converge toward the exact solution if the mesh size h approaches zero while at the same time the number l of decimal places retained in the calculations goes fast enough to infinity. As far as the more specific dependence of the error on h and l is concerned, we emphasize that it corresponds directly to the kind of inhomogeneous terms that appear in (13.58) after a division by h^2 has been performed to reduce the finite differences there to difference quotients which are the true analogues of the derivatives featured in Laplace's equation (13.7). It is of interest, too, that for a fixed choice of l the expected size Ω/h^2 of the round-off error is proportional to the number N of difference equations, or, in other words, of grid points, that have been used.

We close this section with a few comments about an interesting probabilistic interpretation of the Laplace difference equation (13.56). Consider in a given lattice D the solution $u_{m,n}$ of (13.56) which assumes the boundary values 1 at the mesh points of some arc C of ∂D but reduces to zero on the remainder of ∂D. We shall indicate how $u_{m,n}$ can be constructed by means of a *random walk* that is defined in a natural way over the interior grid points of D.

Suppose a particle moves about at random from one mesh point of D to another in such a fashion that as it steps away from any interior grid point the probability of its advancement to each one of the four adjacent grid points is precisely $\frac{1}{4}$. Moreover, require that the particle come to rest whenever it reaches the boundary ∂D of D. Now let $P_{m,n}$ denote the probability that a trajectory of the particle starting from the mesh point (mh, nh) terminate on the boundary arc C of ∂D. We maintain that

$$P_{m,n} = u_{m,n}.$$

Since the particle in our random walk is never allowed to leave ∂D once it has arrived there, we see immediately that $P_{m,n} = 1$ when the mesh point (mh, nh) is located on C, whereas $P_{m,n} = 0$ when that point is located on the remainder of ∂D. Thus the boundary values of $P_{m,n}$ coincide with those of $u_{m,n}$. On the other hand, the probability $P_{m,n}$ that a trajectory emanating from some interior point (mh, nh) of D terminate on C decomposes into the sum of the probabilities that it do so by proceeding first to one or another of the four adjacent grid points. Since the probabilities of the latter events are given by the numbers $P_{m+1,n}/4$, $P_{m,n+1}/4$, $P_{m-1,n}/4$ and $P_{m,n-1}/4$, we conclude that

$$(13.61) \qquad P_{m,n} = \tfrac{1}{4}P_{m+1,n} + \tfrac{1}{4}P_{m,n+1} + \tfrac{1}{4}P_{m-1,n} + \tfrac{1}{4}P_{m,n-1}.$$

Thus $P_{m,n}$ and $u_{m,n}$ both solve the same discrete Dirichlet problem for the Laplace difference equation. Hence they must actually be identical, which is what we originally set out to prove. This probabilistic model of the Dirichlet problem suggests a statistical procedure for computing the answer that is usually referred to as the *Monte Carlo method*.

Finally, we point out that no essential difficulties are involved in formulating valid finite difference approximations to the most general partial differential equations of the elliptic type. Indeed, such a treatment has even more significance for equations with variable coefficients than it has for Laplace's equation, since other methods, such as interpolation by linear combinations of explicit solutions, are not available for them.

EXERCISES

1. Use trigonometric polynomials to derive a closed expression for the solution of Dirichlet's problem for the Laplace difference equation in a rectangle.

2. Show that when the *nine-point difference equation*

$$(13.62) \qquad u_{m,n} = \frac{u_{m+1,n} + u_{m,n+1} + u_{m-1,n} + u_{m,n-1}}{5}$$

$$+ \frac{u_{m+1,n+1} + u_{m+1,n-1} + u_{m-1,n+1} + u_{m-1,n-1}}{20}$$

is used to approximate solutions of Laplace's equation (13.7), the resulting truncation error is of the order of magnitude h^6 instead of h^2.

3. Formulate a finite difference approximation to the first boundary value problem for the biharmonic equation.

4. How can the method of finite differences be used to estimate the principal frequency of a vibrating membrane?

5. Discuss finite difference approximations to the Laplace equation in space.

6. Introduce a Green's function for the Laplace difference equation and develop a probabilistic interpretation of it.

7. Show by means of the von Neumann criterion that the elliptic difference equation (13.10) is unstable for the solution of initial value problems no matter how small the ratio of mesh sizes k/h is chosen. Compare this conclusion with our description of Hadamard's counterexample for the Cauchy problem in Section 4.1.

8. Show that it is a solution of the difference equation (13.56) that minimizes the Dirichlet sum

$$I_D = \sum (u_{m+1,n} - u_{m,n})^2 + \sum (u_{m,n+1} - u_{m,n})^2$$

extended over all possible choices of first differences of the unknown $u_{m,n}$ in a given lattice D, where only the values of $u_{m,n}$ at the boundary ∂D are prescribed. Use this remark to deduce the existence of a unique solution of the Dirichlet problem for (13.56).

9. Use summation by parts to establish[6] an *energy inequality* of the form

$$(13.63) \qquad 2MI_D \leq \sum u_{m,n}^2$$

for solutions $u_{m,n}$ of the Laplace difference equation, where D is a square $|m| \leq N$, $|n| \leq N$, and where the sum on the right is extended over all grid points on the perimeter of the larger square $|m| \leq N + M$, $|n| \leq N + M$.

10. In analogy with *Liouville's theorem*, deduce from (13.63) that any solution of the Laplace difference equation (13.56) that is defined and bounded on a grid covering the entire plane must be a constant.

11. By using energy inequalities to develop equicontinuity properties of the higher differences of a solution of Dirichlet's problem for (13.56) that describe its convergence in the limit as $h \to 0$, prove[7] the existence of a solution of the analogous problem for (13.7). Compare this procedure with Exercise 12.2.17.

5. RELAXATION AND OTHER ITERATIVE METHODS

Although we have convinced ourselves that a unique solution of the Dirichlet problem for Laplace's difference equation exists, we still have to find ways of determining it effectively when the number of mesh points involved is large. Thus our purpose now is to develop numerical methods for treating the unwieldy systems of simultaneous linear equations that occur in more accurate finite difference approximations to boundary value problems.

We shall begin by discussing the procedure known as *relaxation*, which resembles in several respects the Poincaré sweeping out process that led in Section 9.2 to an existence theorem for the Dirichlet problem based on subharmonic functions. Our point of departure is a set of guessed values $u_{m,n}^{(0)}$ of the desired solution $u_{m,n}$ of

$$(13.56) \qquad u_{m,n} = \frac{u_{m+1,n} + u_{m,n+1} + u_{m-1,n} + u_{m,n-1}}{4}$$

which coincide with it on the boundary ∂D of the grid D. What we have in mind is to go through the grid in some reasonably ordered fashion, replacing in succession the value of our guess at each interior mesh point by the average of its four adjacent values. Repetition of this operation infinitely often furnishes a sequence $u_{m,n}^{(j)}$ of approximations that may be expected to converge toward the required exact solution $u_{m,n}$ of the Laplace difference equation as $j \to \infty$.

[6] Cf. Courant-Friedrichs-Lewy 1.
[7] Cf. Epstein 1.

Let $z_1, z_2, z_3, \ldots$ stand for some sequence of mesh points inside D which covers every interior point of that lattice an infinite number of times. Given $u_{m,n}^{(j-1)}$ we define $u_{m,n}^{(j)}$ to be identical with $u_{m,n}^{(j-1)}$ at all points of D except z_j, where it is set equal instead to the average

$$u_{m,n}^{(j)} = \frac{u_{m+1,n}^{(j-1)} + u_{m,n+1}^{(j-1)} + u_{m-1,n}^{(j-1)} + u_{m,n-1}^{(j-1)}}{4}$$

of the values of the previous approximation at the four adjacent grid points. Starting from our original guess $u_{m,n}^{(0)}$ of the solution of the Dirichlet problem, we thus obtain by induction a sequence of improved answers $u_{m,n}^{(j)}$ that will be shown to have the desired limit

(13.64)
$$\lim_{j \to \infty} u_{m,n}^{(j)} = u_{m,n}.$$

The actual relaxation process that is used in practice deviates from the construction we have just outlined in that to improve the rate of convergence a certain amount of overadjustment of the solution is incorporated at each step, following rules suggested intuitively by numerical data already accumulated.

We denote by α and β lower and upper bounds on $u_{m,n}$, so that there is no real loss of generality if we impose the restriction

$$\alpha \leq u_{m,n}^{(0)} \leq \beta$$

on our choice of $u_{m,n}^{(0)}$. Next let $v_{m,n}^{(j)}$ and $w_{m,n}^{(j)}$ stand, respectively, for the sequences of approximations to $u_{m,n}$ found from the procedure for constructing $u_{m,n}^{(j)}$ described above when α and β themselves are substituted for $u_{m,n}^{(0)}$ at the interior grid points of D. From the obvious inequality

$$v_{m,n}^{(0)} \leq u_{m,n}^{(0)} \leq w_{m,n}^{(0)},$$

and from our definition of the successive iterates $u_{m,n}^{(j)}$, $v_{m,n}^{(j)}$ and $w_{m,n}^{(j)}$ by means of an averaging process, we deduce that

(13.65)
$$v_{m,n}^{(j)} \leq u_{m,n}^{(j)} \leq w_{m,n}^{(j)}.$$

It is our intention to prove that as the index j advances toward infinity $v_{m,n}^{(j)}$ always increases or remains unchanged, whereas $w_{m,n}^{(j)}$ always decreases or remains unchanged, while both converge toward $u_{m,n}$ in the limit. Our original assertion (13.64) will then follow because of the estimates (13.65).

We find it convenient to introduce the linear operator

$$(13.66) \quad L[u_{m,n}] = \frac{u_{m+1,n} + u_{m,n+1} + u_{m-1,n} + u_{m,n-1}}{4} - u_{m,n},$$

in terms of which Laplace's difference equation becomes

$$L[u_{m,n}] = 0.$$

Since the boundary values of $v_{m,n}^{(0)}$ and $w_{m,n}^{(0)}$, which are, of course, identical with those of $u_{m,n}$, lie in the interval between α and β , we have

$$(13.67) \qquad L[v_{m,n}^{(0)}] \geq 0, \qquad L[w_{m,n}^{(0)}] \leq 0$$

inside D. In fact the equality signs hold unless at least one of the mesh points concerned is located on ∂D. We maintain that, more generally,

$$(13.68) \qquad L[v_{m,n}^{(j)}] \geq 0, \qquad L[w_{m,n}^{(j)}] \leq 0.$$

When (mh, nh) coincides with the mesh point z_j, our definition of $v_{m,n}^{(j)}$ there becomes

$$(13.69) \qquad v_{m,n}^{(j)} = v_{m,n}^{(j-1)} + L[v_{m,n}^{(j-1)}],$$

whereas

$$v_{m,n}^{(j)} = v_{m,n}^{(j-1)}$$

throughout the rest of D. Hence if (13.68) is fulfilled with j replaced by $j - 1$, then

$$(13.70) \qquad v_{m,n}^{(j)} \geq v_{m,n}^{(j-1)}$$

everywhere. Furthermore, since it is only the value of $v_{m,n}^{(j)}$ at z_j itself that could be raised, we have

$$L[v_{m,n}^{(j)}] \geq 0,$$

too, with the equality sign actually holding at the key mesh point z_j. A similar analysis gives

$$(13.71) \qquad w_{m,n}^{(j)} \leq w_{m,n}^{(j-1)}.$$

The conclusion that (13.68), (13.70) and (13.71) are valid for every choice of the index j now follows from (13.67) and an application of mathematical induction.

Because of the inequalities (13.65) and the monotonicity theorems (13.70) and (13.71), the sequences $v_{m,n}^{(j)}$ and $w_{m,n}^{(j)}$ must approach finite limits $v_{m,n}$ and $w_{m,n}$ as $j \to \infty$. Since each interior mesh point of D is included infinitely often in the sequence z_j, a passage to the limit in (13.69) through values j_μ of the index j with the property

$$z_{j_1} = z_{j_2} = \cdots = z_{j_\mu} = \cdots$$

can be performed to establish that

$$L[v_{m,n}] = 0.$$

Similarly

$$L[w_{m,n}] = 0,$$

and because $v_{m,n}$ and $w_{m,n}$ have the same boundary values as $u_{m,n}$, the uniqueness of the solution of Dirichlet's problem for the Laplace difference equation yields the identity

$$u_{m,n} = v_{m,n} = w_{m,n}.$$

Thus we have

$$\lim_{j \to \infty} v_{m,n}^{(j)} = \lim_{j \to \infty} w_{m,n}^{(j)} = u_{m,n},$$

which, together with (13.65), completes our proof of the convergence (13.64) of the simplified relaxation process.

It is desirable to have a more accurate idea of the rate of convergence to be expected in iterative methods of solution of the Laplace difference equation. For such an analysis it is necessary to consider more specific choices of the sequence of points z_j. Let them therefore be ordered from left to right in each row, and then downward from one row to the next, as in reading some page of a book repeatedly. It follows that the subsequence of approximations

$$(13.72) \qquad u_{m,n}^{(0)}, u_{m,n}^{(N)}, u_{m,n}^{(2N)}, \ldots,$$

where N stands for the total number of interior mesh points in D, has the property that each element is obtained from its predecessor by altering the values at every mesh point just once. Changing notation, we now use the symbols

$$u_{m,n}^{(0)}, u_{m,n}^{(1)}, u_{m,n}^{(2)}, \ldots$$

to indicate the subsequence (13.72) of our original sequence, so that we can write

$$(13.73) \qquad u_{m,n}^{(j)} = \frac{u_{m-1,n}^{(j)} + u_{m,n+1}^{(j)} + u_{m+1,n}^{(j-1)} + u_{m,n-1}^{(j-1)}}{4}$$

for $j = 1, 2, \ldots$. It might be anticipated that (13.73), which is known as the *Gauss-Seidel method,* converges about twice as fast as the less imaginative iteration scheme

$$(13.74) \qquad u_{m,n}^{(j)} = \frac{u_{m-1,n}^{(j-1)} + u_{m,n+1}^{(j-1)} + u_{m+1,n}^{(j-1)} + u_{m,n-1}^{(j-1)}}{4}$$

$$= u_{m,n}^{(j-1)} + L[u_{m,n}^{(j-1)}],$$

which is, however, easier to deal with. Thus, to avoid a lengthy discussion, we shall content ourselves here with a more or less heuristic examination of the rate of convergence of (13.74), which we call the *first order Richardson method*[8] of solving Laplace's equation.

If we divide formula (13.74) through by $h^2/4$, rearrange terms appropriately, and then set $k = h^2/4$, we can express it in the more suggestive form

$$(13.75) \qquad \frac{u_{m,n}^{(j)} - u_{m,n}^{(j-1)}}{k} = \frac{u_{m+1,n}^{(j-1)} - 2u_{m,n}^{(j-1)} + u_{m-1,n}^{(j-1)}}{h^2}$$

$$+ \frac{u_{m,n+1}^{(j-1)} - 2u_{m,n}^{(j-1)} + u_{m,n-1}^{(j-1)}}{h^2}.$$

The latter relation may be interpreted as a finite difference approximation to the heat equation

$$(13.76) \qquad u_t = u_{xx} + u_{yy}$$

based on a three-dimensional lattice of grid points

$$x = mh, \quad y = nh, \quad t = jk.$$

Thus in making a numerical analysis of the Dirichlet problem by means of the iterative scheme (13.74) we are essentially treating an analogous mixed initial and boundary value problem for the heat equation (13.76) and then allowing the time t to become infinite in order to arrive at a steady solution. By marching forward this way with respect to the auxiliary variable t we manage to overcome the difficulties in handling the large system of simultaneous linear difference equations encountered in our original boundary value problem, which could not be solved one at a time by the direct technique of elimination that is available in the case of initial value problems.

To estimate the rate of convergence of the successive approximations (13.74) toward an exact solution $u_{m,n}$ of the Laplace difference equation (13.56), we compare the sequence $u_{m,n}^{(j)}$ with the analogous solution

[8] Cf. Frankel 1.

$u(mh, nh, jk)$ of the heat equation (13.76) which it must approach (cf. Exercise 3.1) in the limit as $h \to 0$. The deviation of the latter function $u(x, y, t)$ from the corresponding solution $u(x, y)$ of Dirichlet's problem for (13.7) has an explicit expansion into a Fourier series (cf. Exercise 11.1.8)

$$(13.77) \qquad u(x, y, t) - u(x, y) = \sum_{i=1}^{\infty} a_i u_i(x, y) e^{-\lambda_i t}$$

of eigenfunctions u_i of the vibrating membrane equation

$$(13.78) \qquad\qquad \Delta u_i + \lambda_i u_i = 0$$

for the region D, with the requirement

$$(13.79) \qquad\qquad u_i = 0$$

imposed along ∂D. The representation (13.77) shows that $u(x, y, t)$ converges toward $u(x, y)$ at an exponential rate defined by the factor $e^{-\lambda_1 t}$, where λ_1 stands for the lowest eigenvalue of (13.78) and (13.79). Therefore we see intuitively that it is reasonable to expect the error in the approximation $u_{m,n}^{(j)}$ to $u_{m,n}$ to fulfill an inequality of the form

$$(13.80) \qquad |u_{m,n}^{(j)} - u_{m,n}| \leq Re^{-\lambda t} = Re^{-\lambda h^2 j/4}$$

for some fixed choice of the positive numbers λ and R.

According to the estimate (13.80) of the rate of convergence of $u_{m,n}^{(j)}$, the number j of iterations needed to reach a prescribed degree of accuracy in the first order Richardson scheme (13.74) grows like the reciprocal of h^2, which is proportional to the total number N of difference equations (13.56) we set out to solve. Hence it becomes desirable to achieve better convergence. This can be accomplished by introducing a more sophisticated iterative procedure that is motivated by the hyperbolic partial differential equation

$$(13.81) \qquad\qquad \epsilon^2 u_{tt} + u_t - u_{xx} - u_{yy} = 0$$

instead of (13.76).

When the parameter ϵ is small enough, Fourier analysis shows that any solution $u(x, y, t)$ of (13.81) with fixed boundary values approaches a steady state as $t \to \infty$, with the transient terms diminishing like $e^{-\lambda^* t}$, where λ^* stands in the relationship

$$\lambda^* = \frac{1 - \sqrt{1 - 4\lambda_1 \epsilon^2}}{2\epsilon^2} = \lambda_1 + O(\epsilon^2)$$

to the eigenvalue λ_1 mentioned above. Now if the finite difference analogue

$$(13.82) \quad \epsilon^2 \frac{u_{m,n}^{(j+1)} - 2u_{m,n}^{(j)} + u_{m,n}^{(j-1)}}{k^2} + \frac{u_{m,n}^{(j+1)} - u_{m,n}^{(j)}}{k}$$

$$- \frac{u_{m-1,n}^{(j)} + u_{m,n+1}^{(j)} + u_{m+1,n}^{(j)} + u_{m,n-1}^{(j)} - 4u_{m,n}^{(j)}}{h^2} = 0$$

of (13.81) is used instead of (13.75) to define successive approximations $u_{m,n}^{(j)}$ to a solution $u_{m,n}$ of Dirichlet's problem for (13.56), then h and k may be taken of the same order of magnitude. Consequently the convergence of the scheme is so rapid that

$$(13.83) \qquad\qquad |u_{m,n}^{(j)} - u_{m,n}| \leq Re^{-\lambda hj}$$

for appropriate positive constants λ and R. It follows that the number j of iterations needed to attain a given accuracy only grows like $1/h$, rather than $1/h^2$, for the process (13.82), which is frequently referred to as the *second order Richardson method* because of the appearance of the second derivative u_{tt} in (13.81).

In closing we wish to call attention to a further and quite practical scheme for solving the Laplace difference equation called *successive overrelaxation*.[9] It is based on the introduction into (13.73) of a systematic *relaxation factor* $\omega > 1$, and, more precisely, is defined by the rule

$$(13.84) \quad u_{m,n}^{(j)} = u_{m,n}^{(j-1)}$$

$$+ \omega \frac{u_{m-1,n}^{(j)} + u_{m,n+1}^{(j)} + u_{m+1,n}^{(j-1)} + u_{m,n-1}^{(j-1)} - 4u_{m,n}^{(j-1)}}{4}.$$

Because of the lack of symmetry in (13.84), overrelaxation can actually be associated with a partial differential equation in three independent variables involving time derivatives of the second order (cf. Exercise 6 below). Consequently we might expect it to converge at the rate indicated in (13.83). The success of these second order methods may be attributed to the fact that two levels of previous approximations are used in determining each iterate. Finally, we stress again that the special systems of difference equations arising in the numerical analysis of boundary value problems are so complicated that they must for the most part be solved with limited accuracy by what amounts to a marching process in the direction of some new independent variable. This limited accuracy results, of course, in an additional contribution to our estimate (13.60) of the error that occurs in the finite difference solution of Dirichlet's problem.

[9] Cf. Young 1.

EXERCISES

1. Show how dice can be used as an analogue computer to solve the Laplace difference equation by means of a Monte Carlo method based on the probabilistic model (13.61). For this purpose what advantage is there in using a triangular grid?

2. Figure 59 defines a specific Dirichlet problem for the Laplace difference equation. Solve this problem exactly by elimination, approximately by relaxation, and experimentally by the Monte Carlo method. Compare the three answers so obtained.

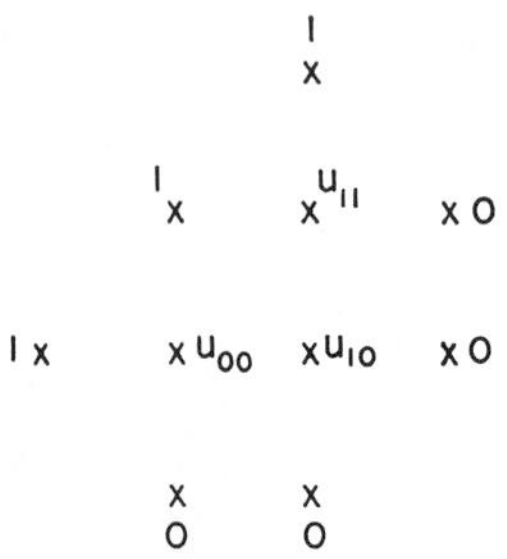

3. Express the eigenfunctions of the finite difference operator (13.66) for a rectangular grid as products of trigonometric functions. Use them to check the statements in the text concerning rates of convergence of the first and second order Richardson and overrelaxation methods of solving the discrete form of Dirichlet's problem.

4. Develop an iterative procedure to solve the nine-point difference equation (13.62).

5. Give a detailed proof based on matrix theory that the inequality (13.80) does, indeed, describe the rate of convergence of the iterative scheme (13.74).

FIG. 59. Example of a Dirichlet problem.

6. Justify more rigorously our contention that the rates of convergence of the second order Richardson and overrelaxation processes are governed by (13.83). In particular, put

$$(13.85) \qquad \omega = \frac{2}{1 + \alpha h}$$

in (13.84) to derive from it the analogous hyperbolic partial differential equation

$$u_{xx} + u_{yy} - u_{xt} + u_{yt} - 2\alpha u_t = 0$$

in the limit as the mesh size h approaches zero.[10] Show that the best choice of the positive parameter α that could be inserted into (13.85) has a connection with the lowest eigenvalue λ_1 of (13.78) and (13.79).

7. Develop a finite difference approximation to the Tricomi problem. How can the resulting system of simultaneous linear equations be handled?

6. EXISTENCE THEOREM FOR THE HEAT EQUATION

It is instructive to use the method of finite differences to establish the existence of solutions of initial and boundary value problems. In such an approach one examines the limit of the solution of a set of difference

[10] Cf. Garabedian 7.

equations as a related mesh size parameter h converges toward zero. In order to show that the limit function so obtained satisfies a corresponding differential equation, it is usually necessary to derive *a priori estimates* of the solution of the difference equations. For this purpose one of the simplest devices to which we can appeal is the maximum principle.

Here we shall present an example of an existence proof of the above kind which is concerned with the solution of a mixed initial and boundary value problem for the heat equation

$$(13.86) \qquad \Delta u = u_{xx} + u_{yy} = u_t.$$

(Cf. Exercise 11.1.8 for a treatment of the same problem by means of eigenfunction expansions.) We shall introduce a difference-differential equation connected with (13.86) which is based on finite differences in the time variable t only.[11] Such a technique has some bearing on more abstract procedures occurring in the theory of semigroups, and, moreover, it can be applied to more general parabolic equations with variable coefficients (cf. Exercises 9 and 10 below).

Consider a sequence of mesh points $t = nh$ along the t-axis, where n stands for an integer and h stands for some positive number. We shall be interested in the *difference-differential equation*

$$(13.87) \qquad \Delta u_{n+1} = \frac{u_{n+1} - u_n}{h},$$

which represents an approximation to the heat equation (13.86). It is reasonable to expect this finite difference scheme to converge in the limit as $h \to 0$ because it is *implicit* in the sense that the Laplacian operates on u_{n+1} rather than on u_n.

Let D be a domain of the (x,y)-plane bounded by a smooth curve ∂D. Our purpose is to establish that there exists in D, and for $t > 0$, a solution $u = u(x, y, t)$ of (13.86) which reduces at $t = 0$ to a prescribed function

$$(13.88) \qquad u(x, y, 0) = f(x, y)$$

over the region D and which satisfies the boundary condition

$$(13.89) \qquad u = 0$$

on ∂D for all $t > 0$. To achieve this purpose we first solve (13.87) in D, subject to the requirement

$$(13.90) \qquad u_0(x \ y) = f(x, y)$$

[11] Cf. Rothe 1, 2.

there and with

(13.91) $u_n = 0$

on ∂D, for $n = 0, 1, 2, \ldots$. Then we show that the rule

(13.92) $u(x, y, nh) = \lim\limits_{h \to 0} u_n(x, y)$

defines the desired solution u of the mixed initial and boundary value problem (13.86), (13.88), (13.89), provided that

$$f = \Delta f = \Delta \Delta f = 0$$

along ∂D.

The solvability of the difference-differential equation (13.87) in conjunction with the initial and boundary conditions (13.90) and (13.91) can be deduced by mathematical induction from results of Section 9.1. Suppose that u_n is known throughout D, which is, of course, true in the special case $n = 0$. Since u_{n+1} vanishes along ∂D, it is determined by a Dirichlet problem in D for the inhomogeneous elliptic partial differential equation

(13.93) $\Delta u_{n+1} - \dfrac{1}{h} u_{n+1} = -\dfrac{1}{h} u_n,$

which is equivalent to (13.87). The existence and uniqueness of u_{n+1} are assured because of the minus sign before the factor multiplying the undifferentiated term on the left in (13.93). It follows that all elements of the sequence $u_0, u_1, u_2, \ldots$ are well defined.

In order to obtain needed estimates of the quantity u_n we shall have to apply the maximum principle repeatedly to solutions v of the partial differential equation

(13.94) $\Delta v - \dfrac{1}{h} v = -\dfrac{1}{h} g,$

of which (13.93) is evidently a special example. Suppose that v vanishes on ∂D. We assert that

(13.95) $\max\limits_{D} |v| \leq \max\limits_{D} |g|.$

Indeed, at a point of D where v assumes its maximum value we have

$$\Delta v \leq 0,$$

whence $v \leq g$ there by virtue of (13.94). Similarly, the inequality $v \geq g$ holds at any minimum of v inside D, and therefore (13.95) follows. In particular, from (13.93) and (13.95) we deduce that

$$\max_{D} |u_{n+1}| \leq \max_{D} |u_n|.$$

Thus the important estimate

$$|u_n| \leq \max_{D} |f|$$

of u_n becomes a consequence of the initial condition (13.90).

Differences $u_n^{(k)}$ of the sequence u_n of arbitrary order k are defined by the formulas

(13.96) $$u_n^{(1)} = u_{n+1} - u_n$$

and

$$u_n^{(k+1)} = u_{n+1}^{(k)} - u_n^{(k)}, \qquad k = 1, 2, \dots .$$

Subtracting equation (13.93) from that same relation with the index n replaced by $n + 1$, we find for $u_n^{(1)}$ the difference-differential equation

$$\Delta u_{n+1}^{(1)} - \frac{1}{h} u_{n+1}^{(1)} = -\frac{1}{h} u_n^{(1)}.$$

Iteration of this procedure yields

(13.97) $$\Delta u_{n+1}^{(k)} - \frac{1}{h} u_{n+1}^{(k)} = -\frac{1}{h} u_n^{(k)}.$$

Thus from the maximum principle we conclude by mathematical induction that

(13.98) $$|u_n^{(k)}| \leq \max_{D} |u_0^{(k)}|.$$

It remains to derive estimates of $u_0^{(1)}$ and $u_0^{(2)}$ so that bounds on $u_n^{(1)}$ and $u_n^{(2)}$ will become available for our proof of the convergence theorem (13.92).

Subtracting $\Delta u_n^{(k)}$ from both sides of (13.97) and transposing the term on the right, we obtain

$$\Delta u_{n+1}^{(k)} - \Delta u_n^{(k)} - \frac{1}{h} [u_{n+1}^{(k)} - u_n^{(k)}] = -\Delta u_n^{(k)},$$

or, in other words,

$$(13.99) \qquad \Delta u_n^{(k+1)} - \frac{1}{h} u_n^{(k+1)} = -\Delta u_n^{(k)}.$$

An application of the maximum principle to (13.99) yields the inequality

$$(13.100) \qquad |u_n^{(k+1)}| \leq h \max_D |\Delta u_n^{(k)}|.$$

In particular, according to the initial condition (13.90) we have

$$(13.101) \qquad |u_0^{(1)}| \leq h \max_D |\Delta f|.$$

For $u_0^{(2)}$, however, (13.100) merely furnishes the result

$$(13.102) \qquad |u_0^{(2)}| \leq h \max_D |\Delta u_0^{(1)}|,$$

which means that we must still proceed to estimate the quantity $\Delta u_0^{(1)}$.

We now subtract $h\,\Delta\Delta u_n^{(k)}$ from both sides of (13.99) and rearrange terms to derive

$$(13.103) \quad \Delta[u_n^{(k+1)} - h\,\Delta u_n^{(k)}] - \frac{1}{h}[u_n^{(k+1)} - h\,\Delta u_n^{(k)}] = -h\,\Delta\Delta u_n^{(k)}.$$

Because Δu_n vanishes on ∂D by virtue of (13.91) and (13.93), we find from the maximum principle that

$$|u_n^{(k+1)} - h\,\Delta u_n^{(k)}| \leq h^2 \max_D |\Delta\Delta u_n^{(k)}|.$$

Inserting this estimate into (13.103), we obtain in turn

$$|\Delta u_n^{(k+1)} - h\,\Delta\Delta u_n^{(k)}| \leq 2h \max_D |\Delta\Delta u_n^{(k)}|,$$

whence

$$(13.104) \qquad |\Delta u_n^{(k+1)}| \leq 3h \max_D |\Delta\Delta u_n^{(k)}|.$$

In particular, when $k = n = 0$ the inequality (13.104) reduces to

$$|\Delta u_0^{(1)}| \leq 3h \max_D |\Delta\Delta f|,$$

which combines with (13.102) to yield

$$(13.105) \qquad |u_0^{(2)}| \leq 3h^2 \max_D |\Delta\Delta f|.$$

From (13.98) and (13.101) we conclude that

(13.106) $$u_n^{(1)} = O(h)$$

uniformly, and similarly (13.98) and (13.105) imply that

(13.107) $$u_n^{(2)} = O(h^2).$$

If the relations (13.106) and (13.107) were meaningful over the entire t-axis, and not just at the mesh points $t = nh$, we could interpret them as Lipschitz conditions implying that u_n and $u_n^{(1)}/h$ are *equicontinuous*.

Suppose that we confine our attention to mesh sizes h of the special form

$$h = \frac{\epsilon}{2^m},$$

where $\epsilon > 0$ is fixed and m ranges over the positive integers. It becomes clear from the proof of Arzela's theorem (cf. Section 10.3) that an increasing sequence $m_1, m_2, m_3, \ldots$ can be found for which the limits

(13.108) $$u(x, y, t) = \lim_{\nu \to \infty} u_{n+1}(x, y)$$

and

(13.109) $$u_t(x, y, t) = \lim_{\nu \to \infty} \frac{u_n^{(1)}(x, y)}{h}$$

exist, where

(13.110) $$t = nh = \frac{n\epsilon}{2^{m\nu}} \geq 0.$$

Moreover, the *a priori* estimates (13.106) and (13.107) enable us to define both the functions (13.108) and (13.109) for arbitrary $t \geq 0$ by continuity, and (13.96) shows that u_t really does represent the partial derivative of u with respect to the variable t.

In order to verify that (13.108) yields a solution u of the heat equation (13.86), we convert the difference-differential equation (13.87) into an integral equation in which a passage to the limit becomes feasible. Let $G = G(x, y; \xi, \eta)$ stand for the Green's function of Laplace's equation

$$\Delta G = 0$$

in D. According to (9.62), the equation (13.87) and the boundary condition (13.91) may be combined in the single formula

(13.111) $$u_{n+1} + \iint_D \frac{u_{n+1} - u_n}{h} G \, d\xi \, d\eta = 0.$$

Allowing h to approach zero here through the sequence of values

$$h = \frac{\epsilon}{2^{mv}}, \qquad v = 1, 2, 3, \ldots,$$

and taking (13.108) to (13.110) into account, we deduce from (13.111) the integrodifferential equation

$$u + \iint_D u_t G \, d\xi \, d\eta = 0$$

for the function u. This result is equivalent to the heat equation (13.86) and the boundary condition (13.89) together. Finally, observe that (13.90) reduces to the initial condition (13.88) in the limit as $h \to 0$.

Our analysis demonstrates the existence of a solution u of the mixed initial and boundary value problem (13.86), (13.88), (13.89) for the heat equation. It is not difficult to derive the somewhat more general convergence theorem (13.92) from the fact that the answer is unique (cf. Exercises 3 and 4 below). We note that the key to our proof was furnished by the *a priori* estimates (13.106) and (13.107) of the solution u_n of the approximating problem (13.87), (13.90), (13.91), which merely involved bounds on the expressions Δf and $\Delta \Delta f$. Extensions and implications of this finite difference technique will be described in the exercises.

EXERCISES

1. Compare the difference-differential equation (13.87) with the implicit finite difference scheme described in Exercise 3.2 for solving the heat equation numerically.

2. From the proof of Arzela's theorem develop a detailed explanation of the convergence of the expressions on the right in (13.108) and (13.109).

3. Use the maximum principle to establish the uniqueness of the solution of the mixed initial and boundary value problem (13.86), (13.88), (13.89).

4. Complete the proof of the convergence theorem (13.92) by means of the result of Exercise 3.

5. If $\Delta \Delta f$ remains bounded throughout the region D, derive the estimate

$$u_n^{(3)} = O(h^3)$$

analogous to (13.106) and (13.107).

6. Solve the mixed initial and boundary value problem (13.86), (13.88), (13.89) by means of an eigenfunction expansion (cf. Exercise 11.1.8). Compare this approach with the method of the text.

7. Apply a Laplace transform with respect to the variable t to reduce the mixed initial and boundary value problem (13.86), (13.88), (13.89) to Dirichlet's problem for a partial differential equation like (13.94).

8. Solve the problem of the text by means of a generalization of the Perron process introduced in Section 9.2.

9. Formulate the procedure of the text in the language of *semigroups*.

10. Develop a generalization[12] of the finite difference method described in the text in order to solve a mixed initial and boundary value problem analogous to (13.86), (13.88), (13.89) for the parabolic partial differential equation

$$L[u] = au_t + b,$$

where L stands for a linear self-adjoint elliptic differential operator in the (x,y)-plane satisfying the maximum principle over the domain D, where a and b represent smooth coefficients depending on the three variables x, y and t, and where a is supposed to have the same sign as the diagonal coefficients from the operator L.

[12] Cf. Rothe 1, 2.

14
Fluid Dynamics

1. FORMULATION OF THE EQUATIONS OF MOTION

Among the most intriguing questions of the theory of partial differential equations are those that arise in fluid dynamics. They are especially interesting and instructive because of the light they throw on nonlinear problems. Therefore we find it appropriate to study here the equations of motion of a perfect, inviscid fluid. Moreover, we shall take the opportunity thus offered to derive these equations from the fundamental principles of mechanics as an illustration of how the equations of mathematical physics are formulated. From the discussion of such matters it will become apparent that good physical intuition concerning fluid flow serves as an important guide in the investigation of partial differential equations.

We shall confine our attention to two-dimensional motion. Let

$$u = u(x, y, t), \qquad v = v(x, y, t)$$

stand for the horizontal and vertical components of velocity of a fluid particle which at the time t is situated at the point (x, y). Similarly, denote by

$$p = p(x, y, t), \qquad \rho = \rho(x, y, t)$$

the pressure and density, respectively, of the fluid. Our first objective is to develop the so-called *Eulerian formulation* of the partial differential equations for u, v, p and ρ which express in mathematical language the physical laws of conservation of mass, momentum and energy.

Consider the quantity of fluid contained in some plane region D bounded by a simple closed curve ∂D. The law of *conservation of mass* asserts that the rate at which fluid pours across ∂D into D must just balance the rate at which the total amount of fluid in D increases with time. The formula

$$\int_{\partial D} \left[\rho u \frac{\partial x}{\partial v} + \rho v \frac{\partial y}{\partial v} \right] ds = \frac{\partial}{\partial t} \iint_D \rho \, dx \, dy$$

expresses this same law in a more precise fashion, where v stands as usual for the inner normal along ∂D, and s represents arc length. Applying the divergence theorem to the integral on the left and simultaneously bringing the differential operator on the right under the sign of integration, we conclude that

$$(14.1) \qquad \iint_D \left[\frac{\partial \rho}{\partial t} + \frac{\partial(\rho u)}{\partial x} + \frac{\partial(\rho v)}{\partial y} \right] dx \, dy = 0.$$

Since we can take the region D to be as small as we please, the integrand in (14.1) must vanish identically. It follows that the law of conservation of mass can be restated in the form of the partial differential equation

$$(14.2) \qquad \frac{\partial \rho}{\partial t} + \frac{\partial(\rho u)}{\partial x} + \frac{\partial(\rho v)}{\partial y} = 0$$

for u, v and ρ, which is called the *continuity equation*.

We shall derive next a formula expressing the law of *conservation of momentum* in the direction of the x-axis when not only viscosity but also all external fields of force, such as gravity, are neglected. The horizontal component of force exerted on a small section D of our fluid by the pressure p along its boundary ∂D is given by the integral

$$-\int_{\partial D} p \, dy = -\iint_D \frac{\partial p}{\partial x} dx \, dy.$$

Suppose that D is allowed to shrink down on a specific particle which describes the trajectory

$$x = x(t), \qquad y = y(t).$$

Since $\dot{x} = u$, the horizontal acceleration $\ddot{x}$ of this particle is equal to

$$\frac{du}{dt} = \frac{\partial u}{\partial t} + u \frac{\partial u}{\partial x} + v \frac{\partial u}{\partial y}.$$

Newton's second law of motion identifies force with mass times acceleration. Therefore a passage to the limit as the area of D approaches zero
yields

$$\frac{du}{dt} = \lim \left[-\iint_D \frac{\partial p}{\partial x} \, dx \, dy \bigg/ \iint_D \rho \, dx \, dy \right],$$

or, finally,

(14.3) $$\frac{\partial u}{\partial t} + u \frac{\partial u}{\partial x} + v \frac{\partial u}{\partial y} + \frac{1}{\rho} \frac{\partial p}{\partial x} = 0.$$

In a similar way we deduce that the law of conservation of momentum in
the direction of the y-axis is equivalent to the partial differential equation

(14.4) $$\frac{\partial v}{\partial t} + u \frac{\partial v}{\partial x} + v \frac{\partial v}{\partial y} + \frac{1}{\rho} \frac{\partial p}{\partial y} = 0.$$

In order to formulate the law of *conservation of energy* as a partial
differential equation for u, v, p and ρ it is necessary to introduce some
concepts from thermodynamics. To be precise, we shall have to deal with
the *entropy S* of our fluid and with the *ratio of specific heats* γ, which is a
constant exceeding unity. The hypothesis will be made that the pressure p
and the density ρ of the fluid are connected with the entropy S by an
equation of state of the special form

(14.5) $$p = A(S)\rho^\gamma,$$

where the function $A = A(S)$ is positive but otherwise arbitrary. It can be
shown (cf. Courant-Friedrichs 1) that energy is conserved if and only if the
entropy S remains constant along each particle path, which means that

(14.6) $$\frac{dS}{dt} = \frac{\partial S}{\partial t} + u \frac{\partial S}{\partial x} + v \frac{\partial S}{\partial y} = 0.$$

Using the equation of state (14.5) to eliminate S from (14.6), we obtain the
final partial differential equation

(14.7) $$\frac{\partial}{\partial t} \frac{p}{\rho^\gamma} + u \frac{\partial}{\partial x} \frac{p}{\rho^\gamma} + v \frac{\partial}{\partial y} \frac{p}{\rho^\gamma} = 0.$$

We collect (14.2), (14.3), (14.4) and (14.7), which are known as *Euler's
equations of motion* for an inviscid fluid, as a system of four simultaneous

nonlinear partial differential equations

(14.8)
$$u_t + uu_x + vu_y + \frac{1}{\rho} p_x = 0,$$

(14.9)
$$v_t + uv_x + vv_y + \frac{1}{\rho} p_y = 0,$$

(14.10)
$$\rho_t + (\rho u)_x + (\rho v)_y = 0,$$

(14.11)
$$\left(\frac{p}{\rho^\gamma}\right)_t + u\left(\frac{p}{\rho^\gamma}\right)_x + v\left(\frac{p}{\rho^\gamma}\right)_y = 0$$

for the four unknowns u, v, p and ρ. For gases the ratio of specific heats γ is not very much greater than 1, and, in particular,

$$\gamma = \tfrac{7}{5} = 1.4$$

for air. Consequently the energy equation (14.11) plays a significant role in most problems of *aerodynamics*. On the other hand, γ is much larger for liquids, and hence it is customary to put $\gamma = \infty$ in the formulation of problems of *hydrodynamics*. Thus in hydrodynamics one can replace the equation of state (14.5) by the more elementary requirement

(14.12)
$$\rho = \text{const.}$$

of *incompressibility*, and one can omit the partial differential equation (14.11) from consideration altogether.

For any flow it is of interest to examine the *circulation*

$$\Gamma = \oint_C (u\,dx + v\,dy)$$

around a closed curve C. When C is allowed to move with the fluid particles, Γ becomes a function of the time t. Using a dot to indicate differentiation with respect to time along each particle path, we find that

$$\dot{\Gamma} = \oint_C (u\,d\dot{x} + v\,d\dot{y}) + \oint_C (\dot{u}\,dx + \dot{v}\,dy).$$

Because
$$\dot{x} = u, \qquad \dot{y} = v,$$

and because the momentum equations (14.8) and (14.9) are equivalent to

$$\dot{u} = -\frac{p_x}{\rho}, \qquad \dot{v} = -\frac{p_y}{\rho},$$

this formula for $\dot{\Gamma}$ reduces to

$$(14.13) \qquad \dot{\Gamma} = \oint_C (u\,du + v\,dv) - \oint_C \frac{p_x\,dx + p_y\,dy}{\rho}$$

$$= \oint_C \frac{du^2 + dv^2}{2} - \oint_C \frac{dp}{\rho} = -\oint_C \frac{dp}{\rho}.$$

For *isentropic flow*, that is, for flow with the property

$$(14.14) \qquad\qquad S \equiv \text{const.},$$

the equation of state (14.5) serves to define ρ as a single-valued function of p, so that the last integral on the right in (14.13) has to vanish. We conclude that the circulation Γ is actually independent of the time t. Moreover, it must be zero if the flow has either started from rest or becomes uniform at large distances. Thus it is natural to study *irrotational flows*, which are those with

$$(14.15) \qquad\qquad \Gamma \equiv 0$$

over every closed loop C. In this connection note that the energy equation (14.11) can be ignored when (14.14) holds, and that for an incompressible fluid (14.15) follows without reference to (14.14).

For any irrotational flow the integral

$$\phi = \int_{(x_0, y_0)}^{(x, y)} (u\,dx + v\,dy)$$

is independent of path and defines a function ϕ of the point (x, y) and of the time t called the *velocity potential*. Clearly

$$(14.16) \qquad\qquad \phi_x = u, \qquad \phi_y = v,$$

whence it follows that

$$(14.17) \qquad\qquad u_y - v_x = \phi_{xy} - \phi_{yx} = 0.$$

Inserting (14.16) and (14.17) into the two momentum equations (14.8) and (14.9), multiplying the first of them by dx and the second by dy, and then adding and integrating, we obtain

$$\int \left[\left(\phi_{tx} + uu_x + vv_x + \frac{1}{\rho}p_x \right) dx \right.$$

$$\left. + \left(\phi_{ty} + uu_y + vv_y + \frac{1}{\rho}p_y \right) dy \right] = 0.$$

A few rearrangements now yield the *Bernoulli equation*

(14.18) $$\phi_t + \frac{q^2}{2} + \int \frac{dp}{\rho} = H,$$

where q stands for the speed

$$q = \sqrt{u^2 + v^2}$$

and H is a function of t alone. The equation of state (14.5) and Bernoulli's law (14.18) enable us to calculate the pressure p and the density ρ when the velocity potential ϕ is known. Finally, observe that in the special case of an incompressible fluid the continuity equation (14.10) reduces to Laplace's equation

$$\phi_{xx} + \phi_{yy} = 0$$

for ϕ.

For steady plane flow, rotational or irrotational, the continuity equation (14.10) shows that there exists a *stream function* $\psi = \psi(x, y)$ such that

(14.19) $$u\rho = \psi_y, \qquad v\rho = -\psi_x.$$

We see that ψ is constant along each particle path, or *streamline*. Thus the entropy S is a function of ψ only, and the equation of state (14.5) can be expressed in the alternate form

(14.20) $$p = A(\psi)\rho^\gamma.$$

For steady rotational flow the Bernoulli law

(14.21) $$\frac{1}{2}q^2 + \frac{\gamma}{\gamma - 1}\frac{p}{\rho} = H(\psi)$$

remains valid along each individual streamline, as can easily be verified by differentiating in the direction of the velocity vector (u, v). On the other hand, for motion that is both steady and irrotational the relations (14.20) and (14.21) can be used to determine ρ as a function of either $\phi_x^2 + \phi_y^2$ or $\psi_x^2 + \psi_y^2$, and ϕ and ψ together satisfy the pair of generalized Cauchy-Riemann equations

(14.22) $$\phi_x = \frac{1}{\rho}\psi_y, \qquad \phi_y = -\frac{1}{\rho}\psi_x,$$

because of (14.16) and (14.19).

Reasonable boundary conditions to impose on the velocity potential ϕ and the stream function ψ for flows past rigid bodies can be formulated in an obvious fashion. No fluid may pass through a solid wall ∂D bordering a fixed region D of plane flow. Hence there is no component of velocity perpendicular to ∂D, and ∂D itself must be a streamline. It follows that the conditions

(14.23) $$\frac{\partial \phi}{\partial \nu} = 0, \qquad \psi = \text{const.}$$

ought to be fulfilled along the boundary ∂D of D. In this connection it should be emphasized, however, that for any fluid with viscosity $\mu > 0$ the velocity has to vanish at a rigid wall, so that (14.23) can only be viewed as a degenerate case of the more stringent requirements

$$u = v = 0.$$

It is essentially because second derivatives of u and v appear in the equations of motion of a viscous fluid that the boundary condition of no tangential slip becomes lost in the limit as $\mu \to 0$. Such a phenomenon is explained by the fact that the passage from inviscid to viscous flow through an asymptotic expansion with respect to μ leads to singular perturbation problems (cf. Exercises 6 and 8 below).

Lack of uniformity in the transition from viscous to frictionless motion may also lead to the appearance of discontinuities in the limiting inviscid flow known as *shock waves*. Here we shall be content to interpret shock waves as surfaces of discontinuity exhibited by weak solutions of Euler's equations of motion (cf. Sections 8.1 and 12.2).

To avoid a lengthy discussion let us consider only the law of conservation of mass

$$\rho_t + (\rho u)_x = 0$$

for one-dimensional flow. Multiplying by a smooth function f that vanishes outside some compact set D and then integrating the result by parts, we arrive at the identity

$$\iint\limits_{D} [\rho f_t + \rho u f_x] \, dx \, dt = 0.$$

Any pair of functions u and ρ satisfying this integral identity for all reasonable choices of f is called a *weak solution* of the continuity equation. Now let u, ρ be a weak solution which has jumps across a curve C describing our shock wave but is continuously differentiable elsewhere. Integrating by parts once more we obtain

$$\int_{C} f[(\rho_2 u_2 - \rho_1 u_1) \, dt - (\rho_2 - \rho_1) \, dx] - \iint\limits_{D} f[\rho_t + (\rho u)_x] \, dx \, dt = 0,$$

where u_1 and ρ_1 stand for the values of the velocity and density on one side of C, while u_2 and ρ_2 stand for the same quantities on the other side of C. The new double integral vanishes because u and ρ are supposed to fulfill Euler's equations everywhere except at C. Since f is arbitrary it follows that the jumps in u and ρ have to obey the shock condition

$$\rho_1\left(u_1 - \frac{dx}{dt}\right) = \rho_2\left(u_2 - \frac{dx}{dt}\right),$$

where dx/dt indicates the speed at which the shock wave C progresses.

The shock condition we have just established states that mass is conserved across the discontinuity curve C. We leave it to the reader to formulate similar conditions expressing the laws of conservation of momentum and energy across a shock wave (cf. Exercises 9 and 10 below). Finally, the fact that weak solutions of an elliptic equation always turn out to be strong ones, too, suggests that shock waves, which reduce to characteristics if they become weak enough, ought to occur primarily in flows governed by partial differential equations of the hyperbolic type.

EXERCISES

1. Show that the *complex potential*

$$\zeta = \phi + i\psi$$

describing steady, irrotational plane motion of a homogeneous incompressible fluid is an analytic function of the complex variable $z = x + iy$. Why does the expression

$$\zeta = z + \frac{1}{z}$$

represent flow past a circle? How can conformal mapping be used to determine other flows?

2. Apply Cauchy's theorem to the contour integral

$$\oint \left(\frac{d\zeta}{dz}\right)^2 dz = -\frac{2}{\rho} \oint (p\,dx - ip\,dy)$$

in order to establish that no force is exerted on a body immersed in a flow of the kind defined in Exercise 1.

3. Show that any steady, irrotational motion of an incompressible fluid in space which exhibits axial symmetry can be described by a *Stokes stream function* ψ satisfying the partial differential equation

$$\psi_{xx} + \psi_{yy} - \frac{1}{y}\,\psi_y = 0.$$

4. Give a more detailed derivation of the Bernoulli equation (14.21) and generalize it to the case of axially symmetric flows.

5. Prove that for steady flow the streamlines are included among the characteristics of the hydrodynamic equations (14.8) to (14.11).

6. Use symmetry arguments and orthogonal transformations to show[1] that the simplest conceivable description of the motion of a viscous incompressible fluid is given by the *Navier-Stokes equations*

$$\rho u_t + \rho u u_x + \rho v u_y + p_x = \mu u_{xx} + \mu u_{yy},$$

$$\rho v_t + \rho u v_x + \rho v v_y + p_y = \mu v_{xx} + \mu v_{yy},$$

$$u_x + v_y = 0,$$

where μ is a constant known as the *coefficient of viscosity*.

7. Discuss viscous flow between two parallel plates by means of special solutions of the Navier-Stokes equations such that

$$v = 0, \qquad p = \text{const.},$$

and such that $u = u(y, t)$ satisfies the heat equation

$$\rho u_t = \mu u_{yy}.$$

8. Introduce polar coordinates r and θ to describe the viscous flow in a wedge $-\alpha < \theta < \alpha$ by seeking solutions of the Navier-Stokes equations with an exclusively radial velocity q given by the formula[2]

$$q = \frac{f(\theta)}{r},$$

where f satisfies the ordinary differential equation

$$\frac{\mu}{\rho} [f''' + 4f'] + 2ff' = 0.$$

Analyze the loss of boundary conditions at the walls of the wedge for inviscid flow by representing f in terms of an elliptic integral of the form

$$\theta = \sqrt{\frac{3\mu}{2\rho}} \int \frac{df}{\sqrt{(f_1 - f)(f_2 - f)(f_3 - f)}}$$

and by examining *boundary layers* where $f(\theta)$ does not converge uniformly in the limit as $\mu \to 0$.

[1] Cf. Lamb 1.
[2] Cf. S. Goldstein 1.

9. Shock waves are discontinuity surfaces across which there are jumps in the flow quantities u, v, p and ρ such that mass, momentum and energy are conserved.[3] Let the subscripts 1 and 2 indicate values of u, v, p and ρ on the two different sides of a shock wave. In the case of one-dimensional flow show that the conservation laws, when put in divergence form, yield the *Rankine-Hugoniot shock conditions*

(14.24)
$$\rho_1 q_1 = \rho_2 q_2,$$

(14.25)
$$\rho_1 q_1^2 + p_1 = \rho_2 q_2^2 + p_2,$$

(14.26)
$$\frac{1}{2} q_1^2 + \frac{\gamma}{\gamma - 1} \frac{p_1}{\rho_1} = \frac{1}{2} q_2^2 + \frac{\gamma}{\gamma - 1} \frac{p_2}{\rho_2},$$

where
$$q_j = u_j - U, \quad j = 1, 2,$$

and where U stands for the speed at which the shock wave itself advances.

10. For two-dimensional steady flow, derive the shock conditions

(14.27)
$$\rho_1 N_1 = \rho_2 N_2,$$

(14.28)
$$L_1 = L_2,$$

(14.29)
$$\rho_1 N_1^2 + p_1 = \rho_2 N_2^2 + p_2,$$

(14.30)
$$\frac{1}{2} (L_1^2 + N_1^2) + \frac{\gamma}{\gamma - 1} \frac{p_1}{\rho_1} = \frac{1}{2} (L_2^2 + N_2^2) + \frac{\gamma}{\gamma - 1} \frac{p_2}{\rho_2}$$

applying across a stationary shock wave, where L and N denote the components of velocity parallel and perpendicular to the shock, respectively.[4]

2. ONE-DIMENSIONAL FLOW

One of the few questions in nonlinear continuum mechanics that yields to a completely satisfactory and quite explicit mathematical analysis is the problem of one-dimensional isentropic flow. Euler's equations of motion reduce in this case to

(14.31)
$$u_t + u u_x + \frac{1}{\rho} p_x = 0$$

and

(14.32)
$$\rho_t + u \rho_x + \rho u_x = 0,$$

together with the equation of state

$$p = A \rho^\gamma,$$

[3] Cf. Courant-Friedrichs 1.
[4] Cf. Courant-Friedrichs 1.

where A is now a constant. The first step in our investigation of one-dimensional flow will be to determine the characteristics of the system of partial differential equations (14.31), (14.32). Fortunately it will turn out that the characteristics suggest new coordinates in terms of which these equations of motion become linear and can even be solved in closed form.

Introducing the positive quantity

$$c^2 = \frac{dp}{d\rho} = A\gamma\rho^{\gamma-1},$$

which is a function of ρ alone, we can eliminate p from the momentum equation (14.31) to obtain

(14.33) $$u_t + uu_x + \frac{c^2}{\rho}\,\rho_x = 0$$

instead. Next we recall (cf. Section 3.5) that a *characteristic* of (14.32) and (14.33) is a curve C in the (x,t)-plane along which those equations do not suffice to determine the normal derivatives of u and ρ when the values of u and ρ themselves are given on C. Differentiating u and ρ along such a curve, we can write

(14.34) $$du = u_x\,dx + u_t\,dt,$$

(14.35) $$d\rho = \rho_x\,dx + \rho_t\,dt.$$

We interpret (14.32) to (14.35) as a system of four simultaneous linear equations for the four unknowns u_x, u_t, ρ_x and ρ_t on C, with du and $d\rho$ prescribed. According to what has been said, this system does not have a unique solution when C is a characteristic, and therefore its determinant must vanish in that case. Thus we conclude that C is a characteristic of (14.32) and (14.33) if and only if it satisfies the ordinary differential equation

(14.36) $$\begin{vmatrix} \rho & 0 & u & 1 \\ u & 1 & \dfrac{c^2}{\rho} & 0 \\ dx & dt & 0 & 0 \\ 0 & 0 & dx & dt \end{vmatrix} = (dx - u\,dt)^2 - c^2\,dt^2 = 0.$$

It becomes evident that the equations (14.32) and (14.33) of one-dimensional gas dynamics constitute a first order system of the *hyperbolic type.* Solving (14.36) for dx/dt, we find that the slopes of the two families of characteristics of these equations are defined by the relation

(14.37)
$$\frac{dx}{dt} = u \pm c.$$

Clearly the characteristics may be viewed as curves across which the derivatives of the flow quantities u and ρ can jump while u and ρ themselves remain continuous. As such they ought to be interpreted physically as *sound waves.* Thus formula (14.37) asserts that any sound wave C progresses at a total velocity dx/dt which differs from the velocity u of the actual fluid particles by the amount $\pm c$. Hence we are justified in referring to c as the *local speed of sound* of the fluid.

Since it is impossible to solve (14.32) and (14.33) for the normal derivatives of u and ρ along a characteristic C, we should be able (cf. Section 3.5) to form from them a linear combination containing only differentiations in the direction of the tangent to C. Indeed, multiplying (14.32) by $\pm c/\rho$ and then adding it to (14.33), we obtain

(14.38)
$$\left[\frac{\partial}{\partial t} + (u \pm c)\frac{\partial}{\partial x}\right]u \pm \frac{c}{\rho}\left[\frac{\partial}{\partial t} + (u \pm c)\frac{\partial}{\partial x}\right]\rho = 0.$$

Now introduce *characteristic coordinates* α and β such that α is variable, but β remains constant, along the characteristics defined by using the plus sign in (14.37), whereas β is variable, but α remains constant, along the characteristics defined by using the minus sign in (14.37). In terms of α and β we can write (14.37) and (14.38) as a *canonical system* of four partial differential equations

(14.39)
$$x_\alpha = (u + c)t_\alpha, \qquad x_\beta = (u - c)t_\beta,$$

(14.40)
$$u_\alpha + \frac{c}{\rho}\rho_\alpha = 0, \qquad u_\beta - \frac{c}{\rho}\rho_\beta = 0$$

of the first order for the four unknowns x, t, u and ρ.

It is remarkable that the physical coordinates x and t do not occur at all in (14.40). We can restate the latter equations in the more convenient form

(14.41)
$$u_\alpha + l_\alpha = 0, \qquad u_\beta - l_\beta = 0$$

based on the function

$$l(\rho) = \int \frac{c}{\rho}\,d\rho = \frac{2\sqrt{A\gamma}}{\gamma - 1}\rho^{(\gamma-1)/2} = \frac{2c}{\gamma - 1}.$$

It follows by direct integration that

$$(14.42) \qquad u + l = 2r, \qquad u - l = -2s,$$

where r is a function of β alone and s is a function of α alone. The quantities r and s, which are called *Riemann invariants*, may serve as characteristic coordinates instead of α and β, provided that neither one is identically constant. Then (14.39) becomes a pair of *linear* partial differential equations

$$(14.43) \qquad x_s = (u + c)t_s, \qquad x_r = (u - c)t_r$$

with coefficients given explicitly by the relations

$$(14.44) \qquad u + c = \frac{\gamma + 1}{2} r + \frac{\gamma - 3}{2} s,$$

$$(14.45) \qquad u - c = -\frac{\gamma - 3}{2} r - \frac{\gamma + 1}{2} s.$$

We have succeeded in reducing the nonlinear equations (14.32) and (14.33) of one-dimensional flow to a system (14.43) of two linear partial differential equations in which the known functions r and s of u and ρ feature as independent variables. Notice now that because the coefficients in the equations of motion involve only u and ρ, and not x and t, we could have achieved essentially the same result by introducing u and ρ directly as the new independent variables. This substitution, which is usually referred to as the *hodograph transformation*, has several important applications in fluid dynamics (cf. Section 3). For the example of (14.32) and (14.33) it leads to the equations

$$(14.46) \qquad x_u - ut_u + \rho t_\rho = 0$$

and

$$(14.47) \qquad -x_\rho + ut_\rho - \frac{c^2}{\rho} t_u = 0,$$

whose linearity is due to the fact that the simple rules

$$u_x = \frac{t_\rho}{J}, \qquad \rho_x = -\frac{t_u}{J},$$

$$u_t = -\frac{x_\rho}{J}, \qquad \rho_t = \frac{x_u}{J}$$

for calculating first partial derivatives permit us to eliminate the Jacobian

$$J = \frac{\partial(x, t)}{\partial(u,\rho)} = x_u t_\rho - t_u x_\rho$$

from the final formulas. We can appeal again, of course, to the method of characteristics to bring (14.46) and (14.47) into the canonical form (14.43).

In order to calculate one-dimensional isentropic flows explicitly we differentiate the relations (14.43) with respect to r and s so as to eliminate the mixed derivative $x_{sr} = x_{rs}$ and derive the second order equation

(14.48) $\qquad 2ct_{rs} + (u_r + c_r)t_s - (u_s - c_s)t_r = 0$

for t alone. Because of (14.44) and (14.45), (14.48) reduces to

(14.49) $\qquad t_{rs} + \dfrac{\lambda}{r + s}(t_r + t_s) = 0,$

where

$$\lambda = \frac{1}{2}\frac{\gamma + 1}{\gamma - 1}.$$

The latter partial differential equation has been discussed in Section 4.4 and Section 5.1. In particular, we recall (cf. Exercise 5.1.9) that its Riemann function $A(r_0, s_0; r, s)$, which is a solution in dependence on the parameters r and s, has the representation

(14.50) $\quad A(r_0, s_0; r, s) = \dfrac{(r_0 + s_0)^{2\lambda}}{(r + s_0)^\lambda (r_0 + s)^\lambda} F\left[\lambda, \lambda; 1; \dfrac{(r - r_0)(s - s_0)}{(r + s_0)(r_0 + s)}\right]$

in terms of the hypergeometric function F.

The Riemann function (14.50) enables us to solve the Cauchy problem for (14.49) in closed form.[5] Moreover, the hodograph transformation takes any initial values of the velocity u and density ρ of our gas which are so assigned at $t = 0$ that r varies monotonically with s into a reasonable curve of Cauchy data in the (r,s)-plane. Consequently the corresponding flow quantities u and ρ at some later time $t > 0$ can be found from specific implicit relations involving the hypergeometric function. However, instead of pursuing this rather complicated matter in detail, we prefer to develop here an interesting physical interpretation of the Riemann function (14.50) itself.

[5] Cf. Riemann 1.

As a preliminary we investigate solutions of the equations of motion (14.32) and (14.33) for which the hodograph transformation breaks down because

$$\frac{\partial(r, s)}{\partial(x, t)} = 0.$$

For such a flow, which is called a *simple wave*, there must exist a relation between the Riemann invariants r and s. Suppose, for example, that r is a function of s. Along any *minus characteristic*, that is, along any characteristic (14.39) on which β is variable, (14.41) and (14.42) show that s remains constant. Hence r, as a function of s, must be constant on that minus characteristic, too. On the other hand, (14.41) and (14.42) imply that r remains constant along each *plus characteristic* (14.39), by which we mean a characteristic where α is variable. It follows that throughout the simple wave under consideration r must be constant, for the relevant region of the (x,t)-plane is covered by plus characteristics that all intersect a specific minus characteristic. Thus, because of the symmetry in the roles played by r and s, we conclude that a simple wave is actually a flow for which either r or s is identically constant. Observe, however, that r and s are both constant only for uniform flow.

For any simple wave in which r is a constant, all the minus characteristics have to be straight lines. Indeed, along each such minus characteristic s remains constant as well as r, and consequently both u and c are constant there. Hence the slope $u - c$ of the characteristic cannot vary, which shows that it is a line. Similarly, in any simple wave where s is identically constant every plus characteristic must be a straight line.

We prove next that any one-dimensional flow adjacent to a region of uniform motion must be a simple wave. First, observe that the curve separating two different flow regimes such as are under discussion here

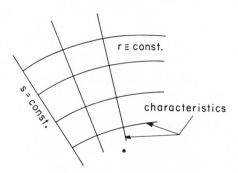

FIG. 60. Simple wave.

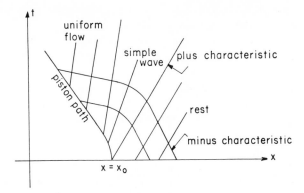

FIG. 61. Piston problem.

has to be a characteristic and is therefore also a straight line, since u and ρ remain constant along it. To be specific, suppose that the curve in question is a minus characteristic. Then families of plus characteristics cross it from the region of uniform motion and penetrate the non-trivial flow. Since r is constant along each plus characteristic, and since all the plus characteristics under consideration come from a region where the value of r is fixed, we conclude that r is identically constant throughout the flow, which must, indeed, be a simple wave. Finally, we emphasize again that all the minus characteristics progressing through such a flow are straight lines.

We are now prepared to describe physical problems in which simple waves occur. Suppose that a narrow pipe parallel to the x-axis contains gas filling the region $x > x_0$ to the right of a piston located at the point $x = x_0$. Assuming a state of rest at the initial time $t = 0$, we seek to determine the one-dimensional flow caused by pulling the piston gradually to the left, away from the gas. We prescribe for the piston a trajectory which accelerates continuously from zero speed at the start until a specific negative velocity is attained and then proceeds uniformly to the left.

Because $u = 0$ and $\rho = $ const. initially, the gas remains at rest throughout the infinite triangular domain of the (x,t)-plane that lies above the positive x-axis and to the right of the plus characteristic emanating from the point $(x_0, 0)$. The flow to the left of that characteristic is adjacent to a region of no motion and must therefore be a simple wave in which s is identically constant and in which all the plus characteristics are straight lines. At the points of intersection of the latter characteristics with the path of the piston it is possible to read off the values of r and of the slope $u + c$ to be associated with them, since u and s are calculable by inspection. In particular, the slopes of those plus characteristics which meet the uniform section of the piston trajectory are all the same, so that beyond the first of them the flow must actually be uniform.

We conclude that when the piston has achieved a fixed speed it is followed down the pipe by a section of fluid which moves uniformly to the left like a solid block but which simultaneously increases in size. Next comes a simple wave, or *rarefaction wave*, through which the gas thins out toward the left. Finally, to the right of the rarefaction wave lies a region of more compressed fluid at rest. Observe that in the limiting case where the piston is brought instantaneously to its ultimate velocity, all the plus characteristics in the rarefaction wave start out from the same initial point $(x_0, 0)$. A flow of the latter kind is referred to as a *centered simple wave*. We call attention to the fact that along each plus characteristic of such a centered simple wave (14.44) yields the explicit relation

$$(14.51) \qquad \frac{x - x_0}{t} = \frac{dx}{dt} = u + c = \frac{\gamma + 1}{2} r + \frac{\gamma - 3}{2} s$$

among r, x and t, where s is understood to represent an auxiliary constant.

Consider now a situation in which gas lies initially at rest in the section of pipe $x_0 < x < x_1$ between a pair of pistons located at the points $x = x_0$ and $x = x_1$. We shall investigate what happens when the pistons are abruptly pulled away from each other at fixed speeds. Our previous analysis shows that a centered simple wave proceeds down the pipe to the right from $x = x_0$, while another centered simple wave moves up the pipe to the left from $x = x_1$. We shall use the Riemann function (14.50) to determine the *interaction* of these two simple waves when they collide in the middle of the pipe.

Before the waves meet there is a region of gas at rest between them which corresponds to a triangle in the (x,t)-plane bounded by the initial segment $x_0 < x < x_1$, together with a pair of characteristics through its ends that intersect at a vertex with the coordinates

$$x = \frac{x_0 + x_1}{2}, \qquad t = t_0 > 0.$$

The continuations of these characteristics are curves along which either r or s remains constant. From our knowledge of the flow in a centered simple wave we shall be able to find t on each of the latter characteristic curves explicitly in terms of either s or r. Thus we shall be led to a *characteristic initial value problem* (cf. Chapter 4) which defines t as a function of the Riemann invariants r and s over a rectangular domain of interaction of the above centered simple waves.

The simple wave centered at $(x_0, 0)$ is bounded from above by the characteristic curve just described on which s is constant. Consequently on this arc (14.51) must hold. According to the differential equation (14.43) for minus characteristics it follows that we have along the arc in question

$$(u - c)t_r = x_r = \frac{\partial}{\partial r}\left[\frac{\gamma + 1}{2} rt + \frac{\gamma - 3}{2} st\right],$$

which reduces in view of (14.45) to

$$t_r + \frac{\lambda}{r + s} t = 0.$$

Letting r_0 and s_0 stand for the values of r and s at the point where the simple waves centered at $(x_0, 0)$ and at $(x_1, 0)$ first collide, we conclude from an integration that

(14.52)
$$t = \left(\frac{r_0 + s_0}{r + s_0}\right)^{\lambda} t_0$$

along the minus characteristic rising from that point. Similarly, along an arc of the opposite plus characteristic rising from the point of collision we have

(14.53)
$$t = \left(\frac{r_0 + s_0}{r_0 + s}\right)^{\lambda} t_0.$$

Between the latter pair of characteristics we can combine the partial differential equation (14.49) with the conditions (14.52) and (14.53) to

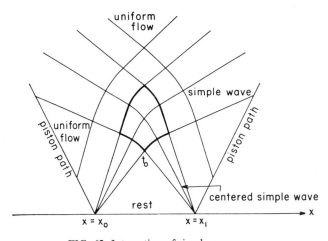

FIG. 62. Interaction of simple waves.

formulate a characteristic initial value problem for t. Its solution is immediately seen to be given in terms of the known Riemann function (14.50) by the remarkable formula

$$(14.54) \qquad\qquad t = t_0 A(r_0, s_0; r, s).$$

The result (14.54) enables us to calculate the interaction of two centered simple waves throughout a curvilinear rectangle bounded by four characteristics. Adjacent to this rectangle on all four sides we see that the flow consists of simple waves. Afterward a region of uniform motion is observed, and then further interactions and reflections of simple waves appear. However, a detailed analysis of the more elaborate configurations that may occur here surpasses the scope of our present discussion.

EXERCISES

1. Show that a linearization of the equations (14.32) and (14.33) of one-dimensional flow based on the assumption that u is small and ρ only undergoes small variations leads to the wave equation as it appears in acoustics.

2. Obtain from (14.54) an expression for the interaction of two centered rarefaction waves that only involves rational functions when the ratio of specific heats γ is given in terms of some positive integer n by the formula

$$\gamma = \frac{2n + 3}{2n + 1}.$$

In this context what is the physical significance of the number n?

3. In the piston problem of Figure 62, verify that the flows appearing on the far sides of the rectangular region of interaction of the two centered rarefaction waves are, indeed, simple waves.

4. Discuss equations for the spherically symmetric motion of a gas, that is, for radial flow depending only on the time and the distance from the origin.[6]

5. If the piston is pushed toward the gas instead of being pulled away from it in the flow described by Figure 61, show that the plus characteristics in the resulting simple wave form an *envelope*. What has this to do with the formation of shock waves?

6. Use the shock conditions (14.24), (14.25) and (14.26) to show that when a piston is pushed at a fixed speed into a gas at rest the resulting flow consists of a shock wave which advances into the stationary fluid, followed by an increasing accumulation of gas in front of the piston which moves forward uniformly like a solid block.[7]

7. Derive (14.43) from (14.46) and (14.47) by the method of characteristics.

[6] Cf. Courant-Friedrichs 1.
[7] Cf. Courant-Friedrichs 1.

3. STEADY SUBSONIC FLOW

The partial differential equations governing steady two-dimensional flow of a compressible fluid have attracted considerable attention because of the important role they play in modern aerodynamics. Much of the progress that has been made in treating them is based on special devices for determining particular solutions or on inverse methods of obtaining flows with properties of physical interest. Here we shall begin our investigation by discussing the case of subsonic flow, which provides a useful example of the difficulties encountered in solving boundary value problems for nonlinear equations of the elliptic type.

We have seen in Section 1 that isentropic, irrotational flow in the (x,y)-plane may be described by a velocity potential $\phi = \phi(x, y)$. Substituting the expressions $u = \phi_x$ and $v = \phi_y$ for the velocity components into the continuity equation (14.10), we obtain for ϕ the second order equation

$$(14.55) \qquad (\rho\phi_x)_x + (\rho\phi_y)_y = 0.$$

The density ρ can be eliminated from (14.55) by means of Bernoulli's law (14.18), which we now write in the form

$$(14.56) \qquad \frac{1}{2}q^2 + \frac{1}{\gamma - 1}c^2 = H,$$

where

$$c^2 = \frac{dp}{d\rho} = \gamma \frac{p}{\rho} = A\gamma\rho^{\gamma-1}$$

still stands for the square of the local speed of sound. Thus

$$\rho_x = \frac{d\rho}{dq}q_x = -\frac{\rho}{c^2}(u\phi_{xx} + v\phi_{yx})$$

and

$$\rho_y = \frac{d\rho}{dq}q_y = -\frac{\rho}{c^2}(u\phi_{xy} + v\phi_{yy}),$$

so that (14.55) reduces to the more specific quasi-linear equation

$$(14.57) \qquad \left(1 - \frac{u^2}{c^2}\right)\phi_{xx} - \frac{2uv}{c^2}\phi_{xy} + \left(1 - \frac{v^2}{c^2}\right)\phi_{yy} = 0.$$

In the case of rotational flow (14.57) must be replaced by an analogous equation for the stream function ψ. For the sake of simplicity we shall suppose here that the constant of integration H on the right in Bernoulli's equation is actually independent of ψ. However, the factor A in the equation of state (14.20) will be allowed to vary from streamline to streamline, which makes it necessary to interpret the derivative $dp/d\rho$ used to define the speed of sound c as a partial differentiation with ψ and S held fixed. These hypotheses enable us to formulate a partial differential equation for ψ from the relation

$$vuu_x + v^2u_y - u^2v_x - uvv_y + \frac{v}{\rho}p_x - \frac{u}{\rho}p_y = 0,$$

which results from multiplying (14.8) by v, multiplying (14.9) by $-u$, and then adding. Indeed, (14.19) and (14.20) give

$$\frac{\psi_y^2}{\rho^3}\psi_{xx} - \frac{2\psi_x\psi_y}{\rho^3}\psi_{xy} + \frac{\psi_x^2}{\rho^3}\psi_{yy} - (\psi_x^2 + \psi_y^2)A'(\psi)\rho^{\gamma-2}$$
$$-\gamma A(\psi)\rho^{\gamma-3}(\psi_x\rho_x + \psi_y\rho_y) = 0,$$

whereupon elimination of ρ and its derivatives by means of Bernoulli's law yields

$$(14.58)\quad \left(1 - \frac{u^2}{c^2}\right)\psi_{xx} - \frac{2uv}{c^2}\psi_{xy} + \left(1 - \frac{v^2}{c^2}\right)\psi_{yy}$$
$$+ \rho^2\frac{q^2 + 2H}{2\gamma}\frac{A'(\psi)}{A(\psi)} = 0.$$

Both (14.57) and (14.58) are quasi-linear partial differential equations of the second order, since their coefficients depend only on the unknown and its first derivatives. Thus the theory we developed in Section 3.4 for the general quasi-linear equation (3.88) applies to them. The two equations have the same discriminant

$$ac - b^2 = \left(1 - \frac{u^2}{c^2}\right)\left(1 - \frac{v^2}{c^2}\right) - \left(\frac{uv}{c^2}\right)^2$$
$$= 1 - \frac{q^2}{c^2},$$

which is a function of the *Mach number*

$$M = \frac{q}{c}.$$

It follows that the type of the equation describing steady flow of a compressible fluid depends on the sign of $1 - M^2$. More precisely, these equations are of the elliptic type in any region of *subsonic flow*, where $M < 1$, whereas they are of the hyperbolic type in any region of *supersonic flow*, where $M > 1$. The subsonic and supersonic regimes therefore turn out to be of quite different structure. Most interesting of all are the *transonic flows* involving a transition from the subsonic to the supersonic regime, for they include both elliptic and hyperbolic regions and lead to problems of mixed type. In this section, however, we deal primarily with the elliptic case of subsonic flow.

Let us confine our attention to irrotational motion, which can be described either by (14.57), or by (14.58) with the last term on the left omitted. We prefer, however, to work with an equivalent pair of equations of the first order for the velocity components u and v directly. We are then able to apply more readily an analogue of the hodograph transformation developed in Section 2 for the purpose of reducing the equations of one-dimensional flow to a linear system.

Eliminating ϕ from (14.57) by means of (14.16), and taking (14.17) into account, we find that u and v must satisfy the generalized Cauchy-Riemann equations

$$(c^2 - u^2)u_x - uv(u_y + v_x) + (c^2 - v^2)v_y = 0,$$

$$u_y - v_x = 0,$$

where

$$c^2 = (\gamma - 1)H - \frac{\gamma - 1}{2}(u^2 + v^2).$$

Because these equations have coefficients depending only on u and v, the experience we gained in Section 2 suggests that they will become linear if we apply to them the hodograph transformation, which now consists in introducing the velocity components u and v as independent variables and considering the physical coordinates x and y as unknowns. Since

$$u_x = \frac{y_v}{J}, \qquad v_x = -\frac{y_u}{J},$$

$$u_y = -\frac{x_v}{J}, \qquad v_y = \frac{x_u}{J},$$

where

$$J = \frac{\partial(x, y)}{\partial(u, v)} = x_u y_v - y_u x_v,$$

we are led to the linear system

(14.59) $$(c^2 - v^2)x_u + uv(x_v + y_u) + (c^2 - u^2)y_v = 0,$$

(14.60) $$x_v - y_u = 0,$$

whose coefficients are actually polynomials in the new independent variables u and v.

It is natural to inquire next under what circumstances we can expect the hodograph transformation to be at least locally one-to-one. That it has this property for all subsonic flows which are not uniform is a consequence of the implicit function theorem, since, as we shall see immediately, the Jacobian J can have only isolated zeros at points (u, v) of the hodograph plane where $M < 1$. Indeed, multiplying (14.59) by x_u and using (14.60) to simplify the result, we derive for J the relation

$$(c^2 - v^2)x_u^2 + 2uvx_ux_v + (c^2 - u^2)x_v^2 = -(c^2 - u^2)J.$$

The quadratic form in x_u and x_v on the left is positive-definite when $M < 1$, since its discriminant must then exceed zero. Under the same hypothesis the coefficient $c^2 - u^2$ occurring on the right has to be positive, from which we conclude that J itself is negative unless

$$x_u = x_v = 0.$$

On the other hand, in that case the differential equations (14.59) and (14.60) give

$$y_u = y_v = 0,$$

too, so that all the partial derivatives x_u, x_v, y_u and y_v must vanish at any root of the Jacobian J in the subsonic region $M < 1$.

Suppose now that $J = 0$ along a curve in the hodograph plane where $M < 1$, which means in particular that the unknowns x and y must remain constant there. Since the system (14.59), (14.60) has no real characteristics in the region $M < 1$, Holmgren's uniqueness theorem (cf. Section 6.1) implies that x and y have to be constants throughout that region and therefore do not correspond to a physically significant flow. In the case of a non-trivial subsonic flow, on the other hand, x and y satisfy linear analytic partial differential equations of the elliptic type. Hence, according to results from Section 5.1, they must themselves be real analytic functions of the two arguments u and v. Thus J, too, is analytic when $M < 1$, and since it does not vanish along any arc it can only have isolated zeros, as we originally asserted.

A similar analysis shows that the inverse Jacobian

$$J^{-1} = \frac{\partial(u, v)}{\partial(x, y)}$$

also possesses only isolated zeros in any region of non-trivial subsonic flow. Thus the hodograph transformation behaves topologically like a conformal mapping and may even be described as an *interior transformation* throughout the subsonic regime. Indeed, for an incompressible fluid the mapping to the hodograph plane actually becomes conformal after an auxiliary reflection has been performed. However, we shall see later that supersonic flows can be found for which the Jacobian J^{-1} vanishes identically. Be this as it may, we proceed here to construct specific examples of the steady flow of a compressible fluid by determining explicit solutions of the linear equations (14.59) and (14.60) in the hodograph plane and then expressing the velocity components u and v afterward as functions of the rectangular coordinates x and y.

Instead of treating the system (14.59), (14.60) directly, we observe that the second equation (14.60) leads to the existence of an integral Φ whose gradient is given by

$$\Phi_u = x, \qquad \Phi_v = y.$$

Substitution into (14.59) yields for Φ the linear partial differential equation

(14.61) $$(c^2 - v^2)\Phi_{uu} + 2uv\Phi_{uv} + (c^2 - u^2)\Phi_{vv} = 0$$

of the second order. Analogous equations for the velocity potential ϕ and for the stream function ψ can also be derived in the hodograph plane. For such a purpose, however, it will turn out to be more convenient to work with the polar coordinates

$$q = \sqrt{u^2 + v^2}, \qquad \theta = \tan^{-1}\frac{v}{u}.$$

In terms of q and θ we can recast the basic relations (14.16) and (14.19) in the more concise form

(14.62) $$dx + i\,dy = \frac{e^{i\theta}}{q}\left(d\phi + i\,\frac{d\psi}{\rho}\right)$$

involving differentials, where ρ is a function of q defined by the Bernoulli law (14.56). We interpret (14.62) to mean that $dz = dx + i\,dy$ is an exact differential in the (u,v)-plane, or, for that matter, even in the (q,θ)-plane. Thus the contour integral

$$\oint \frac{e^{i\theta}}{q}\left(d\phi + i\,\frac{d\psi}{\rho}\right) = \oint \frac{e^{i\theta}}{q}\left[\left(\phi_q + i\,\frac{\psi_q}{\rho}\right)dq + \left(\phi_\theta + i\,\frac{\psi_\theta}{\rho}\right)d\theta\right]$$

is independent of path. Because of the divergence theorem it follows that

$$\frac{\partial}{\partial\theta}\left[\frac{e^{i\theta}}{q}\left(\phi_q + i\,\frac{\psi_q}{\rho}\right)\right] = \frac{\partial}{\partial q}\left[\frac{e^{i\theta}}{q}\left(\phi_\theta + i\,\frac{\psi_\theta}{\rho}\right)\right].$$

The second derivatives $\phi_{q\theta}$ and $\psi_{q\theta}$ clearly cancel out here, and since

$$\frac{d\rho}{dq} = -\frac{\rho q}{c^2}$$

what remains reduces to the generalized Cauchy-Riemann equations

(14.63) $\qquad\qquad\qquad q\psi_q = \rho\phi_\theta$

and

(14.64) $\qquad\qquad\qquad \rho q\phi_q = -(1 - M^2)\psi_\theta$

for ϕ and ψ as functions of q and θ. Equations of the second order for ϕ and ψ separately can now be obtained in the usual way by cross differentiating (14.63) and (14.64) and then eliminating the mixed derivatives $\psi_{q\theta}$ and $\phi_{q\theta}$. We have, for example,

(14.65) $\qquad q^2\psi_{qq} + (1 - M^2)\psi_{\theta\theta} + q(1 + M^2)\psi_q = 0.$

Solutions of (14.65) can be found by the method of separation of variables. Thus the only interesting solution that depends on θ alone is

(14.66) $\qquad\qquad\qquad \psi = k_1\theta,$

where k_1 is a parameter. This corresponds to radial flow, since the angle of inclination θ of the velocity vector remains constant along each streamline. On the other hand, the function

(14.67) $\qquad\qquad\qquad \psi = k_2\int\frac{\rho\,dq}{q}$

of q alone is a solution of (14.65) describing a flow with circular streamlines. Finally, the linear combination

(14.68) $\qquad\qquad\qquad \psi = k_1\theta + k_2\int\frac{\rho\,dq}{q}$

of (14.66) and (14.67) represents flow along a family of spirals. Further flows can be constructed explicitly by seeking solutions of (14.61) or of (14.65) that have the form of a function of q times a function of θ. However, it is not our intention here to go into such matters in detail.

We see that the hodograph method furnishes a convenient inverse construction of flows whose qualitative properties might be of interest in gas dynamics. However, it is not adequate to determine the subsonic flow past a prescribed body. That question leads to a Neumann problem for the velocity potential ϕ and a Dirichlet problem for the stream function ψ in the physical plane. Because these problems are nonlinear they lie beyond the scope of any of the existence theorems we have presented. On the other hand, the techniques of Chapters 8 and 9 do apply to the case of flow so slow that (14.57) and (14.58) reduce to the Laplace equation.

EXERCISES

1. Give a precise formulation of the question of determining the subsonic flow past a prescribed body as a nonlinear Neumann problem for the velocity potential ϕ or as a nonlinear Dirichlet problem for the stream function ψ.

2. Reformulate the partial differential equation (14.61) for Φ in terms of polar coordinates.

3. Show that the integral Φ can be defined in terms of the velocity potential ϕ by means of the *Legendre transformation*

$$\Phi + \phi = xu + yv.$$

Use this transformation to deduce (14.61) directly from (14.57).

4. Verify that the formula

$$\Phi = k_1 \int \frac{dq}{\rho q} - k_2 \theta$$

defines the same flow as (14.68) and discuss its properties in detail.

5. Show that (14.57) is the Euler equation for the double integral extremal problem

(14.69) $$\iint p(\phi_x^2 + \phi_y^2)\, dx\, dy = \text{maximum}$$

in the calculus of variations, where p is the quite specific function of $q^2 = \phi_x^2 + \phi_y^2$ defined by Bernoulli's law.

6. Prove that if the equation of state (14.5) is replaced by the relation

(14.70) $$p = -\frac{\alpha}{\rho} + \beta$$

between p and ρ, where α and β stand for positive constants, then the associated hodograph equations (14.63) and (14.64) reduce to the true Cauchy-Riemann equations for the real and imaginary parts of an analytic function of an appropriate complex variable. How should the parameters α and β be chosen so that (14.70), which is known as the *Karman-Tsien gas law* and which corresponds to a fictitious ratio of specific heats $\gamma = -1$, provides a reasonable approximation to the actual equation of state (14.5) locally? Show that for a Karman-Tsien gas (14.69) becomes the Plateau problem for the determination of minimal surfaces (cf. Section 15.4).

7. Complete the details of our derivation of equation (14.58) in the case of rotational flow.

8. The axially symmetric motion of an incompressible fluid is governed by the equation of Exercise 1.3 for a Stokes stream function $\psi = \psi(x, y)$. Prove that the hodograph transformation to new coordinates

$$u = \frac{1}{y}\,\psi_y, \qquad v = -\frac{1}{y}\,\psi_x$$

is not in general one-to-one even locally for axially symmetric flow by studying the special example

$$\psi = x + xy^2.$$

9. Use the hodograph transformation to prove that the velocity components u and v of any steady, irrotational subsonic flow are real analytic functions of the coordinates x and y.

10. In the case of steady flow we have reduced the system (14.8)–(14.11) of four first order equations to a single equation (14.58) of the second order. In what sense might the two arbitrary functions $A(\psi)$ and $H(\psi)$ appearing in this integration process be interpreted as part of the general solution of the original system?

4. TRANSONIC AND SUPERSONIC FLOW

The behavior of steady supersonic flow is quite unlike that of steady subsonic flow because for $M > 1$ the equations of motion (14.57) and (14.58) are of the hyperbolic type and possess real characteristics. Indeed, across such characteristic curves, which are often referred to as *Mach lines*, the partial derivatives of the velocity components u and v, of the pressure p and of the density ρ need not be continuous, whereas for subsonic irrotational flow each of those quantities was seen to be not only continuous but even analytic. Thus the problem of determining supersonic flows is not at all similar to what we are familiar with in the subsonic case. However, we shall find that it is in some respects simpler to handle because of the possibility that several different flows may be fitted together along the Mach lines.

The characteristics of the equation

$$(14.57) \qquad \left(1 - \frac{u^2}{c^2}\right)\phi_{xx} - \frac{2uv}{c^2}\,\phi_{xy} + \left(1 - \frac{v^2}{c^2}\right)\phi_{yy} = 0$$

for the velocity potential ϕ are two families of curves satisfying the ordinary differential equation

$$(14.71) \qquad (c^2 - u^2)\,dy^2 + 2uv\,dy\,dx + (c^2 - v^2)\,dx^2 = 0,$$

which can also be expressed in the form

$$c^2(dx^2 + dy^2) = (u\,dy - v\,dx)^2.$$

Letting s denote the arc length along such a Mach line, we conclude that

$$(14.72) \qquad c^2 = \left(u\frac{dy}{ds} - v\frac{dx}{ds}\right)^2 = q^2 \sin^2 \omega,$$

where ω is the angle between the direction of the flow and the Mach line. Since (14.72) implies that the magnitude of ω, which is called the *Mach angle*, is the same for both the characteristics (14.71) through a given point, it becomes clear that the angle between those characteristics is bisected by the velocity vector at their intersection. Finally, note that because of (14.72) the Mach lines are nearly perpendicular to the flow when the Mach number M barely exceeds unity, whereas they become parallel to the flow as $M \to \infty$.

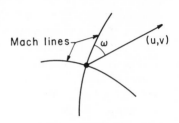

FIG. 63. Mach lines.

Since the characteristics (14.71) are precisely those curves across which the higher derivatives of the velocity potential ϕ can exhibit jumps, they are invariant under changes of the independent variables and must, in particular, map into the characteristics of (14.61) in the hodograph plane. The latter characteristics satisfy the ordinary differential equation

$$(14.73) \qquad (c^2 - u^2)\,du^2 - 2uv\,du\,dv + (c^2 - v^2)\,dv^2 = 0,$$

or

$$c^2(du^2 + dv^2) = (u\,du + v\,dv)^2.$$

It follows that at their intersections these Mach lines cut the radius vector (u, v) at equal angles which coincide with the complement of the Mach angle ω. Observe, too, that in the hodograph plane the characteristics are fixed families of curves, whereas in the physical plane they depend on the specific flow under consideration.

We now introduce the *critical speed* c_* and the *escape speed* q_*, which are defined by the relations

$$\frac{1}{2}c_*^2 + \frac{1}{\gamma - 1}c_*^2 = H, \qquad \frac{1}{2}q_*^2 = H.$$

In terms of c_* and q_* we can put Bernoulli's law (14.56) into the two equivalent forms

$$(14.74) \qquad \mu^2 q^2 + (1 - \mu^2)c^2 = c_*^2, \qquad \mu = \sqrt{\frac{\gamma - 1}{\gamma + 1}},$$

and

$$(14.75) \qquad q^2 + \frac{2}{\gamma - 1}c^2 = q_*^2.$$

From (14.74) we conclude that a flow is subsonic or supersonic according to whether q is smaller or larger than the critical speed c_*. On the other hand, (14.75) implies that q can in no circumstances exceed the escape speed q_*. Thus in the hodograph plane the circle $q < c_*$ corresponds to subsonic flows, while supersonic flows map into the annulus $c_* < q < q_*$. Moreover, the characteristics (14.73) are seen to be curves in the latter region which have cusps at the inner boundary $q = c_*$, where $M = 1$, but which just touch the outer boundary $q = q_*$, where $M = \infty$. Incidentally, transonic flows are associated with domains of the hodograph plane overlapping the parabolic circle $q = c_*$, which is also referred to as the *sonic line* (cf. Section 12.1).

The Jacobian of the hodograph transformation may vanish identically for supersonic flow. Indeed, in the supersonic case the equations of motion are quite analogous to the hyperbolic system governing one-dimensional isentropic flow, for their coefficients involve only the unknowns u and v and not the independent variables x and y. Consequently the methods developed in Section 2 are applicable to them. For example, we can introduce characteristic coordinates α and β which reduce (14.71) and (14.73) to the first order canonical system

$$(u^2 - c^2)y_\alpha = (uv + c^2\sqrt{M^2 - 1})x_\alpha,$$

$$(u^2 - c^2)y_\beta = (uv - c^2\sqrt{M^2 - 1})x_\beta,$$

$$(u^2 - c^2)u_\alpha + (uv - c^2\sqrt{M^2 - 1})v_\alpha = 0,$$

$$(u^2 - c^2)u_\beta + (uv + c^2\sqrt{M^2 - 1})v_\beta = 0$$

similar to (14.39), (14.40).

Solutions of the above system such that

$$\frac{\partial(u, v)}{\partial(x, y)} \equiv 0$$

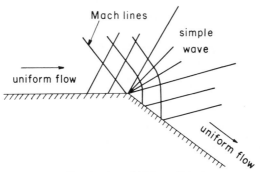

FIG. 64.   Prandtl-Meyer flow.

are again called *simple waves*. A repetition of the analysis of Section 2 shows that a simple wave maps into an arc of one characteristic in the hodograph plane, and that in the physical plane the opposite family of characteristics is composed exclusively of straight lines. In particular, a *centered simple wave* is a supersonic flow for which one of the two families of Mach lines consists of rays emanating from a single point. Such a flow is sometimes also called a *Prandtl-Meyer expansion fan*. By fitting it along two radial Mach lines to a pair of appropriate uniform flows we can construct the supersonic flow around the outside of a corner. This flow is a direct analogue of the one-dimensional time-dependent motion that results from pulling a piston abruptly and uniformly out of a tube of gas that was originally at rest.

It is not feasible to use simple waves to construct the supersonic flow inside a corner because the associated Mach lines would form an envelope. The situation here is connected with that encountered in the case of one-dimensional flow in front of a piston that accelerates into a gas at rest (cf. Exercises 2.5 and 2.6). Therefore we are motivated to seek a solution involving a stationary shock wave across which the conservation laws (14.27) to (14.30) apply. The algebra of those relations shows that if the shock wave is a straight line in the (x,y)-plane, then a constant state in front of it is converted into another constant state behind. Thus as a uniform flow crosses such a stationary shock it is bent through a fixed angle and becomes a different uniform flow behind. The shock condition (14.28) asserts that the component of velocity L parallel to the shock front experiences no jump, and, because of the physics of the problem, we adopt the convention that the side of the wave on which the normal component of velocity N is largest should feature as the front of the shock wave.[8] It follows that the direction of the flow is turned toward the shock front as the fluid passes through.

[8] Cf. Courant-Friedrichs 1.

We observe that each streamline in the motion just described consists of a pair of line segments joined at the shock wave so as to form an obtuse angle there. If we conceive of one such streamline as the juncture of two semi-infinite rigid walls, we see that our flow turns inside the corner between those walls by passing through a suitable shock front. If we adjoin to the latter motion its reflected image in the leading wall, which we suppose afterward to have been removed, we obtain a symmetric flow past the infinite wedge bounded by the remaining wall and its reflection. This flow may be viewed as the simplest model of a supersonic stream impinging on the sharp leading edge of a profile. At the edge are attached two shock waves, and we therefore speak of the motion in question as a flow with *attached shock wave*.

In supersonic flow past any blunt body a *detached shock wave*, or *bow wave*, appears at some distance upstream. This wave is analogous to the attached shock front just described, but it also has features in common with the shock wave that penetrates a pipe filled with gas into which a piston has been accelerated. A detached shock wave is necessarily curved, and a portion at least of the flow behind it is subsonic, since that is the case behind a normal shock front. Indeed, algebraic manipulation of the shock conditions (14.27) to (14.30) shows that

$$N_1 N_2 = c_*^2 - \mu^2 L^2,$$

whence

$$\sqrt{L_2^2 + N_2^2} = N_2 < c_*$$

whenever

$$L = L_1 = L_2 = 0, \qquad N_1 > c_*.$$

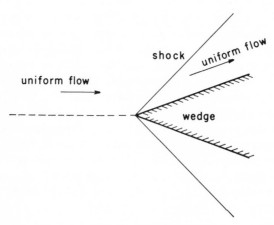

FIG. 65.   Attached shock.

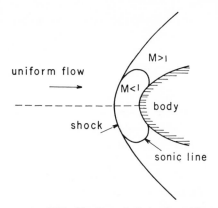

uniform flow

M>I

M<I

body

shock

sonic line

FIG. 66.   Detached shock problem.

The most interesting detached shock waves, of course, are those behind which both subsonic and supersonic regions appear, so that a transonic regime is involved.

A difficulty in the detached shock problem that is perhaps even harder to overcome than the complications connected with mixed type arises from our lack of knowledge concerning the shape and location of the bow wave itself, which must be found as part of the answer. We have thus to deal with what is commonly referred to as a *free boundary problem*, which simply means a problem whose solution is defined over a domain that is not given in advance.

In the present example we suppose that the flow in front of the shock wave is uniform and supersonic. Consequently the shock conditions (14.27) to (14.30) serve to specify the four quantities u, v, p and ρ just behind the shock as functions of its slope $dy/dx = N_1/L_1$ alone. Thus the Cauchy data can be calculated along the shock front in terms of its geometry. To be more precise, the law of conservation of energy (14.30) asserts that H remains constant. Hence the equation of motion (14.58) based on a stream function ψ is applicable behind the free boundary, even though the flow is no longer isentropic there. The law of conservation of mass implies that ψ is continuous across the shock, too, but the normal derivative $\partial \psi/\partial v$ and the coefficient $A = A(\psi)$ can only be evaluated when the slope dy/dx of the bow wave is known. It is natural to think of the relation here between $\partial \psi/\partial v$ and dy/dx as an extra free boundary condition which might suffice to determine the position of the detached shock wave at least along the arc where it is adjacent to a region of subsonic flow.

As we have just formulated it, the detached shock problem is far too difficult to treat in a satisfactory way. However, examples of flow with a bow wave can be generated by solving an *inverse problem* in which it is the detached shock wave rather than the actual body that is prescribed. Let us assume that the shock wave has a representation.

$$x = x(y)$$

in terms of a given real analytic function $x(y)$. The foregoing analysis then shows that the flow behind the shock can be found as the solution of a Cauchy problem for an analytic system of partial differential equations, with initial data defined analytically along what before played the role of the free boundary but is now a known curve. The Cauchy-Kowalewski theorem (cf. Section 1.2) assures us of the existence of a power series solution of this analytic Cauchy problem at least locally, regardless of the type of the equations involved. On the other hand, in Section 16.1 we shall describe a more practical method of handling analytic initial value problems for equations of mixed type like (14.58) which exploits complex substitutions similar to those encountered in Sections 3.1 and 3.4 and in Section 12.2.

A transonic flow problem which yields to more direct analysis than does the detached shock problem arises in the study of a subsonic stream of compressible fluid that emerges from a narrowing channel to form a supersonic free jet. We shall indicate how in the case of irrotational motion the latter question can be formulated[9] as a Tricomi or Frankl problem in the hodograph plane (cf. Section 12.1).

To be precise, consider a channel in the left half-plane $x < 0$ which is symmetric with respect to the x-axis, which is open at the right end, and which is bounded from above and below by a pair of rays that emanate from points z and $-z$ on the y-axis and diverge toward the left, cutting out a positive angle at infinity. We suppose that fluid moves steadily from left to right in the channel at subsonic speed and then issues forth into the right half-plane $x > 0$ as a free jet surrounded by gas of relatively negligible density. Because of its low inertia, the surrounding gas is assumed to be at rest in a state of constant pressure. Therefore along the boundaries of the jet, which are two streamlines of unknown shape and location, the pressure of the moving fluid must remain at a constant value exactly balancing the exterior pressure (cf. Section 5 below). From Bernoulli's law and the equation of state it follows that the speed is constant along these free boundaries, too. We are interested in the case where its value q_0 there is supersonic.

In order to accelerate from specifically subsonic speeds on the walls of the channel to a fixed supersonic speed along the free streamlines bounding the jet, our fluid must turn through corners at the points z and $-z$ where it

---

[9] Cf. Frankl 1.

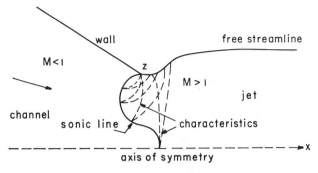

FIG. 67. Supersonic jet.

separates from the walls. In fact, at each of these points of separation the flow behaves locally like a Prandtl-Meyer expansion fan, so that the free streamlines make obtuse angles there with the two rays composing the channel boundary. Moreover, a sonic line crosses the channel at its throat and joins z to $-z$. On the left side of the sonic line the flow is subsonic, and its speed actually approaches zero at the far end of the channel. To the right of the sonic line is situated the free jet, in which the velocity is supposed, of course, to be supersonic.

With the foregoing remarks in mind we intend to examine carefully the image of the flow in the hodograph plane. Because of the symmetry of the motion with respect to the x-axis it will suffice to consider only the upper half of the fluid, which simplifies the geometry significantly.

The upper half of the subsonic region of our flow maps onto a circular sector in the lower half of the hodograph plane because of the fixed direction of the velocities on each channel wall and the fixed magnitude c_* of the velocities on the sonic line. On the other hand, the free boundary of the jet must be transformed into arcs of a circle in the hodograph plane whose radius q_0 exceeds the critical speed c_*. Since the speed has to jump from c_* to q_0 at the point of separation z, that point itself must correspond to an entire curve of velocities. Our earlier analysis of simple waves can be applied to the flow near z because it behaves asymptotically like a Prandtl-Meyer expansion fan. We conclude that the image of z is actually a characteristic which climbs from the lower end of the circular arc bounding the subsonic sector mentioned above until it reaches an arc of the circle of radius q_0 that is turned further toward the u-axis. The latter arc is the map of the free streamline. Unfortunately, at its remaining end we are not quite sure how to specify additional curves that might serve to complete the boundary of our image of the supersonic jet.

Since shock waves may appear anyhow at some distance to the right

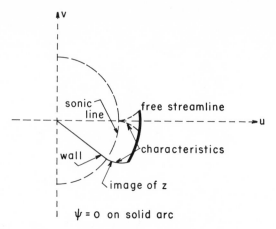

FIG. 68. Frankl problem in the hodograph plane.

within the jet itself, we are only justified in studying the flow inside the channel and near its throat by the hodograph method. We observe that for reasons of symmetry two cusped pairs of characteristics touch the sonic line at its intersection with the x-axis. We shall establish that the flow to the left of these Mach lines can be found by solving a Frankl problem.

Indeed, according to what was said previously, the corresponding region of the (u,v)-plane must consist of a sector of the circle $q \le c_*$, plus the part of the annulus $c_* < q < q_0$ located between the two closest characteristics that emanate from the ends of the circular arc bounding that sector. Furthermore, the stream function ψ reduces to a constant along each curve on the boundary of the latter region except the characteristic issuing from the point where the sonic line meets the u-axis, which corresponds to one of the above Mach lines that rises from a cusp at the x-axis and forks to the left. Thus it turns out that in the domain of the hodograph plane just described, ψ is the solution of a Frankl problem for the linear partial differential equation (14.65).

Note that q_0 should be no larger than the distance from the origin to the intersection of the two characteristics bounding the hyperbolic portion of the region in which the above Frankl problem has been formulated. Moreover, in the limiting case where q_0 has that extreme value we obtain a Tricomi problem for the stream function ψ rather than a Frankl problem. Finally, observe that in the curvilinear triangle bounded by the image of the free streamline and the pair of Mach lines forming a cusp at the intersection of the sonic line with the u-axis, ψ can now be determined by solving a Goursat problem (cf. Section 4.2). However, farther out in the jet we might expect to encounter shock waves, and therefore it is of no avail to attempt continuing the solution of the problem beyond the present stage by such techniques.

EXERCISES

1. Show by explicit integration that in the hodograph plane the Mach lines (14.73) are *epicycloids* obtained as the trajectories of a point on the perimeter of a wheel of diameter $q_* - c_*$ which rolls around on the outside of the disc $q \leq c_*$.

2. From the shock conditions (14.27) to (14.30) deduce that points just behind a stationary shock wave placed in a uniform stream map into the so-called *shock polar*

$$v^2 = \frac{(q_1 - u)^2 \, (u - c_*^2/q_1)}{(1 - \mu^2)q_1 + c_*^2/q_1 - u}$$

in the hodograph plane, where q_1 stands for the speed of the flow in front of the shock.

3. Give an example of a supersonic flow that is not analytic in its dependence on the physical coordinates x and y.

4. Compare the characteristic surfaces of time-dependent subsonic flow with those of time-dependent supersonic flow.

5. Show that there are in general two shock waves, the *weak shock* and the *strong shock*, that turn a given uniform supersonic flow through a prescribed angle.

6. Combine an attached shock wave with a Prandtl-Meyer flow to construct a symmetrical example of supersonic flow past a semi-infinite bullet composed of a finite wedge fitted onto the end of a semi-infinite strip.

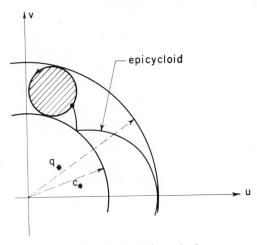

FIG. 69. Mach lines in the hodograph plane.

7. Replace the equation of state (14.5) by a new relation between p and ρ such that the equation (14.65) for ψ in the (q,θ)-plane can be reduced to Tricomi's equation (12.1).

8. Work out in detail all the formulas needed for an explicit statement of the analytic Cauchy problem connected with the inverse determination of the rotational flow behind a detached shock wave whose equation is

$$x = \sqrt{M_1^2 - 1}\ \sqrt{y^2 + 1}.$$

9. Formulate the detached shock problem for axially symmetric flow.

10. Use the hodograph method to prove that in the limiting case $q_0 = c_*$ of a sonic free jet, the only non-trivial flow involved is subsonic because the bulk of the jet moves uniformly forward at the speed of sound like a solid block.[10]

5. FLOWS WITH FREE STREAMLINES

In the previous section we became familiar with two free boundary problems of gas dynamics, one of which was concerned with shock waves and the other with jets. Here we intend to discuss in much greater detail the free streamline problems of classical hydrodynamics. Because we shall be dealing with an incompressible fluid, the system of partial differential equations (14.22) for the velocity potential ϕ and the stream function ψ will now reduce to the true Cauchy-Riemann equations

$$\phi_x = \psi_y, \qquad \phi_y = -\psi_x,$$

which assert that the complex potential

$$\zeta = \phi + i\psi$$

is an analytic function of the complex variable

$$z = x + iy.$$

This simplification in the equations of motion will enable us to solve a special class of free boundary problems explicitly. The device we shall appeal to for that purpose is the *hodograph transformation*, which, after a reflection in the u-axis, becomes a conformal mapping

$$w = u - iv = \frac{d\zeta}{dz}$$

[10] Cf. Sedov 1.

in the case of irrotational flow of an incompressible fluid. Before proceeding further with the mathematical analysis of such problems, however, we wish to explain briefly the more essential aspects of their physical background and significance.

For a liquid Bernoulli's law (14.18) assumes the form

$$(14.76) \qquad \frac{1}{2}q^2 + \frac{p}{\rho} = \text{const.},$$

which shows that the pressure p must decrease as the speed q increases. In fact, p will fall to the vapor pressure p_* of the liquid when q reaches a certain escape speed q_*. Hence q has to stay below that value in the liquid, which must boil, or *cavitate*, in regions where the flow would otherwise tend to become excessively rapid. We suppose that the vapor generated by this cavitation process has negligible inertia, so that it can be treated as a gas at rest in which the pressure is everywhere equal to p_*. We shall be interested in steady flows of liquid that include streamlines bounding cavities of vapor of the latter kind. Such free streamlines are called *slip streams* because the particles on one side of them move at a different speed than those on the other. Unlike a shock wave, however, a slip stream is a curve of discontinuity across which there is no actual mass flow. Consequently the conditions for equilibrium of a slip stream reduce to the statement that the pressure on the two sides of it must balance, which simply means that the pressure has to remain continuous across it. In particular, the liquid pressure p along a free streamline bounding a vapor cavity must be constantly equal to the vapor pressure p_*.

A jet of water flowing steadily into a region filled with air leads to a quite analogous situation. Assuming that the air is of negligible inertia relative to the water, and supposing that the air is everywhere at atmospheric pressure, we find that the free boundaries of the jet are streamlines along which the water has that same constant pressure. A glance at Bernoulli's law (14.76) shows that for both the case of a cavity and the case of a jet the extra requirement along the free streamlines can be expressed in terms of velocities by the equation

$$q = \text{const.}$$

This free boundary condition compensates for the fact that the shape and location of the free streamlines are not given in advance but must be determined as part of the flow. With the above quite general description in mind, we shall next formulate in a precise fashion several more specific free boundary problems, and we shall supply evidence that they are well posed by exhibiting closed solutions for particular examples.

Consider a monotonic arc joining some point on the negative x-axis to the point $-h + ik$ in the second quadrant, and let C_+ denote the curve composed of that arc plus its reflection in the x-axis. We want to construct a potential flow ζ past C_+ which is uniform at large distances in the sense that

$$\zeta \sim Uz$$

as $z \to \infty$, where U represents a positive coefficient. More specifically, we shall suppose that the central streamline of the flow forks over C_+ and then separates from the endpoints $-h + ik$ and $-h - ik$ to form a pair of free streamlines Γ bounding a vapor cavity Ω behind C_+. The simplest model of this kind, called the *Helmholtz flow*, occurs when both branches of Γ extend to infinity. A more sophisticated model is obtained, on the other hand, by introducing the reflection C_- of C_+ in the y-axis as an additional fixed boundary and assuming that the free streamlines Γ reattach to C_- at the points $h + ik$ and $h - ik$, so that the motion has symmetry in the y-axis as well as in the x-axis. In the latter configuration, which we shall refer to as the *Riabouchinsky flow*, the reflected curve C_- can be thought of as an idealization of the wake that actually appears behind the steam cavity Ω created by a stream of water passing over C_+ and boiling as it attempts to turn around the sharp edges at $-h + ik$ and $-h - ik$.

It has become apparent that the principal difficulty in determining a Helmholtz or Riabouchinsky flow is that the very region D_z in which the equations of motion are to be solved is not even known to begin with.

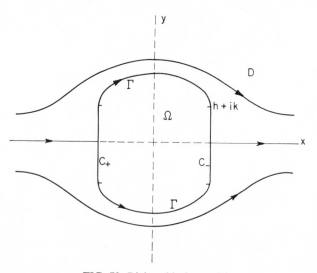

FIG. 70. Riabouchinsky model.

In the special case where the fixed boundary C_+ consists of a finite set of line segments we shall contrive, however, to overcome this difficulty by examining the conformal mappings ζ and w of the domain D_z in the physical plane into the plane of the complex potential and into the reflection of the hodograph plane. Because they turn out to be polygons, the respective images D_ζ and D_w of D_z in the ζ-plane and in the w-plane can be located exactly. The latter information enables us to calculate the required flow by working out the conformal transformation

$$(14.77) \qquad w = F(\zeta)$$

of D_ζ onto D_w explicitly and then using the ordinary differential equation

$$(14.78) \qquad \frac{d\zeta}{dz} = F(\zeta),$$

based on the fact that $w = d\zeta/dz$, to find ζ as a function of z. Thus it is possible to derive a representation in closed form for the free streamline flow past a polygonal obstacle.

We intend to discuss in detail the elementary example of a fixed boundary C_+ that consists of a single vertical segment joining the points of separation $-h + ik$ and $-h - ik$. The upper half of the Helmholtz flow past such a plate is mapped by the complex potential onto the upper half $\psi > 0$ of the ζ-plane, provided that we put

$$\psi = 0$$

along the central streamline. The hodograph of that same region is the intersection of a circle about the origin with the first quadrant; therefore it corresponds to the sector of the circle

$$|w| < q_*$$

in the w-plane cut out by the fourth quadrant. Indeed, the velocity is horizontal along the interval of the x-axis to the left of $-h$, and $-h$ itself is a stagnation point, which shows that this interval maps into the segment $0 < u < U$ of the u-axis. As we climb along C_+ from $-h$ to $-h + ik$ the velocities are all vertical, and they correspond to the interval $0 < v < q_*$ of the v-axis. Finally, along the free streamline Γ leaving the point of separation $-h + ik$, the velocity has fixed magnitude q_* and turns through a right angle, ending up in the horizontal direction with speed $U = q_*$ at infinity. Thus the three different portions of the boundary of the upper half of our flow in the physical plane map onto the three edges of the circular sector in the hodograph plane referred to above.

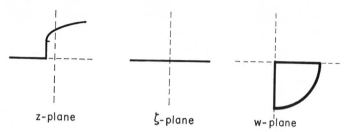

FIG. 71. Maps of the Helmholtz flow.

Since (14.77) must transform the half-plane $\psi > 0$ conformally onto the sector of the circle $|w| < q_*$ located in the fourth quadrant, we obtain by standard procedures the specific expression

$$(14.79) \qquad w = F(\zeta) = q_* \frac{\sqrt{\zeta - \phi_0} - i\epsilon}{\sqrt{\zeta - \phi_0 + \epsilon^2}}$$

for the mapping function F, where ϕ_0 and $\epsilon > 0$ are real constants. Substituting (14.79) into (14.78) and integrating, we find that

$$(14.80) \quad z = \int \frac{d\zeta}{F(\zeta)} = \frac{1}{q_*} \sqrt{\zeta + \epsilon^2} \left(\sqrt{\zeta} + 2i\epsilon \right) + \frac{\epsilon^2}{q_*} \log \frac{\sqrt{\zeta} - \sqrt{\zeta + \epsilon^2}}{i\epsilon} - h,$$

where we have set $\phi_0 = 0$ for the sake of simplicity. Comparing the values of z at $\zeta = -\epsilon^2$ and at $\zeta = 0$, which correspond, respectively, to the stagnation point $z = -h$ and the separation point $z = -h + ik$, we derive the relation

$$k = \frac{\epsilon^2}{q_*} \frac{4 + \pi}{2}$$

between ϵ and k. Formulas (14.79) and (14.80) constitute the desired representation of the Helmholtz flow past a vertical flat plate. Thus we have demonstrated how the hodograph method can be exploited to solve certain free boundary problems explicitly.

To calculate the Riabouchinsky flow past a flat plate C_+ we shall resort to a somewhat more specialized technique that only applies when the fixed boundaries are composed exclusively of vertical or horizontal line segments but which is more direct than the hodograph method in such cases. Our analysis will be based on an investigation of the analytic function

$$(14.81) \qquad g(z) = \frac{1}{q_*^2} \int \zeta'(z)^2 \, dz,$$

which is connected with the force exerted by the fluid on the boundary of the flow region (cf. Exercise 1.2).

Suppose that the boundary has a parametric representation

$$z = z(s)$$

in terms of its arc length s, and let differentiation with respect to s be denoted by a dot. The derivative $w = \zeta'(z)$ is a complex number whose magnitude coincides with the speed q and whose direction is conjugate to that of the flow, which means that

$$\zeta'(z) = q\overline{\dot{z}(s)}$$

at each boundary point, since

$$|\dot{z}|^2 = \dot{z}\overline{\dot{z}} = \dot{x}^2 + \dot{y}^2 = 1.$$

Because of Bernoulli's law (14.76), it follows that

$$(14.82) \qquad g(z) = \frac{1}{q_*^2}\int q^2 \dot{\overline{z}}^2\, dz = \frac{1}{q_*^2}\int q^2\, d\bar{z}$$

$$= \frac{2i}{q_*^2}\int (p - p_0)(dy + i\, dx),$$

where to simplify matters we have taken $\rho = 1$, and where p_0 stands for the stagnation pressure. Observe that the integral on the right represents the complex conjugate of a hydrodynamic force applied at the boundary.

Along the free streamlines Γ we have $q \equiv q_*$, so that (14.82) reduces to

$$g(z) = \int d\bar{z} = \bar{z}$$

there, provided that the constant of integration is selected appropriately. This leads us to introduce the pair of analytic functions

$$(14.83) \qquad \Phi(z) = z + g(z)$$

and

$$(14.84) \qquad \Psi(z) = z - g(z),$$

which have on Γ the values

$$(14.85) \qquad \Phi = z + \bar{z} = 2x$$

and

(14.86)
$$\Psi = z - \bar{z} = 2iy.$$

Since

$$g'(z) = \frac{\zeta'(z)^2}{q_*^2} = \frac{w^2}{q_*^2}$$

is real on the x-axis and on both the vertical line segments C_+ and C_-, we find that

$$\text{Im}\,\{\Phi\} = \text{const.}, \qquad \text{Im}\,\{\Psi\} = \text{const.}$$

along the x-axis, whereas

(14.87)
$$\text{Re}\,\{\Phi\} = \text{const.}, \qquad \text{Re}\,\{\Psi\} = \text{const.}$$

along C_+ and along C_- separately. Finally, note that Φ and Ψ are regular throughout the flow region D and that

$$\Phi \sim \left[1 + \frac{U^2}{q_*^2}\right]z, \qquad \Psi \sim \left[1 - \frac{U^2}{q_*^2}\right]z$$

as $z \to \infty$.

Our remarks concerning the analytic functions Φ and Ψ show that they map the upper half $\psi > 0$ of the Riabouchinsky flow past a flat plate conformally onto infinite polygonal domains. The image of $\psi > 0$ in the Φ-plane is an upper half-plane from which a symmetrically located rectangle has been removed. The two vertical sides of this rectangle correspond to C_+ and C_-, in view of (14.87), whereas the top of the rectangle corresponds to the free streamline Γ, because of (14.85). On the other hand, the image of $\psi > 0$ in the Ψ-plane is an upper half-plane slit along a segment of the imaginary axis. The relations (14.86) and (14.87) imply that C_+, Γ and C_- all map into the slit.

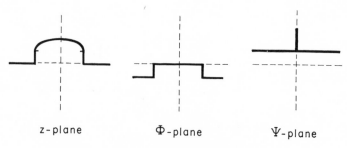

z-plane Φ-plane Ψ-plane

FIG. 72. Maps of the Riabouchinsky flow.

We can now use the Schwarz-Christoffel transformation (cf. Exercise 7.3.7) to map the upper half $\psi > 0$ of the ζ-plane conformally onto the above regions in the Φ-plane and the Ψ-plane. Thus we find that

$$(14.88) \qquad \Phi = \left[\frac{1}{U} + \frac{U}{q_*^2}\right] \int_0^\zeta \frac{\sqrt{t^2 - b^2}}{\sqrt{t^2 - a^2}}\, dt$$

and

$$(14.89) \qquad \Psi = \left[\frac{1}{U} - \frac{U}{q_*^2}\right] \int_0^\zeta \frac{t\, dt}{\sqrt{t^2 - a^2}} + \text{const.}$$

$$= \left[\frac{1}{U} - \frac{U}{q_*^2}\right] \left[\sqrt{\zeta^2 - a^2} - \sqrt{b^2 - a^2}\right] + 2ik,$$

where the singularities at $\zeta = \pm a$ correspond to the stagnation points $z = \pm h$ and where the singularities at $\zeta = \pm b$ correspond to the points of separation $z = \pm h + ik$, with $a > b > 0$.

From (24.83) and (14.84) we see that

$$z = \frac{\Phi + \Psi}{2},$$

whence (14.88) and (14.89) yield the explicit expression

$$(14.90) \quad z = \left[\frac{1}{2U} + \frac{U}{2q_*^2}\right] \int_0^\zeta \frac{\sqrt{t^2 - b^2}}{\sqrt{t^2 - a^2}}\, dt$$

$$+ \left[\frac{1}{2U} - \frac{U}{2q_*^2}\right]\left[\sqrt{\zeta^2 - a^2} - \sqrt{b^2 - a^2}\right] + ik$$

for z in terms of the complex potential ζ. The parameters a and b may be determined from the auxiliary conditions

$$2h = \left[\frac{1}{U} + \frac{U}{q_*^2}\right] \int_0^b \frac{\sqrt{b^2 - t^2}}{\sqrt{a^2 - t^2}}\, dt$$

and

$$2k = \left[\frac{1}{U} + \frac{U}{q_*^2}\right] \int_b^a \frac{\sqrt{t^2 - b^2}}{\sqrt{a^2 - t^2}}\, dt + \left[\frac{1}{U} - \frac{U}{q_*^2}\right]\sqrt{a^2 - b^2}.$$

Thus our conformal mapping technique based on the force integral (14.81) furnishes a closed expression for the Riabouchinsky flow past a vertical plate. Note that our direct appeal to the functions Φ and Ψ avoids the intermediate integration of the ordinary differential equation (14.78) occurring in the hodograph method and makes it almost immediately evident that the answer should merely involve an elliptic integral.

The examples (14.80) and (14.90) suggest that the free streamline problems of classical hydrodynamics which we have described here are well posed. In Section 15.2 we shall present further evidence to that effect by developing a quite general proof of the existence of Riabouchinsky flows, provided merely that the arc of the symmetrical fixed boundary C_+ located in the second quadrant consists of a monotonically ascending curve with non-parametric representations

$$(14.91) \qquad\qquad y = y(x), \qquad x = x(y).$$

Under the same hypothesis we proceed to establish next a theorem which asserts that the solution of the latter problem is also unique. Our demonstration hinges on a comparison argument[11] that exploits an extended form of the maximum principle which appeared in Exercise 7.1.7.

For a given choice of the fixed boundaries C_+ and C_- suppose that two Riabouchinsky flows can be found with stream functions ψ_1 and ψ_2 that are not identical, and let Γ_1 and Γ_2 stand for the arcs in the upper half-plane of the different free streamlines associated with ψ_1 and ψ_2. We denote by D_1 and D_2, respectively, the regions of the upper half-plane that are bounded by Γ_1 and Γ_2 together with portions of C_+ and C_- and two semi-infinite segments of the x-axis. It is understood, of course, that D_j is the domain where $\psi_j > 0$ and that

$$\psi_j = 0$$

on ∂D_j for $j = 1, 2$. To deal with the possibility that the free boundaries Γ_1 and Γ_2 may cross each other, we shall contrive to compare the normal derivatives of ψ_1 and ψ_2 along them by raising one of the flow regions D_1 or D_2 until it just contains the other, so that the maximum principle can be applied.

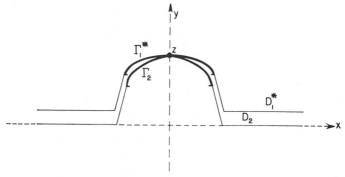

FIG. 73. Uniqueness proof.

[11] Cf. Gilbarg 1, Lavrentiev 1.

Let D_1 be translated upwards until it is just included within D_2, if that is not already the case. Denote by D_1^* and Γ_1^* the translated versions of D_1 and Γ_1, and let ψ_1^* stand for the correspondingly transformed stream function ψ_1. In view of the hypothesis (14.91) concerning the fixed boundaries C_+ and C_-, there must be a point z where the free streamline Γ_2 just touches Γ_1^*. Now observe that ψ_2 is defined throughout D_1^* and is positive there. The maximum principle shows that the harmonic function

$$\psi = \psi_2 - \psi_1^*$$

has to be positive in D_1^*, too, because $\psi_2 \geq \psi_1^*$ on ∂D_1^*, and ψ_1^* and ψ_2 have the same asymptotic behavior

$$\psi_1^* \sim Uy, \qquad \psi_2 \sim Uy$$

at infinity but are not identical. On the other hand, ψ vanishes at the point of contact z of the two free boundaries Γ_1^* and Γ_2. According to Exercise 7.1.7 it follows that

$$\frac{\partial \psi}{\partial \nu} > 0$$

at z, where ν indicates the common inner normal to Γ_1^* and Γ_2 there. Hence

(14.92)
$$\frac{\partial \psi_2}{\partial \nu} > \frac{\partial \psi_1^*}{\partial \nu},$$

which means that the respective speeds q_1 and q_2 of the flows ψ_1 and ψ_2 along the free streamlines Γ_1 and Γ_2 obey the strict inequality

(14.93)
$$q_2 > q_1.$$

By raising D_2 until it is just contained within D_1 and then comparing normal derivatives of the relevant stream functions at a point where the translated free streamline Γ_2 is tangent to Γ_1, we find conversely that

(14.94)
$$q_1 > q_2.$$

Since the relations (14.93) and (14.94) are contradictory, we conclude that the free streamlines Γ_1 and Γ_2 could not have been distinct in the first place and that the Riabouchinsky flow past C_+ and C_- must actually be unique. Our proof depends on the lemma (14.92) concerning comparison of free streamline speeds, and it therefore requires us to assume that Γ_1 and Γ_2 are appropriately differentiable curves which separate smoothly from the fixed boundaries C_+ and C_-. Finally, note that the strict monotonicity of the principal arc of C_+ in the second quadrant played an important role in providing a true point of contact between the translated free streamlines so that no indeterminate speeds on the fixed boundaries could feature in the key inequalities (14.93) and (14.94).

EXERCISES

1. Use the Schwarz-Christoffel transformation (cf. Exercise 7.3.7) to derive (14.79).
2. Show that (14.80) is a limiting case of (14.90).
3. Use the hodograph method to establish that any infinite cavity is shaped asymptotically like a parabola.
4. Use the hodograph method and the Schwarz-Christoffel transformation to find a closed expression for the free streamlines bounding a plane jet, directed parallel to the x-axis, which issues from an orifice between two semi-infinite segments of the y-axis that form the walls of a reservoir covering the left half-plane. Calculate the *discharge coefficient* for this flow, which is defined to be the ratio of the width of the jet at infinity to the length of the interval of the y-axis comprising the orifice.
5. By means of the hodograph transformation investigate the flow of Exercise 4 in the case of a compressible fluid, too.
6. Consider the axially symmetric irrotational motion of an incompressible fluid that flows from the exterior of a semi-infinite pipe of circular cross section into a jet turning back into the pipe, which we shall refer to as *Borda's mouthpiece*.[12] The discharge coefficient C_d of Borda's mouthpiece is defined to be the ratio of the area of the jet at infinity to the area of the circular cross section of the pipe. Prove that

$$C_d = \frac{1}{2}$$

by introducing a force integral analogous to (14.81) and appealing to the law of conservation of momentum.
7. Describe the Riabouchinsky model of axially symmetric flow past a circular disc. Use the fact that the square of the speed is a subharmonic function to establish that the free boundary of this flow appears as a convex curve in the meridian plane.
8. Use the conformal mappings (14.83) and (14.84) to prove that any plane free streamline is an analytic curve. By means of the same device show that the separation of a free streamline from any sufficiently regular fixed boundary is smooth.
9. Prove that the equation of any analytic curve in the z-plane can be expressed in the form

(14.95) $$\bar{z} = g(z),$$

[12] Cf. Lamb 1.

where g is a complex analytic function regular throughout some neighborhood of that curve (cf. Section 7.3). Verify that the complex potential ζ of the flow for which (14.95) features as a free streamline is defined by the formula

$$\zeta(z) = q_* \int^z \sqrt{g'(z)} \, dz,$$

where q_* stands for the constant value of the speed along the free boundary.

10. Use the Riemann function (5.37) to obtain an integral formula in the complex domain (cf. Section 16.2) for the solution of arbitrary analytic Cauchy problems for the elliptic partial differential equation found in Exercise 1.3 to govern the Stokes stream function ψ of any axially symmetric flow. Thus establish that an axially symmetric flow for which the analytic curve (14.95) appears as a free streamline in the meridian plane is defined in that plane by the expression[13]

$$\psi(z, \bar{z}) = \operatorname{Re} \left\{ \frac{q_*}{2i} \int_{z_0}^{z} F\left(\frac{[z - t][\bar{z} - g(t)]}{[z - g(t)][\bar{z} - t]} \right) \sqrt{[z - g(t)][\bar{z} - t]g'(t) \, dt} \right\},$$

where q^* is the speed on the free boundary, z_0 stands for a fixed point there, and F is the hypergeometric function

$$F(w) = F\left(-\frac{1}{2}, -\frac{1}{2}; 1; w \right).$$

11. For any Riabouchinsky flow in the plane, verify that the free streamline speed q_* exceeds the speed at infinity U.

12. Let C denote the closed curve composed of the pair of arcs C_+ and C_- that feature as fixed boundaries in the Riabouchinsky model, together with two horizontal line segments joining their endpoints. Consider a family of Riabouchinsky flows, symmetric in the x-axis and in the y-axis, whose free streamlines separate from and then reattach to the horizontal portions of C. Next, let C_+ and C_- be extended both upward and downward by adjoining vertical segments to their endpoints, and consider the family of symmetrical Riabouchinsky flows whose fixed boundaries consist of such extensions of C_+ and C_-. By means of the comparison theorem (14.92), prove that under the

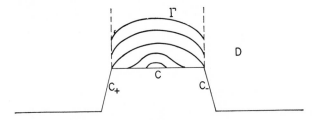

FIG. 74. Family of Riabouchinsky flows.

[13] Cf. Garabedian 3.

hypothesis (14.91) the cavity occurring in the families of Riabouchinsky flows just described depends continuously and monotonically on the dimensionless ratio q_*/U of the free streamline speed to the speed at infinity.[14]

6. MAGNETOHYDRODYNAMICS

We shall take up here a few special topics in magnetohydrodynamics that help to indicate the role of the theory of partial differential equations in plasma physics. What motivates us is the problem of containment of an ionized gas by a strong magnetic field, which is of interest in thermonuclear research. In one idealization that has proved to be quite useful the gas, or plasma, is treated as a perfect conductor. Therefore our principal aim will be to investigate the flow of a perfectly conducting fluid in the presence of a magnetic field.

Magnetohydrodynamics involves coupling Maxwell's equations (cf. Abraham-Becker 1) with the classical equations of gas dynamics. The motion of a perfectly conducting fluid tends to induce electric currents affecting the surrounding magnetic field, which in turn exerts a Lorentz force on the fluid. In order to obtain a mathematical model that is at once reasonable and tractable, we shall neglect displacement currents and electrostatic forces. Consequently we have to ignore the charge, too. Similarly, the effects of viscosity and heat conduction will be omitted. Within the framework of these simplifications, the basic physical principles governing magnetohydrodynamic flow are Faraday's law of induction and Newton's second law of motion, together with the laws of conservation of mass and energy.

Let E and B stand for the electric and magnetic field vectors in space with the rectangular coordinates x, y, z, and let J and u stand for the electric current and the flow velocity. We employ standard vector notation and, more specifically, we set

$$\nabla = \begin{pmatrix} \dfrac{\partial}{\partial x} \\[2mm] \dfrac{\partial}{\partial y} \\[2mm] \dfrac{\partial}{\partial z} \end{pmatrix}, \quad B = \begin{pmatrix} B_1 \\ B_2 \\ B_3 \end{pmatrix}, \quad u = \begin{pmatrix} u_1 \\ u_2 \\ u_3 \end{pmatrix}.$$

Units will be selected so that the magnetic permeability reduces to 1, and a dot will be used to indicate differentiation with respect to the time t.

[14] Cf. Garabedian-Spencer 1.

As far as the pressure p, the density ρ and the entropy S are concerned, we shall adopt the principles of Section 1, including the equation of state (14.5) and the formula $c^2 = dp/d\rho$ for the speed of sound c. According to Ohm's law, our assumption that the fluid is a perfect conductor takes the form

$$(14.96) \qquad E = B \times u,$$

which we shall interpret as a rule for calculating the electric field E when B and u are known. In particular, viewed from a frame of reference moving with the fluid E vanishes. We agree to ignore both the charge density and Poisson's equation for it. Since the displacement current is to be neglected, too, we can omit the term involving $\dot{E}$ from Maxwell's equations to obtain the expression

$$J = \nabla \times B$$

for the current J. The *Lorentz force* per unit volume of fluid therefore becomes

$$(14.97) \qquad J \times B = -B \times (\nabla \times B).$$

The Maxwell equations which we shall have occasion to deal with more directly are

$$(14.98) \qquad \nabla \cdot B = 0$$

and Faraday's law of induction

$$(14.99) \qquad \dot{B} + \nabla \times E = 0.$$

We eliminate E from the latter relation by means of (14.96) to obtain

$$\dot{B} + \nabla \times (B \times u) = 0$$

instead. Putting

$$\frac{d}{dt} = \frac{\partial}{\partial t} + u \cdot \nabla$$

and taking the identity

$$\nabla \times (B \times u) = (\nabla \cdot u)B + (u \cdot \nabla)B - (\nabla \cdot B)u - (B \cdot \nabla)u$$

into account, together with (14.98), we find that (14.99) reduces to

$$(14.100) \qquad \frac{dB}{dt} + (\nabla \cdot u)B - (B \cdot \nabla)u = 0,$$

which is evidently Galilean invariant. In vector form, Newton's second law of motions yields

(14.101) $$\rho \frac{du}{dt} + \nabla p + B \times (\nabla \times B) = 0,$$

where the last term on the left represents the Lorentz force (14.97) exerted on the fluid by the magnetic field. Finally, the continuity equation is

(14.102) $$\frac{c^2}{\rho} \frac{d\rho}{dt} + c^2 \nabla \cdot u = 0,$$

and the law of conservation of energy gives

(14.103) $$\frac{dS}{dt} = 0.$$

In view of (14.99) we have

$$\frac{\partial}{\partial t}(\nabla \cdot B) = \nabla \cdot \dot{B} = -\nabla \cdot (\nabla \times E) = 0,$$

since the divergence of a curl is always zero. Hence we are permitted to interpret (14.98) as an initial condition rather than as an actual equation of motion. The remaining relations (14.100) to (14.103) constitute a system of eight partial differential equations of the first order for B, u, ρ and S, which comprise in all a set of eight scalar unknown functions of the variables x, y, z and t. In magnetohydrodynamics (14.100) to (14.103) are usually referred to as the *Lundquist equations*. We maintain that they are a symmetric hyperbolic system in the sense that they can be expressed in the form

(14.104) $$A_0 U_t + A_1 U_x + A_2 U_y + A_3 U_z = 0,$$

where U indicates a vector of unknowns and where the A_j are symmetric matrices of coefficients depending on U, with A_0 positive-definite (cf. Section 3.5). This will be verified in Exercise 1 below.

Cauchy's problem is well posed for the symmetric hyperbolic system (14.104) when the initial data are prescribed on a space-like manifold (cf. Section 12.2). This observation may be viewed as evidence confirming that the Lundquist equations have been formulated in a reasonable way, despite the omission of displacement currents. Of interest in connection with the initial value problem for (14.104) are, of course, its characteristics, which represent magnetohydrodynamic wave fronts. Let $C(t)$ stand for the surface in (x,y,z)-space consisting of those points of a characteristic manifold that correspond to the time t. We indicate by ν the unit normal to $C(t)$, and we introduce the characteristic speed

$$b = \frac{d\nu}{dt},$$

which specifies the rate at which the wave front $C(t)$ advances in the normal direction. Let

$$b = u \cdot \nu \pm a,$$

where $u \cdot \nu$ is the normal component of the velocity of the fluid and $a > 0$. We shall proceed to calculate a, which describes the speed of the wave relative to that of the flow.

A characteristic of equation (14.104) is a manifold along which the normal derivative of U cannot be determined from the equation and the values of U itself (cf. Chapter 3). Therefore the matrix of coefficients multiplying the normal derivative of U in (14.104) must have determinant zero on a characteristic. Since the normal to the characteristic associated with the surface $C(t)$ has components of the form x_ν, y_ν, z_ν and b, we conclude that

$$|bA_0 + x_\nu A_1 + y_\nu A_2 + z_\nu A_3| = 0.$$

To evaluate this requirement more explicitly, we choose coordinates so that ν becomes parallel to the x-axis, whence

$$x_\nu = 1, \qquad y_\nu = z_\nu = 0.$$

Moreover, because we wish merely to calculate a, it will suffice to treat the special case in which u vanishes at the point of $C(t)$ under consideration. We thus derive for a the equation

$$\begin{vmatrix} a & 0 & 0 & 0 & 0 & 0 & 0 & 0 \\ 0 & a & 0 & B_2 & -B_1 & 0 & 0 & 0 \\ 0 & 0 & a & B_3 & 0 & -B_1 & 0 & 0 \\ 0 & B_2 & B_3 & \rho a & 0 & 0 & c^2 & 0 \\ 0 & -B_1 & 0 & 0 & \rho a & 0 & 0 & 0 \\ 0 & 0 & -B_1 & 0 & 0 & \rho a & 0 & 0 \\ 0 & 0 & 0 & c^2 & 0 & 0 & \dfrac{c^2 a}{\rho} & 0 \\ 0 & 0 & 0 & 0 & 0 & 0 & 0 & a \end{vmatrix} = 0.$$

Elementary manipulation of the above determinant yields the quartic equation

$$c^2 a^2 (\rho a^2 - B_1^2)[\rho a^4 - (\rho c^2 + B_1^2 + B_2^2 + B_3^2)a^2 + c^2 B_1^2] = 0$$

for a^2. Denoting the normal component of the magnetic field by B_ν instead of B_1 and putting

$$B^2 = B_1^2 + B_2^2 + B_3^2,$$

we arrive at the more transparent formula

(14.105) $c^2 a^2 (\rho a^2 - B_\nu^2)[\rho a^4 - (\rho c^2 + B^2)a^2 + c^2 B_\nu^2] = 0.$

This is a polynomial equation of the eighth degree whose roots are all real. The non-negative roots of the first two factors are

$$a = 0, \qquad a = \sqrt{B_\nu^2/\rho}.$$

We shall denote by a_f and a_s, respectively, the larger and the smaller positive roots of the last factor, and we shall refer to them as the *fast* and *slow characteristic speeds*.

Obviously a_f and a_s satisfy the relation

$$(a^2 - c^2)(\rho a^2 - B_\nu^2) = a^2(B^2 - B_\nu^2) \geq 0,$$

and therefore

$$a_s \leq c \leq a_f,$$

$$a_s \leq \sqrt{B_\nu^2/\rho} \leq a_f.$$

Thus the fast and slow speeds at which magnetohydrodynamic waves can propagate straddle the speed of sound c and the root $\sqrt{B_\nu^2/\rho}$ of (14.105). In this connection the case where the magnetic field B becomes parallel to the direction of propagation ν is of special interest, for then

$$\sqrt{B_\nu^2/\rho} = \sqrt{B^2/\rho}.$$

The quantity appearing on the right here is known as the *Alfvén speed*, and for an incompressible fluid the waves associated with it are called *Alfvén waves*.

In a somewhat different context we wish to describe a problem of *hydromagnetic equilibrium* in which there occur free boundaries quite analogous to those of Section 5. In an equilibrium configuration $u \equiv 0$, and therefore we must also take $E \equiv 0$ in order to be consistent with our assumption (14.96) of a perfectly conducting fluid. Thus for equilibrium the Lundquist equations (14.100) to (14.103) reduce to the system

$$(14.106) \qquad\qquad \nabla \cdot B = 0,$$

$$(14.107) \qquad\qquad \nabla p + B \times (\nabla \times B) = 0$$

of four partial differential equations for the four unknowns B_1, B_2, B_3 and p. Since we shall be interested in deriving boundary conditions to be imposed at the interface between two distinct media, our first aim will be to convert (14.106) and (14.107) into integral relations that will help us to study the jumps of B and p across such a surface of discontinuity.

An application of the divergence theorem to (14.106) over an arbitrary region D shows that

$$(14.108) \qquad\qquad \iint_{\partial D} B \cdot \nu \, d\sigma = 0,$$

where $d\sigma$ stands for the area element on the closed surface ∂D bounding D. A rearrangement of (14.107) gives

$$\nabla(p + \tfrac{1}{2}B^2) - (B \cdot \nabla)B = 0,$$

to which we add an appropriate multiple of (14.106) to obtain

$$\nabla(p + \tfrac{1}{2}B^2) - (B \cdot \nabla)B - (\nabla \cdot B)B = 0.$$

Another application of the divergence theorem therefore yields

$$(14.109) \qquad\qquad \iint_{\partial D} [(p + \tfrac{1}{2}B^2)\nu - (B \cdot \nu)B] \, d\sigma = 0.$$

Introducing the *Maxwell stress tensor*

$$T = \begin{pmatrix} T_{11} & T_{12} & T_{13} \\ T_{21} & T_{22} & T_{23} \\ T_{31} & T_{32} & T_{33} \end{pmatrix},$$

where

$$T_{ij} = B_i B_j - \tfrac{1}{2}\delta_{ij}B^2,$$

we can restate the latter equation in the form

$$(14.110) \qquad\qquad \iint_{\partial D} (p\nu - T \cdot \nu) \, d\sigma = 0.$$

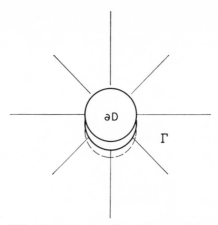

FIG. 75. Derivation of boundary conditions.

The physical meaning of (14.110) is simply that the total force exerted on D by the fluid pressure p and the magnetic field B is zero.

Now consider the interface Γ between two different fluids that are in equilibrium. Let D be a thin disc whose broader flat surfaces are located on opposite sides of Γ, while its narrower curved edge forms a band cutting through Γ. Both (14.108) and (14.109) are valid for such a domain because we can view the discontinuities of B and p across Γ as the result of a limit process performed on corresponding quantities for a single medium whose properties change rapidly, but continuously, in a neighborhood of Γ. We may eliminate from (14.108) and (14.109) the contributions involving integration over the narrow band of ∂D perpendicular to Γ by allowing D to collapse into that interface. Since the two sides of ∂D that are left can be taken as small as we please, the jumps of the integrands in (14.108) and (14.109) across Γ must vanish. It follows, in particular, that both $B \cdot \nu$ and $p + \frac{1}{2}B^2$ remain continuous as we pass through the interface between two perfectly conducting fluids in equilibrium.

We shall be concerned with a perfectly conducting fluid that is contained inside a vacuum by a magnetic field. It will be assumed that no current or field is present in the fluid, which takes the form of a *figure of equilibrium* throughout which the pressure p_0 is constant. Of course, no current or fluid pressure exists in the vacuum. Hence the vacuum magnetic field B satisfies the partial differential equations

(14.111) $$\nabla \cdot B = 0, \quad \nabla \times B = 0.$$

Although surface currents may occur at the interface Γ separating the fluid from the vacuum, the expressions $B \cdot \nu$ and $p + \frac{1}{2}B^2$ must, as we have seen, be continuous across Γ. Thus the vacuum field fulfills the boundary condition

(14.112)' $$B \cdot \nu = 0$$

on Γ because B vanishes inside the fluid. In addition we have the balance

(14.113) $$\tfrac{1}{2}B^2 = p_0$$

between the *magnetic pressure* $\tfrac{1}{2}B^2$ on Γ and the constant value p_0 of the fluid pressure, since p reduces to zero in the vacuum (cf. Exercise 6 below). The requirement (14.113) can be interpreted as a free boundary condition that might serve to determine the shape and location of the interface Γ in a specific example.

It is instructive to examine in more detail the case where B depends only on x and y, and

$$B_3 \equiv 0.$$

Under these hypotheses (14.111) reduces to the Cauchy-Riemann equations for

(14.114) $$w = B_1 - iB_2,$$

which is therefore an analytic function of the complex variable $x + iy$. It is natural to introduce the complex potential

$$\zeta = \phi + i\psi = \int w(dx + i\,dy),$$

which is also analytic. After a suitable choice of the additive constant implicit in the definition of ψ, the boundary conditions (14.112) and (14.113) holding on Γ acquire the simple form

(14.115) $$\psi = 0, \qquad |w|^2 = 2p_0 = \text{const.}$$

Thus the determination of plane figures of equilibrium for a perfectly conducting fluid contained by a vacuum magnetic field is a free boundary problem of precisely the kind we studied in Section 5.

A specific model that is of practical importance occurs when the magnetic field (14.114) has the form of a quadripole

$$w = x + iy + \sum_{n=1}^{\infty} \frac{a_n}{(x + iy)^{2n+1}}$$

at infinity but is regular elsewhere. In that case the fluid is found to lie in a plane region of finite extent bounded by four concave arcs terminating in

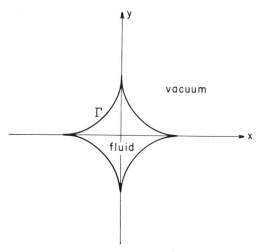

FIG. 76. Cusped geometry.

cusps at the coordinate axes. Such a figure is usually said to have *cusped geometry*.[15] An application of the hodograph method explained in Section 5 to the present example shows that its free boundary Γ is the hypocycloid

$$(14.116) \qquad x^{2/3} + y^{2/3} = \left(\frac{32p_0}{9}\right)^{1/3}.$$

In Section 15.3 we shall pursue our investigation of this model further and shall develop a minimum energy principle in order to prove that cusped figures like (14.116) are in *stable* equilibrium.

EXERCISES

1. Verify that the Lundquist equations (14.100) to (14.103) may be expressed as a symmetric hyperbolic system (14.104) by determining the relevant matrices of coefficients A_j explicitly.

2. Establish the explicit formula (14.116) for the free boundary of a cusped figure of equilibrium by means of the hodograph method or the conformal mapping technique based on (14.81).

3. Use the hodograph method to find a one-parameter family of plane figures of equilibrium analogous to (14.116) contained by a surrounding magnetic field

$$B_1 - iB_2 = \left(\frac{\partial}{\partial y} + i\frac{\partial}{\partial x}\right) \log \left| \frac{(x + iy)^2 - (h - ih)^2}{(x + iy)^2 - (h + ih)^2} \right| + \cdots$$

[15] Cf. Grad 1.

that has four symmetrically located poles of equal strengths but alternating signs instead of a single quadripole.

4. Show that a plane magnetic field with the asymptotic expansion

$$w \sim e^{ix-y}$$

at infinity holds in place a figure of equilibrium whose free boundary consists of a picket fence of inverted cycloids.

5. Formulate the cusped geometry model of a perfectly conducting spindle of fluid that is contained in equilibrium by an axially symmetric magnetic field.

6. Interpret the magnetic pressure $\frac{1}{2}B^2$ acting on the free surface of a figure of equilibrium composed of perfectly conducting fluid as a Lorentz force

$$K \times \frac{B}{2} = \frac{1}{2}(v \times B) \times B = -\frac{1}{2}B^2 v$$

due to a surface current $K = v \times B$ and an average magnetic field $(B + 0)/2$ present in the interface between the fluid and the surrounding vacuum.

7. Show that the total potential energy M of the physical system consisting of a perfectly conducting liquid held in a vacuum by a magnetic field is

(14.117) $$M = \frac{1}{2} \iiint_D B^2 \, dx \, dy \, dz,$$

where the region D represents the vacuum.

8. Develop an analogy between the equations (14.106) and (14.107) describing hydromagnetic equilibrium and Euler's equations of motion for the steady flow of an inviscid, incompressible fluid.

9. Compare the Euler equations for two-dimensional, time-dependent flow with the Lundquist equations when B, u, ρ and S are independent of z and

$$B_1 \equiv B_2 \equiv u_3 \equiv 0.$$

10. Develop a theory of simple waves for the Lundquist equations.

11. Define what is meant by a characteristic conoid for the Lundquist equations and plot a graph describing its geometry in terms of the angle between the field vector B and the direction of propagation v.

12. Show that in formulating the equilibrium equations (14.106) and (14.107) it is not necessary to make assumptions about neglecting the displacement current and the charge because the electric field E vanishes identically anyway.

15

Free Boundary Problems

1. HADAMARD'S VARIATIONAL FORMULA

It has become apparent that most boundary value problems of mathematical physics are too difficult to solve in closed form, and even the information provided by approximate or numerical treatments does not always turn out to be completely satisfactory. Thus it is important to study the qualitative properties of solutions. Perturbation methods prove to be quite a useful tool in any such analysis.

As an example of what is involved here we shall be interested in examining the variation of the Green's function of Laplace's equation (cf. Section 7.2) under perturbations of its domain of definition. More specifically, we intend to derive the so-called *Hadamard variational formula*, which represents the first order term in the expansion of the Green's function in powers of a small parameter describing an infinitesimal displacement of the boundary of the domain.[1] This formula has many applications in the calculus of variations, and it suggests the answers to a variety of extremal problems that are basic to an understanding of the behavior of solutions of linear elliptic partial differential equations.

One of the most important applications of Hadamard's variational formula occurs in the theory of the free boundary problems of hydrodynamics and of plasma physics that were introduced in Sections 14.5 and 14.6. It will be established presently that every constant pressure free streamline can be characterized as an extremal within an appropriate family of curves bounding flow regions whose shape is to be adjusted so that the Dirichlet integral defining the kinetic energy of the corresponding motion becomes a minimum. In this context it is Hadamard's formula

[1] Cf. Hadamard 1.

that explains why the pressure ought to be constant along the extremal free boundary. In practice the minimum energy principle to which we have just referred furnishes both a method for proving the mathematical existence and uniqueness of free streamline flows (cf. Section 2) and a procedure for testing the stability of figures of equilibrium in magneto-hydrodynamics (cf. Section 3).

We start by giving a more or less heuristic derivation of Hadamard's variational formula for the harmonic Green's function G of a plane domain D. Our argument will be rigorously justifiable only if the boundary ∂D of D consists of an analytic simple closed curve. Let D^* be the region whose boundary ∂D^* is obtained by shifting ∂D an infinitesimal distance

$$(15.1) \qquad \delta \nu = \epsilon \rho(s)$$

along its inner normal ν, where ϵ is a small number and ρ is a real analytic function of the arc length s on ∂D. It is to be understood that points of ∂D where $\delta \nu$ is negative are moved in the direction of the outer normal instead of the inner normal. Our objective will be to find the first order term δG in the expansion of the harmonic Green's function G^* of D^* in powers of the parameter ϵ.

Consider the identity

$$(15.2) \qquad \iint\limits_{D-D^*} \nabla G \cdot \nabla G^* \, dx_1 \, dx_2 = \int_{\partial D} \frac{\partial G}{\partial \nu} \frac{\partial G^*}{\partial \nu} \delta \nu \, ds + O(\epsilon^2),$$

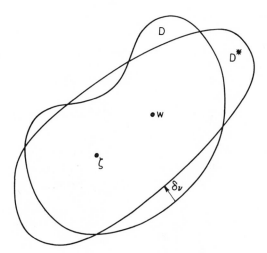

FIG. 77. Variation of the domain.

which is a consequence of (15.1) and the fact that G vanishes on ∂D, so that

$$\frac{\partial G}{\partial s} = 0$$

there. It is assumed that the integral on the left in (15.2) is evaluated over $D^* - D$ instead of $D - D^*$, but with a reversed sign, at points associated with negative values of the normal displacement $\delta \nu$ of ∂D. Moreover, we suppose that ϵ is so small that the analytic extensions of G and G^* beyond the boundary curves ∂D and ∂D^*, which can be found from the Schwarz principle of reflection (cf. Chapter 7), are defined in a fixed region including the closures of both D and D^*.

We use complex numbers z, ζ, w, $\cdots$ to indicate points inside the domain D, with, for example, $z = x_1 + ix_2$, and we adopt the normalization

$$G(z, \zeta) = \frac{1}{2\pi} \log \frac{1}{|z - \zeta|} + \cdots$$

of the singularity of the Green's function. Thus Green's theorem gives

$$\iint_D \nabla G \cdot \nabla G^* \, dx_1 \, dx_2 = -\int_{\partial D} G \frac{\partial G^*}{\partial \nu} \, ds + G(w, \zeta) = G(w, \zeta)$$

and

$$\iint_{D^*} \nabla G \cdot \nabla G^* \, dx_1 \, dx_2 = -\int_{\partial D^*} G^* \frac{\partial G}{\partial \nu} \, ds + G^*(\zeta, w) = G^*(\zeta, w),$$

where

$$G = G(z, \zeta), \qquad G^* = G^*(z, w).$$

Comparing these results with (15.2), we conclude that

$$G^*(\zeta, w) - G(\zeta, w) = -\int_{\partial D} \frac{\partial G}{\partial \nu} \frac{\partial G^*}{\partial \nu} \, \delta \nu \, ds + O(\epsilon^2)$$

because of the symmetry of G. Since the difference between G^* and G is of the order of magnitude ϵ, we can replace the normal derivative $\partial G^*/\partial \nu$ on the right by $\partial G/\partial \nu$ to establish the announced Hadamard formula

$$(15.3) \qquad \delta G(\zeta, w) = -\int_{\partial D} \frac{\partial G(z, \zeta)}{\partial \nu} \frac{\partial G(z, w)}{\partial \nu} \, \delta \nu \, ds$$

for the first variation δG of the Green's function G.

A more rigorous mathematical analysis of Hadamard's formula requiring fewer assumptions about the boundary ∂D can be made by introducing the notion of *interior variations*[2] of the domain D. Let

$$S_j = S_j(x_1, x_2), \qquad j = 1, 2,$$

stand for a pair of functions possessing piece-wise continuous partial
derivatives of a conveniently high order in some neighborhood of the
closure of D. For small enough choices of the parameter ϵ the transformation

$$(15.4) \qquad\qquad x_j^* = x_j + \epsilon S_j, \qquad j = 1, 2,$$

is one-to-one in D and maps that region onto a nearby domain D^*.
We refer to (15.4) as an interior variation because it is defined inside D as
well as on the boundary curves ∂D, where, incidentally, it behaves quite
smoothly. Our intention is to develop the Green's function G^* of D^* as a
perturbation series in powers of ϵ. For that purpose we shall use the
change of variables (15.4) to express it as a function of x_1 and x_2 over the
original region D, where it will be found to satisfy a linear elliptic partial
differential equation with coefficients involving ϵ. The variation of G^*
under changes in these coefficients will suffice to specify in a precise
fashion how the Green's function depends on its domain of definition.

It will facilitate our calculations to employ the complex notation

$$(15.5) \qquad\qquad z^* = z + \epsilon F(z, \bar{z})$$

for the interior variation (15.4), where

$$(15.6) \qquad\qquad F(z, \bar{z}) = S_1(x_1, x_2) + i S_2(x_1, x_2).$$

Similarly, we put

$$\zeta^* = \zeta + \epsilon F(\zeta, \bar{\zeta}), \qquad w^* = w + \epsilon F(w, \bar{w}).$$

Our aim is to determine a partial differential equation in D for the transformed Green's function

$$g(z, \zeta; \epsilon) = G^*(z^*, \zeta^*),$$

whose principal arguments are the real and imaginary parts x_1 and x_2 of the
variable z.

Since Laplace's equation

$$\Delta G^* = 0$$

[2] Cf. Garabedian-Schiffer 2, Schiffer 2.

is the Euler equation for an extremal problem of the form

$$\iint \left[\sum_{j=1}^{2} \left(\frac{\partial G^*}{\partial x_j^*} \right)^2 \right] dx_1^* \, dx_2^* = \text{minimum}$$

in the calculus of variations, g must satisfy the Euler equation of the transformed problem

$$\iint \left[\sum_{j=1}^{2} \left(\sum_{k=1}^{2} \frac{\partial g}{\partial x_k} \frac{\partial x_k}{\partial x_j^*} \right)^2 \right] \frac{\partial(x_1^*, x_2^*)}{\partial(x_1, x_2)} \, dx_1 \, dx_2 = \text{minimum}.$$

Standard procedures show that the latter equation is

(15.7) $L_\epsilon[g] = 0,$

where L_ϵ represents the self-adjoint differential operator defined by

(15.8) $L_\epsilon[g] = \sum_{k,l=1}^{2} \frac{\partial}{\partial x_k} \left(A_{kl} \frac{\partial g}{\partial x_l} \right),$

with coefficients

$$A_{kl} = \frac{\partial(x_1^*, x_2^*)}{\partial(x_1, x_2)} \sum_{j=1}^{2} \frac{\partial x_k}{\partial x_j^*} \frac{\partial x_l}{\partial x_j^*}$$

depending on the parameter ϵ. Note, incidentally, that

(15.9) $\begin{vmatrix} A_{11} & A_{12} \\ A_{21} & A_{22} \end{vmatrix} = 1.$

We intend to derive a Fredholm integral equation (cf. Section 10.1) for g which will enable us to expand it in a Neumann series each of whose terms is multiplied by a successively higher power of ϵ. The integral equation in question will be based on a parametrix (cf. Section 5.3) for the linear elliptic partial differential equation (15.7). For present purposes we shall need the inverse matrix of elements $a_{jk} = a_{jk}(z)$ such that

$$\sum_{k=1}^{2} a_{jk} A_{kl} = \delta_{jl}.$$

Let us introduce the quadratic form

(15.10) $\Gamma(z, \zeta) = \sum_{j,k=1}^{2} a_{jk}(\zeta)(x_j - \xi_j)(x_k - \xi_k),$

where ξ_1 and ξ_2 stand for the real and imaginary parts of ζ. We denote by $\alpha = \alpha(z, \zeta)$ a fixed function of the two points z and ζ in D which possesses continuous derivatives of an appropriate order, which fulfills the boundary condition $\alpha(z, \zeta) = 0$ when either z or ζ lies on ∂D, but which has the value

$$\alpha(\zeta, \zeta) = \frac{1}{4\pi}$$

when z coincides with ζ inside D. The expression

$$P_\epsilon(z, \zeta) = \alpha(z, \zeta) \log \frac{1}{\Gamma(z, \zeta)}$$

then defines the required parametrix P_ϵ.

Direct computation shows that

$$(15.11) \qquad L_\epsilon[P_\epsilon] = O\left(\frac{1}{r}\right),$$

where $r = |z - \zeta|$ is the distance between the points z and ζ. Thus, in particular, the quantity on the left in (15.11) is integrable over the domain D. In a similar way it is found that

$$(15.12) \qquad L_\epsilon[P_\epsilon] - L_0[P_0] = O\left(\frac{\epsilon}{r}\right).$$

Finally, we shall also have occasion in what follows to exploit the estimate

$$(15.13) \qquad L_\epsilon[G - P_0] - L_0[G - P_0] = O\left(\frac{\epsilon}{r}\right),$$

which is obviously valid for reasonable choices of the factor α.

For the self-adjoint differential operator (15.8) Green's identity (5.65) takes the form

$$(15.14) \qquad \iint_D \{v L_\epsilon[u] - u L_\epsilon[v]\}\, dx_1\, dx_2 = \int_{\partial D} B_\epsilon[u, v]\, ds,$$

where

$$B_\epsilon[u, v] = \sum_{k,l=1}^{2} A_{kl} \left\{ u\, \frac{\partial v}{\partial x_k} \frac{\partial x_l}{\partial \nu} - v\, \frac{\partial u}{\partial x_k} \frac{\partial x_l}{\partial \nu} \right\}.$$

We put

$$u = G(z, \zeta) - P_0(z, \zeta) + P_\epsilon(z, \zeta)$$

and

$$v = g(z, w; \epsilon)$$

in (15.14), after removing from D small circles about the singular points ζ and w which are later allowed to disappear. The limits of the contributions from these circles may be evaluated in a standard fashion (cf. the derivation of (5.68) in Section 5.2), with note made that g has the same logarithmic infinity as P_ϵ and that

$$\begin{vmatrix} a_{11} & a_{12} \\ a_{21} & a_{22} \end{vmatrix} = 1$$

because of (15.9). We are thus led to the representation

$$(15.15) \qquad g(\zeta, w; \epsilon) = G(w, \zeta) - P_0(w, \zeta) + P_\epsilon(w, \zeta)$$

$$+ \iint\limits_{D} g(z, w; \epsilon) L_\epsilon[G - P_0 + P_\epsilon]\, dx_1\, dx_2$$

of the transformed Green's function g, since the equation (15.7) holds throughout D and the boundary conditions

$$g = G = P_\epsilon = P_0 = 0$$

apply along ∂D.

We interpret (15.15) as a Fredholm integral equation for g with the known kernel

$$K(z, \zeta; \epsilon) = L_\epsilon[G(z, \zeta) - P_0(z, \zeta) + P_\epsilon(z, \zeta)].$$

The relations (15.12) and (15.13) provide the simple restriction

$$K = O\left(\frac{\epsilon}{r}\right)$$

on the growth of K, which shows that it is integrable and that the Neumann series (cf. Section 10.1) obtained by iterating (15.15) has to converge for every sufficiently small value of the parameter ϵ. Moreover, each successive term of the series contains the factor ϵ more times than does its predecessor. We conclude, in particular, that

$$g - G = O(\epsilon).$$

Hence we can replace g by G on the right in (15.15) to derive the perturbation formula

$$(15.16) \quad G^*(\zeta^*, w^*) - G(\zeta, w) = P_\epsilon(w, \zeta) - P_0(w, \zeta)$$

$$+ \iint\limits_{D} G(z, w) K(z, \zeta; \epsilon)\, dx_1\, dx_2 + O(\epsilon^2)$$

for the Green's function, where on the left the definition of g and the symmetry of G have been taken into account.

Our purpose will be to simplify (15.16) and to develop from it a rigorous verification of the Hadamard variational formula (15.3). Let

$$\delta A_{jk} = \frac{\partial A_{jk}}{\partial \epsilon} \epsilon,$$

where the derivative on the right is to be evaluated at $\epsilon = 0$. An application of Green's theorem shows that

$$(15.17) \quad \iint_D G(z, w) K(z, \zeta; \epsilon) \, dx_1 \, dx_2 = G(\zeta, w)$$

$$- \iint_D \sum_{j,k=1}^{2} A_{jk} \frac{\partial G}{\partial x_j} \frac{\partial}{\partial x_k} [G - P_0 + P_\epsilon] \, dx_1 \, dx_2$$

$$= G(w, \zeta) - \iint_D \sum_{j=1}^{2} \frac{\partial G}{\partial x_j} \frac{\partial}{\partial x_j} [G - P_0 - P_\epsilon] \, dx_1 \, dx_2$$

$$- \iint_D \sum_{j,k=1}^{2} \delta A_{jk} \frac{\partial G}{\partial x_j} \frac{\partial}{\partial x_k} [G - P_0 + P_\epsilon] \, dx_1 \, dx_2 + O(\epsilon^2),$$

since $A_{jk} = \delta_{jk}$ when $\epsilon = 0$. On the other hand, an additional integration by parts gives

$$\iint_D \sum_{j=1}^{2} \frac{\partial G}{\partial x_j} \frac{\partial}{\partial x_j} [G - P_0 + P_\epsilon] \, dx_1 \, dx_2 = G(w, \zeta) - P_0(w, \zeta) + P_\epsilon(w, \zeta),$$

and it is evident that

$$\iint_D \sum_{j,k=1}^{2} \delta A_{jk} \frac{\partial G}{\partial x_j} \frac{\partial}{\partial x_k} [P_\epsilon - P_0] \, dx_1 \, dx_2 = O(\epsilon^2).$$

Therefore (15.16) combines with (15.17) to yield the more elegant variational formula

$$(15.18)$$

$$G^*(\zeta^*, w^*) - G(\zeta, w) = - \iint_D \sum_{j,k=1}^{2} \delta A_{jk} \frac{\partial G(z, \zeta)}{\partial x_j} \frac{\partial G(z, w)}{\partial x_k} \, dx_1 \, dx_2 + O(\epsilon^2),$$

from which the parametrix P_ϵ has been eliminated.

By direct calculation we find that

$$\frac{\partial x_k}{\partial x_j^*} = \delta_{jk} - \epsilon \frac{\partial S_k}{\partial x_j} + O(\epsilon^2)$$

and

$$\frac{\partial(x_1^*, x_2^*)}{\partial(x_1, x_2)} = 1 + \epsilon \left\{\frac{\partial S_1}{\partial x_1} + \frac{\partial S_2}{\partial x_2}\right\} + O(\epsilon^2),$$

whence

(15.19) $$\delta A_{jk} = \epsilon \left\{\delta_{jk}\left(\frac{\partial S_1}{\partial x_1} + \frac{\partial S_2}{\partial x_2}\right) - \frac{\partial S_j}{\partial x_k} - \frac{\partial S_k}{\partial x_j}\right\}.$$

Substituting (15.19) into (15.18) and rearranging terms appropriately, we obtain

(15.20) $$G^*(\zeta^*, w^*) - G(\zeta, w) = \epsilon \iint_D \sum_{j,k=1}^{2} T_{jk} \frac{\partial S_j}{\partial x_k} dx_1 dx_2 + O(\epsilon^2),$$

where

$$T_{jk} = T_{jk}(z; \zeta, w) = \frac{\partial G(z, \zeta)}{\partial x_j}\frac{\partial G(z, w)}{\partial x_k} + \frac{\partial G(z, w)}{\partial x_j}\frac{\partial G(z, \zeta)}{\partial x_k}$$

$$- \delta_{jk}\left\{\frac{\partial G(z, \zeta)}{\partial x_1}\frac{\partial G(z, w)}{\partial x_1} + \frac{\partial G(z,\zeta)}{\partial x_2}\frac{\partial G(z, w)}{\partial x_2}\right\}$$

is a generalization of Maxwell's stress tensor (cf. Section 14.6). Because G is a harmonic function of the variables x_1 and x_2, the identities

$$\sum_{k=1}^{2} \frac{\partial T_{jk}}{\partial x_k} = 0, \quad j = 1, 2,$$

are seen to hold throughout D. Thus an integration by parts enables us to bring (15.20) into the form

(15.21)

$$G^*(\zeta^*, w^*) - G(\zeta, w) = \epsilon\sum_{j=1}^{2}\left[\frac{\partial G(\zeta, w)}{\partial \xi_j}S_j(\xi_1, \xi_2) + \frac{\partial G(\zeta, w)}{\partial u_j}S_j(u_1, u_2)\right]$$

$$-\epsilon\int_{\partial D}\sum_{j,k=1}^{2}T_{jk}(z; \zeta, w)S_j(x_1, x_2)\frac{\partial x_k}{\partial \nu}ds + O(\epsilon^2)$$

involving only a boundary integral, where u_1 and u_2 stand for the real and imaginary parts of w, and where the residual terms on the right evaluated at the points ζ and w result from the singularities of the tensor T_{jk} there.

According to Taylor's theorem we have

$$G^*(\zeta^*, w^*) - G^*(\zeta, w)$$

$$= \epsilon \sum_{j=1}^{2} \left[\frac{\partial G^*(\zeta, w)}{\partial \xi_j} S_j(\xi_1, \xi_2) + \frac{\partial G^*(\zeta, w)}{\partial u_j} S_j(u_1, u_2) \right] + O(\epsilon^2).$$

Moreover, G^* can be replaced by G on the right, because of (15.21). Therefore (15.21) actually reduces to the simpler formula

$$(15.22) \qquad \delta G = -\epsilon \int_{\partial D} \sum_{j,k=1}^{2} T_{jk} S_j \frac{\partial x_k}{\partial \nu} \, ds$$

for the first order perturbation

$$\delta G = \epsilon \lim_{\epsilon \to 0} \frac{G^*(\zeta, w) - G(\zeta, w)}{\epsilon}$$

of the Green's function G. It is natural to ascribe to (15.22) the physical interpretation that the variation δG in potential energy, viewed as work done, must be proportional to a force, represented by the stress tensor T_{jk}, times the displacement defined by the vector (S_1, S_2). Finally, if we insert the expression

$$\rho = S_1 \frac{\partial x_1}{\partial \nu} + S_2 \frac{\partial x_2}{\partial \nu}$$

for the normal shift of the boundary curves ∂D into (15.1) and exploit the boundary condition $G = 0$ when calculating T_{jk} along ∂D, we find that (15.22) is equivalent to Hadamard's variational formula (15.3).

The importance of our derivation of the perturbation formula (15.3) by means of the Fredholm integral equation (15.15) is that it requires only a certain degree of differentiability of the boundary curves ∂D and of their infinitesimal displacement (15.1), rather than analyticity of those quantities. Moreover, the Neumann series solution of (15.15) enables us to expand the Green's function G^* of the varied domain D^* as a power series in ϵ whose radius of convergence is positive. Thus we have at our disposal a method for determining higher variations of the Green's function. The main significance of perturbation formulas like (15.3) and (15.20) in practice is their usefulness for the investigation of maximum and minimum problems in the calculus of variations. In this connection we wish to close by describing a technique that makes it feasible to apply interior variations (15.4) to an extremal domain D whose boundary ∂D does not fulfill any smoothness hypotheses whatever.

To motivate the discussion we introduce the complex notation (15.6) into (15.20) and obtain

$$(15.23) \quad G^*(\zeta^*, w^*) - G(\zeta, w)$$

$$= \mathrm{Re} \left\{ 8\epsilon \iint_D \frac{\partial G(z, \zeta)}{\partial z} \frac{\partial G(z, w)}{\partial z} \frac{\partial F(z, \bar{z})}{\partial \bar{z}} \, dx_1 \, dx_2 \right\} + O(\epsilon^2),$$

because

$$4 \, \mathrm{Re} \left\{ \frac{\partial G(z, \zeta)}{\partial z} \frac{\partial G(z, w)}{\partial z} \right\} = T_{11} = -T_{22},$$

$$4 \, \mathrm{Im} \left\{ \frac{\partial G(z, \zeta)}{\partial z} \frac{\partial G(z, w)}{\partial z} \right\} = -T_{12} = -T_{21},$$

and

$$2 \frac{\partial F}{\partial \bar{z}} = \frac{\partial S_1}{\partial x_1} - \frac{\partial S_2}{\partial x_2} + i \left(\frac{\partial S_1}{\partial x_2} + \frac{\partial S_2}{\partial x_1} \right).$$

In particular, when F is an analytic function of the complex variable $z = x_1 + ix_2$, the Cauchy-Riemann equations

$$\frac{\partial F}{\partial \bar{z}} = 0$$

convert (15.23) into an extraordinary verification of the invariance of the harmonic Green's function G under conformal mapping. This suggests that we may be able to avoid difficulties in assessing the effect of the interior variation

$$(15.5) \qquad z^* = z + \epsilon F(z, \bar{z})$$

in the neighborhood of an irregular boundary ∂D by choosing[3] the function F to be complex analytic there, but by at the same time allowing it to possess poles or other singularities inside the domain D.

Let t be a fixed point in the interior of D, and let R be a positive number smaller than the distance from t to ∂D. We shall investigate in detail the special case of (15.5) where F is the continuous function defined by

$$(15.24) \qquad F(z, \bar{z}) = \begin{cases} \dfrac{1}{z - t}, & |z - t| \geq R, \\[2ex] \dfrac{\bar{z} - \bar{t}}{R^2}, & |z - t| \leq R. \end{cases}$$

[3] Cf. Schiffer 1.

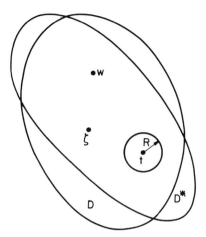

FIG. 78. Interior variation.

We emphasize that the resulting perturbation of the domain D actually consists in an application of the quite specific transformation

$$(15.25) \qquad\qquad z^* = z + \frac{\epsilon}{z - t}$$

to the boundary ∂D, where we may take the parameter ϵ to be complex.

Observe that the mapping (15.25) does not involve R and is conformal near ∂D, but exhibits a pole at the point t inside D. Since (15.24) can be approximated by twice continuously differentiable functions with first partial derivatives that converge in the mean, (15.23) is valid for the present example. Hence

$$G^*(\zeta^*, w^*) - G(\zeta, w)$$
$$= \frac{1}{\pi R^2} \iint\limits_{|z-t|\le R} \mathrm{Re}\left\{8\pi\epsilon\, \frac{\partial G(z, \zeta)}{\partial z}\, \frac{\partial G(z, w)}{\partial z}\right\} dx_1\, dx_2 + O(|\epsilon|^2),$$

and because the expression on the right must be independent of the radius R of the circle over which the integral there is extended, we can let $R \to 0$ to arrive at the *Schiffer variational formula*

$$(15.26) \quad G^*(\zeta^*, w^*) - G(\zeta, w) = \mathrm{Re}\left\{8\pi\epsilon\, \frac{\partial G(t, \zeta)}{\partial t}\, \frac{\partial G(t, w)}{\partial t}\right\} + O(|\epsilon|^2).$$

We maintain that (15.26) holds without any restrictions concerning the boundary ∂D of D other than that it should be composed of a finite number of continua which do not degenerate to isolated points. Indeed, in the case of the variation (15.24) the linear partial differential equation (15.7) reduces to Laplace's equation for $|z - t| > R$, and, similarly, the quadratic form (15.10) becomes equal to the square of the distance $r = |z - \zeta|$ if $|\zeta - t| > R$. Consequently the Fredholm integral equation (15.15) for g merely involves integration over the circle $|z - t| \leq R$ when R is smaller than the distance between the points t and ζ. Note, too, that the use of Green's identity (15.14) in deriving (15.15) can be justified here through a conformal transformation to an intermediate coordinate system in which the boundary becomes analytic. Thus our development of the varied Green's function G^* into a power series in ϵ only entails estimates over a closed subdomain of D. It follows that not only (15.16) but also (15.20) and (15.23) are applicable. Therefore our proof of (15.26) goes through with only minor modifications for extremal domains in the calculus of variations whose boundaries are not known in advance to be smooth.

EXERCISES

1. Show that the first variation δA of the area A of the domain D under a normal shift (15.1) of the boundary curve ∂D is given by the simple formula

(15.27) $$\delta A = - \int_{\partial D} \delta v \, ds.$$

2. Derive Hadamard's variational formula (15.3) from the observation that the boundary values of δG are approximately equal to $-(\partial G/\partial v)\delta v$.

3. Establish that the first variation of the Dirichlet integral

$$||u||^2 = 4 \iint_D \frac{\partial u}{\partial z} \frac{\partial u}{\partial \bar{z}} dx_1 \, dx_2$$

under an infinitesimal transformation of variables (15.5) is

$$\delta ||u||^2 = \mathrm{Re} \left\{ - 8\epsilon \iint_D \left(\frac{\partial u}{\partial z} \right)^2 \frac{\partial F}{\partial \bar{z}} dx_1 \, dx_2 \right\}.$$

4. Use the method of interior variation to establish in a rigorous way the Hadamard formula

$$\delta G = - \epsilon \int_{\partial D} \cdots \int \sum_{j,k=1}^{n} T_{jk} \, S_j \, \frac{\partial x_k}{\partial v} \, d\sigma,$$

analogous to (15.22), for the first variation δG of the harmonic Green's function of an n-dimensional domain D.

5. Prove that the transformation (15.4) has to be one-to-one if the functions S_1 and S_2 are continuously differentiable and the parameter ϵ is small enough.

6. Under what hypotheses are the estimates (15.11), (15.12) and (15.13) valid near the boundary ∂D of the domain of D? Use your answer to this question to reconfirm that Hadamard's formula (15.3) is applicable to examples in which the curves ∂D and the normal displacement $\delta\nu$ possess continuous derivatives of a certain finite order but are not analytic.

7. Derive the formula[4]

$$(15.28) \qquad \delta^2 G(\zeta, w) = -2 \iint_D \nabla \delta G(z, \zeta) \cdot \nabla \delta G(z, w)\, dx_1\, dx_2$$

$$- \int_{\partial D} \frac{\partial G(z, \zeta)}{\partial \nu} \frac{\partial G(z, w)}{\partial \nu} (\delta\nu)^2\, \kappa\, ds$$

for the second variation $\delta^2 G$ of the harmonic Green's function of a plane domain D under the normal shift (15.1) of the boundary ∂D, where κ stands for the curvature of ∂D.

8. Show that Hadamard's formula (15.3) remains valid for the Green's function G of the linear elliptic partial differential equation

$$(15.29) \qquad\qquad \Delta G + cG = 0,$$

provided that the coefficient c is a negative real analytic function of the variables x_1 and x_2 in the closure of the domain D.

9. If the first eigenfunction u_1 of the vibrating membrane problem defined by (11.4) and (11.5) is normalized so that

$$\iint_D u_1^2\, dx_1\, dx_2 = 1,$$

verify that the first variation of the corresponding lowest eigenvalue λ_1 of the domain D under an infinitesimal displacement (15.1) of ∂D is given by the formula

$$(15.30) \qquad \delta\lambda_1 = \int_{\partial D} \left(\frac{\partial u_1}{\partial \nu}\right)^2 \delta\nu\, ds.$$

Explain by means of (15.30) why the Faber-Krahn inequality (11.60) is reasonable.

10. Use (15.30) to derive the representation

$$\lambda_1 = -\frac{1}{4} \int_{\partial D} \left(\frac{\partial u_1}{\partial \nu}\right)^2 \frac{\partial r^2}{\partial \nu}\, ds$$

of the eigenvalue λ_1.

[4] Cf. Garabedian-Schiffer 2.

11. Suppose that $t_1, \ldots, t_m$ are interior points of the domain D and that $\epsilon_1, \ldots, \epsilon_m$ are parameters dominated by the positive number ϵ. Show that the infinitesimal transformation

$$z^* = z + \sum_{j=1}^{m} \frac{\epsilon_j}{z - t_j}$$

maps the curves ∂D onto the boundary ∂D^* of a domain D^* whose Green's function G^* is given by the variational formula

(15.31)

$$G^*(\zeta^*, w^*) - G(\zeta, w) = 8\pi \, \mathrm{Re} \left\{ \sum_{j=1}^{m} \epsilon_j \frac{\partial G(t_j, \zeta)}{\partial t_j} \frac{\partial G(t_j, w)}{\partial t_j} \right\} + O(\epsilon^2)$$

analogous to (15.26).

12. If the plane domain D includes the point at infinity, then the harmonic Green's function whose pole is located there has an asymptotic expansion of the form

$$G(z, \infty) = \frac{1}{2\pi} \log |z| + \frac{\gamma}{2\pi} + O\left(\frac{1}{|z|}\right).$$

The exponential $e^{-\gamma}$ of the constant γ appearing on the right is referred to as the *logarithmic capacity* of the curves ∂D bounding D (cf. Exercise 9.2.14). Suppose that ∂D consists of a single continuum containing two prescribed points z_1 and z_2. Use an ordinary differential equation based on (15.31) to show[5] that under these circumstances the capacity of ∂D becomes a minimum when that continuum reduces to the line segment joining z_1 to z_2.

13. Establish the variational formula

$$\delta N(\zeta, w) = \int_{\partial D} [\nabla N(z, \zeta) \cdot \nabla N(z, w) - c(z)N(z, \zeta)N(z, w)] \, \delta v \, ds$$

for the Neumann's function of the partial differential equation (15.29).

14. Let D be an infinite domain of the (x,y)-plane bounded by a curve ∂D that encloses an obstacle Ω possessing symmetry in the x-axis. Let ψ be the stream function of a potential flow past Ω, by which we mean that ψ should be harmonic in D, should vanish on ∂D, and should have an asymptotic expansion of the form

$$\psi = y - \frac{ay}{x^2 + y^2} + \cdots$$

at infinity. Let A stand for the area of Ω, and introduce the *virtual mass* (cf. Exercise 11.3.10)

$$M = \iint_D (\nabla \psi - \nabla y)^2 \, dx \, dy.$$

Show that

(15.32) $2\pi a = M + A.$

[5] Cf. Schiffer 2.

Prove that under the variation (15.1) of D we have[6]

(15.33)
$$\delta a = \frac{1}{2\pi} \int_{\partial D} (\nabla \psi)^2 \, \delta v \, ds$$

and

(15.34)
$$\delta M = \int_{\partial D} [(\nabla \psi)^2 - 1] \, \delta v \, ds.$$

15. Use the Hadamard variational formulas (15.3), (15.30) and (15.33) to show that the logarithmic capacity $e^{-\gamma}$, the principal eigenvalue λ_1 and the virtual mass coefficient a are all monotonic domain functionals.

16. The quantity e^γ formed from the coefficient γ in the expansion

$$G(z, 0) = \frac{1}{2\pi} \log \frac{1}{|z|} + \frac{\gamma}{2\pi} + O(|z|)$$

about the origin of the Green's function of a finite, simply connected region D is sometimes referred to as its *inner mapping radius*. If A stands for the area of D, show that the dimensionless product $Ae^{-2\gamma}$ becomes a minimum when D is a circle centered at the origin. More specifically, use the Hadamard variational formula (15.3) to establish[7] that the value of $Ae^{-2\gamma}$ associated with either the union or the intersection of D with a circle of radius R about the origin depends monotonically on R.

17. Consider the region D^* found by performing an interior variation (15.5) on the infinite domain D of Exercise 14, and let M^*, a^* and A^* denote the virtual mass and related quantities associated with D^*. By appealing to the procedures in the text, prove[8] that

(15.35)
$$A^* - A = 2 \, \text{Re} \left\{ \epsilon \iint_\Omega \frac{\partial F}{\partial z} \, dx \, dy \right\} + O(|\epsilon|^2)$$

and

(15.36)
$$a^* - a = -\frac{4}{\pi} \, \text{Re} \left\{ \epsilon \iint_D \left(\frac{\partial \psi}{\partial z}\right)^2 \frac{\partial F}{\partial \bar{z}} \, dx \, dy \right\} + O(|\epsilon|^2).$$

Exploit the fact (cf. Section 8.1) that the virtual mass M^* is actually the minimum of the Dirichlet integral of all continuously differentiable functions in D^* that assume the boundary values $-y$ along ∂D^* to derive more directly the inequality

(15.37)
$$M^* - M \le \iint_{D^*} (\nabla \psi^{**} - \nabla y)^2 \, dx \, dy - \iint_D (\nabla \psi - \nabla y)^2 \, dx \, dy$$
$$= \text{Re} \left\{ \epsilon \iint_D \left[2 \frac{\partial F}{\partial z} - 8 \left(\frac{\partial \psi}{\partial z}\right)^2 \frac{\partial F}{\partial \bar{z}} \right] dx \, dy \right\} + O(|\epsilon|^2),$$

[6] Cf. Garabedian-Spencer 1.
[7] Cf. Garabedian-Schiffer 4.
[8] Cf. Garabedian-Lewy-Schiffer 1.

where ψ^{**} represents, not the true stream function ψ^* for D^*, but the transform of ψ defined in a more elementary way by (15.5) and the relation

$$\psi^{**}(x^*, y^*) = \psi(x, y).$$

Observe that (15.37), which comprises one part of a variational formula for the virtual mass resulting from (15.35) and (15.36), is applicable when ∂D is merely known to be a simple closed curve having no special smoothness properties.

2. EXISTENCE OF FLOWS WITH FREE STREAMLINES

In Section 14.5 we discussed at some length potential flows with free streamlines. Here we shall return to such matters in order to establish an existence and uniqueness theorem for the Riabouchinsky model describing plane motion of a liquid past an obstacle that generates a vapor cavity. Our analysis will be based on an extremal problem for the stream function which exhibits the free streamlines as curves shaped so that the virtual mass of the flow is minimized for a fixed area of the cavity.[9] The method of proof we shall develop can be viewed as an illustration of the role that the calculus of variations plays more generally in the theory of free boundary problems for partial differential equations of the elliptic type.

Let the complex number $-h + ik$ indicate a point in the second quadrant of the (x,y)-plane, and consider an arc with non-parametric representations

$$(15.38) \qquad\qquad y = y(x), \qquad x = x(y)$$

which ascends monotonically from a point on the negative x-axis until it reaches $-h + ik$. We denote by C_+ the curve consisting of the arc (15.38) plus its reflection in the x-axis, and we denote by C_- the reflection of C_+ in the y-axis. Let Γ_0 stand for the pair of horizontal line segments joining $-h + ik$ to $h + ik$ and $-h - ik$ to $h - ik$, and let C_0 stand for the four semi-infinite vertical line segments that rise from $\pm h + ik$ and descend from $\pm h - ik$. We denote by Ω_0 the closure of the curvilinear rectangle bounded by C_+, C_- and Γ_0, and we denote by Ω_1 the closure of the infinite curvilinear strip bounded by C_+, C_- and C_0. We shall be interested in potential flows past obstacles Ω which are symmetric in the x-axis and which contain Ω_0, but which are at the same time contained within Ω_1. The letter Γ will be used to indicate those arcs of the boundary ∂D of the complement D of Ω which are not located on either $\partial \Omega_0$ or $\partial \Omega_1$.

---

[9] Cf. Riabouchinsky 1, Friedrichs 1, Garabedian-Lewy-Schiffer 1.

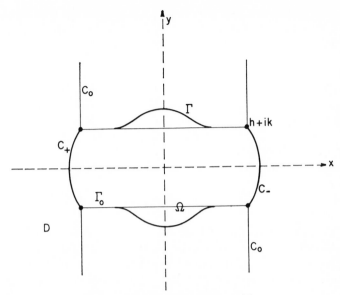

FIG. 79. Riabouchinsky flow model.

Consider the flow past Ω whose stream function ψ satisfies the partial differential equation

(15.39) $\Delta\psi = 0$

throughout the region D, satisfies the boundary condition

(15.40) $\psi = 0$

along the curve ∂D and along the x-axis, and satisfies the normalization requirement

(15.41) $\psi = y - \dfrac{ay}{x^2 + y^2} + \cdots$

at infinity. Let A stand for the area of Ω, which is intended to play the role of a vapor cavity following the obstacle C_+. We recall that Ω is supposed to be a subset of the strip Ω_1 and is supposed to contain the rectangular body Ω_0. Subject to these restrictions, we ask that Ω be shaped so that

(15.42) $2\pi a - \lambda A = \text{minimum},$

where $\lambda > 1$ is a prescribed constant and a is the coefficient defined by (15.41). The identity (15.32) shows that in terms of the virtual mass

$$M = \iint\limits_{D} (\nabla\psi - \nabla y)^2 \, dx \, dy$$

of the flow ψ, the extremal problem (15.42) acquires the alternate form

(15.43) $$M - \sigma A = \text{minimum},$$

where

$$\sigma = \lambda - 1 > 0$$

is the so-called *cavitation parameter*.

We assert that the unrestricted arcs Γ bounding a domain D that solves the minimum problem (15.42) have to be free streamlines (cf. Section 14.5) along which the constant pressure condition

(15.44) $$(\nabla \psi)^2 = \lambda$$

is fulfilled. To start our analysis, we verify this contention under the preliminary hypothesis that an extremal flow for (15.42) exists possessing boundary curves Γ smooth enough to justify an application of the Hadamard variational formulas (15.27) and (15.33). For such a configuration the expression occurring on the left in (15.42) must be stationary under every normal shift (15.1) of the free arcs Γ. Since A is the area of Ω rather than of D in the present context, it follows that the variational equation

(15.45) $$2\pi \, \delta a - \lambda \, \delta A = \int_\Gamma [(\nabla \psi)^2 - \lambda] \, \delta v \, ds = 0$$

holds for arbitrary choices of δv on Γ. Thus the integrand in (15.45) must vanish identically, which means that the free boundary condition (15.44) is, indeed, satisfied along Γ in the case of a smoothly bounded extremal domain.

The equations (15.39), (15.40) and (15.44) fulfilled by the stream function ψ are recognized to be those encountered in the classical free boundary problems of hydrodynamics that we described in Section 14.5. Our current investigation of existence questions for the Riabouchinsky model of cavity flow centers about the minimum problem (15.43), which we have introduced in order to determine the free streamlines Γ. Now observe that the virtual mass M represents the kinetic energy of a flow $\psi - y$ resulting from motion of the obstacle Ω through liquid that is at rest at infinity. Moreover, the product σA is a measure of the potential energy needed to confine a suitable amount of vapor in the cavity Ω. Consequently (15.45) asserts that the difference between the kinetic energy and the potential energy must be stationary for actual free streamline flows. Thus what we are dealing with is an analogue of Hamilton's principle (cf. Section 2.5). In the present case (15.43) shows that the difference of energies even becomes a minimum, which is consistent with the fact that the motion is steady. It should be mentioned that phenomena of this kind are also encountered in connection with physical interpretations of Dirichlet's principle (cf. Section 8.1).

Note that for each value of the cavitation parameter $\sigma > 0$, the cavity Ω produced by the extremal problem (15.43) has to have a different size and shape. We shall find later on that there is a special choice σ_0 of σ for which the free streamlines Γ bounding Ω separate from the ends $-h \pm ik$ of the profile C_+ and join them to the ends $h \pm ik$ of C_-. For $\sigma < \sigma_0$ it will turn out that the points of separation of Γ located on Γ_0, whereas for $\sigma > \sigma_0$ they will be situated on C_0 (cf. Figure 74 in Chapter 14).

We see that according to Dirichlet's principle the virtual mass M can be identified as the minimum

$$M = \min_{\tilde{\psi}} \iint_D (\nabla \tilde{\psi})^2 \, dx \, dy$$

of the Dirichlet integral among all functions $\tilde{\psi}$ that are continuously differentiable in D and have the boundary values

$$\tilde{\psi} = -y$$

along ∂D. Thus it is possible to reformulate (15.43) as a more general free boundary problem

$$(15.46) \qquad \iint_D (\nabla \tilde{\psi})^2 \, dx \, dy - \sigma \iint_\Omega dx \, dy = \text{minimum}$$

in the calculus of variations with respect to this wider class of competing functions $\tilde{\psi}$. For the applications it is desirable to have both (15.43) and (15.46) at our disposal.

We shall establish next that for the minimum problem (15.46) we may confine our attention to obstacles Ω whose intersection with any line parallel to the x-axis or the y-axis is composed of not more than one segment. In other words, we maintain that the boundary ∂D of the extremal domain D should consist in each quadrant of a monotonic curve with non-parametric representations (15.38), except, perhaps, for the occurrence of horizontal or vertical linear sections. This assertion is a consequence of the theorem that the virtual mass M decreases, or at least does not increase, under Steiner symmetrization of Ω in the x-axis or the

y-axis, which was proved in Exercise 11.3.10. To define the symmetrization processes referred to, we merely represent the streamlines

(15.47)
$$\psi(x, y) = \text{const.}$$

in the form

(15.48)
$$y = y_j(x), \qquad j = 0, 1, \ldots, 2m,$$

or

(15.49)
$$x = x_j(y), \qquad j = 0, 1, \ldots, 2n,$$

and if either of the expressions on the right is multiple-valued, so that $m > 0$ or $n > 0$, we replace the corresponding level curves (15.47) by appropriate alternating sums

(15.50)
$$y = \sum_{j=0}^{2m} (-1)^j y_j(x),$$

or

(15.51)
$$x = \sum_{j=0}^{2n} (-1)^j x_j(y).$$

The procedure we have just outlined is obviously meaningful if the boundary curve ∂D and the other streamlines (15.47) each have at most a finite number of inflection points, but the construction remains valid for more general domains also because they can be obtained as limiting cases of regions satisfying the stated hypothesis. Notice, furthermore, that in view of the asymptotic expansion (15.41) of ψ, the representations (15.48) and (15.49) must be single-valued at large distances from Ω. Hence the symmetrization processes (15.50) and (15.51) only alter the Dirichlet integral M over a region of finite extent. However, a final application of Dirichlet's principle leads in general to a completely new stream function for the symmetrized domain.

Since Steiner symmetrization leaves the area A of Ω invariant, we see that it either diminishes or does not change the energy expression $M - \sigma A$ on the left in (15.43). Moreover, the closed sets Ω_0 and Ω_1 which have been introduced to furnish constraints on Ω are already symmetrized because of the assumption (15.38) concerning C_+ and C_-. It follows that the symmetrized version of an admissible obstacle Ω can be put into competition for the minimum problem (15.43), too, and actually yields as small a value of the quantity to be minimized. Since Steiner symmetrization (15.51) in the y-axis transforms a figure which has already been symmetrized in the x-axis into one retaining that property, we conclude that no generality is lost if we consider in the first place in (15.43) only cavities Ω which have been symmetrized with respect to both coordinate axes. In particular, our analysis shows that the level curves (15.47) of an extremal stream function cannot be altered by further symmetrization, which means that they must be monotonic in each quadrant.

We now introduce the letters ψ_n, a_n, etc., to describe a minimal sequence of flows such that

$$(15.52) \qquad \lim_{n \to \infty} \{2\pi a_n - \lambda A_n\} = \inf \{2\pi a - \lambda A\}.$$

We should like to be able to construct from ψ_n a subsequence of stream functions that converges toward an extremal flow for the problem (15.42). As a preliminary step we observe that Steiner symmetrization of Ω_n in the x-axis or in the y-axis yields another minimal sequence. Thus we may assume that ψ_n is symmetrized to begin with, which leads to a quite essential simplification in the discussion to follow.

In order to ensure that (15.52) makes sense, it is necessary to find an *a priori* lower bound on the expression appearing there and to ascertain that the obstacles Ω_n do not become large without limit as $n \to \infty$. Since the variable portion of any admissible body Ω is confined to the strip $-h \le x \le h$ we evidently have

$$A \le 4hd + \mu,$$

where d stands for the height of Ω and μ indicates a positive constant. Moreover, in Exercise 1.15 it was established that the virtual mass coefficient a is a monotonic functional of the flow region D. Hence the value of a associated with the (symmetrized) obstacle Ω exceeds that corresponding to the vertical line segment

$$(15.53) \qquad x = 0, \qquad -d \le y \le d,$$

which is a subset of Ω. Since the stream function ψ_d for the flow past (15.53) is defined by the explicit formula

$$\psi_d = \mathrm{Im} \{\sqrt{z^2 + d^2}\} = y - \frac{d^2}{2} \frac{y}{x^2 + y^2} + \cdots,$$

we conclude from (15.41) that

$$a \ge \frac{d^2}{2}.$$

Consequently

$$2\pi a - \lambda A \ge \pi d^2 - 4\lambda hd - \lambda \mu \ge -\frac{4\lambda^2 h^2}{\pi} - \lambda \mu,$$

which shows that the quantity on the left grows without limit as $d \to \infty$. It follows that the bodies Ω_n comprising our minimal sequence have to remain bounded as $n \to \infty$.

The properties of Ω_n which we have established so far imply that the closed curves ∂D_n are monotonic and uniformly bounded in each quadrant. Hence it is possible to deduce from Helly's convergence theorem that the sequence ∂D_n includes a convergent subsequence. However, the same result is seen more conclusively in a new coordinate system

$$X = \frac{x - y}{\sqrt{2}}, \qquad Y = \frac{x + y}{\sqrt{2}}$$

obtained by performing a rotation through $-45°$. In terms of the variables X and Y let $Y_n = Y_n(X)$ be the non-parametric representation of the arc of ∂D_n located in the first quadrant of the (x,y)-plane. Our assertion concerning the monotonicity of that arc becomes, in the (X, Y)-plane, a Lipschitz condition

$$|Y_n(X'') - Y_n(X')| \leq |X'' - X'|$$

on the function Y_n. Thus the sequence Y_n is equicontinuous (cf. Section 10.3), and by Arzela's theorem it must therefore contain a subsequence that converges uniformly to a Lipschitz continuous limit. This limit defines a closed curve ∂D bounding a region D which we anticipate will turn out to be the extremal domain for the minimum problem (15.42).

For the sake of simplicity we may suppose that it is actually the original sequence of domains D_n itself that converges toward D in the sense just described. Thus each closed subdomain of D must lie inside all but a finite number of the regions D_n. We contend that the functions $\psi_n - y$ are equicontinuous in any such closed subdomain. Since the Dirichlet integral of $\psi_n - y$ over D_n is uniformly bounded, our assertion can be deduced from the estimate (7.103). It is also a consequence of the Harnack inequalities (7.61) and the fact that ψ_n is positive in the upper half-plane and negative in the lower half-plane. We conclude that there exists a subsequence of ψ_n that converges uniformly toward a harmonic function ψ in each closed subdomain of D. We shall endeavor to establish that this limit ψ is the stream function of a flow past the complement Ω of D and that it solves the extremal problem (15.42).

There is no loss of generality if we assume, as in the case of D_n, that the sequence ψ_n itself converges toward ψ. It follows that the coefficient a in the asymptotic expansion (15.41) of ψ is given by

$$(15.54) \qquad\qquad a = \lim_{n \to \infty} a_n.$$

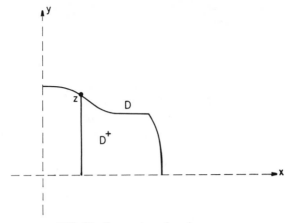

FIG. 80. Comparison domain.

From the uniform convergence of the boundary curves ∂D_n we also conclude that the area A of the limiting obstacle Ω satisfies

$$A = \lim_{n \to \infty} A_n.$$

According to (15.52) we have therefore only to prove that ψ vanishes on the boundary ∂D of the domain D in order to complete our verification that it defines a solution of (15.42).

To obtain the desired boundary condition on ψ at, say, a point z of ∂D in the first quadrant, we compare ψ_n with an auxiliary harmonic function, or barrier, ψ_n^+. Let z_n denote a point on ∂D_n such that

(15.55)
$$\lim_{n \to \infty} z_n = z.$$

We choose ψ_n^+ to be the stream function of the normalized flow past the vertical line segment connecting z_n with its complex conjugate $\bar{z}_n$. From the maximum principle it is clear that in the intersection of the first quadrant with D_n we have

(15.56)
$$0 < \psi_n < \psi_n^+,$$

since the corresponding flow regions D_n and D_n^+ stand in the relationship $D_n \subset D_n^+$. Furthermore, in view of (15.55) the limit

$$\psi^+ = \lim_{n \to \infty} \psi_n^+$$

exists and defines a flow past the vertical line segment connecting z with $\bar{z}$. Thus we deduce from (15.56) that

(15.57) $$0 \leq \psi \leq \psi^{+},$$

throughout the portion of D located in the first quadrant. On the other hand, as the boundary point z is approached from the interior of D one sees directly that $\lim \psi^{+} = 0$. Therefore the desired boundary condition $\psi = 0$ on the stream function ψ along the arc of ∂D in the first quadrant is a consequence of (15.57).

Our analysis thus far demonstrates the existence of a solution ψ of the three equivalent free boundary problems (15.42), (15.43) and (15.46) in the calculus of variations. Moreover, the corresponding extremal body Ω has been shown to have the property that it is symmetrized in both the x-axis and the y-axis, which means that in each quadrant its boundary curve ∂D is continuous and monotonic. Our next objective will be to deduce from (15.46) that the constant pressure requirement (15.44) must be fulfilled along those arcs Γ of ∂D which do not coincide with Γ_0, C_0, C_+ or C_-. In this sense Γ will turn out to consist of free streamlines, so that what we have actually accomplished is to establish the existence of a family of Riabouchinsky flows depending on the cavitation parameter σ.

Denote by z_0 a point located on an arc of the extremal free boundary Γ. We intend to construct an interior variation

(15.5) $$z^{*} = z + \epsilon F(z, \bar{z})$$

of D in the neighborhood of z_0 which leaves the auxiliary closed sets Ω_0 and Ω_1 invariant, and which can therefore be applied to (15.46) in order to show that Γ is analytic and satisfies (15.44). Let ρ_1 and ρ_2 be numbers in the range $0 < \rho_1 < \rho_2$ which are so small that no point of the line segments Γ_0 and C_0 lies inside the closed circle

(15.58) $$|z - z_0| \leq \rho_2.$$

We use the letter ω to indicate an infinitely differentiable function with the properties

$$\omega = \begin{cases} 1, & |z - z_0| < \rho_1, \\ 0, & |z - z_0| > \rho_2, \end{cases}$$

and we let t stand for any point with $|t - z_0| < \rho_1$ which is not situated on Γ. Finally, choose $R > 0$ small enough so that the disc

(15.59) $$|z - t| \leq R$$

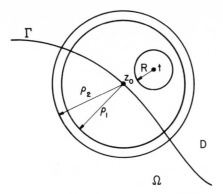

FIG. 81. Variation of the free boundary.

intersects neither Γ nor the circumference of the circle of radius ρ_1 about z_0. We define a continuous function F describing the desired interior variation (15.5) of D by means of the specific rules

(15.60)
$$F(z, \bar{z}) = \begin{cases} \dfrac{\omega}{z - t}\,, & |z - t| \geq R, \\[2ex] \dfrac{\bar{z} - \bar{t}}{R^2}\,, & |z - t| \leq R. \end{cases}$$

Since the infinitesimal transformation given by formulas (15.5) and (15.60) does not shift points outside the circle (15.58), it leaves Γ_0, C_0, C_+ and C_- invariant. Thus it does not alter either Ω_0 or Ω_1, and consequently we can apply it to the extremal domain D for the problem (15.46). Let D^* and Ω^* denote the images of D and Ω found in this fashion, and let M^* and A^* stand for the values of the virtual mass and the area associated with Ω^*. Since D solves the minimum problem (15.46), we obviously have

$$M^* - M - \sigma[A^* - A] \geq 0.$$

Substituting the variational formulas (15.35) and (15.37) into the latter inequality, we derive the fundamental relation

(15.61) $\operatorname{Re} \left\{ \epsilon \displaystyle\iint_D \left[2 \dfrac{\partial F}{\partial z} - 8 \left(\dfrac{\partial \psi}{\partial z} \right)^2 \dfrac{\partial F}{\partial \bar{z}} \right] dx\, dy \right.$

$$\left. - 2\sigma\epsilon \iint_\Omega \dfrac{\partial F}{\partial z}\, dx\, dy \right\} + O(|\epsilon|^2) \geq 0,$$

where F is understood to indicate the function (15.60). Because the combined area of D and Ω remains unchanged under the interior variation (15.5), we also have

$$\iint_D \frac{\partial F}{\partial z}\, dx\, dy + \iint_\Omega \frac{\partial F}{\partial z}\, dx\, dy = 0.$$

Consequently (15.61) can be brought into the more convenient form

$$(15.62)\quad \mathrm{Re}\left\{4\epsilon \iint_D \left(\frac{\partial \psi}{\partial z}\right)^2 \frac{\partial F}{\partial \bar z}\, dx\, dy + \lambda\epsilon \iint_\Omega \frac{\partial F}{\partial z}\, dx\, dy\right\} + O(|\epsilon|^2) \leq 0.$$

Since we may choose the complex parameter ϵ to be as small as we please in (15.62), it becomes evident that

$$(15.63)\quad 4\iint_D \left(\frac{\partial \psi}{\partial z}\right)^2 \frac{\partial F}{\partial \bar z}\, dx\, dy + \lambda\iint_\Omega \frac{\partial F}{\partial z}\, dx\, dy = 0.$$

We shall denote by $\delta(t)$ the function which is equal to 1 inside D but vanishes outside D. Also, let D_ρ and Ω_ρ stand for the intersections of D and Ω, respectively, with the annulus

$$\rho_1 < |z - z_0| < \rho_2,$$

and let Ω_R be the part of Ω located inside the circle

$$(15.64)\quad |z - z_0| < \rho_1,$$

but with the disc (15.59) of radius R removed. Inserting the explicit expression (15.60) for F into (15.63), we derive the more useful variational condition

$$(15.65)\quad \frac{4\delta(t)}{R^2} \iint_{|z-t|\leq R} \left(\frac{\partial \psi}{\partial z}\right)^2 dx\, dy - \lambda\iint_{\Omega_R} \frac{dx\, dy}{(z-t)^2} + \alpha(t) = 0,$$

where

$$\alpha(t) = 4\iint_{D_\rho} \left(\frac{\partial \psi}{\partial z}\right)^2 \frac{\partial \omega}{\partial \bar z}\frac{dx\, dy}{z-t} + \lambda\iint_{\Omega_\rho} \frac{\partial}{\partial z}\frac{\omega}{z-t}\, dx\, dy$$

is an analytic function of t in the neighborhood (15.64) of z_0. The integrals on the left in (15.65) are actually independent of R. Hence we can let $R \to 0$ to obtain

$$(15.66)\quad \frac{4\delta(t)}{\lambda}\left(\frac{\partial \psi}{\partial t}\right)^2 = \frac{1}{\pi}\iint_\Omega \frac{dx\, dy}{(z-t)^2} + \beta'(t),$$

where β' is the derivative of another analytic function of t defined throughout the circle (15.64), and where the improper integral on the right is to be interpreted in the sense of the Cauchy principal value.

We introduce the function

$$(15.67) \qquad g(t) = -\frac{4}{\lambda} \int \left(\frac{\partial \psi}{\partial t}\right)^2 dt,$$

which is analytic in the flow region D. The Poisson equation (5.93) implies that

$$\frac{\partial^2}{\partial \bar{t} \, \partial t} \iint_\Omega \log |z - t| \, dx \, dy = \frac{\pi}{2}$$

at interior points t of the set Ω. Consequently we have

$$\frac{\partial}{\partial \bar{t}} \frac{1}{\pi} \iint_\Omega \frac{dx \, dy}{z - t} = \delta(t) - 1,$$

or

$$\frac{\partial}{\partial \bar{t}} \left\{ \frac{1}{\pi} \iint_\Omega \frac{dx \, dy}{z - t} + [1 - \delta(t)] \bar{t} \right\} = 0,$$

whenever t is not situated on the curve ∂D. In these circumstances we also find that

$$\frac{\partial}{\partial t} \left\{ \frac{1}{\pi} \iint_\Omega \frac{dx \, dy}{z - t} + [1 - \delta(t)] \bar{t} \right\} = \frac{1}{\pi} \iint_\Omega \frac{dx \, dy}{(z - t)^2}.$$

Thus an integration of (15.66) leads to the identity

$$(15.68) \qquad \delta(t) g(t) + [1 - \delta(t)] \bar{t} + \frac{1}{\pi} \iint_\Omega \frac{dx \, dy}{z - t} + \beta(t) = 0$$

for appropriate choices of the additive constants implicit in our definitions of β and g.

Both β and the double integral occurring in (15.68) are continuous across the arc of the free boundary Γ located inside the circle (15.64) around the point z_0. Hence the remaining terms $\delta(t) g(t)$ and $[1 - \delta(t)] \bar{t}$ in (15.68) must have jumps across Γ which just cancel each other out. We conclude that the analytic function g has a continuous limit as we approach Γ from the interior of D, and, more specifically, that its boundary values on Γ are given by the important formula

$$(15.69) \qquad g(t) = \bar{t}.$$

From this result we shall deduce that Γ consists of analytic arcs along which the constant pressure condition (15.44) holds (cf. Exercise 14.5.8).

For t in D we set

$$\Phi(t) = t + g(t)$$

and

$$\Psi(t) = t - g(t).$$

Because of (15.69) we find that on Γ the analytic functions Φ and Ψ satisfy the boundary conditions

$$\Phi(t) = t + \bar{t} = 2\operatorname{Re}\{t\}$$

and

$$\Psi(t) = t - \bar{t} = 2i \operatorname{Im}\{t\}.$$

Therefore Φ maps Γ onto a horizontal line segment and Ψ maps Γ onto a vertical line segment. From the Schwarz principle of reflection (cf. Section 7.2) it follows that Φ and Ψ can be extended analytically across the free boundary curve Γ as functions of the complex potential $\zeta = \phi + i\psi$ of our flow. Hence the variable

$$t = \tfrac{1}{2}[\Phi + \Psi]$$

is also an analytic function of ζ along Γ, which must therefore consist of analytic arcs. Differentiation of (15.67) and (15.69) is now permissible on Γ and yields there the relation

$$(15.70) \qquad 4\left(\frac{\partial \psi}{\partial t}\right)^2 = -\lambda g'(t) = -\lambda \dot{\bar{t}}^2,$$

where the dot indicates a derivative with respect to arc length. The free boundary condition (15.44) is an immediate consequence of (15.70).

The fact that the extremal body Ω for (15.43) is invariant under symmetrization in the x-axis and the y-axis shows that the free boundary Γ is composed of precisely two connected arcs, one lying above Γ_0 in the upper half-plane and the other lying below Γ_0 in the lower half-plane. Our variational procedure has served to establish that both are free streamlines. Therefore the minimum problem (15.43) yields a one-parameter family of Riabouchinsky flows past profiles consisting of C_+ and C_-, plus segments from either Γ_0 or C_0. According to the results of Section 14.5 each flow thus obtained is uniquely determined by our choice of the cavitation parameter σ. Furthermore, we have indicated in Exercise 14.5.12 that the dependence of the free boundary Γ on σ is continuous and monotonic (cf. Figure 74). Hence there must be just one value σ_0 of σ which corresponds to a Riabouchinsky flow with free streamlines Γ that separate from the endpoints $-h \pm ik$ of the profile C_+ and join them to the endpoints $h \pm ik$ of C_-. It is the existence of this particular flow that we wanted to prove in the first place.

In summary, we have found that the minimum energy principle (15.43) generates flows with constant pressure free streamlines. By introducing suitable normalizations we have succeeded in showing that, for a given profile C_+ satisfying the hypothesis (15.38), the symmetrically located points of separation of Γ can be prescribed through a proper selection of the cavitation parameter σ. Thus we have been able to obtain a quite satisfactory existence and uniqueness theorem for a classical free boundary problem of hydrodynamics and have demonstrated that it is well posed in so far as the formulation that has been presented here is concerned. The implications of this analysis for more general free boundary problems in the theory of partial differential equations are apparent.

EXERCISES

1. Show that the parameter λ appearing in (15.45) plays the role of a Lagrange multiplier in a reformulation of the extremal problem (15.42) which asks that the virtual mass coefficient a be minimized while the area A is held fixed.

2. Complete all details of the proof that a convergent minimal sequence of domains D_n for the extremal problem (15.42) can be found, that the corresponding stream functions ψ_n are equicontinuous in each closed subregion of the limit domain D, and that (15.54) holds.

3. Apply Green's theorem to show that

$$\frac{\partial}{\partial t} \iint_\Omega \frac{dx\,dy}{z - t} = \iint_\Omega \frac{dx\,dy}{(z - t)^2}$$

at interior points t of the set Ω.

4. Give a more detailed proof of the assertion that a value σ_0 of σ exists for which the free boundary Γ associated with (15.43) separates from the endpoints of C_+.

5. Show that if C_+ is an analytic arc, the curve ∂D has a continuously turning tangent at the point of separation $-h + ik$ of the free boundary Γ specified in Exercise 4. What is the asymptotic behavior of ψ near that point?

6. Apply a limit process to the Riabouchinsky model, with $\sigma \to 0$, in order to establish the existence of a Helmholtz flow past C_+ followed by an infinite cavity Ω.

7. Generalize the method of the text to develop existence and uniqueness theorems for jets.

8. Extend the minimum principle (15.43) to include examples of free streamline flows under the influence of gravity.[10]

9. Develop a minimum problem for the analogue of the virtual mass coefficient a in space whose solution leads to *vortex sheets*, i.e., surfaces on which the normal component of velocity vanishes and across which the tangential component of velocity experiences a jump in direction but not in magnitude.[11]

10. Use the following symmetrization argument (cf. Exercise 11.3.13) to prove that there cannot exist two distinct solutions ψ_1 and ψ_2 of the free boundary problem (15.43). Let $y = y_1(x)$ and $y = y_2(x)$ represent the two streamlines $\psi_1(x, y) = K$ and $\psi_2(x, y) = K$, and let ψ_θ be the function which has the value K on the intermediate curve

(15.71) $$y = \theta y_1(x) + (1 - \theta)y_2(x),$$

where θ is some number in the interval $0 \leq \theta \leq 1$. Introduce the integrals

$$M(\theta) = \iint_{D_\theta} (\nabla \psi_\theta - \nabla y)^2 \, dx \, dy, \qquad A(\theta) = \iint_{\Omega_\theta} dx \, dy$$

over the region D_θ in which (15.71) defines ψ_θ and over the complement Ω_θ of D_θ. Show that the inequality

$$M\left(\frac{\theta_1 + \theta_2}{2}\right) \leq \frac{M(\theta_1) + M(\theta_2)}{2}$$

holds because symmetrization diminishes the Dirichlet integral. Conclude[12] that $M(\theta) - \sigma A(\theta)$ is a convex function of θ, so that it cannot achieve minimum values at both $\theta = 0$ and $\theta = 1$ unless $\psi_1 \equiv \psi_2$.

11. Use the representation formula of Exercise 14.5.10 to show that any constant pressure free streamline in axially symmetric flow must be an analytic curve.

12. Prove the existence of axially symmetric Riabouchinsky flows past a prescribed axially symmetric obstacle in space by means of a generalization[13] of the minimum energy principle (15.43).

13. Apply the Hadamard variational formulas for a and A in the case of an infinitesimal magnification in order to establish that the drag

$$\tau = \tfrac{1}{2} \int_{C_+} [\lambda - (\nabla \psi)^2] \, dy$$

on a vertical line segment C_+ immersed in a plane Riabouchinsky flow is given in terms of the virtual mass by the expression[14]

$$2h\tau = \sigma A - M.$$

[10] Cf. Garabedian-Spencer 1.
[11] Cf. Garabedian-Schiffer 2.
[12] Cf. Friedrichs 1, Garabedian 11.
[13] Cf. Garabedian-Lewy-Schiffer 1.
[14] Cf. Garabedian 6.

14. The total drag τ on a body Ω immersed in a potential flow ψ is defined by the integral

$$\tau = \tfrac{1}{2} \int_{\partial D} [1 - (\nabla \psi)^2] \, dy.$$

Show that the variation δM in virtual mass due to an infinitesimal translation $\delta v = -\epsilon \, dy/ds$ of Ω is given by the rule $\delta M = 2\tau\epsilon$, which may be interpreted to mean that work equals force times displacement. Use this result to explain why τ vanishes (cf. Exercise 14.1.2).

3. HYDROMAGNETIC STABILITY

We resume here the discussion begun at the close of Section 14.6 concerning figures of equilibrium of a perfectly conducting fluid held in place by a vacuum magnetic field B. We shall be concerned with the question of the stability of such figures, which is of importance for the problem of confining a plasma in a strong magnetic field.

We assume that the fluid is incompressible and does not move, so that it remains in a state of constant pressure p and occupies a region Ω of fixed volume A. The magnetic field B is supposed to vanish throughout Ω, but in the vacuum D surrounding Ω it is a non-trivial solution of the Maxwell equations

(15.72) $$\nabla \cdot B = 0$$

and

(15.73) $$\nabla \times B = 0.$$

At the interface Γ separating Ω from D we have to impose the customary requirement

$$B \cdot v = 0$$

related to (15.72), together with the free boundary condition

(15.74) $$\tfrac{1}{2}B^2 = p,$$

which asserts that the magnetic pressure from the field just balances the fluid pressure from Ω. We shall be interested in determining under what circumstances the equilibrium of a configuration of this kind is stable.[15]

In the examples that we intend to treat, the magnetic field B will be generated by a system of perfectly conducting rigid coils. No energy can flow across the surface of the coils, for the tangential components of the

[15] Cf. Berkowitz-Grad-Rubin 1, Blank 1.

electric field E must necessarily vanish there, whence the same is true of the normal component of the Poynting vector $E \times B$. Now consider different shapes which the fluid Ω might take, and let us inquire whether the corresponding vacuum fields B exhibit any invariant features. It is to be observed that no tangential electric field can be induced around any closed circuit at the surface of the coils. Therefore Faraday's law of induction implies that the magnetic flux through such a circuit cannot be altered by changes in the form of Ω. We conclude that each magnetic flux H thus associated with the coils is a constant that must be held fixed under any disturbance to which we may subject the perfectly conducting liquid Ω during our stability analysis.

We shall discuss in detail the special case where Ω and D are actually regions of the (x,y)-plane and where B lies in that plane and depends only on the two variables x and y. Assume to begin with that the coils generating B reduce to four lines which are perpendicular to the (x,y)-plane and intersect it at the points w, $\bar{w}$, $-w$ and $-\bar{w}$, with

$$w = h + ih$$

located in the first quadrant. The field B is supposed to have poles of equal strengths, but with alternating signs, at w, $\bar{w}$, $-w$ and $-\bar{w}$. Thus if we prescribe for its area A a small enough value, Ω should turn out to be a cusped figure centered at the origin and bounded by four concave arcs Γ which terminate at points of tangency to the coordinate axes. The explicit form of these free boundary curves has been found in Exercise 14.6.3.

The partial differential equation (15.72) shows that there exists a vector potential $(0, 0, \psi)$ with the property

$$B = \nabla \times (0, 0, \psi).$$

The remaining Maxwell equation (15.73) then asserts that the scalar function ψ is a solution of Laplace's equation

$$\Delta\psi = 0$$

throughout the vacuum D. The level curves

$$\psi(x, y) = \text{const.}$$

of ψ coincide with the *lines of force* of the magnetic field B; in particular, we can pick the constant of integration implicit in the definition of ψ so that the boundary condition

$$\psi = 0$$

is satisfied along Γ. We choose to normalize the strength of B in such a way that ψ has the asymptotic behavior

$$\psi = \log \left| \frac{z^2 - \bar{w}^2}{z^2 - w^2} \right| + \cdots$$

at the singular points w, $\bar{w}$, $-w$ and $-\bar{w}$. Under these hypotheses ψ is uniquely determined by the value which we assign to the area A of Ω, and A becomes in turn a monotonically increasing function of the pressure p of our perfectly conducting fluid. Finally, we note that in terms of ψ and p the free boundary condition (15.74) along Γ can be expressed in the form

(15.75) $(\nabla\psi)^2 = 2p$

analogous to (15.44).

Before proceeding to investigate the stability of the cusped figure of equilibrium Ω associated with the harmonic function ψ that we have just described, we introduce a slight modification of the boundary value problem defining ψ which enables us to avoid dealing with divergent integrals later on. The modification consists in replacing the line coils through the points w, $\bar{w}$, $-w$ and $-\bar{w}$ by thicker perfectly conducting coils whose cross sections are the simply connected regions located inside the four closed level curves C given by

(15.76) $\psi(x, y) = \pm H,$

where H is some large positive number. This formulation of the problem leads, of course, to a figure of the same shape as before for the fluid Ω, which is still held in place by the same field B. Observe, incidentally, that the constant H appearing in (15.76) plays the role of an invariant magnetic flux to be identified with the new configuration.

We shall accept here the principle that a physical system is in stable equilibrium whenever its potential energy is a minimum compared with that of any other system from an appropriate admissible class (cf. Exercise 7 below). For example, a ball sitting in a bowl is known to be in stable equilibrium when it lies at the bottom of the bowl. In order to apply this principle to the figure of equilibrium Ω, we have first to find an expression for the potential energy naturally associated with such a system, or with any neighboring configuration of incompressible fluid which has the same total mass, or area,

$$A = \iint_{\Omega} dx\, dy.$$

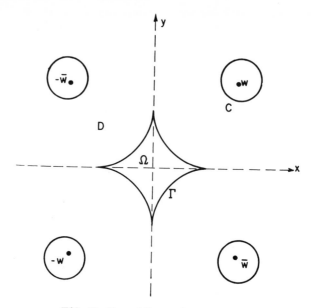

FIG. 82. Cusped figure of equilibrium.

Since both the fluid filling Ω and the coils C generating B are supposed to be perfect conductors, we can treat the internal energies connected with them as absolute constants. Thus it is only the energy

$$M = \tfrac{1}{2} \iint_{D} B^2 \, dx \, dy$$

of the vacuum magnetic field B that we need to consider, so that the potential energy of our hydromagnetic model reduces to the Dirichlet integral

$$M = \tfrac{1}{2} \iint_{D} (\nabla \psi)^2 \, dx \, dy$$

of the scalar function ψ.

We denote by Ω^* a region occupied by perfectly conducting incompressible fluid, not necessarily in equilibrium, which is confined by a plane magnetic field B^* located in the vacuum D^* between Ω^* and our original conducting coils C. Since B^* might vary with time, the only requirements we are permitted to impose on it are the Maxwell equation

(15.77) $\nabla \cdot B^* = 0$

in D^* and the corresponding boundary condition

$$B^* \cdot \nu = 0$$

along C and along the free boundary Γ^* of Ω^*. As in the case of (15.72), we deduce from (15.77) the existence of a scalar function ψ^* in D^* such that

$$B^* = \nabla \times (0, 0, \psi^*),$$

with

(15.78) $\psi^* = 0$

on Γ^* and

(15.79) $\psi^* = \pm H^*$

on C. Our observations above concerning the invariance of the magnetic flux H indicate that we ought to put

(15.80) $H^* = H;$

similarly, the area A^* of Ω^* is supposed to satisfy the constraint

(15.81) $A^* = A.$

The energy

$$M^* = \tfrac{1}{2} \iint\limits_{D^*} B^{*2} \, dx \, dy = \tfrac{1}{2} \iint\limits_{D^*} (\nabla \psi^*)^2 \, dx \, dy$$

of the magnetic field B^* is the right potential energy to associate with the figure Ω^*, which has been introduced for the purpose of a comparison with Ω. According to our previous remarks we have only to show that under the constraints (15.80) and (15.81) we have

(15.82) $M^* \geq M$

in order to prove that Ω is in stable equilibrium. Because of Dirichlet's principle (cf. Section 8.1), (15.82) will follow quite generally if we are able to establish its validity in the special case where ψ^* fulfills the Laplace equation

$$\Delta \psi^* = 0$$

in addition to the boundary conditions (15.78) and (15.79). Thus it suffices to confine our attention to the latter situation. Finally, we mention that our verification of (15.82) will only succeed when A is chosen to be so small that the field strength $|\nabla \psi|$ increases initially as we leave the cusps of Γ and enter D along the x-axis or the y-axis.

We discuss next some useful concepts that apply equally well to both ψ and ψ^*. For reasons of symmetry we know that the same electric current I must be carried by each of the four coils composing C, and it can be expressed as a line integral

$$I = -\int_{C_+} \frac{\partial \psi}{\partial \nu}\, ds$$

over the perimeter C_+ of the coil situated in the first quadrant. From Green's theorem and (15.76) we easily derive the relationship

$$M = -\frac{1}{2} \int_C \psi \frac{\partial \psi}{\partial \nu}\, ds = 2HI$$

between the energy M, the magnetic flux H and the current I. Now let

$$\phi = \frac{1}{I} \int \frac{\partial \psi}{\partial \nu}\, ds$$

indicate the harmonic function conjugate to ψ, renormalized so that its periods around the coils C reduce to ± 1. Clearly

$$(15.83) \qquad (\nabla \phi)^2 = \frac{1}{I^2} (\nabla \psi)^2,$$

and therefore we have

$$(15.84) \qquad \iint_D (\nabla \phi)^2\, dx\, dy = \frac{2M}{I^2} = \frac{8H^2}{M}\,.$$

For the normalized conjugate

$$\phi^* = \frac{1}{I^*} \int \frac{\partial \psi^*}{\partial \nu}\, ds$$

of the harmonic function ψ^* associated with the fluid Ω^* we obtain a result

$$\iint_{D^*} (\nabla \phi^*)^2\, dx\, dy = \frac{8H^2}{M^*}$$

quite analogous to (15.84). Hence the desired inequality (15.82) will follow if we can establish that

$$(15.85) \qquad \iint_D (\nabla \phi)^2\, dx\, dy \geq \iint_{D^*} (\nabla \phi^*)^2\, dx\, dy.$$

A new application of Dirichlet's principle can be brought into play in order to show that for our proof of (15.85) we need not restrict ϕ^* to be harmonic but can pick it to be any comparison function ϕ^{**} in D^* possessing the same periods ± 1 around the coils C as does ϕ itself (cf. Exercise 8.1.8). Indeed, the difference

$$\phi_1 = \phi^{**} - \phi^*$$

is single-valued, and since ϕ^* satisfies the natural boundary condition

$$\frac{\partial \phi^*}{\partial \nu} = 0$$

along C and Γ^*, we find that

$$\iint_{D^*} \nabla \phi^* \cdot \nabla \phi_1 \, dx \, dy = -\int_{C + \Gamma^*} \phi_1 \frac{\partial \phi^*}{\partial \nu} \, ds = 0.$$

Consequently

$$(15.86) \quad \iint_{D^*} (\nabla \phi^{**})^2 \, dx \, dy = \iint_{D^*} (\nabla \phi^* + \nabla \phi_1)^2 \, dx \, dy$$

$$= \iint_{D^*} (\nabla \phi^*)^2 \, dx \, dy + 2 \iint_{D^*} \nabla \phi^* \cdot \nabla \phi_1 \, dx \, dy + \iint_{D^*} (\nabla \phi_1)^2 \, dx \, dy$$

$$= \iint_{D^*} (\nabla \phi^*)^2 \, dx \, dy + \iint_{D^*} (\nabla \phi_1)^2 \, dx \, dy \geq \iint_{D^*} (\nabla \phi^*)^2 \, dx \, dy,$$

so that for the verification of (15.85) it suffices to derive a more elementary estimate of the form

$$(15.87) \quad \iint_{D} (\nabla \phi)^2 \, dx \, dy \geq \iint_{D^*} (\nabla \phi^{**})^2 \, dx \, dy.$$

From (15.83) and the free boundary condition (15.75) we deduce that

$$(15.88) \quad (\nabla \phi)^2 = \frac{2p}{I^2}$$

on Γ. For our proof of (15.87) we shall make essential use of the more significant fact that

$$(15.89) \quad (\nabla \phi)^2 \geq \frac{2p}{I^2}$$

throughout some fixed subregion of D surrounding Γ. In order to be sure of the validity of (15.89), however, we must restrict our discussion to cases where a small enough value has been assigned to the area A of Ω. Then the explicit results about the magnetic field B obtained in Exercise 14.6.3 show not only that $(\nabla\phi)^2$ increases initially as we enter D from Γ along the coordinate axes but also that all four of the arcs composing Γ are concave, which means that they have positive curvature

$$\kappa > 0.$$

On the other hand, because ϕ is harmonic it satisfies the differential equation

(15.90)
$$\frac{\partial}{\partial \nu}(\nabla\phi)^2 = 2\kappa(\nabla\phi)^2$$

along Γ. Therefore $(\nabla\phi)^2$ increases monotonically for a short distance along each of the inner normals to Γ, so that (15.89) becomes a consequence of the free boundary condition (15.88).

We proceed to extend the definition of ϕ smoothly into the fluid region Ω in such a way that

(15.91)
$$(\nabla\phi)^2 \leq \frac{2p}{I^2}$$

there. To achieve (15.91) we have merely to find an extension with the property that $(\nabla\phi)^2$ decreases as we move into Ω along the outer normals to Γ. The necessary construction can be accomplished by merely requiring that ϕ remain constant on each such normal. From the knowledge that Γ is concave it then follows that $(\nabla\phi)^2$ decreases in the desired fashion, for the normals involved spread apart as we enter Ω. Our extension of ϕ thus satisfies (15.91). Moreover, it yields a single-valued result in each quadrant of the (x,y)-plane separately, although jumps in the derivatives of ϕ develop across the segments of the coordinate axes contained inside Ω. No discontinuities in ϕ itself will occur, however, because it is an even function of both x and y in the vacuum domain D.

In order to establish the stability estimate (15.87) we choose ϕ^{**} to coincide in D^* with the extended version of ϕ that has just been introduced. We recall in this connection that ϕ^{**} was not necessarily supposed to be harmonic. Now because $\phi^{**} = \phi$ in the intersection D^*D of D^* with D, we have

(15.92)
$$\iint\limits_{D^*D} (\nabla\phi)^2\, dx\, dy = \iint\limits_{D^*D} (\nabla\phi^{**})^2\, dx\, dy.$$

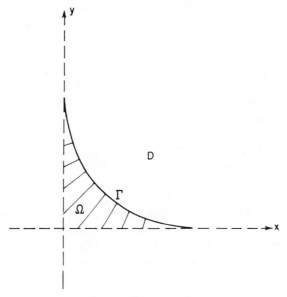

FIG. 83. Normals construction.

On the other hand, the constraint (15.81) on A^* implies that the area of $D - D^*$ is precisely equal to the area of $D^* - D$. At this stage we shall impose one additional hypothesis concerning how much D^* is allowed to deviate from D by requiring that $D - D^*$ be included in the neighborhood of Γ where the inequality (15.89) holds. Finally, observe that (15.91) applies throughout $D^* - D$. From these considerations it follows that

$$\iint\limits_{D-D^*} (\nabla\phi)^2\, dx\, dy \geq \iint\limits_{D-D^*} \frac{2p}{I^2}\, dx\, dy$$

$$= \iint\limits_{D^*-D} \frac{2p}{I^2}\, dx\, dy \geq \iint\limits_{D^*-D} (\nabla\phi^{**})^2\, dx\, dy,$$

which combines with (15.92) to complete our proof of (15.87).

The equivalent inequalities (15.82) and (15.87) show that the hydromagnetic figure of equilibrium Ω is stable even with reference to other shapes Ω^* of the same incompressible fluid that differ from it globally. The essential criterion for stability is seen from our proof to be that the free boundary Γ be concave. This fundamental criterion has numerous applications to other models in magnetohydrodynamics.

We observe here that our specific analysis based on the extremal property (15.82) of the potential energy M is closely connected with the earlier minimum problem (15.43) from which we developed an existence theorem for classical free streamline flows. In the context of more general free boundary problems we may view (15.43) and (15.82) as almost identical results in the calculus of variations. It should be emphasized, however, that the estimate (15.82) of the magnetic energy M is by itself of far more direct significance for plasma physics than is the extremal problem (15.43) for hydrodynamics, since the virtual mass plays a relatively unimportant role in engineering practice.

We conclude our study of free boundary problems with the comment that our derivation of the inequality (15.82) may be interpreted as a proof of uniqueness for the solution of the corresponding cusped geometry model. It is of interest to compare an argument of this kind with the variational method for establishing uniqueness of free streamlines that was described in Exercise 2.10. Since the new uniqueness proof is based on a natural boundary condition and on extension of the velocity potential along normals that spread out into Ω, it turns out, however, only to be applicable to flows with concave free streamlines, such as those appearing in the theory of finite wakes.

EXERCISES

1. Derive (15.90).
2. Make a comparison of the extremal problems (15.42), (15.46) and (15.87), which are based, respectively, on a power series coefficient a, a stream function $\bar{\psi}$, and a velocity potential ϕ. What relation has this comparison to the question of finding upper and lower bounds on domain functionals defined by means of the Dirichlet integral?
3. Use a Hadamard formula like (15.28) for the second variation of the magnetic energy M to verify the stability of the cusped figure of equilibrium Ω described in the text.
4. Derive the inequality (cf. Exercise 1.17)

$$\delta M \geq \frac{1}{2}\int_\Gamma \left(\frac{\partial\psi}{\partial\nu}\right)^2 \delta\nu \, ds,$$

which contains part of the statement of Hadamard's variational formula for the domain functional M, by applying the continuation of ϕ along normals that was described in the text to an estimate of the form (15.86).
5. For the proof of (15.87) replace our extension of ϕ into Ω in which it was specified to be constant along each of the outer normals to Γ by a construction[16] based on solving the first order partial differential equation

$$(\nabla\phi)^2 = \frac{2p}{I^2}$$

[16] Cf. Blank 1.

of geometrical optics in Ω, with initial values prescribed on Γ according to the original definition of ϕ there (cf. Section 2.4).

6. If we replace the constraint (15.80) of constant magnetic flux H by the requirement

$$I^* = I$$

of fixed current I in our formulation of the cusped geometry model, show that the correct expression to use for the potential energy of the resulting system becomes

$$\tilde{M} = M - 4HI = -M$$

instead of M. Apply our stability analysis to this new configuration.

7. Combine the law of conservation of energy with suitable energy inequalities to explain why the minimum potential energy criterion (15.82) for stability of a hydromagnetic figure of equilibrium implies that the figure is stable in the sense of an initial value problem, too, with the kinetic energy remaining bounded for all time.

8. Establish the stability of an infinite plane figure of equilibrium that is bounded by a picket fence composed of a sequence of cycloids (cf. Exercise 14.6.4).

9. Formulate the cusped geometry model of a figure of equilibrium corresponding to an axially symmetric vacuum magnetic field (cf. Exercise 14.6.5). Verify its stability.

10. Use an inequality like (15.87) to develop a uniqueness theorem for plane irrotational flow of an incompressible fluid past a convex body that is followed by a finite wake bounded by two concave free streamlines.

11. Give a rigorous derivation of Hadamard's variational formula (15.3) by establishing upper and lower bounds on the Green's function $G^*(\zeta, w)$ analogous to the inequalities found in Exercises 4 and 1.17.

4. THE PLATEAU PROBLEM

When a closed contour made of wire is dipped into a soap solution, a thin film of soap is formed within the wire. The area of the resulting surface will be a relative minimum among all surfaces stretched within the wire, since surface tension acts to bring the film into a position of stable equilibrium. Simple examples show, however, that these so-called *minimal surfaces* are in general not unique and do not always correspond to an absolute minimum of the surface area (cf. Exercise 2 below). The problem of finding a minimal surface spanned inside a given contour is known as *Plateau's problem.*[17]

The principal aim of this section will be to prove an existence theorem for Plateau's problem in the case of a closed curve in three-dimensional

[17] Cf. Courant 2, Radó 1.

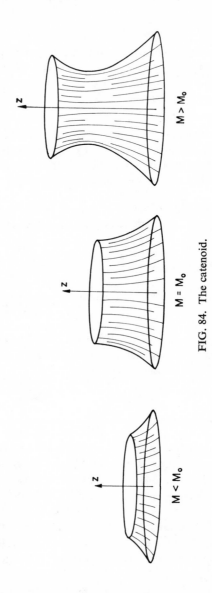

FIG. 84. The catenoid.

space. In addition to its obvious interest from the point of view of soap film experiments, the problem of least area turns out to be instructive and fruitful in other respects as well. On the one hand, it provides a typical example of a nonlinear Dirichlet problem; on the other, it exhibits some features of a free boundary problem. Certain questions in gas dynamics are connected with Plateau's problem (cf. Exercise 14.3.6 and Exercise 6 below), while Riemann's mapping theorem appears as a corollary to our existence theorem in the special case where the given curve lies in a plane.

Our treatment of Plateau's problem involves three main ideas. First the introduction of characteristic coordinates leads in a natural way to a formulation in terms of analytic functions of a complex variable and to an alternate extremal problem connected with Dirichlet's principle. Then the extremal problem is solved by the direct method of the calculus of variations. Finally the variational technique introduced in Section 1 is used to show that the solution of the alternate extremal problem defines a smooth surface of least area.

We shall begin with a discussion of the *non-parametric form* of Plateau's problem. If a smooth surface S is represented in (x,y,z)-space in the non-parametric form $z = u(x, y)$, then the area A of S is given by

$$(15.93) \qquad A = \iint\limits_{D} \sqrt{1 + u_x^2 + u_y^2}\, dx\, dy,$$

where D is the projection of S onto the (x,y)-plane. If the function $u(x, y)$ is to represent a minimal surface, the first variation of the area A must vanish, which implies that u satisfies the *Euler equation*

$$\frac{\partial}{\partial x}\left(\frac{u_x}{\sqrt{1 + u_x^2 + u_y^2}}\right) + \frac{\partial}{\partial y}\left(\frac{u_y}{\sqrt{1 + u_x^2 + u_y^2}}\right) = 0.$$

This reduces to the nonlinear elliptic partial differential equation

$$(15.94) \qquad (1 + u_y^2)u_{xx} - 2u_x u_y u_{xy} + (1 + u_x^2)u_{yy} = 0.$$

Hence the desired function u should be sought as the solution of a non-linear Dirichlet problem for equation (15.94).

The main limitation of the non-parametric form of Plateau's problem is that it excludes surfaces that cannot be represented by a single-valued function of x and y. A typical example is the *catenoid*, that is, the minimal surface formed by a soap film stretched between two circular rings, one located in the plane $z = 0$ and the other in the plane $z = M$. When M is

602   FREE BOUNDARY PROBLEMS

sufficiently small, the minimal surface is single-valued. However, for some value $M = M_0$ the soap film meets the plane $z = M_0$ at right angles. For $M > M_0$ the minimal surface becomes double-valued and the non-parametric form of the problem has no solution (cf. Exercise 1 below). This illustrates the fact that solutions of a nonlinear partial differential equation may not exist in the case of large boundary data.

Although the non-parametric formulation of the problem is inadequate for a general treatment, we shall use equation (15.94) to arrive at the more useful *parametric form* in a quite natural way. By the method of characteristics introduced in Section 3.4 we shall put equation (15.94) into canonical form. We now regard $x, y, u, p = u_x$ and $q = u_y$ as five unknown functions of new (complex) characteristic independent variables α and β. Since the equation of the characteristics of (15.94) is

$$(1 + q^2)\, dy^2 + 2pq\, dy\, dx + (1 + p^2)\, dx^2 = 0,$$

and since α is a characteristic coordinate, we may write

$$(1 + q^2)y_\alpha^2 + 2pq y_\alpha x_\alpha + (1 + p^2)x_\alpha^2 = 0.$$

It follows that

$$x_\alpha^2 + y_\alpha^2 + (px_\alpha + qy_\alpha)^2 = 0,$$

and because $u_\alpha = px_\alpha + qy_\alpha$ we obtain

$$(15.95) \qquad x_\alpha^2 + y_\alpha^2 + u_\alpha^2 = 0.$$

Similarly we have

$$(15.96) \qquad x_\beta^2 + y_\beta^2 + u_\beta^2 = 0.$$

It is natural to regard the triple of functions (x, y, u) as a parametric representation of our surface in terms of the parameters α and β. Since equation (15.94) is elliptic, the characteristic coordinates α and β are actually complex conjugates (cf. Sections 3.4 and 16.2). It is therefore convenient to make the following change in notation: replace α by $z = x + iy$, β by $\bar{z} = x - iy$, and the triple (x, y, u) by the vector $\mathbf{r} = (u, v, w)$. Then the surface is represented parametrically in (u,v,w)-space by the position vector $\mathbf{r} = (u, v, w)$, which is a function of the parameters z and $\bar{z}$. In the new notation equation (15.95) reads

$$(15.97) \qquad \mathbf{r}_z^2 = u_z^2 + v_z^2 + w_z^2 = 0,$$

of which equation (15.96) becomes the complex conjugate.

Differentiating (15.97) with respect to $\bar{z}$ we obtain

$$\mathbf{r}_z \cdot \mathbf{r}_{z\bar{z}} = 0.$$

Because $\Delta\mathbf{r} = 4\mathbf{r}_{z\bar{z}}$ this yields

(15.98) $\mathbf{r}_z \cdot \Delta\mathbf{r} = 0,$ $\mathbf{r}_{\bar{z}} \cdot \Delta\mathbf{r} = 0.$

However, according to Exercise 3.4.6 we have

$$\begin{vmatrix} \Delta u & \Delta v & \Delta w \\ u_z & v_z & w_z \\ u_{\bar{z}} & v_{\bar{z}} & w_{\bar{z}} \end{vmatrix} = 0,$$

which implies that $\Delta\mathbf{r}$ is a linear combination of $\mathbf{r}_z$ and $\mathbf{r}_{\bar{z}}$. The latter assertion is compatible with (15.98) only if

(15.99) $\Delta\mathbf{r} = (\Delta u, \Delta v, \Delta w) = 0.$

Equation (15.99) is the desired canonical form of equation (15.94).

Note that whereas the original equation (15.94) was nonlinear, the canonical system (15.99) consists of three linear equations. Thus we see that in parametric form Plateau's problem consists in finding three harmonic functions u, v, w satisfying (15.97), together with appropriate boundary conditions. The parametric problem has the advantage geometrically of admitting surfaces that intersect parallels to the w-axis more than once.

The representation of a minimal surface in terms of harmonic functions leads immediately to the classical *Weierstrass representation* in terms of analytic functions. If the three harmonic functions u, v, w are considered to be the real parts

$$u = \text{Re}\,\{\xi(z)\}, \quad v = \text{Re}\,\{\eta(z)\}, \quad w = \text{Re}\,\{\zeta(z)\}$$

of three analytic functions ξ, η, ζ, then (15.97) becomes

(15.100) $\xi'(z)^2 + \eta'(z)^2 + \zeta'(z)^2 = 0.$

Equation (15.100) shows that the three functions ξ, η, ζ are not independent. If one of them is introduced as a new independent variable, then of the remaining two precisely one is still arbitrary. Hence the totality of all minimal surfaces depends on just one arbitrary analytic function of a complex variable. In fact, in Exercise 4 it is shown that any minimal surface may be represented in the form

(15.101)
$$\begin{cases} u = \text{Re}\left\{\int (1 - \zeta^2)F(\zeta)\,d\zeta\right\}, \\[2mm] v = \text{Re}\left\{\int i(1 + \zeta^2)F(\zeta)\,d\zeta\right\}, \\[2mm] w = \text{Re}\left\{\int 2\zeta F(\zeta)\,d\zeta\right\}, \end{cases}$$

where $F(\zeta)$ is some analytic function of the complex variable ζ. The formulas (15.101), which are known as the *formulas of Weierstrass,* define the general solution of the equation (15.94) of minimal surfaces.

Let us discuss next the differential geometry of the parametric problem, which is connected with conformal mapping. Separating (15.97) into its real and imaginary parts and using the usual notation

$$e = u_x^2 + v_x^2 + w_x^2,$$

$$f = u_x u_y + v_x v_y + w_x w_y,$$

$$g = u_y^2 + v_y^2 + w_y^2,$$

we obtain the conditions

(15.102) $e - g = 0, \quad f = 0$

on the *first fundamental form*

$$ds^2 = e\, dx^2 + 2f\, dx\, dy + g\, dy^2$$

of the surface. These are precisely the requirements that the coordinates x and y be *isometric*. Consequently our mapping of a region in the (x,y)-plane onto the minimal surface is conformal. In the special case where $w \equiv 0$ equations (15.102) merely express the fact that $u + iv$ is a complex analytic function of $x + iy$ or of $x - iy$.

The harmonic property of the functions u, v, w suggests a possible relationship between Dirichlet's principle and the problem of least area. It is convenient to define the Dirichlet integral of a vector function $\mathbf{r} = (u, v, w)$ over a domain D as

$$\|\mathbf{r}\|^2 = \tfrac{1}{2} \iint\limits_{D} (e + g)\, dx\, dy$$

$$= \tfrac{1}{2} \iint\limits_{D} (u_x^2 + u_y^2 + v_x^2 + v_y^2 + w_x^2 + w_y^2)\, dx\, dy.$$

On the other hand, the area A of a surface S parametrized by the vector function $\mathbf{r}$ of x and y is

$$A = \iint\limits_{D} \sqrt{eg - f^2}\, dx\, dy$$

$$= \iint\limits_{D} \sqrt{\left[\frac{\partial(u, v)}{\partial(x, y)}\right]^2 + \left[\frac{\partial(v, w)}{\partial(x, y)}\right]^2 + \left[\frac{\partial(w, u)}{\partial(x, y)}\right]^2}\, dx\, dy.$$

Since

$$\sqrt{eg - f^2} \leq \sqrt{eg} \leq \frac{e + g}{2},$$

with equality holding if and only if $e = g$ and $f = 0$, we have

$$A \leq \|\mathbf{r}\|^2$$

for arbitrary parametrizations $\mathbf{r}$ of the surface S. Furthermore,

$$A = \|\mathbf{r}\|^2$$

whenever (15.102) is satisfied. Although the area A of a surface S is independent of the choice of parameters, the Dirichlet integral $\|\mathbf{r}\|^2$ is not; it is in fact found to be a minimum for isometric parameters. Thus when we grant that isometric parameters can be introduced on every surface, the problem of minimizing Dirichlet's integral and the problem of least area become equivalent.

From Exercise 8.3.10 it follows that isometric parameters can always be introduced on surfaces which are sufficiently smooth. This amounts to solving the Beltrami equations (3.42). We shall rely on the results of Exercise 8.3.10 mentioned above to secure a complete equivalence between the problem of minimizing Dirichlet's integral and the problem of least area. Otherwise our treatment of the Plateau problem is self-contained. In particular, we shall only need Dirichlet's principle in the case of a circle, which is a relatively elementary result (cf. Exercise 8.1.4).

We come now to a more precise formulation of the version of the Plateau problem that we intend to solve here. Let Γ be a given rectifiable Jordan curve in three-dimensional (u,v,w)-space and let D be the open unit disc in the (x,y)-plane. Our aim is to find a vector $\mathbf{r}_0$ which is harmonic in D, satisfies condition (15.97) and describes a surface S spanned through Γ in such a way that ∂D is mapped one-to-one and continuously onto Γ. Our proof of the existence of the desired vector $\mathbf{r}_0$ will consist of two parts. First we shall solve the problem of minimizing Dirichlet's integral $\|\mathbf{r}\|^2$ by the direct method of the calculus of variations, using an equicontinuity argument. Then we shall show that the extremal vector $\mathbf{r}_0$ satisfies the conformality condition (15.97).

Our class U of admissible vectors shall consist of all functions $\mathbf{r}$ that are continuous in $D + \partial D$, are piecewise smooth in D, define a continuous one-to-one mapping of ∂D onto Γ, and have uniformly bounded Dirichlet integrals. The assumption that Γ is rectifiable ensures that U is not empty, provided the upper bound M on Dirichlet's integral is taken sufficiently large (cf. Exercise 8 below). Thus our extremal problem is to find a vector $\mathbf{r}_0$ in U for which

$$(15.103) \qquad \|\mathbf{r}_0\|^2 = d = \min \|\mathbf{r}\|^2.$$

Because of Dirichlet's principle for a circle (cf. Exercise 8.1.4), we may restrict our attention to those functions in U that are harmonic. Indeed, the effect of replacing any admissible vector by the harmonic vector having the same boundary values on ∂D is to decrease the value of the Dirichlet integral. Our class of admissible functions U was chosen to include functions that are not harmonic only in order to facilitate our variational procedure later on.

Let $\mathbf{r}_n$ be a minimizing sequence of admissible harmonic vectors for which

$$\lim_{n \to \infty} \|\mathbf{r}_n\|^2 = d.$$

Since the Dirichlet integral is invariant under conformal mapping, we may normalize the sequence by the following *three point condition*. Let P_1, P_2, P_3 be three distinct points on ∂D and let P_1', P_2', P_3' be three distinct points on Γ. We simply suppose that each of the functions $\mathbf{r}_n$ carries the three points P_1, P_2, P_3 into P_1', P_2', P_3', respectively. No generality is lost in this way because the unit circle can be mapped conformally onto itself so that any three boundary points go into P_1, P_2, P_3.

It should be noted that although each function $\mathbf{r}_n$ maps P_1, P_2, P_3 onto P_1', P_2', P_3', the nature of the mapping on the rest of the boundary remains free. Hence each function $\mathbf{r}_n$ defines, in general, a different correspondence between ∂D and Γ. Since $\mathbf{r}_n$ is harmonic, its Dirichlet integral is a minimum among all functions having the same boundary values. Now, for harmonic functions $\mathbf{r}$ Dirichlet's integral $\|\mathbf{r}\|^2$ becomes a functional of only the boundary correspondence. Hence we may take the point of view that our minimizing sequence is a sequence of boundary correspondences for which Dirichlet's integral approaches a minimum. In this sense what we are dealing with is more or less a free boundary problem.

We need the following lemma to show that our minimizing sequence of boundary correspondences contains a convergent subsequence:

LEMMA.   *Consider the class of admissible functions in U which are normalized by the three point condition. The boundary values of these functions are equicontinuous. The corresponding harmonic functions on D form a compact set.*

The second statement follows from the first because of Arzela's theorem (cf. Section 10.3) and the maximum principle. Therefore the only difficulty is to establish the equicontinuity of the normalized boundary values of the vectors **r**. Let an arbitrary $\epsilon > 0$ be given, and pick any point Q_0 on ∂D. Our aim is to find a $\delta > 0$ depending only on ϵ such that whenever the two points Q, Q_0 lie on ∂D with

(15.104) $|Q - Q_0| < \delta,$

then their images $\mathbf{r}(Q)$, $\mathbf{r}(Q_0)$ on Γ satisfy

(15.105) $|\mathbf{r}(Q) - \mathbf{r}(Q_0)| < \epsilon.$

Recall that the diameter of an arc is defined to be the supremum of the distance between any two points on the arc. The Jordan curve Γ clearly has the following property. Given $\epsilon > 0$, there exists a $\tau > 0$ such that whenever A', B' are two points on Γ with $|A' - B'| < \tau$, then one of the two arcs of Γ connecting A' and B' has diameter less than ϵ. This smaller arc will contain at most one of the points P'_j used in the three point normalization, provided τ is taken less than the minimum distance among all the points P'_j.

Consider a circle of radius R about Q_0 intersecting ∂D at points A and B. Suppose for the moment that for some value of R the images $A' = \mathbf{r}(A)$, $B' = \mathbf{r}(B)$ of the points A, B satisfy $|A' - B'| < \tau$. By taking R less than a suitable number $\delta > 0$ we can make the arc AQ_0B of ∂D contain at most

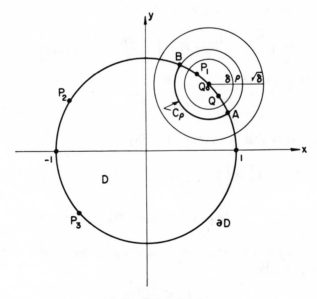

FIG. 85. Estimating L_ρ.

one of the points P_j and hence correspond to the smaller arc of Γ connecting A' and B'. Then any point Q on ∂D for which (15.104) holds will have an image $\mathbf{r}(Q)$ on Γ satisfying (15.105).

In order to prove the existence of the required values of R and δ, we shall estimate the length L_R of the image of the arc C_R that D has in common with the circle of radius R about Q_0. In a polar coordinate system about Q_0 we have the inequality

$$\frac{1}{2} \iint_D \left\{ u_R^2 + v_R^2 + w_R^2 + \frac{1}{R^2} (u_\theta^2 + v_\theta^2 + w_\theta^2) \right\} R \, dR \, d\theta \leq M$$

for the Dirichlet integral, from which it follows in particular that

$$\iint_D (u_\theta^2 + v_\theta^2 + w_\theta^2) \frac{dR}{R} \, d\theta \leq 2M.$$

If we set

$$F(R) = \int_{C_R} (u_\theta^2 + v_\theta^2 + w_\theta^2) \, d\theta,$$

then certainly

$$\int_\delta^{\sqrt{\delta}} F(R) \frac{dR}{R} \leq 2M$$

for any δ in the interval $0 < \delta < 1$. However, by the mean value theorem there is a ρ with $\delta < \rho < \sqrt{\delta}$ for which

$$\int_\delta^{\sqrt{\delta}} F(R) \frac{dR}{R} = F(\rho) \int_\delta^{\sqrt{\delta}} \frac{dR}{R} = \frac{1}{2} F(\rho) \log \frac{1}{\delta}.$$

Therefore we obtain

$$F(\rho) \leq \frac{4M}{\log 1/\delta}.$$

Now observe that Schwarz's inequality gives

$$L_\rho^2 = \left\{ \int_{C_\rho} \sqrt{u_\theta^2 + v_\theta^2 + w_\theta^2} \, d\theta \right\}^2$$

$$\leq 2\pi \int_{C_\rho} (u_\theta^2 + v_\theta^2 + w_\theta^2) \, d\theta = 2\pi F(\rho).$$

Hence to any $\delta > 0$ there corresponds a ρ in the interval $\delta < \rho < \sqrt{\delta}$ with the property

$$L_\rho^2 \leq \frac{8\pi M}{\log 1/\delta}.$$

Because the length L_p of the image of C_p exceeds the distance between the endpoints A' and B', and because $L_p \to 0$ as $\delta \to 0$, we can find a $\delta > 0$ depending only on ϵ such that $L_p < \tau$ and such that (15.105) holds whenever Q satisfies (15.104). This completes the proof of the equicontinuity lemma.

Next we apply the lemma to our minimizing sequence $\mathbf{r}_n$ of harmonic vectors. It enables us to choose a subsequence that converges uniformly in D to a harmonic limit function $\mathbf{r}_0$. In view of the lower semicontinuity of Dirichlet's integral (cf. Section 8.2) we obtain

$$\|\mathbf{r}_0\|^2 \leq \underline{\lim} \|\mathbf{r}_n\|^2 = d,$$

where $\mathbf{r}_n$ is now supposed to indicate the convergent subsequence. However, since $\mathbf{r}_0$ is an admissible function we have

$$d \leq \|\mathbf{r}_0\|^2,$$

too, which means that $\mathbf{r}_0$ is the desired solution of the extremal problem (15.103).

We still have to show that the extremal vector $\mathbf{r}_0$ satisfies the conformality condition (15.97). It is at this stage that we make use of the variational technique introduced in Section 1. We shall derive here a particular variational formula for the Dirichlet integral that is especially helpful in the case of the Plateau problem (cf. Exercise 1.3).

Let the unit disc D be mapped onto itself by the infinitesimal transformation

(15.106) $$z^* = z + \epsilon F(z, \bar{z}; \epsilon, \bar{\epsilon}).$$

We shall allow ϵ to be complex and F to be piecewise continuously differentiable. We define a transformed vector $\mathbf{r}^*$ in the z^*-plane by the rule

$$\mathbf{r}^*(z^*) = \mathbf{r}(z),$$

which gives a function that is piecewise continuously differentiable, but not necessarily harmonic. It is our intent to estimate the varied Dirichlet integral $\|\mathbf{r}^*\|^2$. Thus we must calculate the first variation $\delta \|\mathbf{r}\|^2$ of Dirichlet's integral.

We shall first determine the quantity

$$\delta \|u\|^2 = \delta \iint_D |\nabla u|^2 \, dx \, dy$$

by a conceptual argument. Since this variation ought to be linear in F_z and $F_{\bar{z}}$, we may write

$$\delta \|u\|^2 = \epsilon \iint_D [aF_{\bar{z}} + bF_z]\, dx\, dy.$$

If the mapping (15.106) is conformal, which means that $F_{\bar{z}} = 0$, then the Dirichlet integral is invariant. Consequently the coefficient b must vanish identically. On the other hand, since the integrand $|\nabla u|^2 = 4u_z u_{\bar{z}}$ is quadratic in the first derivatives of u, the coefficient a ought to be a quadratic form in u_z and $u_{\bar{z}}$, too. Hence

$$\delta \|u\|^2 = \epsilon \iint_D [\alpha u_z^2 + 2\beta u_z u_{\bar{z}} + \gamma u_{\bar{z}}^2] F_{\bar{z}}\, dx\, dy,$$

where α, β, γ are constants.

If we restrict F to those functions which vanish on ∂D, then an integration by parts gives

$$\delta \|u\|^2 = -\epsilon \iint_D [\alpha u_z^2 + 2\beta u_z u_{\bar{z}} + \gamma u_{\bar{z}}^2]_{\bar{z}} F\, dx\, dy.$$

For such functions F the first variation of Dirichlet's integral has to vanish whenever u is harmonic, whereas our formula implies that

$$\alpha u_z^2 + 2\beta u_z u_{\bar{z}} + \gamma u_{\bar{z}}^2$$

is a complex analytic function of z. Since only the analyticity of the term αu_z^2 follows from the fact that u is harmonic, we conclude that $\beta = \gamma = 0$. Therefore we have

$$\delta \|u\|^2 = \operatorname{Re}\left\{ \epsilon \iint_D \alpha u_z^2 F_{\bar{z}}\, dx\, dy \right\}.$$

A more detailed calculation (cf. Exercise 1.3) actually shows that

$$\delta \|u\|^2 = -8 \operatorname{Re}\left\{ \epsilon \iint_D u_z^2 F_{\bar{z}}\, dx\, dy \right\}.$$

Repeating the above argument for v and w and recalling that the Dirichlet integral of the vector function $\mathbf{r} = (u, v, w)$ was defined to be one half the sum of the Dirichlet integrals of its components, we obtain the variational formula

(15.107) $$\delta \|\mathbf{r}\|^2 = -4 \operatorname{Re}\left\{ \epsilon \iint_D [u_z^2 + v_z^2 + w_z^2] F_{\bar{z}}\, dx\, dy \right\}.$$

We shall use this formula to show that the extremal vector $\mathbf{r}_0 = (u, v, w)$ satisfies the conformality condition (15.97). Since $\mathbf{r}_0$ minimizes the Dirichlet integral, the first variation $\delta \|\mathbf{r}_0\|^2$ must vanish. What we want to do now is to pick an infinitesimal transformation (15.106) which alters the boundary correspondence in such a way that the vanishing of the first variation reduces to condition (15.97).

Let $z_0 \neq 0$ be an arbitrary point of D and choose $\rho > 0$ small enough so that the disc E where $|z - z_0| \leq \rho$ is contained in D. For small complex values of the parameter ϵ we put $z_1 = z_0 + \epsilon$ and define the mapping (15.106) as follows. In $D - E$ we set

$$(15.108) \qquad z^* = z \frac{z - z_1}{1 - \bar{z}_1 z} \frac{1 - \bar{z}_0 z}{z - z_0}$$

$$= z \frac{1 - \bar{z}_0 z}{1 - \bar{z}_1 z} \left[1 - \frac{\epsilon}{z - z_0} \right],$$

while in E we put

$$(15.109) \qquad z^* = z \frac{1 - \bar{z}_0 z}{1 - \bar{z}_1 z} \left[1 - \frac{\epsilon}{\rho^2} (\bar{z} - \bar{z}_0) \right].$$

The transformation given by (15.108) and (15.109) is continuous across the boundary of E, since that boundary has the equation

$$z - z_0 = \frac{\rho^2}{(\bar{z} - \bar{z}_0)}.$$

The mapping (15.108) is conformal in $D - E$, and consequently the first variation of the Dirichlet integral there must be zero. Thus the only contribution to the first variation comes from (15.109). From (15.107) we obtain

$$\delta \|\mathbf{r}_0\|^2 = 4 \, \text{Re} \left\{ \epsilon \iint_E [u_z^2 + v_z^2 + w_z^2] \frac{z}{\rho^2} \, dx \, dy \right\} = 0.$$

Because the argument of ϵ is arbitrary and because of the mean value theorem for harmonic functions (cf. Section 7.2), it follows that

$$[u_z^2 + v_z^2 + w_z^2]z \bigg|_{z=z_0} = \frac{1}{\pi \rho^2} \iint_E [u_z^2 + v_z^2 + w_z^2]z \, dx \, dy = 0.$$

Therefore

$$u_z^2 + v_z^2 + w_z^2 \equiv 0$$

throughout D, which is exactly the conformality condition (15.97).

We have now completed all three steps in our proof of the fundamental

THEOREM.   *In the case of a rectifiable Jordan curve, the Plateau problem can always be solved.*

More specifically, what we have constructed is a minimal surface having the topological structure of a disc. This surface has least area among all simply connected surfaces through the given Jordan curve. However, there may be other surfaces of smaller area through the same curve, and even the simply connected minimal surface may not be unique (cf. Exercise 2 below). Whenever our minimal surface has a non-parametric representation it can be interpreted as a solution of Dirichlet's problem for the nonlinear elliptic partial differential equation (15.94). However, the non-parametric problem does not always have a solution (cf. Exercises 1 and 14.).

An important special case of the preceding theorem arises when the curve Γ lies in the plane $w = 0$. Then it is clear that the surface of least area spanned inside Γ is the region it bounds in the (u,v)-plane. Our method thus produces a solution of the special form $(u, v, 0)$. In particular, $u + iv$ is an analytic function of $x + iy$ and defines a conformal mapping of the unit disc onto the plane region bounded by Γ. Hence we obtain the Riemann mapping theorem (cf. Section 7.3) as a corollary of the existence theorem for Plateau's problem. More precisely, any simply connected region in the plane that is bounded by a rectifiable Jordan curve Γ can be mapped conformally onto the interior of the unit disc D. Moreover, any three points on Γ can be made to correspond to any three points on the circle ∂D, and the mapping is one-to-one and continuous at the boundary.

EXERCISES

1. Find an ordinary differential equation for the axially symmetric minimal surface connecting two coaxial circles. Show that the answer is a catenoid (rotated catenary), and discuss its dependence on parameters. In particular, exhibit two solutions, one of which is stable and the other unstable.

2. Combine two circular arcs with two arcs of a catenary to form a Jordan curve through which at least two different simply connected minimal surfaces can be spanned.

3. Prove that in the special case where $w \equiv 0$ equations (15.102) imply that $u + iv$ is a complex analytic function of $x + iy$ or of $x - iy$.

4. Establish the Weierstrass representation formulas (15.101). Discuss them for the special case of a polygonal boundary curve Γ in the light of the Schwarz-Christoffel transformation (cf. Exercise 7.3.7).

5. Verify that the choices $F = ia/\zeta^2$ and $F = a/\zeta^2$ for the function F in the Weierstrass formulas (15.101) yield a right helicoid and a catenoid, respectively. Show that $F = 1 - \zeta^{-4}$ leads to a one-sided surface.

6. Develop the connection between minimal surfaces and the steady flow of a Karman-Tsien gas (cf. Exercise 14.3.6). Use the hodograph transformation to

reduce the equation (15.94) for minimal surfaces to Laplace's equation. What has this to do with the mapping of the surface onto the unit sphere by means of its normals?

7. Consider a minimal surface whose boundary Γ is an analytic curve. Use the Weierstrass formulas (15.101) to develop rules[18] for extending the surface analytically beyond its boundary (cf. Section 16.4).

8. Use Fourier expansions of u, v, w as functions of the arc length along Γ to show that a finite Dirichlet integral $\|\mathbf{r}\|^2 < \infty$ can be achieved whenever Γ is rectifiable (cf. Exercise 8.1.3).

9. Consider the Jordan curve Γ defined in terms of spherical coordinates (r, θ, ϕ) by the formulas $r = \sin\theta$ and $\phi = \cot^5\theta$, with $0 \leq \theta < \pi$. Show that Γ is analytic everywhere except at one point, where it has a well defined tangent. Prove that any simply connected surface spanned inside Γ has infinite area. What is the length of Γ?

10. The estimates used to establish the equicontinuity lemma of the text imply the continuity of any conformal mapping at a boundary composed of Jordan arcs. By means of similar estimates[19] discuss the assumption of boundary values in Dirichlet's principle for the case of two independent variables (cf. Section 8.3).

11. Show that in the analysis of the text the parameter domain D may be taken as a half-plane instead of a circle. What advantages has this choice of D in determining a specific infinitesimal transformation (15.106) to replace (15.108) and (15.109)?

12. Explain why the derivation of the conformality condition (15.97) from the variational formula (15.107) becomes easier if we suppose that the Riemann mapping theorem is already known.

13. Show that every minimal surface spanned through a Jordan curve Γ must lie within the convex hull of Γ. Interpret this result as a generalization of the maximum principle.

14. Let D be a convex region of the (x,y)-plane and let Γ be a curve in space which has a one-to-one projection onto the boundary of D. Use Steiner symmetrization (cf. Section 11.3) to establish that the simply connected minimal surface through Γ has a non-parametric representation $z = u(x, y)$ and is unique (cf. Exercises 7.1.15 and 2.10). From these results formulate an existence theorem concerning Dirichlet's problem for the nonlinear elliptic partial differential equation (15.94).

15. Show that the formula $z^* = z + \epsilon/(z - z_0)$ gives a conformal mapping of the region $|z - z_0| > \rho > \sqrt{|\epsilon|}$ onto the exterior of an ellipse, whereas $z^* = z + \epsilon(\bar{z} - \bar{z}_0)/\rho^2$ defines an affine transformation of the circle $|z - z_0| < \rho$ onto the interior of the same ellipse.

[18] Cf. Lewy 2.
[19] Cf. Courant 2.

16

Partial Differential Equations in the Complex Domain

1. CAUCHY'S PROBLEM FOR ANALYTIC SYSTEMS

We have already had occasion quite often in this book to discuss partial differential equations in the complex domain. Such techniques were used at the very start of Chapter 3 to reduce an elliptic equation to canonical form and in Chapter 5 to construct fundamental solutions. They arose again in Section 6.2 in connection with Cauchy's problem, and in Section 7.3 in connection with the method of images. In Section 12.2 we even based an existence theorem for symmetric hyperbolic systems on complex substitutions. Here we intend to make a broader study of the results that can be achieved by extending solutions of analytic partial differential equations into the domain of complex values of their independent variables. Combined with the method of finite differences, this theory has important applications to free boundary problems of fluid dynamics and of magnetohydrodynamics (cf. Exercises 14.5.10 and 15.2.11, and 11 below).

To start with we shall show how to transform incorrectly set Cauchy problems for analytic systems of arbitrary type into well posed initial value problems for symmetric hyperbolic systems in a larger number of independent variables (cf. Section 1). Then we shall refine the procedure in the case of a single equation of the second order in two independent variables by taking advantage of the method of characteristics (cf. Section 2). The latter method will also enable us to establish the analyticity of solutions of nonlinear analytic elliptic equations in two independent variables (cf. Section 3). Finally, for elliptic equations in two independent variables we shall develop reflection rules that will permit us to extend solutions analytically across curves on which they satisfy analytic boundary conditions (cf. Section 4).

Let us consider a quasi-linear system of m partial differential equations

(16.1)
$$\frac{\partial u}{\partial t} = \sum_{j=1}^{n} A_j \frac{\partial u}{\partial x_j} + B$$

of the first order in $n + 1$ independent variables $x_1, \ldots, x_n$ and t, where

$$u = u(x, t) = \begin{pmatrix} u_1 \\ \cdot \\ \cdot \\ \cdot \\ u_m \end{pmatrix}$$

denotes a column vector of unknown functions, where the coefficients

$$A_j = A_j(x, t, u), \qquad j = 1, \ldots, n,$$

are m-by-m matrices which depend analytically on $x = (x_1, \ldots, x_n)$, t and u, and where $B = B(x, t, u)$ is a given column vector whose elements are also analytic functions of x, t and u. Actually we have shown in Section 1.2 that any system of partial differential equations can be brought into the canonical form (16.1), provided that the hyperplane $t = 0$ is not a characteristic. We shall be interested in solving the Cauchy problem for (16.1) with analytic initial values

(16.2)
$$u(x, 0) = f(x)$$

prescribed at $t = 0$.

The Cauchy-Kowalewski theorem, which was proved in Section 1.2, asserts that there exists a convergent power series solution of (16.1) and (16.2) in the neighborhood of each point where the data of the problem are analytic. According to the principles of analytic continuation, this power series defines a solution u of (16.1) and (16.2) for complex as well as for real values of the independent variables x and t. For our purposes it will suffice to keep the time coordinate t real, but we shall replace the space coordinates $x_1, \ldots, x_n$ by corresponding complex variables $z_j = x_j + iy_j$. Since u is an analytic function, the partial derivatives

(16.3)
$$\frac{\partial u}{\partial z_j} = \frac{1}{2} \left(\frac{\partial u}{\partial x_j} - i \frac{\partial u}{\partial y_j} \right)$$

can be evaluated in the customary fashion as limits of difference quotients based on complex increments of $z_1, \ldots, z_n$ with arbitrary arguments.

Moreover, it is evident that the n Cauchy-Riemann equations

$$(16.4) \qquad \frac{\partial u}{\partial \bar{z}_j} = \frac{1}{2}\left(\frac{\partial u}{\partial x_j} + i\frac{\partial u}{\partial y_j}\right) = 0$$

are fulfilled.

In terms of the complex variables $z_1, \ldots, z_n$ we can express the system of partial differential equations (16.1) in the form

$$(16.5) \qquad \frac{\partial u}{\partial t} = \sum_{j=1}^{n} A_j \frac{\partial u}{\partial z_j} + B.$$

The identity (16.5) holds throughout some region of the $(2n + 1)$-dimensional Euclidean space spanned by the real coordinates $x_1, \ldots, x_n$, $y_1, \ldots, y_n$ and t. Similarly, analytic continuation of (16.2) into the complex domain yields the extended initial condition

$$(16.6) \qquad u(z, 0) = f(z)$$

over a $(2n)$-dimensional section of the hyperplane $t = 0$ in space of dimension $2n + 1$, where we have put $z = (z_1, \ldots, z_n)$. Now observe that (16.4) and (16.5) comprise an overdetermined system of $m(n + 1)$ complex partial differential equations for the m complex-valued unknown functions $u_1, \ldots, u_m$. Our aim[1] is to find a combination of these relations which reduces to a symmetric hyperbolic system of precisely m complex equations for u.

Let $\bar{A}'_j$ stand for the transpose of the complex conjugate of the matrix A_j. Multiplying the Cauchy-Riemann equations (16.4) by $\bar{A}'_j$ and adding the results to (16.5), we obtain

$$\frac{\partial u}{\partial t} = \sum_{j=1}^{n} A_j \frac{\partial u}{\partial z_j} + \sum_{j=1}^{n} \bar{A}'_j \frac{\partial u}{\partial \bar{z}_j} + B.$$

Formulas (16.3) and (16.4) show that this is equivalent to

$$(16.7) \qquad \frac{\partial u}{\partial t} = \sum_{j=1}^{n} \frac{A_j + \bar{A}'_j}{2}\frac{\partial u}{\partial x_j} + \sum_{j=1}^{n} \frac{A_j - \bar{A}'_j}{2i}\frac{\partial u}{\partial y_j} + B,$$

which constitutes the desired symmetric hyperbolic system. To verify that (16.7) is of the stated type, we merely note that the coefficients $(A_j + \bar{A}'_j)/2$ and $(A_j - \bar{A}'_j)/2i$ appearing on the right are all Hermitian matrices (cf. Section 3.5). Thus through a preliminary analytic continuation into the

[1] Cf. Garabedian 9.

complex domain we have been able to convert the quite general Cauchy problem (16.1), (16.2) for m equations in $n + 1$ independent variables into the more reasonable initial value problem (16.6), (16.7) for a symmetric hyperbolic system of $2m$ real equations in $2n + 1$ real independent variables. Because the latter question is correctly set (cf. Section 12.2), such an analysis explains in a new and enlightening way why the Cauchy-Kowalewski theorem is applicable to analytic systems of arbitrary type.

The characteristic surfaces for the hyperbolic system (16.7) are real manifolds of dimension $2n$. By familiar rules (cf. Section 6.1) they serve to define the domain of dependence in the initial hyperplane $t = 0$ where data must be known if we are to determine a solution of that system at some specific point. Even when the point at issue lies in the real domain

$$y = (y_1, \ldots, y_n) = 0,$$

the corresponding domain of dependence may be complex. In fact, we can expect it in general to be so unless the original system (16.1) is symmetric hyperbolic to begin with (cf. Section 12.2). In the latter case we obtain zero coefficients

$$\frac{A_j - \bar{A}'_j}{2i} = 0$$

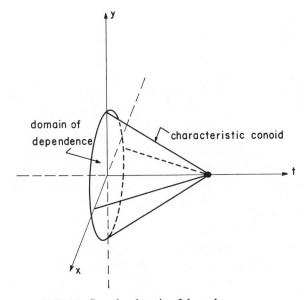

FIG. 86. Complex domain of dependence.

in (16.7) for the partial derivatives of u with respect to the imaginary parts y_j of the variables z_j in the real domain; hence the relevant characteristic conoids degenerate to manifolds of lower dimension there and the associated domains of dependence become strictly real.

The most remarkable feature about the complex system (16.7) is that it is symmetric hyperbolic regardless of the type of the analytic system (16.1) from which it was derived. Moreover, the characteristics of (16.7) are not necessarily related in a direct way to those of (16.1), which might be either real or imaginary. When (16.1) does possess imaginary characteristics, we cannot consider the Cauchy problem to be reasonable for it. That (16.6) and (16.7) nevertheless define a well posed initial value problem is explained by the fact that the data (16.2) must be extended into the complex domain in such cases to give meaning to (16.6). The analytic continuation of the data is itself an unstable process, since small perturbations in the prescribed function f for real choices of x may result in large changes in its extension to complex arguments z (cf. Section 4.1). These instabilities are actually what causes the original problem (16.1), (16.2) to be incorrectly set for a system of, say, the elliptic type.

Our next objective will be to show that for complex analytic data every solution u of (16.6), (16.7) is itself analytic and therefore represents a solution of the system (16.1), (16.2), too. To this end we introduce the quantities

$$w^{(j)} = \begin{pmatrix} w_1^{(j)} \\ \cdot \\ \cdot \\ \cdot \\ w_m^{(j)} \end{pmatrix} = \frac{\partial u}{\partial \bar{z}_j}, \qquad j = 1, \ldots, n,$$

and proceed to derive for them a symmetric hyperbolic system of mn linear homogeneous partial differential equations of the first order. The corresponding initial values of the $w^{(j)}$ are seen to vanish because of the Cauchy-Riemann equations

$$(16.8) \qquad \frac{\partial f}{\partial \bar{z}_j} = 0, \qquad j = 1, \ldots, n,$$

satisfied by the prescribed vector f. Consequently a standard uniqueness theorem (cf. Section 6.1) will identify $w^{(j)}$ as the zero solution

$$(16.9) \qquad w^{(j)} = 0$$

of a homogeneous Cauchy problem. According to the definition of $w^{(j)}$, (16.9) is equivalent to the Cauchy-Riemann equations

$$(16.10) \qquad \frac{\partial u}{\partial \bar{z}_j} = 0, \qquad j = 1, \ldots, n,$$

which comprise one statement of the required complex analyticity of u in its dependence on z.

Symmetric hyperbolic differential equations for $w^{(1)}, \ldots, w^{(n)}$ are obtained by applying the differential operators $\partial/\partial \bar{z}_l$ to (16.7). Since the A_j and B are supposed to be analytic functions of z and u, we have

$$\frac{\partial A_j}{\partial \bar{z}_k} = \frac{\partial B}{\partial \bar{z}_k} = 0, \quad j, k = 1, \ldots, n,$$

and

$$\frac{\partial A_j}{\partial \bar{u}_k} = \frac{\partial B}{\partial \bar{u}_k} = 0, \quad j = 1, \ldots, n; k = 1, \ldots, m.$$

Therefore the equation found for $w^{(l)}$ is

(16.11)
$$\frac{\partial w^{(l)}}{\partial t} = \sum_{j=1}^{n} \frac{A_j + \bar{A}_j'}{2} \frac{\partial w^{(l)}}{\partial x_j} + \sum_{j=1}^{n} \frac{A_j - \bar{A}_j'}{2i} \frac{\partial w^{(l)}}{\partial y_j}$$
$$+ \sum_{j=1}^{n} \left[\frac{\partial \bar{A}_j'}{\partial \bar{z}_l} + \sum_{k=1}^{m} \frac{\partial \bar{A}_j'}{\partial \bar{u}_k} \frac{\partial \bar{u}_k}{\partial \bar{z}_l} \right] w^{(j)}$$
$$+ \sum_{j=1}^{n} \sum_{k=1}^{m} \frac{\partial A_j}{\partial u_k} \frac{\partial u}{\partial z_j} w_k^{(l)} + \sum_{k=1}^{m} \frac{\partial B}{\partial u_k} w_k^{(l)}.$$

Similarly, differentiation of (16.6) with respect to $\bar{z}^l$ yields the initial condition

(16.12)
$$w^{(l)}(z, 0) = 0,$$

in view of (16.8). The relations (16.11) and (16.12) together define a linear homogeneous Cauchy problem for the combined set of vectors $w^{(1)}, \ldots, w^{(n)}$, provided that we consider the coefficients appearing in (16.11) to be known.

One verifies by inspection that zero is a solution of (16.11) and (16.12). Since (16.11) constitutes a symmetric hyperbolic system of mn equations for the individual components of the quantities $w^{(l)}$, it follows from the uniqueness results of Chapter 6 that (16.9) is valid. Therefore u satisfies the Cauchy-Riemann equations (16.10) and must be analytic in its dependence on $z_1, \ldots, z_n$. In particular, it fulfills (16.1) as well as (16.7). Thus we see that the complex substitution $z_j = x_j + iy_j$ leading to the symmetric hyperbolic system (16.7) furnishes a practical tool for the construction of solutions of an analytic system (16.1) of arbitrary type. In this connection observe that (16.7) might even be solved by the method of finite differences (cf. Exercise 13.2.6). Finally, note that we have made no essential use of analyticity in the time variable t.

It is of interest to present examples of analytic partial differential equations for which there are several different ways of converting Cauchy's problem into properly set hyperbolic initial value problems by extension into the complex domain.[2] Consider the case of a quasi-linear equation

$$(16.13) \qquad au_{tt} + 2bu_{tx} + cu_{xx} + d = 0$$

of the second order, where a, b, c and d are real analytic functions of the five arguments x, t, u, u_x and u_t which satisfy the ellipticity requirement

$$(16.14) \qquad ac - b^2 > 0.$$

Let analytic Cauchy data

$$(16.15) \qquad u(x, 0) = f(x), \qquad u_t(x, 0) = g(x)$$

be prescribed at $t = 0$. Instead of bringing (16.13) into the canonical form (16.1) and then reducing it to a symmetric hyperbolic system (16.7), we shall arrive at a hyperbolic problem by making the complex substitution $z = x + iy$ directly.

Because of (16.14) there is no loss of generality if we suppose that

$$a = a(x, t, u, u_x, u_t) \equiv 1.$$

The remaining coefficients b and c will be represented in terms of their real and imaginary parts

$$b = b_1 + ib_2, \qquad c = c_1 + ic_2.$$

The analytic continuation

$$u = u(x + iy, t) = u_1 + iu_2$$

of the power series solution of (16.13) and (16.15) into a three-dimensional domain with the rectangular coordinates x, y and t fulfills the Cauchy-Riemann equations

$$(16.16) \qquad u_{\bar{z}} = \tfrac{1}{2}(u_x + iu_y) = 0$$

and Laplace's equation

$$(16.17) \qquad u_{xx} + u_{yy} = 0.$$

Our construction of u will be based on the fact that it also solves the linear combination

$$(16.18) \quad u_{tt} + 2b_1u_{tx} + 2b_2u_{ty} + c_1u_{xx} + c_2u_{xy} - \lambda(u_{xx} + u_{yy}) + d = 0$$

[2] Cf. Garabedian 9.

of (16.13), (16.16) and (16.17), where λ is an arbitrary parameter. By keeping y small enough we can arrange that $|b|$ and $|c|$ stay within fixed bounds; in such a range it is possible to pick $\lambda > 0$ so large that (16.18) becomes of the hyperbolic type with t featuring as the time-like variable.

Analytic extension of (16.15) into the complex domain yields the initial conditions

(16.19) $$u(z, 0) = f(z), \qquad u_t(z, 0) = g(z).$$

Formulas (16.18) and (16.19) together constitute a well posed Cauchy problem for the determination of the complex unknown function u. Moreover, this assertion holds for a variety of large positive values of λ. Thus we obtain a one-parameter family of hyperbolic problems equivalent to (16.13), (16.15), each of which furnishes a different construction of u. Since the characteristic conoids for (16.18) are altered by variations in λ, the corresponding domains of dependence in the initial plane $t = 0$ change with λ, too. Consequently the quite simple method we have described here for calculating u has the disadvantage that it does not single out a specific minimal complex domain where data will always be needed for the evaluation of u at a given point. Finally, note that in solving (16.18) we deal with a hyperbolic equation in three independent variables, whereas our ultimate interest is only in that part of the answer defined over the real plane of the original problem (16.13), (16.15). The method of characteristics will enable us to overcome some of these difficulties in the next section.

EXERCISES

1. Derive from (16.7) a real system of $2m$ partial differential equations for the real and imaginary parts of u, and show that the matrices of coefficients appearing in the real system are symmetric because the corresponding matrices of coefficients in (16.7) are Hermitian.

2. Show that the complex substitution $z = x + iy$ transforms Laplace's equation

(16.20) $$u_{tt} + u_{xx} = 0$$

into the wave equation

$$u_{tt} - u_{yy} = 0.$$

Derive from this result the expression

$$u = \text{Re}\, \{f(t + ix)\}$$

for the general solution of (16.20), where f stands for an arbitrary analytic function of the complex variable $t + ix$.

3. Show that the initial value problem (16.1), (16.2) is equivalent to (16.6), (16.7) even when the coefficients A_j and B do not depend analytically on the time variable t.

4. Prove that when the data involved are complex analytic, every solution u of the Cauchy problem (16.18), (16.19) satisfies the Cauchy-Riemann equations (16.16).

5. Generalize the method we introduced in order to solve the Cauchy problem (16.13), (16.15) to the case of an analytic partial differential equation of the second order in any number of independent variables.

6. Discuss Hadamard's counterexample concerning the Cauchy problem for Laplace's equation (cf. Section 4.1) in the light of our reduction of the general analytic system (16.1) to a symmetric hyperbolic system (16.7) in the complex domain.

7. Compare our use here of the complex substitution $z_j = x_j + iy_j$ with the method we developed in Section 6.2 for the solution of the wave equation in three or more independent variables.

8. In Section 12.2 we established that in the case of analytic data Cauchy's problem for a *linear* symmetric hyperbolic system (16.1) can be solved in the large. Use (16.11) to show that the solution in question has to be an analytic function of its arguments in the large.

9. Use (16.7) in conjunction with the abstract procedure outlined at the close of Section 12.2 to prove the Cauchy-Kowalewski theorem without introducing power series expansions.

10. Apply Exercise 13.2.6 to (16.7) in order to develop a finite difference scheme for the numerical solution of the general analytic Cauchy problem (16.1), (16.2).

11. The *inverse detached shock problem* consists in finding the steady flow behind a curved shock wave that is prescribed analytically (cf. Section 14.4). Formulate this question for both two-dimensional and three-dimensional flow as a special case of the analytic Cauchy problem (16.1), (16.2), and then use complex substitutions to develop a numerical procedure for its solution based on the symmetric hyperbolic system (16.7).

12. Show how inverse solutions of more general free boundary problems can be found by the method suggested in Exercise 11.

13. Discuss Cauchy's problem for the Tricomi equation

$$yu_{xx} + u_{yy} = 0$$

with initial data prescribed analytically along the line $y = x$.

14. The steady motion of an electron beam is governed by equations in which no pressure term appears because the electrons in the beam do not exhibit random movement. The equations in question are $\nabla \times E = 0$, $\nabla \cdot E = \rho$, $(u \cdot \nabla)u = E$, $\nabla \cdot (\rho u) = 0$, where E stands for the electrostatic field, ρ stands for the charge density, and u stands for the electron velocity.[3] It is of interest

---

[3] Cf. Harker 1.

to determine the shape of electrodes for an electron gun designed to guide a prescribed beam in a given channel. For an axially symmetric hollow beam whose boundary is defined by analytic expressions, formulate this physical problem as an incorrectly set analytic Cauchy problem. Show how the problem can nevertheless be solved numerically by the method of Exercise 10, or, in other words, by reduction to a symmetric hyperbolic problem of the type (16.6), (16.7).

15. In the case of a linear system (16.1) with entire coefficients A_j and B and with entire initial data (16.2), combine the method of this section with that of Section 12.2 in order to establish the existence of a global solution of Cauchy's problem even when (16.1) is not of the hyperbolic type.

16. Consider an analytic symmetric hyperbolic system (16.1) in the case $n = 1$ of just two independent variables x and t. Suppose that the initial data (16.2) can be represented as the boundary values of a complex analytic function $f(z)$ that is regular in the upper half-plane $y > 0$. Under these hypotheses prove that for each t the solution $u(x, t)$ of the Cauchy problem (16.1), (16.2) can be represented as the boundary values of a complex analytic function of $z = x + iy$ that is regular for small enough positive values of y.

17. Show how to convert the forward heat equation into the backward heat equation by means of a complex substitution.

2. CHARACTERISTIC COORDINATES

In the previous section we solved seemingly unstable analytic initial value problems in two steps, the first of which consisted in an extension of the data into the complex domain, and the second of which amounted to a standard treatment of Cauchy's problem for a hyperbolic system. Essential for the success of such a construction is that the analytic continuation of the data can be performed in the large. In particular, any instability in the original problem must necessarily appear in the first step of our process, which can fortunately be carried out by inspection in many of the more important applications of the method.

One disadvantage of our procedure so far is that it converts a problem in only $n + 1$ variables into one in no less than $2n + 1$ variables. In practice, the large number of independent variables occurring in the final problem may lead to forbidding difficulties. However, for the special case of a quasi-linear equation of the second order in just two independent variables we shall show here how complex substitutions can be made that lead to a Cauchy problem for a hyperbolic system also involving only two independent variables.[4] This is achieved by the introduction of characteristic coordinates (cf. Section 3.4), which have the effect of giving to the third

[4] Cf. Garabedian-Lieberstein 1.

variable that featured in our original approach the role of an auxiliary parameter rather than a principal variable. Thus the new procedure can be handled in a more elementary way numerically, since stable two-dimensional initial value problems are relatively easy to handle by the method of finite differences (cf. Section 13.2). Moreover, an analysis based on characteristic coordinates makes it possible to study the dependence of the singularities of the solution of Cauchy's problem on the corresponding singularities of the extension of the initial data into the complex domain.

As a preliminary example let us study the Laplace equation

$$(16.21) \qquad u_{tt} + u_{xx} = 0.$$

The extended equation (16.18) to be associated with (16.21) in the three-dimensional space of the real variable t and the complex variable $z = x + iy$ is obviously

$$(16.22) \qquad u_{tt} + u_{xx} - \lambda(u_{xx} + u_{yy}) = 0.$$

When λ lies in the range $\lambda > 1$, the partial differential equation (16.22) is seen to be of the hyperbolic type with t appearing as the time-like variable. However, our interest will be in the limiting case where $\lambda = 1$, so that (16.22) reduces to the wave equation

$$(16.23) \qquad u_{tt} - u_{yy} = 0$$

in two-dimensional space (cf. Exercise 1.2). In the latter situation the characteristic conoids

$$\frac{(x - x_0)^2}{\lambda - 1} + \frac{y^2}{\lambda} = (t - t_0)^2$$

for (16.22) degenerate into a pair of infinite wedges bounded by the characteristics

$$x = x_0, \qquad y = \pm(t - t_0)$$

of (16.23).

Observe that for (16.22) the complex domain of dependence

$$\frac{(x - x_0)^2}{\lambda - 1} + \frac{y^2}{\lambda} \leq t_0^2$$

in the initial plane $t = 0$ corresponding to the real point (x_0, t_0) is an ellipse that encloses the line segment

$$x = x_0, \qquad -t_0 \leq y \leq t_0, \qquad t = 0,$$

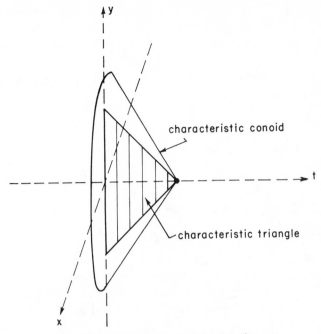

FIG. 87. Characteristics in the complex domain.

which in turn comprises the domain of dependence for (16.23) associated with the same point. Thus (16.23) leads to a minimal domain of dependence carrying data for the determination of $u(x_0, t_0)$. In this connection it is clear that $t + y$ and $t - y$ are characteristic coordinates for (16.23). Within the framework of the substitution $z = x + iy$, we can ascribe the same property to $t - iz$ and $t + iz$ for Laplace's equation (16.21) in the complex domain. Therefore the introduction of characteristic coordinates suggests itself as an appropriate tool for the reduction of more general quasi-linear elliptic equations of the second order to hyperbolic systems. Furthermore, we might expect the latter to involve just two independent variables, since x merely features as a parameter in the wave equation (16.23).

We are now prepared to discuss the quasi-linear equation

$$(16.24) \qquad au_{xx} + 2bu_{xy} + cu_{yy} + d = 0$$

in the plane of the two real independent variables x and y. This equation has precisely the same form as (3.88), for which we introduced characteristic coordinates ξ and η in Section 3.4. Our interest here will be primarily in the case $ac - b^2 > 0$ where (16.24) is of the elliptic type, and we shall assume that the coefficients a, b, c and d are real analytic functions of their five arguments x, y, u and

$$p = u_x, \qquad q = u_y.$$

Our ultimate concern will be with the Cauchy problem for (16.24), with analytic data

(16.25) $$u = u(s), \quad p = p(s), \quad q = q(s)$$

assigned along an analytic arc

(16.26) $$x = x(s), \qquad y = y(s),$$

where, of course, u must be subject to the compatibility requirement

(16.27) $$du = p \, dx + q \, dy.$$

In Section 3.4 we explained how to convert (3.88) into a canonical hyperbolic system of five partial differential equations (3.93), (3.95), (3.96) and (3.97) of the first order for the unknowns x, y, u, p and q as functions of the pair of characteristic coordinates ξ and η. Recognizing that the quantities

$$\lambda_+ = \frac{b + i\sqrt{ac - b^2}}{a}, \qquad \lambda_- = \frac{b - i\sqrt{ac - b^2}}{a}$$

become conjugate complex numbers in the elliptic case, we can carry out exactly the same transformation of variables for analytic solutions u of (16.24). It follows that the analytic extension of u into the four-dimensional domain of complex values of the original independent variables x and y must satisfy the canonical system

(16.28) $$y_\xi - \lambda_+ x_\xi = 0, \qquad y_\eta - \lambda_- x_\eta = 0,$$

(16.29) $$p_\xi + \lambda_- q_\xi + \frac{d}{a} x_\xi = 0,$$

(16.30) $$p_\eta + \lambda_+ q_\eta + \frac{d}{a} x_\eta = 0,$$

(16.31) $$u_\xi - p x_\xi - q y_\xi = 0,$$

where ξ and η may represent either real or complex arguments. When the characteristic coordinates ξ and η are real, the system (16.28) to (16.31) is seen to be of the hyperbolic type. It is our intention to reduce the unstable Cauchy problem (16.25), (16.26) for the elliptic equation (16.24) to a well posed initial value problem for (16.28) to (16.31) over the real (ξ,η)-plane by continuing the data analytically into the complex domain.

Since it is permissible to replace ξ by any function of ξ and η by any function of η in the canonical system of partial differential equations (16.28) to (16.31), there is no loss of generality if we suppose that the Cauchy data of our problem are assigned along the line

$$(16.32) \qquad\qquad \xi + \eta = 0.$$

Afterward we are still free to prescribe the initial values of one of the five unknowns x, y, u, p or q, for example of x, arbitrarily in dependence on the coordinate

$$\sigma = \xi - \eta$$

orthogonal to

$$\tau = \xi + \eta.$$

However, the four remaining unknowns then become determinate functions of σ on (16.32) defined by the initial curve of data (16.25), (16.26). The rule specifying the select unknown x here in terms of σ need not even be analytic. Indeed, we shall see that it merely determines what amounts to a path of integration associated with the differential equations (16.28) to (16.31).

When the new independent variables σ and τ are permitted to assume complex values

$$\sigma = \sigma_1 + i\sigma_2, \qquad \tau = \tau_1 + i\tau_2,$$

it is convenient to view their imaginary parts σ_2 and τ_2 as auxiliary parameters in the solution of the hyperbolic system (16.28) to (16.31). On the other hand, if the dependence of the initial data (16.25), (16.26) on σ is assigned in a natural way so that

$$(16.33) \qquad\qquad x = x(\sigma)$$

becomes a real analytic function of σ_2 for $\sigma_1 = 0$, then by reason of symmetry the real (x,y)-plane corresponds to the (σ_2,τ_1)-plane in the four-dimensional space with coordinates σ_1, σ_2, τ_1 and τ_2. Thus in the (σ_2,τ_1)-plane the system (16.28) to (16.31) has an analogy to the Laplace equation (16.21), whereas it behaves over the (σ_1,τ_1)-plane more like the wave equation (16.23). To calculate the real solution u of the real Cauchy problem (16.24) to (16.26) we therefore put

$$\tau_2 = 0$$

and solve (16.28) to (16.31) in the (σ_1,τ_1)-plane for a variety of choices of the parameter σ_2.

In order to determine u over a full region of the real (x,y)-plane it is necessary to solve the two-dimensional system (16.28) to (16.31) for a complete interval of values of σ_2. Each hyperbolic problem identified with an individual choice of σ_2 generates only a single curve of results in the real (x,y)-plane, which are found by setting

$$\sigma_1 = 0.$$

As σ_2 varies, these paths sweep out the required plane region.

The geometry of the construction of u defined above is best visualized in the three-dimensional space of the coordinates τ_1, σ_1 and σ_2, where it has a direct analogy with our discussion of Laplace's equation (16.21) based on analyzing the wave equation (16.23) in the space spanned by t, y and x. The answer to each separate initial value problem for the hyperbolic system (16.28) to (16.31) is obtained in a characteristic triangle parallel to the (σ_1,τ_1)-plane. However, a one-parameter family of these characteristic triangles, which depend on σ_2, must be introduced to cover a three-dimensional region of the complex domain including a section of the (σ_2,τ_1)-plane where the real solution of our original Cauchy problem (16.24) to (16.26) is specified. Note that the characteristic triangles correspond to two-dimensional surfaces in the four-dimensional domain of complex values of the independent variables x and y over which the quasi-linear equation (16.24) is of the hyperbolic type.

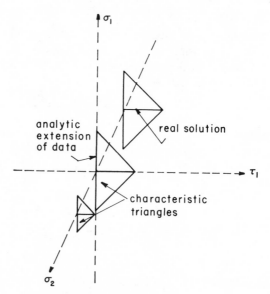

FIG. 88. Method of characteristics.

The method of solving the unstable Cauchy problem (16.24) to (16.26) that we have just outlined is more readily adaptable to numerical computation than is the previous treatment based on the symmetric hyperbolic system (16.7). The new procedure merely involves the integration of a one-parameter family of two-dimensional hyperbolic equations, whereas our earlier technique required analysis of a three-dimensional initial value problem, which is in general more difficult to handle. In both approaches it is necessary, of course, to perform an analytic continuation of the data of the problem into the complex domain. When this key step cannot be carried out in the large, we do not expect to find an answer in the large, either. The latter phenomenon is what explains our success in treating the original question, which was not correctly set.

A more specific relationship actually exists between the singularities of the analytic continuation of the Cauchy data (16.25), (16.26) in the complex domain and the singularities of the solution u of the corresponding Cauchy problem. For complex values of the coordinates ξ and η, the characteristics

$$(16.34) \qquad\qquad \xi = \text{const.}, \qquad \eta = \text{const.}$$

of equation (16.24) become two-dimensional surfaces within the four-dimensional complex domain into which we have extended u. A singularity of the initial data located at a point on the surface (16.32) can propagate along both the characteristics (16.34) passing through that point. The solution u of (16.24) ought therefore to exhibit singularities of a similar kind at the pair of points where these two characteristics intersect the real (x,y)-plane.

An example will serve to indicate what we have in mind here. Consider the linear elliptic partial differential equation

$$(16.35) \qquad\qquad u_{tt} + u_{xx} = Pu,$$

where the coefficient $P = P(x, t)$ is supposed to be an entire function of its arguments. We seek an analytic solution $u = u(x, t)$ of the Cauchy problem for (16.35) with initial data defined by the explicit formulas

$$(16.36) \qquad\qquad u(x, 0) = 0$$

and

$$(16.37) \qquad\qquad u_t(x, 0) = \frac{\epsilon}{\delta^2 + x^2},$$

where δ and ϵ are small real parameters. Our general theory shows that the complex substitution $z = x + iy$ can be used to represent (16.35) in the hyperbolic form

(16.38) $$u_{tt} - u_{yy} = Pu.$$

Furthermore, analytic continuation leads to the new expressions

(16.39) $$u(iy, 0) = 0$$

and

(16.40) $$u_t(iy, 0) = \frac{\epsilon}{\delta^2 - y^2}$$

for the initial conditions (16.36) and (16.37) within the section $x = 0$ of the complex domain.

Although the real initial values (16.36), (16.37) for (16.35) are quite regular, it is apparent that the corresponding data (16.39), (16.40) possess poles at the two points

$$x = 0, \quad y = \pm\delta, \quad t = 0$$

in the complex domain. These singularities must propagate along the four characteristic lines

(16.41) $$x = 0, \quad y \pm t = \pm\delta$$

associated with (16.38). Consequently the solution u of the Cauchy problem (16.35) to (16.37) must become infinite at the points

(16.42) $$x = 0, \quad t = \pm\delta$$

where the characteristics (16.41) meet the real plane $y = 0$. On the other hand, for $x \neq 0$ the initial values

$$u(x + iy, 0) = 0,$$

$$u_t(x + iy, 0) = \frac{\epsilon}{\delta^2 + (x + iy)^2}$$

remain finite, so that u is defined over the whole (y,t)-plane in that case. Thus the singularities of the Cauchy data (16.36), (16.37) in the complex domain, together with the geometry of the characteristics there, indicate precisely where we can expect the real solution u of (16.35) to exist, even in the large.

It becomes clear again why Cauchy's problem is incorrectly set for the elliptic equation (16.35). For an arbitrary choice of $\delta > 0$ the data (16.36), (16.37) can be made as small as we please by taking ϵ sufficiently small. However, the corresponding solution u of (16.35) will become infinite at the points (16.42), which have only a distance of δ units from the initial line $t = 0$. Thus u does not approach zero with the data (16.36),

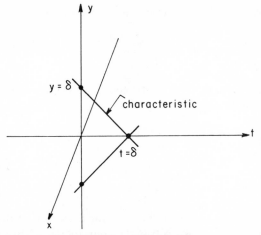

FIG. 89. Propagation of singularities.

(16.37) when $\epsilon \to 0$. Our analysis shows, on the other hand, that solutions do depend continuously on the global extension of their initial values into the complex domain, and that their singularities can be located there in a quite satisfactory fashion.

EXERCISES

1. Precisely why should the (σ_2, τ_1)-plane referred to in the text correspond to the real (x,y)-plane?

2. Show that the requirement (16.27) is needed to ensure that a solution of the first order system (16.28) to (16.31) yields a solution of the second order equation (16.24).

3. Use a uniqueness theorem to prove that any solution u of (16.28) to (16.31), considered as a function of the complex variables x and y, has to satisfy the Cauchy-Riemann equations when the associated initial data (16.25), (16.26) are analytic in the complex domain.

4. Under what circumstances can we expect that the inverse transformation from the characteristic coordinates ξ and η to the original independent variables x and y should have a non-trivial Jacobian

$$\frac{\partial(x, y)}{\partial(\xi, \eta)} \neq 0?$$

5. What simplifications are possible in the theory of the text for the case of the nonlinear elliptic equation

$$u_{xx} + u_{yy} = F(x, y, u, u_x, u_y),$$

where F is supposed to be a real analytic function of its five arguments?

6. In Exercise 3.4.3 we introduced real coordinates

$$\alpha = \frac{\xi + \eta}{2}, \qquad \beta = \frac{\xi - \eta}{2i}$$

in terms of which (16.28) to (16.31) could be reduced to a real elliptic system in the canonical form

$$(16.43) \qquad x_{\alpha\alpha} + x_{\beta\beta} + \cdots = 0, \quad y_{\alpha\alpha} + y_{\beta\beta} + \cdots = 0,$$

$$u_{\alpha\alpha} + u_{\beta\beta} + \cdots = 0, \quad p_{\alpha\alpha} + p_{\beta\beta} + \cdots = 0,$$

$$q_{\alpha\alpha} + q_{\beta\beta} + \cdots = 0.$$

Use this canonical system as the basis for a solution of the analytic Cauchy problem for (16.24) which is patterned directly on our treatment of Laplace's equation (16.21).

7. Develop a finite difference scheme for the numerical solution of the analytic Cauchy problem (16.24) to (16.26) that is based on the hyperbolic system (16.28) to (16.31).

8. Use the method of Exercise 7 to solve the inverse detached shock problem (cf. Exercise 1.11) for plane or axially symmetric flow.[5]

9. Compare the respresentation given in Exercise 14.5.10 of the axially symmetric flow corresponding to a prescribed free streamline with the method described in the text for the solution of analytic Cauchy problems for an elliptic equation. What applications has this construction to the inverse free boundary problems of magnetohydrodynamics?

10. Let $A(z, z^*; \zeta, \zeta^*)$ denote the Riemann function for the self-adjoint linear elliptic partial differential equation (16.35), where

$$z = t + ix, \qquad z^* = t - ix,$$

and t and x are permitted to be complex variables. Indicating by a bar the actual complex conjugate of a complex number, show that the integral formula

$$(16.44) \qquad u(\zeta, \bar\zeta) = u(\zeta, \zeta^*) + u(\bar\zeta^*, \bar\zeta) - u(\bar\zeta^*, \zeta^*)A(\bar\zeta^*, \zeta^*; \zeta, \bar\zeta)$$

$$- \int_{\zeta^*}^{\zeta} u(z, \zeta^*) \frac{\partial}{\partial z} A(z, \zeta^*; \zeta, \bar\zeta)\, dz - \int_{\zeta^*}^{\bar\zeta} u(\bar\zeta^*, z^*) \frac{\partial}{\partial z^*} A(\bar\zeta^*, z^*; \zeta, \bar\zeta)\, dz^*$$

represents the solution of a characteristic initial value problem for (16.35) in the complex domain (cf. Section 4.4). Interpret (16.44) as an expression[6] for the general solution of (16.35).

[5] Cf. Garabedian-Lieberstein 1.
[6] Cf. Bergman 2, Vekua 1.

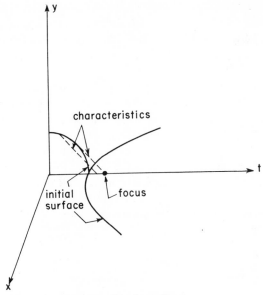

FIG. 90. Branch point at focus.

11. Solve Cauchy's problem for the Laplace equation (16.21) with initial data

$$u = 0, \qquad \frac{\partial u}{\partial v} = 1$$

assigned along the equilateral hyperbola $t^2 - x^2 = 1$. In terms of the geometry of the characteristics of (16.21), explain why the answer has a branch point at the focus of the hyperbola. Does the solution of the same problem for equation (16.35) have a branch point there, too?

12. Show that a conformal mapping of the (σ_2, τ_1)-plane which leaves the σ_2-axis invariant corresponds to a transformation of the characteristic coordinates ξ and η that preserves the initial line (16.32).

13. Show that an assignment (16.33) of initial values which is not analytic in its dependence on the variable σ_1 leads to a path of results in the real (x,y)-plane which is not analytic, either. Prove that (16.33) can be chosen so that the latter path coincides with an arbitrary prescribed curve.

3. ANALYTICITY OF SOLUTIONS OF NONLINEAR ELLIPTIC EQUATIONS

In Section 5.1 we have used the integral representation (5.28) of solutions of a linear analytic elliptic partial differential equation in two independent variables to establish that they have to be real analytic functions of their arguments. The proof depended on the analyticity of the fundamental solution appearing in (5.28), which was actually constructed in the complex domain. Heuristically speaking, the analyticity property of the solutions results from the fact that their singularities always propagate along characteristics, which turn out to be imaginary in the case of an elliptic equation. Motivated by these ideas, we shall devote our attention here to proving that any sufficiently differentiable solution of a *nonlinear* analytic elliptic equation in two independent variables has to be analytic, too. Our demonstration[7] will be based on hyperbolic initial value problems that enable us to extend a given solution into the complex domain directly and then to show that the extension fulfills the Cauchy-Riemann equations there (cf. Exercise 1.8). To simplify matters we shall only present the details of our analysis for the special nonlinear equation

$$(16.45) \qquad u_{xx} + u_{yy} = F(x, y, u, u_x, u_y),$$

where F is assumed to be an entire function of its five arguments. Generalizations will be described afterward in the exercises.

In order to find the hyperbolic equations needed to extend functions $u = u(x, y)$ satisfying (16.45) into the complex domain, we shall suppose at first that an analytic solution u is known to us there. Let us use the notation

$$x = x_1 + ix_2, \qquad y = y_1 + iy_2$$

to indicate the real and imaginary parts of the independent variables x and y, and let us introduce the differential operators $\partial/\partial x$, $\partial/\partial \bar{x}$, $\partial/\partial y$ and $\partial/\partial \bar{y}$. The analyticity of u in the complex domain means that the Cauchy-Riemann equations

$$(16.46) \qquad \frac{\partial u}{\partial \bar{x}} = 0, \qquad \frac{\partial u}{\partial \bar{y}} = 0$$

are fulfilled continuously. It follows that u is also a solution of the two Laplace equations

$$(16.47) \qquad \frac{\partial^2 u}{\partial x_1^2} + \frac{\partial^2 u}{\partial x_2^2} = 0$$

and

$$(16.48) \qquad \frac{\partial^2 u}{\partial y_1^2} + \frac{\partial^2 u}{\partial y_2^2} = 0.$$

Our intention is to combine (16.46) to (16.48) with (16.45) in such a way as to obtain a hyperbolic equation for u as a function of x_1 and y_2 alone or as a function of y_1 and x_2 alone (cf. Sections 1 and 2).

[7] Cf. Lewy 1.

We deduce immediately from (16.45) that

$$(16.49) \qquad u_{x_1 x_1} + u_{y_1 y_1} = F(x, y, u, u_{x_1}, u_{y_1})$$

holds in a four-dimensional region with the coordinates x_1, x_2, y_1 and y_2. An application of (16.46) and (16.48) therefore yields

$$(16.50) \qquad u_{x_1 x_1} - u_{y_2 y_2} = F(x, y, u, u_{x_1}, -iu_{y_2}).$$

Similarly,

$$(16.51) \qquad - u_{x_2 x_2} + u_{y_1 y_1} = F(x, y, u, -iu_{x_2}, u_{y_1})$$

because of (16.46) and (16.47). Both (16.50) and (16.51) are differential equations of the hyperbolic type. We shall see that they serve, respectively, to define u for non-trivial values of y_2 and of x_2 without specific appeal to the concept of analytic continuation. To this end we aim to exploit the second Cauchy-Riemann equation (16.46) as an initial condition of the form

$$(16.52) \qquad u_{y_2} = iu_{y_1}$$

at $y_2 = 0$; similarly, we shall interpret the first equation (16.46) as an initial condition at $x_2 = 0$ of the form

$$(16.53) \qquad u_{x_2} = iu_{x_1}.$$

Now suppose that we are given a solution $u(x, y)$ of the nonlinear elliptic equation (16.45) possessing continuous partial derivatives of the third order over a two-dimensional region D of the real (x,y)-plane. Our purpose is to establish that u must actually be an analytic function of the variables x and y throughout D. For the proof we shall first extend u into an intermediate three-dimensional domain by means of (16.50) and (16.52), and then we shall define it over a full four-dimensional region of the complex domain by means of (16.51) and (16.53). Afterward the extended function will be shown to satisfy the Cauchy-Riemann equations (16.46) identically, whence its analyticity will follow.

Let $u(x_1, y_1, y_2)$ stand for the solution of the hyperbolic equation

$$(16.54) \qquad u_{x_1 x_1} - u_{y_2 y_2} = F(x_1, y_1 + iy_2, u, u_{x_1}, -iu_{y_2})$$

which fulfills the initial conditions

$$(16.55) \qquad u(x_1, y_1, 0) = u(x_1, y_1),$$

$$(16.56) \qquad u_{y_2}(x_1, y_1, 0) = iu_{y_1}(x_1, y_1)$$

at points of D. For each fixed choice of the parameter y_1, the two-dimensional Cauchy problem (16.54) to (16.56) determines $u(x_1, y_1, y_2)$ in

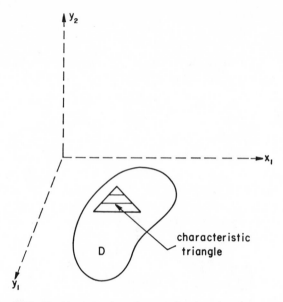

FIG. 91. Extension into the complex domain.

every sufficiently small triangle parallel to the (x_1, y_2)-plane whose base lies in D and whose two remaining sides are characteristics of (16.54). Thus the method of successive approximations developed in Section 4.2 can be used to define u throughout some neighborhood of D within the three-dimensional space spanned by x_1, y_1 and y_2. We shall prove next that the extension $u(x_1, y_1, y_2)$ so obtained is an analytic function of the complex variable $y = y_1 + iy_2$ in that neighborhood.

What has to be shown is that the Cauchy-Riemann equation

$$(16.57) \qquad u_{\bar{y}}(x_1, y_1, y_2) = 0$$

holds even when $y_2 \neq 0$. First, observe that it is permissible to differentiate $u(x_1, y_1, y_2)$ with respect to y_1 as well as with respect to x_1 and y_2 because of the differentiability of a solution of Cauchy's problem in its dependence on auxiliary parameters (cf. Section 4.2) and because of our hypothesis concerning the differentiability of $u(x_1, y_1)$ in D. We intend to exploit this remark in order to derive a homogeneous initial value problem for the quantity

$$w = u_{\bar{y}}$$

which will enable us to conclude that it is identically zero (cf. Section 1).

An application of the differential operator $\partial/\partial\bar{y}$ to (16.54) yields the linear differential equation

(16.58) $\qquad w_{x_1x_1} - w_{y_2y_2} = F_u w + F_p w_{x_1} - iF_q w_{y_2}$

for w, where, as usual, $p = u_x$, $q = u_y$. Note that the analyticity of F has been taken into account in calculating the right-hand side of (16.58). The initial conditions (16.55) and (16.56) imply that

(16.59) $\qquad w(x_1, y_1, 0) = \dfrac{1}{2} u_{y_1}(x_1, y_1) + \dfrac{i}{2} u_{y_2}(x_1, y_1, 0) = 0.$

However, to evaluate w_{y_2} at $y_2 = 0$ we have to appeal to (16.49) and (16.54) there. Thus we observe that

(16.60) $\qquad w_{y_2}(x_1, y_1, 0) = \dfrac{1}{2} u_{y_1y_2} + \dfrac{i}{2} u_{y_2y_2}$

$\qquad\qquad\qquad = \dfrac{1}{2} u_{y_1y_2} + \dfrac{i}{2} [u_{x_1x_1} - F(x_1, y_1, u, u_{x_1}, -iu_{y_2})]$

because of (16.54), whereas according to (16.49) we have

(16.61) $\qquad u_{x_1x_1} - F(x_1, y_1, u, u_{x_1}, u_{y_1}) = -u_{y_1y_1}.$

Using (16.59) to combine (16.60) with (16.61), we finally arrive at the initial condition

(16.62) $\qquad w_{y_2}(x_1, y_1, 0) = -\dfrac{i}{2}\dfrac{\partial}{\partial y_1}(u_{y_1} + iu_{y_2})$

$\qquad\qquad\qquad = -iw_{y_1}(x_1, y_1, 0) = 0.$

Since zero is an obvious solution of the linear homogeneous Cauchy problem (16.58), (16.59), (16.62), a standard uniqueness theorem (cf. Section 4.2) shows that

(16.63) $\qquad w(x_1, y_1, y_2) = 0$

throughout the region where $u(x_1, y_1, y_2)$ has been defined. We conclude that u satisfies the Cauchy-Riemann equation (16.57) and must be an analytic function of the variable $y = y_1 + iy_2$. It remains to extend u similarly into the domain of complex values of $x = x_1 + ix_2$ so that we can establish its analyticity as a function of both of the arguments x and y simultaneously.

For suitable fixed choices of x_1 and y_2 let $u(x_1, x_2, y_1, y_2)$ be the solution of

(16.64) $-u_{x_2 x_2} + u_{y_1 y_1} = F(x_1 + ix_2, y_1 + iy_2, u, -iu_{x_2}, u_{y_1})$

such that

(16.65) $$u(x_1, 0, y_1, y_2) = u(x_1, y_1, y_2)$$

and

(16.66) $$u_{x_2}(x_1, 0, y_1, y_2) = iu_{x_1}(x_1, y_1, y_2).$$

The initial value problem (16.64) to (16.66) serves to determine

$$u = u(x_1, x_2, y_1, y_2)$$

in a four-dimensional region containing the domain of definition of $u(x_1, y_1, y_2)$. In order to verify that the extended function is actually analytic, we wish to prove that the Cauchy-Riemann equations

(16.67) $$u_{\bar{y}}(x_1, x_2, y_1, y_2) = 0,$$

(16.68) $$u_{\bar{x}}(x_1, x_2, y_1, y_2) = 0$$

hold. For this purpose we formulate homogeneous Cauchy problems for the quantities

$$w = u_{\bar{y}}, \qquad \zeta = u_{\bar{x}}$$

which imply that they vanish identically.

Applying the operators $\partial/\partial\bar{y}$ and $\partial/\partial\bar{x}$ to (16.64), we find that

(16.69) $$-w_{x_2 x_2} + w_{y_1 y_1} = F_u w - iF_p w_{x_2} + F_q w_{y_1}$$

and

(16.70) $$-\zeta_{x_2 x_2} + \zeta_{y_1 y_1} = F_u \zeta - iF_p \zeta_{x_2} + F_q \zeta_{y_1}.$$

Moreover, for $x_2 = 0$ the operator $\partial/\partial\bar{y}$ can be used to convert the initial conditions (16.65) and (16.66) into

(16.71) $$w(x_1, 0, y_1, y_2) = w(x_1, y_1, y_2) = 0$$

and

(16.72) $$w_{x_2}(x_1, 0, y_1, y_2) = iw_{x_1}(x_1, y_1, y_2) = 0,$$

because of (16.63). Thus w must coincide with the trivial solution

$$w(x_1, x_2, y_1, y_2) = 0$$

of the linear homogeneous Cauchy problem (16.69), (16.71), (16.72), whence (16.67) follows.

As in our derivation of (16.59), we deduce from (16.65) and (16.66) that

$$(16.73) \quad \zeta(x_1, 0, y_1, y_2) = \frac{1}{2} u_{x_1}(x_1, y_1, y_2) + \frac{i}{2} u_{x_2}(x_1, 0, y_1, y_2) = 0.$$

Also, we can subtract (16.64) from (16.54) to obtain

$$u_{x_1 x_1} + u_{x_2 x_2} - u_{y_1 y_1} - u_{y_2 y_2} = 0$$

when $x_2 = 0$. In view of (16.67) this gives in turn

$$u_{x_1 x_1} + u_{x_2 x_2} = 0,$$

and therefore

$$(16.74) \quad \zeta_{x_2}(x_1, 0, y_1, y_2) = \frac{1}{2} u_{x_1 x_2} + \frac{i}{2} u_{x_2 x_2}$$

$$= \frac{1}{2} u_{x_1 x_2} - \frac{i}{2} u_{x_1 x_1} = -i \zeta_{x_1}(x_1, 0, y_1, y_2) = 0.$$

Since the linear homogeneous initial value problem (16.70), (16.73), (16.74) has the unique solution

$$\zeta(x_1, x_2, y_1, y_2) = 0,$$

our proof of (16.68) is now complete.

The combined Cauchy-Riemann equations (16.67) and (16.68) are enough to establish that $u(x_1, x_2, y_1, y_2)$ is an analytic function of the two complex variables x and y throughout some region of four-dimensional space surrounding the real domain D where the original solution $u(x, y)$ of the nonlinear elliptic partial differential equation (16.45) was defined. Thus when F is analytic we have demonstrated the real analyticity of any sufficiently smooth function u that satisfies (16.45) by extending it into the complex domain as a complex-valued solution of the hyperbolic equations (16.54) and (16.64). If F is linear, it is not difficult to find the precise form of the domain of existence of the extension, too, although for that purpose our construction must be modified so as to eliminate its excessive dependence on the particular choice of a coordinate system (cf. Exercise 2 below).

EXERCISES

1. In the notation of formula (16.44), derive the integral representation

$$u(\zeta, \zeta^*) = u(\zeta, \bar{\zeta}) - \int_{\bar{\zeta}*}^{\zeta} \left[u(z, \bar{z}) \frac{\partial A(z, \bar{z}; \zeta, \zeta^*)}{\partial z} \, dz + A(z, \bar{z}; \zeta, \zeta^*) \frac{\partial u(z, \bar{z})}{\partial \bar{z}} \, d\bar{z} \right]$$

for the extension of a solution u of the linear elliptic equation (16.35) into the complex domain. Verify that the expression on the right is independent of the path of integration running from ζ^* to ζ in the real domain.

2. Show that the values of a solution u of (16.45) and of its gradient along a short enough arc in the real domain whose shape is arbitrary suffice to determine the analytic extension of u at a point in the complex domain where the imaginary characteristics of (16.45) through the ends of that arc intersect. In the case of a linear equation (16.45), use this result to find the precise four-dimensional region of the complex domain where the extension of any solution defined over a given real domain D must exist.

3. When the coefficients a, b, c and d are real analytic functions of their five arguments x, y, u, p and q, show that any sufficiently differentiable solution of the quasi-linear elliptic equation (16.24) must be analytic by making a transformation to the canonical system (16.43) and then imitating the procedure described in the text. Carry out the same proof using the characteristic system (16.28) to (16.31) instead of (16.43), and discuss the role played by corresponding curves of initial data in the real domain.

4. Why is it necessary to obtain a non-trivial Jacobian

$$\frac{\partial(x, y)}{\partial(\xi, \eta)} \neq 0$$

in Exercise 3, and how is this achieved?

5. Compare the analysis of the text with Exercise 1.8.

6. Establish the analyticity of smooth solutions of the quasi-linear elliptic equation (16.13) by extending them into the complex domain as solutions of the hyperbolic equation (16.18) in three independent variables, with the positive parameter λ chosen so small that the imaginary coordinate y becomes time-like.[8]

7. Use a real integral equation of the form

(16.75)

$$u(s, t) = U(s, t) + \frac{1}{2\pi} \iint F(x, y, u, u_x, u_y) \log \sqrt{(x - s)^2 + (y - t)^2} \, dx \, dy,$$

where U stands for a harmonic function, to prove that (16.45) has only analytic solutions when F is analytic. What can be said when F is not analytic but does possess continuous partial derivatives of some specific order?

8. Exploit Cauchy's theorem to ascribe a meaning to (16.75) in the complex domain, with the original region of integration replaced by an appropriate complex surface whose boundary lies in the real domain (cf. Section 6.2). Use the resulting integral equation to define the extension of u into the complex domain by the method of successive approximations. Show that this technique actually reduces to the procedure of the text when the surface of integration is allowed to collapse onto a characteristic triangle.

[8] Cf. Garabedian 10.

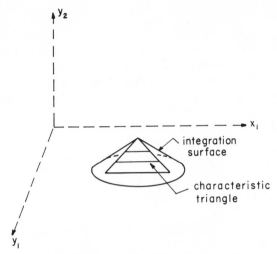

FIG. 92. Complex surface of integration.

9. By introducing a suitable parametrix (cf. Section 5.3), derive an integral equation like (16.75) for the solutions of the quasi-linear elliptic equation (16.24) in the complex domain, and thus establish their analyticity.

10. Show that all sufficiently smooth solutions of a quasi-linear analytic elliptic partial differential equation of the second order in three or more independent variables have to be analytic[9] by extending them into the complex domain as solutions of an integral equation there which is analogous to (16.75) and is based on an appropriate parametrix but does not otherwise involve the imaginary characteristics of the equation (cf. Section 6.4).

11. Through extension into the complex domain, derive *a priori* estimates on the higher derivatives of bounded solutions of an analytic elliptic partial differential equation.

4. REFLECTION

Extension into the complex domain not only provides a tool for establishing the analyticity of solutions of an analytic elliptic partial differential equation in the interior of their domain of definition but also furnishes a method to prove their analyticity at the boundary in the case where they fulfill analytic boundary conditions. Indeed, mixed initial and boundary value problems for the hyperbolic system into which an elliptic equation of the second order can be converted in the complex domain enable us to establish specific reflection rules for the analytic continuation

[9] Cf. Hopf 2.

of solutions satisfying given boundary conditions.[10] The most elementary example of this phenomenon, which is essentially restricted to equations in two independent variables, has already appeared in Section 7.2 in the form of Schwarz's reflection principle for harmonic functions in the plane. We shall present a new proof of that principle here which will suggest how its generalization to quasi-linear equations of the elliptic type can be accomplished.

Consider a real solution u of Laplace's equation

$$(16.76) \qquad u_{xx} + u_{yy} = 0$$

in a region D of the right half-plane $x > 0$ whose boundary ∂D includes a segment C of the y-axis. We shall suppose that u is subject to the boundary condition

$$(16.77) \qquad u(0, y) = 0$$

on C, and, moreover, we shall assume for the time being that u is analytic in some neighborhood of C. Thus we know that its analytic extension

$$u = u(x_1, x_2, y_1, y_2)$$

satisfies the wave equation

$$(16.78) \qquad u_{x_1 x_1} - u_{y_2 y_2} = 0$$

in the domain of complex values of the independent variables $x = x_1 + ix_2$ and $y = y_1 + iy_2$.

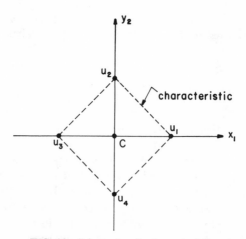

FIG. 93. Schwarz's reflection principle.

[10] Cf. Lewy 5.

In Exercise 4.1.2 it was established that the values u_1, u_2, u_3 and u_4 of a solution of (16.78) at the consecutive corners of a square whose sides are characteristic lines

$$y_2 \pm x_1 = \text{const.}$$

must fulfill the difference equation

(16.79) $u_1 - u_2 + u_3 - u_4 = 0.$

Let us select as the four corners in question the points

$$(h, 0, k, 0), \quad (0, 0, k, h), \quad (-h, 0, k, 0), \quad (0, 0, k, -h),$$

respectively, within the plane $x_2 = 0$, $y_1 = k$ of the four-dimensional space spanned by x_1, x_2, y_1 and y_2. Since analytic continuation of the boundary condition (16.77) implies that

$$u_2 = u(0, 0, k, h) = 0$$

and

$$u_4 = u(0, 0, k, -h) = 0,$$

the relation (16.79) reduces in this case to $u_1 = -u_3$, or, in other words, to

(16.80) $u(h, 0, k, 0) = -u(-h, 0, k, 0).$

In terms of our original real harmonic function $u(x, y)$ the statement (16.80) becomes equivalent to the classical Schwarz principle of reflection

(16.81) $u(h, k) = -u(-h, k),$

which merely asserts that u is an odd function of its first argument x.

We shall now establish a result analogous to (16.81) for solutions of the nonlinear elliptic equation

(16.82) $u_{xx} + u_{yy} = F(x, y, u, u_x, u_y),$

where F is once more assumed to be analytic. Let u be a regular solution of (16.82) in the plane region D described above, and suppose that it, too, satisfies the boundary condition (16.77) along the segment C of the y-axis. We make the additional hypotheses that the first partial derivatives p and q of u are uniformly continuous in a one-sided neighborhood of C located in the half-plane $x > 0$ and that the limiting values of x, y, u, p and q on C lie inside the domain of regularity of F. In these circumstances we will show how to continue u analytically across the y-axis so that it becomes a solution of (16.82) in a certain region containing C in its interior. However, we shall see that the extended region may not include all points of the reflection D' of D in the y-axis when the equation (16.82) is not linear.

In our analysis we shall keep $x = x_1$ real but shall allow $y = y_1 + iy_2$ to be complex. Thus u will be continued into a three-dimensional cross section of the complex domain. The boundary condition (16.77) will enable us to perform the complex extension of u far enough so that on returning to the real domain we shall find it specified for negative values of x. The rule thus obtained for the reflection of u across the y-axis will be based on a two-dimensional hyperbolic equation in which the variable y_1 merely features as a parameter.

We want to determine $u(x_1, y_1, y_2)$ as the solution of a mixed initial and boundary value problem for the hyperbolic equation

$$(16.83) \qquad u_{x_1 x_1} - u_{y_2 y_2} = F(x_1, y_1 + iy_2, u, u_{x_1}, -iu_{y_2}).$$

The Cauchy data

$$(16.84) \qquad u(x_1, y_1, 0) = u(x_1, y_1),$$

$$(16.85) \qquad u_{y_2}(x_1, y_1, 0) = iu_{y_1}(x_1, y_1)$$

will be assigned at points of D, where, of course, $x_1 > 0$, and the boundary condition

$$(16.86) \qquad u(0, y_1, y_2) = 0$$

will be imposed in the complex y-plane, because of (16.77). The Cauchy problem given by formulas (16.83) to (16.85) alone defines $u(x_1, y_1, y_2)$ in a family of characteristic triangles that sweep out the intersection of the wedge

$$(16.87) \qquad -x_1 \leq y_2 \leq x_1$$

with a sufficiently small neighborhood of the segment C. This preliminary extension of u has already been investigated in Section 3, and the discussion there shows that the Cauchy-Riemann equation

$$(16.88) \qquad u_{\bar{y}}(x_1, y_1, y_2) = 0$$

holds in the wedge (16.87). What remains is to construct a continuation of u beyond that wedge.

Let us consider the hyperbolic equation (16.83) for fixed admissible choices of the parameter y_1. Since $u(x_1, y_1, y_2)$ is now known on the characteristic line segments

$$(16.89) \qquad y_2 = \pm x_1, \qquad x_1 \geq 0$$

bounding (16.87), and since its values along the y_2-axis are given by the boundary condition (16.86), it may be determined as the solution of a Goursat problem (cf. Section 4.2) in each of the angular regions

$$(16.90) \qquad\qquad y_2 \geq x_1 \geq 0$$

and

$$(16.91) \qquad\qquad y_2 \leq -x_1 \leq 0.$$

Thus we can find it by the method of successive approximations throughout at least some sufficiently small neighborhood of C within the complex half-space $x_1 \geq 0$. However, we have yet to establish that it satisfies the Cauchy-Riemann equation (16.88) inside the wedges (16.90) and (16.91).

Since u has been constructed by means of a Cauchy problem on the near side of the characteristics (16.89), but by means of a Goursat problem on the far side, we need to verify that its first partial derivatives do not exhibit jumps across those lines. For the proof we introduce the characteristic coordinates

$$\xi = x_1 + y_2, \qquad \eta = x_1 - y_2,$$

which bring (16.83) into the canonical form

$$(16.92) \quad 4u_{\xi\eta} = F\left(\frac{\xi + \eta}{2}, y_1 + i\frac{\xi - \eta}{2}, u, u_\xi + u_\eta, -iu_\xi + iu_\eta\right).$$

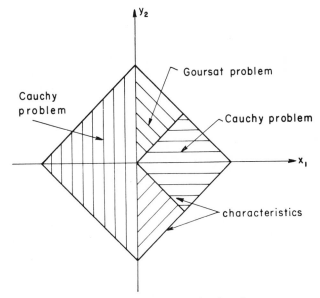

FIG. 94. Reflection via the complex domain.

The tangential derivative u_ξ along the characteristic $y_2 = x_1$ is uniquely defined because u is; consequently formula (16.92) represents an ordinary differential equation for the normal derivative u_η there. Hence not only u_ξ but also u_η ought to remain continuous across the line $y_2 = x_1$ provided only that its initial value at the origin

$$(16.93) \qquad\qquad \xi = \eta = 0$$

has that same property. This is, of course, the case because of our hypothesis that all the first partial derivatives of u have continuous limiting values on C and because of the compatibility of the requirements (16.85) and (16.86) at the crucial point (16.93). A similar argument shows that the first derivatives of u are continuous across the opposite characteristic $y_2 = -x_1$, too.

In order to prove that u satisfies the Cauchy-Riemann equation (16.88) in the intersection of the wedges (16.90) and (16.91) with some neighborhood of C, we introduce once more the quantity

$$w = u_{\bar y}.$$

Since we now know that the first partial derivatives of u remain continuous across the characteristic segments (16.89), and since (16.88) has already been established in the region (16.87) located between those two segments, it becomes clear that on the characteristics (16.89) themselves we must have

$$(16.94) \qquad\qquad w = 0.$$

An application of the operator $\partial/\partial\bar y$ to the boundary condition (16.86) shows that (16.94) is valid for $x_1 = 0$ also. Finally, differentiation of (16.83) yields the hyperbolic equation

$$w_{x_1 x_1} - w_{y_2 y_2} = F_u w + F_p w_{x_1} - i F_q w_{y_2}$$

for w in the interior of the wedges (16.90) and (16.91). Thus w solves a linear homogeneous Goursat problem, which means that it must actually vanish according to the usual uniqueness argument. We conclude that the Cauchy-Riemann equation (16.88) is fulfilled throughout a one-sided neighborhood of C within the half-space $x_1 > 0$.

For each sufficiently small positive value of x_1, the Cauchy-Riemann equation (16.88) implies that $u(x_1, y_1, y_2)$ is an analytic function of the complex variable $y = y_1 + iy_2$. It follows that the same statement must hold for the partial derivative $u_{x_1}(x_1, y_1, y_2)$. Because the limit

$$(16.95) \qquad u_{x_1}(0, y_1, y_2) = \lim_{x_1 \to 0, x_1 > 0} u_{x_1}(x_1, y_1, y_2)$$

exists uniformly, the normal derivative $u_{x_1}(0, y_1, y_2)$ is also seen to be an analytic function of y throughout some region of the complex y-plane containing the line segment C. Combined with (16.86), this result shows that the initial values at $x_1 = 0$ of both u and u_{x_1} are analytic functions of y. Consequently we can appeal to the procedures of Section 1 or of Section 2 to construct u for negative values of x_1 as the solution of an analytic Cauchy problem for (16.82).

More precisely, in any small enough characteristic triangle

$$-x_1 - h \leq y_2 \leq x_1 + h, \qquad x_1 \leq 0$$

associated with a suitable fixed choice of the parameter y_1, we can determine $u(x_1, y_1, y_2)$ as the solution of Cauchy's problem for the hyperbolic equation (16.83) with initial data at $x_1 = 0$ given by (16.86) and (16.95). The extension $u(x_1, y_1, y_2)$ so obtained fulfills the Cauchy-Riemann equation (16.88) in the left half-space $x_1 < 0$ because of the by now familiar linear homogeneous Cauchy problem that is applicable there to the quantity $w = u_{\bar{y}}$. In particular, if we set

(16.96) $\qquad u(x_1, y_1) = u(x_1, y_1, 0)$

in the real left half-plane $x_1 < 0$, $y_2 = 0$, we find that

(16.97) $\quad u_{y_1}(x_1, y_1) = -iu_{y_2}(x_1, y_1, 0), u_{y_1 y_1}(x_1, y_1) = -u_{y_2 y_2}(x_1, y_1, 0).$

Substitution of the expressions (16.97) into (16.83) proves that the reflected function $u(x_1, y_1)$ specified by the rule (16.96) furnishes a real solution of the elliptic partial differential equation

$$u_{x_1 x_1} + u_{y_1 y_1} = F(x_1, y_1, u, u_{x_1}, u_{y_1})$$

in an extended region of the real domain containing both D and C in its interior.

When the underlying equation (16.82) is linear with entire coefficients, the reflection rule (16.96), which is based on a combination of two Cauchy problems and a Goursat problem, defines an analytic continuation of our original solution $u = u(x, y)$ into a domain of the left half-plane $x < 0$ that includes all points of the mirror image D' of D in the y-axis, except perhaps for those which cannot be connected to C by horizontal line segments from D'. The real data from D needed to extend u are located in turn on the reflected images of segments of the latter kind. Even in the nonlinear case, formula (16.96) serves to extend u a short distance across the y-axis and thus establishes its analyticity at points of ∂D where the boundary condition (16.77) holds. Finally, observe that when C is a curved analytic arc instead of a line segment, a rule analogous to (16.96) can be derived for the analytic continuation of u beyond D by performing a preliminary conformal transformation that maps C onto a line segment. Such generalizations of the Schwarz reflection principle will be described in the exercises below.

We close with a few remarks concerning the reflection of a solution u of Laplace's equation (16.76) across an analytic arc C on which it obeys a nonlinear analytic boundary condition

$$(16.98) \qquad \frac{\partial u}{\partial v} = \phi(x, y, u).$$

Denote by v the conjugate of the harmonic function u, and consider the corresponding complex analytic function

$$f(z) = u + iv,$$

where $z = x + iy$. Suppose that the equation of C has been expressed in the form

$$(16.99) \qquad \bar{z} = g(z),$$

where g is regular in some neighborhood of C. We find it convenient to introduce a complex representation

$$(16.100) \qquad \overline{f'(z)} = \Phi[z, \bar{z}, f(z), \overline{f(z)}, f'(z)]$$

of the boundary condition (16.98). Inserting (16.99) into (16.100), we derive along C the complex analytic relation

$$(16.101) \qquad \tilde{f}'(g(z)) = \Phi[z, g(z), f(z), \tilde{f}(g(z)), f'(z)],$$

where $\tilde{f}$ is the analytic function defined by the rule

$$(16.102) \qquad \tilde{f}(\bar{z}) = \overline{f(z)}.$$

According to the basic principles of analytic continuation, formula (16.101) can be viewed as an ordinary differential equation in the complex domain which determines $\tilde{f}$ uniquely, provided that an auxiliary initial condition of the form

$$\tilde{f}(\bar{z}_0) = \overline{f(z_0)}$$

is imposed at a point z_0 on C. Now observe that the two points z and $\overline{g(z)}$ lie on opposite sides of C. It follows that when $f(z)$ is given on one side of C, the equation (16.101) serves to specify $\tilde{f}(\bar{z})$ for values of z on the other side of C. With this new definition of f, the identity (16.102) yields the rule we have been seeking for the reflection of f, and therefore also of u, across the analytic arc (16.99). Note, finally, that the proof becomes especially simple in the case $g(z) = z$ where C coincides with the real axis.

EXERCISES

1. Why is $u(x_1, y_1, 0)$ real for negative values of x_1?
2. Find a rule for reflecting solutions u of the biharmonic equation

$$\Delta \Delta u = 0$$

across an analytic arc (16.99) where they satisfy the boundary conditions

$$u = \frac{\partial u}{\partial \nu} = 0.$$

Compare your result with the method of images described in Section 7.3.

3. Consider a solution $u = u(x, t)$ of the quasi-linear elliptic equation (16.13) which satisfies the boundary condition

$$u(x, 0) = 0$$

along a segment of the x-axis. Construct an analytic continuation of u from the upper half-plane $t > 0$ into the lower half-plane $t < 0$ by solving a mixed initial and boundary value problem for the hyperbolic equation (16.18) with λ taken small and positive and then solving a Cauchy problem for that same equation with λ taken large and positive.[11]

4. Let u be a solution of the quasi-linear elliptic equation (16.24) in D which fulfills the boundary condition (16.77) on C. Reflect u across C by making a preliminary transformation to the canonical system (16.43) and then imitating the procedure of the text. Generalize this construction to the case where C is a curved analytic arc.

5. Instead of (16.43), use the hyperbolic system (16.28) to (16.31) to solve the problem of Exercise 4. Show how to associate with each point in the left half-plane a reflected image in the right half-plane with respect to the equation (16.24) by considering the geometry of the relevant characteristics in the complex domain. Prove that the value of u at a point in the left half-plane is determined by data in the right half-plane located on an arbitrary curve joining the reflected image of that point to the segment C of the y-axis.

6. By appealing to the result of Exercise 5 concerning arbitrary paths of data, verify that when the equation (16.82) is linear and has entire coefficients the analytic extension across the y-axis of a solution u satisfying (16.77) is defined without restriction throughout the whole reflected image D' of the region D where it was originally specified.

[11] Cf. Garabedian 10.

7. Give a detailed discussion of the reflection rule defined by formulas (16.101) and (16.102). Use the method of successive approximations to generalize that rule to the case of solutions u of equation (16.82) that satisfy a nonlinear boundary condition of the form (16.98).

8. Obtain the analytic extension of a solution u of (16.82) and (16.77) across the y-axis by considering u as an analytic function of one complex variable in some characteristic plane through C in the complex domain and by reflecting it there according to rules analogous to (16.101) and (16.102).

9. Derive from (16.44) the representation

(16.103)

$$u(\zeta, \bar{\zeta}) = A(\bar{\zeta}^*, \zeta^*; \zeta, \bar{\zeta})u(\bar{\zeta}^*, \zeta^*) + 2 \operatorname{Re}\left\{ \int_{\bar{\zeta}^*}^{\zeta} A(z, \zeta^*; \zeta, \bar{\zeta}) \frac{\partial u(z, \zeta^*)}{\partial z} dz \right\}$$

for solutions u of the linear elliptic equation (16.35). Use (16.103) to show how the rule for reflecting u across an analytic arc (16.99) where it vanishes can be obtained from the classical Schwarz principle of reflection for the analytic function

$$f(\zeta) = \int_{\bar{\zeta}^*}^{\zeta} A(z, \zeta^*; \zeta, g(\zeta)) \frac{\partial u(z, \zeta^*)}{\partial z} dz,$$

whose real part also vanishes along (16.99) when ζ^* is any fixed point of that arc.[12]

10. Explain why singularities in the reflection of a solution of (16.82) across the curve (16.99) can be expected to occur at the branch points of the analytic function g, which are referred to as *foci* of the curve in question. Use the example of the equilateral hyperbola

$$\bar{z} = \sqrt{2 - z^2}$$

as an illustration of this phenomenon (cf. Exercise 2.11).

11. Establish the analyticity of any sufficiently smooth arc along which a harmonic function u fulfills the free boundary conditions

$$u = 0, \qquad \frac{\partial u}{\partial v} = 1$$

by exploiting the technique of reflection (cf. Section 15.2).

[12] Cf. Garabedian 4, Lewy 5.

Bibliography

M. Abraham and R. Becker
1. *The classical theory of electricity and magnetism.* New York, Hafner, 1949. (Pages 333, 548.)

S. Agmon
1. Boundary value problems for equations of mixed type. *Atti del Convegno Internazionale sulle Equazioni Lineari alle Derivate Parziali, Trieste,* 1954, pp. 54–68. Roma, Edizioni Cremonese, 1955. (Page 433.)

S. Agmon, L. Nirenberg, and M. H. Protter
1. A maximum principle for a class of hyperbolic equations and applications to equations of mixed elliptic-hyperbolic types. *Comm. Pure Appl. Math.,* Vol. 6 (1953), pp. 455–470. (Page 433.)

L. V. Ahlfors
1. *Complex analysis* (Internatl. Ser. Pure Appl. Math.). New York, McGraw-Hill, 1953. (Pages 252, 264, 333.)
2. Conformality with respect to Riemannian metrics. *Ann. Acad. Sci. Fenn.,* Ser. AI, No. 206 (1955), pp. 1–22. (Pages 67, 70.)

L. Asgeirsson
1. Über eine Mittelwertseigenschaft von Lösungen homogener linearer partieller Differentialgleichungen 2. Ordnung mit konstanten Koeffizienten. *Math. Ann.,* Vol. 113 (1936), pp. 321–346.

S. Banach
1. *Théorie des opérations linéaires* (Monografje matematyczne, Vol. 1). Warsaw, Subwencji Funduszu kultury narodowej, 1932. (Page 313.)

H. Bateman
1. *Partial differential equations of mathematical physics.* New York, Dover, 1944.

S. Bergman
1. Über die Entwicklung der harmonischen Funktionen der Ebene und des Raumes nach Orthogonalfunktionen. *Math. Ann.,* Vol. 86 (1922), pp. 238–271.
2. *Integral operators in the theory of linear partial differential equations* (Erg. d. Math. u. Grenzgebiete, Vol. 23). Berlin, Springer, 1961. (Pages 135, 632.)

651

S. Bergman and M. Schiffer

1. A representation of Green's and Neumann's functions in the theory of partial differential equations of second order. *Duke Math. J.*, Vol. 14 (1947), pp. 609–638. (Page 256.)

2. *Kernel functions and elliptic differential equations in mathematical physics* (Pure Appl. Math., Vol. 4). New York, Academic Press, 1953.

J. Berkowitz, H. Grad, and H. Rubin

1. Magnetohydrodynamic stability. *Proceedings of the Second United Nations International Conference on the Peaceful Uses of Atomic Energy*, Geneva, 1958, Vol. 31, pp. 177–189. (Page 589.)

I. B. Bernstein, E. A. Frieman, M. D. Kruskal, and R. M. Kilsrud

1. An energy principle for hydromagnetic stability problems. *Proc. Roy. Soc. London Ser. A*, Vol. 244 (1958), pp. 17–40.

S. Bernstein

1. Sur la nature analytique des solutions des équations aux dérivées partielles du seconde ordre. *Math. Ann.*, Vol. 59 (1904), pp. 20–76.

L. Bers

1. *Mathematical aspects of subsonic and transonic gas dynamics* (Surveys Appl. Math., Vol. 3). New York, Wiley, 1958.

L. Bers and L. Nirenberg

1. On linear and non-linear elliptic boundary value problems in the plane. *Atti del Convegno Internazionale sulle Equazioni Lineari alle Derivate Parziali, Trieste*, 1954, pp. 141–167. Rome, Edizioni Cremonese, 1955. (Page 309.)

G. Birkhoff and S. MacLane

1. *A survey of modern algebra.* New York, Macmillan, 1953. (Pages 72, 95, 282, 370.)

G. Birkhoff and E. H. Zarantonello

1. *Jets, wakes, and cavities* (Appl. Math. Mech., Vol. 2). New York, Academic Press, 1957.

A. A. Blank

1. Stability of magneto-fluid free boundaries against finite perturbation. *Relativistic Fluid Mechanics and Magnetohydrodynamics.* New York, Academic Press, 1963. (Pages 589, 598.)

S. Bochner

1. Über orthogonale Systeme analytischer Funktionen. *Math. Zeit.*, Vol. 14 (1922), pp. 180–207.

R. C. Buck

1. *Advanced calculus* (Internatl. Ser. Pure Appl. Math.). New York, McGraw-Hill, 1956. (Pages 128, 161, 162, 213, 252, 290.)

R. Caccioppoli

1. Sui teoremi d'esistenza di Riemann. *Annali Scuola Norm. Sup. Pisa*, Vol. 7 (1938), pp. 177–187. (Page 317.)

R. V. Churchill

1. *Fourier series and boundary value problems.* New York, McGraw-Hill, 1941. (Pages 3, 5, 77, 259.)

E. A. Coddington and N. Levinson

1. *Theory of ordinary differential equations* (Internatl. Ser. Pure Appl. Math.). New York, McGraw-Hill, 1955. (Pages 323, 334.)

Paul J. Cohen
1. The non-uniqueness of the Cauchy problem. *Tech. Report* No. 93, Stanford University, 1960.

R. Courant
1. *Differential and integral calculus.* New York, Interscience, 1937. (Pages 24, 45, 197, 200.)
2. *Dirichlet's principle, conformal mapping, and minimal surfaces* (Pure Appl. Math., Vol. 3). New York, Interscience, 1950. (Pages 309, 599, 613.)

R. Courant and K. O. Friedrichs
1. *Supersonic flow and shock waves* (Pure Appl. Math., Vol. 1). New York, Interscience, 1948. (Pages 502, 509, 518, 529.)

R. Courant, K. O. Friedrichs, and H. Lewy
1. Über die partiellen Differenzengleichungen der mathematischen Physik. *Math. Ann.*, Vol. 100 (1928), pp. 32–74. (Pages 346, 467, 485.)

R. Courant and D. Hilbert
1. *Methoden der mathematischen Physik*, Vol. 2 (Grundlehren Math. Wiss., Vol. 12). Berlin, Springer, 1937. (Pages 8, 49, 203, 211, 400.)
2. *Methods of mathematical physics*, Vol. 1. New York, Interscience, 1953. (Pages 187, 273, 278, 394, 397, 403, 406, 412.)
3. *Methods of mathematical physics*, Vol. 2. New York, Interscience, 1961. (Pages 224, 331, 346, 364, 448, 449.)

R. Courant, E. Isaacson and M. Rees
1. On the solution of nonlinear hyperbolic differential equations by finite differences. *Comm. Pure Appl. Math.*, Vol. 5 (1952), pp. 243–255. (Page 471.)

R. Courant and P. Lax
1. On nonlinear partial differential equations with two independent variables. *Comm. Pure Appl. Math.*, Vol. 2 (1949), pp. 255–273. (Pages 97, 122.)

E. DeGiorgi
1. Sulla differenziabilità e l'analiticità delle estremali degli integrali multipli regolari. *Mem. Accad. Sci. Torino, Cl. Sci. Fis. Mat. Nat.*, Vol. 3 (1957), pp. 25–43. (Page 309.)

J. Douglas
1. Solution of the problem of Plateau. *Trans. Amer. Math. Soc.*, Vol. 33 (1931), pp. 263–321.

G. F. D. Duff
1. *Partial differential equations* (Math. Expositions, No. 9). Toronto, University of Toronto Press, 1956. (Page 224.)

B. Epstein
1. *Partial differential equations, an introduction.* New York, McGraw-Hill, 1962. (Pages 480, 485.)

G. Faber
1. Beweis, dass unter allen homogenen Membranen von gleicher Fläche und gleicher Spannung die kreisförmige den tiefsten Grundton gibt. *Sitzungsber. d. Bayr. Akad. d. Wiss.*, (1923), pp. 169–172. (Page 413.)

G. Fichera
1. Sull'esistenza delle funzioni potenziali nei problemi della fisica matematica. *Atti Accad. Naz. Lincei, Rend., Cl. Sci. Fis. Mat. Nat.*, Vol. 2 (1947), pp. 527–532. (Page 288.)

654 BIBLIOGRAPHY

G. E. Forsythe and W. R. Wasow

1. *Finite-difference methods for partial differential equations* (Appl. Math. Ser.). New York, Wiley, 1960. (Page 475.)

Y. Fourès-Bruhat

1. Théorème d'existence pour certains systèmes d'équations aux dérivées partielles non linéaires. *Acta Math.*, Vol. 88 (1952), pp. 141–225. (Page 225.)

S. P. Frankel

1. Convergence rates of iterative treatments of partial differential equations. *Math. Tables Aids Comput.*, Vol. 4 (1950), pp. 65–75. (Page 489.)

F. Frankl

1. On the problems of Chaplygin for mixed sub- and supersonic flows. *Bull. Acad. Sci. U.S.S.R.*, Math. Ser., Vol. 9 (1945), pp. 121–143. N.A.C.A. translation, T.M. No. 1155 (1947). (Page 532.)

K. O. Friedrichs

1. Über ein Minimumproblem für Potentialströmungen mit freiem Rande. *Math. Ann.*, Vol. 109 (1934), pp. 60–82. (Pages 574, 588.)

2. Symmetric hyperbolic linear differential equations. *Comm. Pure Appl. Math.*, Vol. 7 (1954), pp. 345–392. (Pages 474, 480.)

3. Symmetric positive linear differential equations. *Comm. Pure Appl. Math.*, Vol. 11 (1958), pp. 333–418. (Page 449.)

K. O. Friedrichs and H. Lewy

1. Über die Eindeutigkeit und das Abhängigkeitsgebiet der Lösungen beim Anfangswertproblem linearer hyperbolischer Differentialgleichungen. *Math. Ann.*, Vol. 98 (1927), pp. 192–204.

P. R. Garabedian

1. The classes L_p and conformal mapping. *Trans. Amer. Math. Soc.*, Vol. 69 (1950), pp. 392–415. (Pages 275, 317, 322.)

2. A partial differential equation arising in conformal mapping. *Pac. J. Math.*, Vol. 1 (1951), pp. 485–524. (Pages 266, 275.)

3. An example of axially symmetric flow with a free surface. *Studies in mathematics and mechanics presented to Richard von Mises*, New York, Academic Press, 1954, pp. 149–159. (Page 547.)

4. Applications of analytic continuation to the solution of boundary value problems. *J. Ratl. Mech. Anal.*, Vol. 3 (1954), pp. 383–393. (Pages 268, 650.)

5. An integral equation governing electromagnetic waves. *Quarterly Appl. Math.*, Vol. 12 (1955), pp. 428–433. (Page 344.)

6. Calculation of axially symmetric cavities and jets. *Pac. J. Math.*, Vol. 6 (1956), pp. 611–684. (Page 588.)

7. Estimation of the relaxation factor for small mesh size. *Math. Tables Aids Comput.*, Vol. 10 (1956), pp. 183–185. (Page 492.)

8. Partial differential equations with more than two independent variables in the complex domain. *J. Math. Mech.*, Vol. 9 (1960), pp. 241–271. (Pages 192, 212.)

9. Stability of Cauchy's problem in space for analytic systems of arbitrary type. *J. Math. Mech.*, Vol. 9 (1960), pp. 905–914. (Pages 616, 620.)

10. Analyticity and reflection for plane elliptic systems. *Comm. Pure Appl. Math.*, Vol. 14 (1961), pp. 315–322. (Pages 640, 649.)

11. Proof of uniqueness by symmetrization. *Studies in mathematical analysis and related topics: Essays in honor of George Pólya*. Stanford, Stanford University Press, 1962, pp. 126–127. (Page 588.)

P. R. Garabedian, H. Lewy, and M. Schiffer
1. Axially symmetric cavitational flow. *Ann. Math.*, Vol. 56 (1952), pp. 560–602. (Pages 573, 574, 588.)

P. R. Garabedian and H. M. Lieberstein
1. On the numerical calculation of detached bow shock waves in hypersonic flow. *J. Aero. Sci.*, Vol. 25 (1958), pp. 109–118. (Pages 623, 632.)

P. R. Garabedian and M. Schiffer
1. On existence theorems of potential theory and conformal mapping. *Ann. Math.*, Vol. 52 (1950), pp. 164–187. (Pages 288, 296, 323.)
2. Convexity of domain functionals. *J. Anal. Math.*, Vol. 2 (1952–53), pp. 281–368. (Pages 561, 571, 588.)
3. On a double integral variational problem. *Can. J. Math.*, Vol. 6 (1954), pp. 441–446. (Page 311.)
4. On estimation of electrostatic capacity. *Proc. Amer. Math. Soc.*, Vol. 5 (1954), pp. 206–211. (Page 573.)

P. R. Garabedian and M. Shiffman
1. On solution of partial differential equations by the Hahn-Banach theorem. *Trans. Amer. Math. Soc.*, Vol. 76 (1954), pp. 288–299. (Page 321.)

P. R. Garabedian and D. C. Spencer
1. Extremal methods in cavitational flow. *J. Ratl. Mech. Anal.*, Vol. 1 (1952), pp. 359–409. (Pages 420, 548, 573, 588.)

L. Garding and J. Leray
1. Book on partial differential equations, to appear. (Page 224.)

P. Germain and R. Bader
1. Sur quelques problèmes relatifs à l'équation de type mixte de Tricomi. *O.N.E.R.A. Pub.* No. 54 (1952), pp. 1–57. (Page 433.)

D. Gilbarg
1. Uniqueness of axially symmetric flows with free boundaries. *J. Ratl. Mech. Anal.*, Vol. 1 (1952), pp. 309–320. (Page 544.)

H. Goldstein
1. *Classical mechanics.* Cambridge, Addison-Wesley, 1950. (Pages 45, 49.)

S. Goldstein
1. *Modern developments in fluid dynamics* (Oxford Eng. Sci. Ser.). Oxford, Clarendon, 1938. (Page 508.)

E. Goursat
1. *Cours d'analyse mathématique*, Vol. 3. Paris, Gauthier-Villars, 1933.

H. Grad
1. Containment in cusped plasma systems. *Progress in Nuclear Energy*, Ser. 11, Vol. 2, New York, Pergamon Press, 1963, pp. 189–200. (Page 556.)

A. Haar
1. Über das Plateausche Problem. *Math. Ann.*, Vol. 97 (1927), pp. 124–158.

J. Hadamard
1. Mémoire sur le problème d'analyse relatif à l'équilibre des plaques élastiques encastrées. *Mémoires présentés par divers savants à l'Académie des Sciences*, Vol. 33 (1908), pp. 1–128. (Page 558.)
2. *Le problème de Cauchy et les équations aux dérivées partielles linéaires hyperboliques.* Paris, Hermann et Cie., 1932.

3. *Lectures on Cauchy's problem in linear partial differential equations.* New York, Dover, 1952. (Pages 108, 152, 166, 168, 203, 217, 223.)

K. J. Harker

1. Determination of electrode shapes for axially symmetric electron guns. *J. Appl. Phys.*, Vol. 31 (1960), pp. 2165–2170. (Page 622.)

G. Hellwig

1. *Partielle Differentialgleichungen* (Math. Leitfäden). Stuttgart, Teubner, 1960. (Page 303.)

G. Herglotz

1. Über die Berechnung retardierter Potentiale. *Gött. Nachr.*, 1904, pp. 459–556.

D. Hilbert

1. *Grundzüge einer allgemeinen Theorie der linearen Integralgleichungen* (Fortschritte d. Math. Wiss. in Mon., Heft 3). Leipzig, Teubner, 1912. (Pages 168, 223.)

E. Hopf

1. Elementare Bemerkungen über die Lösungen partieller Differentialgleichungen zweiter Ordnung vom elliptischen Typus. *Sitzungsber. d. Preuss. Akad. d. Wiss.*, Vol. 19 (1927), pp. 147–152. (Page 232.)
2. Über den Funktionalen, insbesondere den analytischen Charakter der Lösungen elliptischer Differentialgleichungen zweiter Ordnung. *Math. Zeit.*, Vol. 34 (1931), pp. 194–233. (Pages 309, 641.)

L. Hörmander

1. *Linear partial differential operators* (Grundlehren d. Math. Wiss., Vol. 116). New York, Academic Press, 1963. (Page 456.)

E. L. Ince

1. *Ordinary differential equations.* New York, Dover, 1944. (Pages 18, 20, 35.)

J. D. Jackson

1. *Classical electrodynamics.* New York, Wiley, 1962. (Page 243.)

F. John

1. *Plane waves and spherical means applied to partial differential equations* (Tracts Pure Appl. Math.). New York, Interscience, 1955. (Pages 161, 224.)

J. Keller, R. Lewis, and B. Seckler

1. Asymptotic solution of some diffraction problems. *Comm. Pure Appl. Math.*, Vol. 9 (1956), pp. 207–265. (Pages 40, 44, 167.)

O. D. Kellogg

1. *Foundations of potential theory.* New York, Dover, 1953. (Pages 149, 173, 296, 324, 334, 339, 347.)

M. Kline

1. Asymptotic solutions of Maxwell's equations involving fractional powers of the frequency. *Comm. Pure Appl. Math.*, Vol. 8 (1955), pp. 595–614. (Page 40.)

E. Krahn

1. Über eine von Rayleigh formulierte Minimaleigenschaft des Kreises. *Math. Ann.*, Vol. 94 (1925), pp. 97–100. (Page 413.)

H. Lamb

1. *Hydrodynamics.* New York, Dover, 1945. (Pages 508, 546.)

L. D. Landau and E. M. Lifshitz

1. *Fluid mechanics.* Reading, Addison-Wesley, 1959.

M. A. Lavrentiev
1. On certain properties of univalent functions and their application to wake theory. *Mat. Sbornik*, Vol. 46 (1938), pp. 391–458. (Pages 238, 544.)

M. A. Lavrentiev and A. Bitsadze
1. On the problem of equations of mixed type. *Doklady Akad. Nauk SSSR*, Vol. 70 (1950), pp. 373–376. (Page 432.)

P. Lax
1. A remark on the method of orthogonal projections. *Comm. Pure Appl. Math.*, Vol. 4 (1951), pp. 457–464. (Page 288.)
2. On the existence of Green's function. *Proc. Amer. Math. Soc.*, Vol. 3 (1952), pp. 526–531. (Pages 317, 321.)
3. On Cauchy's problem for hyperbolic equations and the differentiability of solutions of elliptic equations. *Comm. Pure Appl. Math.*, Vol. 8 (1955), pp. 615–633. (Pages 444, 445, 449.)
4. Asymptotic solutions of oscillatory initial value problems. *Duke Math. J.*, Vol. 24 (1957), pp. 627–646. (Page 457.)

M. Lees
1. The solution of positive-symmetric hyperbolic systems by difference methods. *Proc. Amer. Math. Soc.*, Vol. 12 (1961), pp. 195–202. (Page 474.)

O. Lehto
1. Anwendung orthogonaler Systeme auf gewisse funktionentheoretische Extremal- und Abbildungsprobleme. *Ann. Acad. Sci. Fenn.*, Ser. AI, No. 59 (1949), pp. 1–51. (Page 288.)

J. Leray
1. Le problème de Cauchy pour une équation linéaire à coefficients polynomiaux. *C. R. Acad. Sci. Paris*, Vol. 242 (1956), pp. 953–959.

E. E. Levi
1. Sulle equazioni lineari totalmente ellittiche alle derivate parziali. *Rend. Circ. Mat. Palermo*, Vol. 24 (1907), pp. 275–317. (Pages 168, 223.)

H. Lewy
1. Neuer Beweis des analytischen Charakters der Lösungen elliptischer Differentialgleichungen. *Math. Ann.*, Vol. 101 (1929), pp. 609–619. (Page 634.)
2. On the boundary behavior of minimal surfaces. *Proc. Nat. Acad. Sci. U.S.A.*, Vol. 37 (1951), pp. 103–110. (Page 613.)
3. A note on harmonic functions and a hydrodynamical application. *Proc. Amer. Math. Soc.*, Vol. 3 (1952), pp. 111–113.
4. An example of a smooth linear partial differential equation without solution. *Ann. Math.*, Vol. 66 (1957), pp. 155–158. (Pages 452, 456.)
5. On the reflection laws of second order differential equations in two independent variables. *Bull. Amer. Math. Soc.*, Vol. 65 (1959), pp. 37–58. (Pages 642, 650.)

L. Lichtenstein
1. Über den analytischen Charakter der Lösungen regulärer zweidimensionaler Variationsprobleme. *Bull. Acad. Cracovie*, 1912, pp. 915–941. (Page 309.)

A. E. H. Love
1. *A treatise on the mathematical theory of elasticity.* New York, Dover, 1944. (Page 239.)

S. Lundquist
1. On the stability of magnetohydrostatic fields. *Phys. Rev.*, Vol. 83 (1951), pp. 307–311.

W. Magnus and F. Oberhettinger

1. *Formulas and theorems for the special functions of mathematical physics.* New York, Chelsea, 1949.

S. G. Mikhlin

1. *Linear integral equations* (Russian Mon. Texts Advanced Math. Phys., Vol. 2). Delhi, Hindustan, 1960. (Pages 349, 359.)

W. E. Milne

1. *Numerical solution of differential equations.* New York, Wiley, 1953.

C. Miranda

1. *Equazioni alle derivate parziali di tipo ellittico* (Erg. d. Math. u. Grenzgebiete, Vol. 2). Berlin, Springer, 1955.

C. S. Morawetz

1. A uniqueness theorem for Frankl's problem. *Comm. Pure Appl. Math.*, Vol. 7 (1954), pp. 697–703. (Pages 425, 426.)
2. Note on a maximum principle and a uniqueness theorem for an elliptic-hyperbolic equation. *Proc. Roy. Soc.*, Ser. A, Vol. 236 (1956), pp. 141–144. (Page 432.)
3. A weak solution for a system of equations of elliptic-hyperbolic type. *Comm. Pure Appl. Math.*, Vol. 11 (1958), pp. 315–331. (Page 449.)

C. B. Morrey

1. Multiple integral problems in the calculus of variations and related topics. *U. Cal. Publ. Math.*, Vol. 1 (1943), pp. 1–130. (Page 309.)

C. Müller

1. *Grundprobleme der mathematischen Theorie elektromagnetischer Schwingungen* (Grunglehren Math. Wiss.). Berlin, Springer, 1957. (Page 347.)

M. E. Munroe

1. *Introduction to measure and integration.* Cambridge, Addison-Wesley, 1953. (Pages 281, 295, 301, 351.)

N. I. Muskhelishvili

1. *Some basic problems of the mathematical theory of elasticity.* Groningen, P. Noordhoff, 1953.

J. Nash

1. Continuity of solutions of parabolic and elliptic equations. *Amer. J. Math.*, Vol. 80 (1958), pp. 931–954. (Page 309.)

Z. Nehari

1. *Conformal mapping.* New York, McGraw-Hill, 1952. (Pages 265, 273, 442.)

C. Neumann

1. *Untersuchungen über das logarithmische und Newton'sche Potential.* Leipzig, Teubner, 1877. (Pages 348, 351.)

L. Nirenberg

1. On elliptic partial differential equations. *Annali. Scuola Norm. Sup. Pisa*, Vol. 13 (1959), pp. 1–48. (Page 167.)

B. Noble

1. *Methods based on the Wiener-Hopf technique for the solution of partial differential equations* (Internatl. Ser. Mon. Pure Appl. Math., Vol. 7). New York, Pergamon Press, 1958.

G. G. O'Brien, M. A. Hyman, and S. Kaplan
1. A study of the numerical solution of partial differential equations. *J. Math. Phys.*, Vol. 29 (1951), pp. 223–251.

O. Perron
1. Eine neue Behandlung der Randwertaufgabe für $\Delta u = 0$. *Math. Zeit.*, Vol. 18 (1923), pp. 42–54. (Page 325.)

I. G. Petrovsky
1. *Lectures on partial differential equations.* New York, Interscience, 1954. (Page 78.)
1. *Lectures on the theory of integral equations.* Rochester, Graylock Press, 1957. (Pages 349, 357.)

A. Pliś
1. Non-uniqueness in Cauchy's problem for differential equations of elliptic type. *J. Math. Mech.*, Vol. 9 (1960), pp. 557–562. (Page 456.)

F. Pockels
1. *Über die partielle Differentialgleichung $\Delta u + k^2 u = 0$ und deren Auftreten in der mathematischen Physik.* Leipzig, Teubner, 1891.

G. Pólya and G. Szegö
1. *Isoperimetric inequalities in mathematical physics* (Ann. Math. Studies, No. 27). Princeton, Princeton University Press, 1951. (Pages 410, 413, 414, 416.)

M. H. Protter
1. Uniqueness theorems for the Tricomi problem. *J. Ratl. Mech. Anal.*, Vol. 2 (1953), pp. 107–114. (Pages 425, 432.)

E. F. Prym
1. Zur Integration der Differenzialgleichung $\dfrac{\partial^2 u}{\partial x^2} + \dfrac{\partial^2 u}{\partial y^2} = 0$. *Journal für die reine und angewandte Mathematik*, Vol. 73 (1871), pp. 340–364. (Page 288.)

T. Radó
1. *On the problem of Plateau* (Erg. d. Math. u. Grenzgebiete, Vol. 2). Berlin, Springer, 1933. (Page 599.)

F. Rellich
1. Über das asymptotische Verhalten der Lösungen von $\Delta u + \lambda u = 0$ in unendlichen Gebieten. *Jahresber. Deutsch. Math. Verein*, Vol. 53 (1943), pp. 57–65. (Page 239.)

D. Riabouchinsky
1. Sur un problème de variation. *C. R. Acad. Sci. Paris*, Vol. 185 (1927), pp. 840–841. (Page 574.)

B. Riemann
1. Über die Fortpflanzung ebener Luftwellen von endlicher Schwingungsweite. *Gesammelte Werke*, New York, Dover, 1953. (Pages 130, 513.)

M. Riesz
1. L'intégrale de Riemann-Liouville et le problème de Cauchy. *Acta Math.*, Vol. 81 (1949), pp. 1–223. (Page 224.)

F. Riesz and B. Sz.-Nagy
1. *Functional analysis.* New York, F. Ungar, 1955. (Pages 316, 447.)

E. Rothe

1. Zweidimensionale parabolische Randwertaufgaben als Grenzfall eindimensionaler Randwertaufgaben. *Math. Ann.*, Vol. 102 (1930), pp. 650–670. (Pages 493, 499.)
2. Über die Wärmeleitungsgleichung mit nichtkonstanten Koeffizienten im räumlichen Falle. *Math. Ann.*, Vol. 104 (1931), pp. 340–354. (Pages 493, 499.)

A. Rubinowicz

1. Über die Eindeutigkeit der Lösung der Maxwellschen Gleichungen. *Phys. Zeit.*, Vol. 27 (1926), pp. 707–710. (Page 189.)

J. Schauder

1. Das Anfangswertproblem einer quasilinearen hyperbolischen Differentialgleichung zweiter Ordnung in beliebiger Anzahl von unabhängigen Veränderlichen. *Fund. Math.*, Vol. 24 (1935), pp. 213–246. (Page 442.)

M. Schiffer

1. Variation of the Green function and theory of the p-valued functions. *Amer. J. Math.*, Vol. 65 (1943), pp. 341–360. (Page 568.)
2. Hadamard's formula and variation of domain-functions. *Amer J. Math.*, Vol. 68 (1946), pp. 417–448. (Pages 561, 572.)

E. Schmidt

1. Zur Theorie der linearen und nicht linearen Integralgleichungen. *Math. Ann.*, Vol. 63 (1907), pp. 433–476 and Vol. 64 (1907), pp. 161–174.

L. I. Sedov

1. *Two-dimensional problems of hydrodynamics and aerodynamics.* Moscow, Gosudarstv. Izdat. Tech.-Teor. Lit., 1950. (Page 536.)

I. N. Sneddon

1. *Elements of partial differential equations* (Internatl. Ser. Pure Appl. Math.). New York, McGraw-Hill, 1957.

S. L. Sobolev

1. A new method for the solution of the Cauchy problem for linear hyperbolic equations. *Mat. Sbornik*, Vol. 43 (1936), pp. 39–71.

A. Sommerfeld

1. *Partial differential equations in physics.* New York, Academic Press, 1949. (Pages 256, 344.)
2. *Electrodynamics* (Lect. Theor. Phys., Vol. 3). New York, Academic Press, 1952. (Pages 82, 84, 203.)

M. H. Stone

1. *Linear transformations in Hilbert space and their applications to analysis* (Amer. Math. Soc., Colloquium Publ., Vol. 15). New York, Amer. Math. Soc., 1932. (Pages 279, 280, 281.)

G. Szegö

1. Über orthogonale Polynome, die zu einer gegebenen Kurve der komplexen Ebene gehören. *Math. Zeit.*, Vol. 9 (1921), pp. 218–270. (Page 275.)

F. Tricomi

1. Sulle equazioni lineari alle derivate parziali di 2° ordine, di tipo misto. *Atti Accad. Naz. Lincei, Rend.*, Vol. 14 (1923), pp. 133–247. (Page 423.)

R. S. Varga

1. *Matrix iterative analysis.* Englewood Cliffs, Prentice-Hall, 1962.

I. N. Vekua
1. *New methods for solving elliptic equations.* Moscow, Gosudarstv. Izdat. Tech.-Teor. Lit., 1948. (Page 632.)

V. Volterra
1. Sur les vibrations des corps élastiques isotropes. *Acta Math.*, Vol. 18 (1894), pp. 161–232. (Page 224.)

H. Weber
1. *Die partiellen Differential-gleichungen der mathematischen Physik.* Braunschweig, F. Vieweg, 1910–19.

H. F. Weinberger
1. An isoperimetric inequality for the N-dimensional free membrane problem. *J. Ratl. Mech. Anal.*, Vol. 5 (1956), pp. 633–636. (Page 421.)

A. Weinstein
1. The singular solutions and the Cauchy problem for generalized Tricomi equations. *Comm. Pure Appl. Math.*, Vol. 7 (1954), pp. 105–116.

H. Weyl
1. Das asymptotische Verteilungsgesetz der Eigenwerte linearer partieller Differential-gleichungen. *Math. Ann.*, Vol. 71 (1912), pp. 441–479. (Page 403.)
2. The method of orthogonal projection in potential theory. *Duke Math. J.*, Vol. 7 (1940), pp. 411–444.

E. T. Wittaker
1. On the partial differential equations of mathematical physics. *Math. Ann.*, Vol. 57 (1903), pp. 333–355. (Page 204.)

E. T. Wittaker and G. N. Watson
1. *A course of modern analysis.* Cambridge, Cambridge University Press, 1902. (Pages 149, 363.)

D. M. Young
1. Iterative methods for solving partial differential equations of elliptic type. *Trans. Amer. Math. Soc.*, Vol. 76 (1954), pp. 92–111. (Page 491.)

S. Zaremba
1. L'équation biharmonique et une classe remarquable de fonctions fondamentales harmoniques. *Bull. Acad. Çracovie*, 1907, pp. 147–196. (Page 266.)
2. Sopra un teorema d'unicità relativo alla equazione delle onde sferiche. *Atti Accad. Naz. Lincei, Rend.*, Vol. 24 (1915), pp. 904–908.

Index